SCIENCE

A CLOSER LOOK

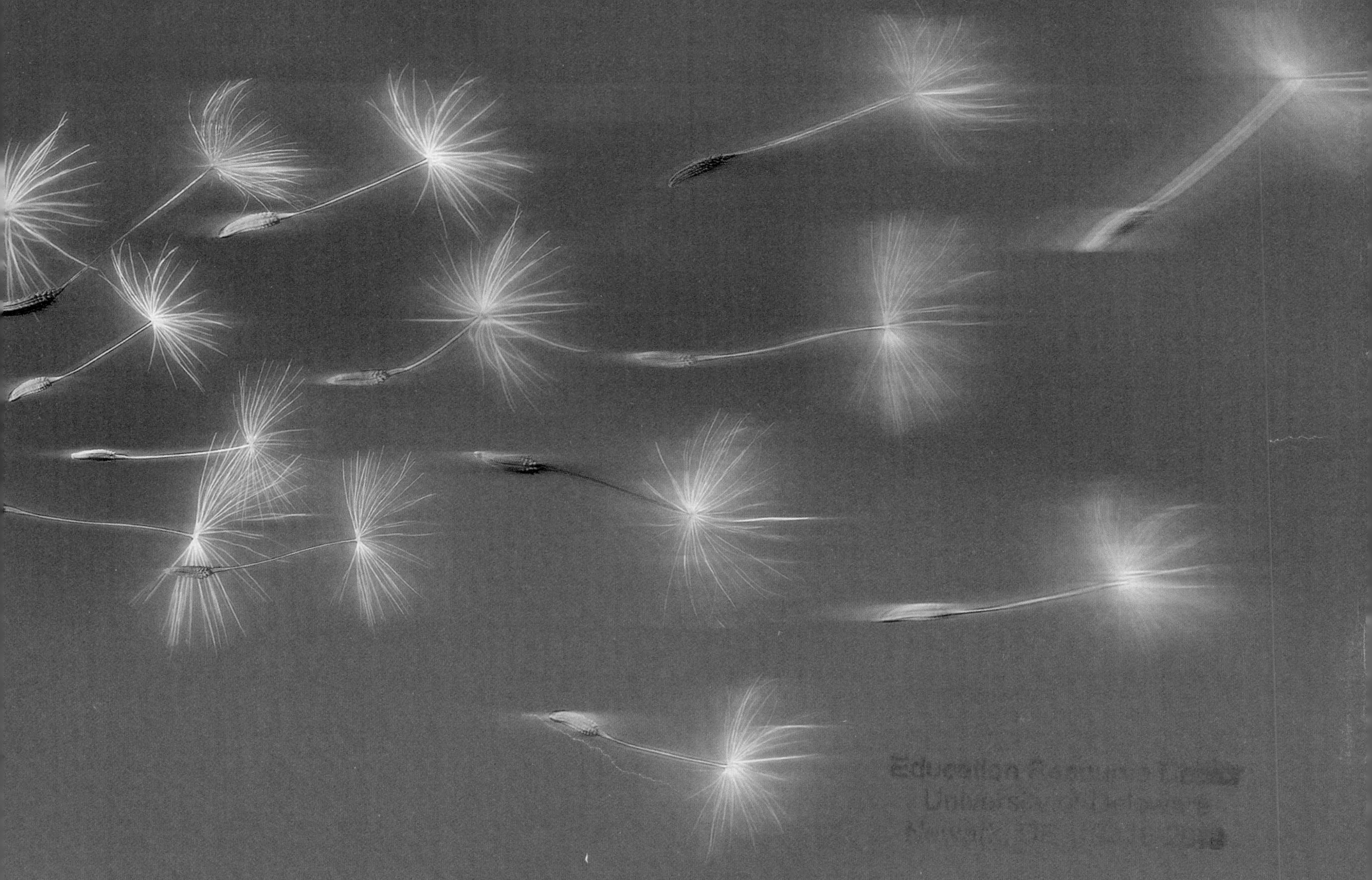

Mc Graw Hill Macmillan/McGraw-Hill

Partnerships

The American Museum of Natural History in New York City is one of the world's preeminent scientific, educational, and cultural institutions, with a global mission to explore and interpret human cultures and the natural world through scientific research, education, and exhibitions. Each year the Museum welcomes around 4 million visitors, including 500,000 schoolchildren in organized field trips. It provides professional development activities for thousands of teachers; hundreds of public programs that serve audiences ranging from preschoolers to seniors; and an array of learning and teaching resources for use in homes, schools, and community-based settings. Visit www.amnh.org for online resources.

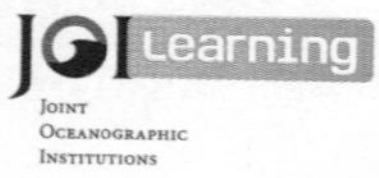

As the education arm of the **National Science Foundation**-funded Joint Oceanographic Institutions, JOI Learning brings the excitement of discovery and the scientific process to your classroom through expedition-based science, math, reading, and social studies activities. JOI Learning and Macmillan/McGraw-Hill have formed a partnership to give teachers and students access to JOI's resources through www.macmillanmh.com.

The National Science Digital Library (NSDL) is funded by the **National Science Foundation** as an online library of resources for science, technology, engineering, and mathematics education. NSDL has partnered with Macmillan/McGraw-Hill to provide teaching and learning resources that will help develop teachers' content knowledge in the topic and skill areas addressed at each grade level.

RFB&D  learning through listening

Students with print disabilities may be eligible to obtain an accessible, audio version of the pupil edition of this textbook. Please call Recording for the Blind & Dyslexic at 1-800-221-4792 for complete information.

The McGraw-Hill Companies A

Macmillan/McGraw-Hill

Send all inquiries to:
Macmillan/McGraw-Hill
8787 Orion Place
Columbus, OH 43240-4027

ISBN: 978-0-02-287975-4 *(Teacher Edition)*
MHID: 0-02-287975-7 *(Teacher Edition)*
ISBN: 978-0-02-288004-0 *(Kindergarten Flipbook)*
MHID: 0-02-288004-6 *(Kindergarten Flipbook)*

Printed in the United States of America.

1 2 3 4 5 6 7 8 9 10 RJE/LEH 15 14 13 12 11 10 09

Program Authors

Dr. Jay K. Hackett
Professor Emeritus of Earth Sciences
University of Northern Colorado
Greeley, CO

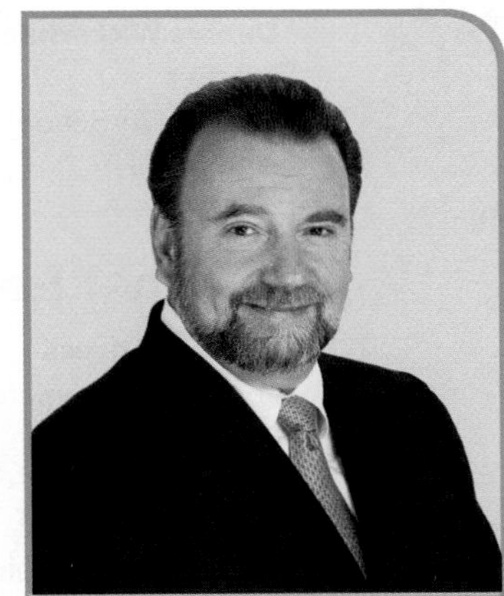

Dr. Richard H. Moyer
Professor of Science Education and Natural Sciences
University of Michigan–Dearborn
Dearborn, MI

Dr. JoAnne Vasquez
Elementary Science Education Consultant
NSTA Past President
Member, National Science Board and NASA Education Board

Mulugheta Teferi, M.A.
Principal, Gateway Middle School
Center of Math, Science, and Technology
St. Louis Public Schools
St. Louis, MO

Dinah Zike, M.Ed.
Dinah Might Adventures LP
San Antonio, TX

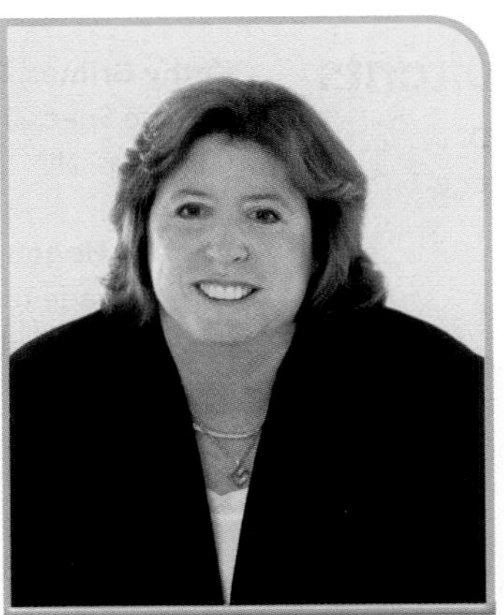

Kathryn LeRoy, M.S.
Chief Officer
Curriculum Services
Duval County Schools, FL

Dr. Dorothy J. T. Terman
Science Curriculum Development Consultant
Former K–12 Science and Mathematics Coordinator
Irvine Unified School District
Irvine, CA

Dr. Gerald F. Wheeler
Executive Director
National Science Teachers Association

Bank Street College of Education
New York, NY

Contributors and Reviewers

Contributing Authors

Dr. Sally Ride
Sally Ride Science
San Diego, CA

Lucille Villegas Barrera, M.Ed.
Elementary Science Supervisor
Houston Independent School District
Houston, TX

American Museum of Natural History
New York, NY

Contributing Writer

Ellen C. Grace, M.S.
Consultant
Albuquerque, NM

Content Consultants

Paul R. Haberstroh, Ph.D.
Mohave Community College
Lake Havasu City, AZ

Timothy Long
School of Earth and Atmospheric Sciences
Georgia Institute of Technology
Atlanta, GA

Rick MacPherson, Ph.D.
Program Director
The Coral Reef Alliance
San Francisco, CA

Hector Córdova Mireles, Ph.D.
Physics Department
California State Polytechnic University
Pomona, CA

Charlotte A. Otto, Ph.D.
Department of Natural Sciences
University of Michigan–Dearborn
Dearborn, MI

Paul Zitzewitz, Ph.D.
Department of Natural Sciences
University of Michigan–Dearborn
Dearborn, MI

Editorial Advisory Board

Deborah T. Boros, M.A.
President, Society of Elementary Presidential Awardees
Second Grade Teacher
Mississippi Elementary
Coon Rapids, MN

Lorraine Conrad
K–12 Coordinator of Science
Richland County School District #2
Columbia, SC

Kitty Farnell
Science/Health/PE Coordinator
School District 5 of Lexington and Richland Counties
Ballentine, SC

Kathy Grimes, Ph.D.
Science Specialist
Las Vegas, NV

Richard Hogen
Fourth Grade Teacher
Rudy Bologna Elementary School
Chandler, AZ

Kathy Horstmeyer
Educational Consultant
Past President, Society of Presidential Awardees
Past Preschool/Elementary NSTA Director
Carefree, AZ and Chester, CT

Jean Kugler
Gaywood Elementary School
Prince Georges County Public Schools
Lanham, MD

Bill Metz, Ph.D.
Science Education Consultant
Fort Washington, PA

Karen Stratton
Science Coordinator K–12
Lexington District One
Lexington, SC

Emma Walton, Ph.D.
Science Education Consultant
NSTA Past President
Anchorage, AK

Debbie Wickerham
Teacher
Findlay City Schools
Findlay, OH

Teacher Reviewers

Barbara Adcock
Pocahontas Elementary
Powhatan, VA

Erma Anderson
Educational Consultant
Needmore, PA

Cathryn Beck-Potter
Chestatee Elementary
Gainesville, GA

Teri Warden Bickmore, M.Ed.
Science Consultant
Midland, MI

Jaime Breedlove
Jane D. Hull Elementary
Chandler, AZ

Jacqueline M. Brown
Cascade Elementary
Atlanta, GA

April M. Bruce
Supervisor for Instruction
Lynchburg City Schools
Lynchburg, VA

Patricia A. Cavanagh
Merrimac Elementary
Holbrook, NY

Meghan Ciacchella
L'Anse Creuse Public Schools
Chesterfield, MI

Gary L. Cooper
Science Department Chair, Biology Teacher
MSD of Pike Township Schools
Indianapolis, IN

Sarah M. D'Agostini
Joseph M. Carkenord Elementary
Chesterfield, MI

Dr. Kelly A. Decker
University of Richmond
Richmond, VA

Frances Pistone DeLuca
Our Lady of Perpetual Help School
South Ozone Park, NY

Wendy DeMers
Hynes Charter School
New Orleans, LA

Kellie DeRango
Washington Elementary
Wauwatosa, WI

Sheri Dudzinski
Marie C. Graham Elementary
Harrison Township, MI

Delores Dalton Dunn
Curriculum Specialist (retired)
Virginia Department of Education
Hanover, VA

Laura A. Edwards
Vickery Creek Elementary
Cumming, GA

Mary Vella Ernat
District Elementary Science Content Leader
Wayne-Westland Community Schools
Westland, MI

Jenny Sue Flannagan
Elementary Science Coordinator
Virginia Beach City Public Schools
Virginia Beach, VA

Marjorie Froberger, M.A.
Anchor Bay Schools
New Baltimore, MI

Clara Mackin Fulkerson
Curriculum Resource Consultant
Nelson County Schools
Bardstown, KY

Lou Gatto
Hunterdon Central District School
Flemington, NJ

Lori Gehrman
Jane D. Hull Elementary
Chandler, AZ

Angela Geibel
Francis A. Higgins Elementary
Chesterfield Township, MI

Lori Gilchrist
Sharon Elementary
Suwanee, GA

Connie Grubbs
Varner Elementary
Powder Springs, GA

Tasha Hamil
Cumming Elementary
Cumming, GA

Nancy Hayes
Educational Consultant
Lemont, IL

Carol Johnson
Jane D. Hull Elementary
Chandler, AZ

Jerry D. Kelley, Ed.S.
Chestatee Elementary
Forsyth, GA

Andrew C. Kemp
Jefferson County Public Schools
Louisville, KY

Heather W. Kemp
Middletown Elementary
Louisville, KY

Tricia Reda Kerr
Science Specialist, EXCEL Program
The Ohio State University
Columbus, OH

Barbara Kingston
Blessed Sacrament School
Jackson Heights, NY

Gina Koger
Carroll County Public School
Westminster, MD

Bonnie Kohler
L'Anse Creuse Public Schools
Harrison Township, MI

Heather LeBlanc
Chestatee Elementary
Gainesville, GA

Larry Lebofsky
Senior Research Scientist
Lunar and Planetary Laboratory
University of Arizona
Tucson, AZ

Richard MacDonald
Science Curriculum Leader
Hampton City Schools
Hampton, VA

Brenda S. Martin
Coal Mountain Elementary
Cumming, GA

Rebecca Martin
Westridge Elementary
Frankfort, KY

Corinne Masters
Natoma Elementary
Natoma, KS

Tiah E. McKinney
Albert Einstein Fellow
National Science Foundation
Arlington, VA

Sharon Meyer
Barnesville Elementary
Barnesville, OH

Janiece Mistich
Tchefuncte Middle School
Mandeville, LA

Anthony Molock
Cascade Elementary
Atlanta, GA

Sandy Morris
Department of Learning Services
Wichita, KS

Terri Oatis-Wilson
Peyton Forest Elementary
Atlanta, GA

Brenda A. Oulsnam
Clayton County Schools (retired)
Jonesboro, GA

Jim Peters
Science Resource Teacher
Carroll County Board of Education
Westminster, MD

Sharon Pinion
Sawnee Elementary
Cumming, GA

Amy Quick
Joseph M. Carkenord Elementary
Chesterfield, MI

Stacey Race
Sharon Elementary
Suwanee, GA

Gloria R. Ramsey
Mathematics/Science Specialist
Memphis City Schools
Memphis, TN

Anna Reitz
Forsyth County Schools
Cumming, GA

Steve A. Rich
Science Coordinator
Georgia Youth Science & Technology Center
Carrollton, GA

Maureen Riordan
Fairway Elementary
Wildwood, MO

Richard Ruiz
Jane D. Hull Elementary
Chandler, AZ

Ruth M. Ruud
Millcreek Township School District
Erie, PA

Sarah Rybarczyk
Joseph M. Carkenord Elementary
Chesterfield, MI

Laura W. Schaefer
Coordinator, School Partnerships
Missouri Botanical Garden
St. Louis, MO

Rhonda Segraves
Settles Bridge Elementary
Suwanee, GA

Ursula M. Sexton
Senior Research Associate/Educational Consultant
WestEd
San Ramon, CA

Rita Jane Shelton
Louisa Middle School
Louisa, KY

Matt Silberglitt
Science Assessment Specialist
Minnesota Department of Education
Roseville, MN

William L. Siletti
Packer Collegiate Institute
Brooklyn, NY

Georgia Ann Smith
Sunflower Elementary
Lenexa, KS

Victoria L. Thom
Baker Elementary
Acworth, GA

Shannon Tribble
Daves Creek Elementary
Cumming, GA

Shirley Whorley
Science Coordinator, K–12
Roanoke City Public Schools
Roanoke, VA

Laura Wilkowski
Science Consultant
Midland, MI

Dr. Sharon Wynstra
Science Coordinator
Rockford Public Schools
Rockford, IL

Brad Yohe
Science Supervisor
Carroll County Public Schools
Westminster, MD

Table of Contents

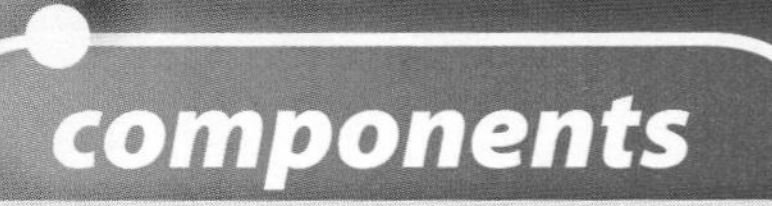

SCIENCE
A CLOSER LOOK

Components	K	1	2	3	4	5	6
Student Edition*		•	•	•	•	•	•
Teacher's Edition	•	•	•	•	•	•	•
Student Edition Unit Big Books		•	•				
Reading Essentials		•	•	•	•	•	•
Reading and Writing*		•	•	•	•	•	•
Math		•	•	•	•	•	•
Activity Lab Book		•	•	•	•	•	•
Visual Literacy*		•	•	•	•	•	•
Assessment		•	•	•	•	•	•
The Human Body and Teacher's Guide	•	•	•	•	•	•	•
Technology A Closer Look and Teacher's Guide	•	•	•	•	•	•	•
Transparencies for Visual Literacy		•	•	•	•	•	•
English Language Learner Teacher's Guide		•	•	•	•	•	•
Vocabulary Cards	•	•	•	•	•	•	•
Key Concept Cards		•	•	•	•	•	•
Leveled Readers and Leveled Reader Teacher's Guide	•	•	•	•	•	•	•
Kindergarten Flipbook	•						
Literature Big Books	•	•	•				
A to Z Activity Book	•						
Science Projects in a Pocket		•	•				
Science Resource Book	•						
Floor Puzzles	•						
Science on the Go	•	•					
Photo Sorting Cards	•	•	•				
Equipment Kits	•	•	•	•	•	•	•
Activity Flipchart		•	•	•	•	•	•
Science Fair Handbook		•	•	•	•	•	•
Online Student Edition		•	•	•	•	•	•
StudentWorks™ Plus CD-ROM		•	•	•	•	•	•
Online Teacher's Edition	•	•	•	•	•	•	•
TeacherWorks™ Plus CD-ROM	•	•	•	•	•	•	•
Exam*View® Assessment Suite* CD-ROM		•	•	•	•	•	•
Classroom Presentation Toolkit CD-ROM		•	•	•	•	•	•
Science Activity DVD		•	•	•	•	•	•
Science Songs Audio CD	•	•	•				
PuzzleMaker CD-ROM		•	•	•	•	•	•
Operation: Science Quest CD-ROM		•	•	•	•	•	•
Professional Development for Science, The Master Teacher Series	•	•	•	•	•	•	•
Teacher's Desk Reference	•	•	•	•	•	•	•
Companion Web site	•	•	•	•	•	•	•

Pre-K Components: Teacher's Edition, Flipbook, Big Science Readers, Photo Cards, Science Song Posters, and Science Songs CD

* also available in Spanish

SCIENCE
A CLOSER LOOK

A wealth of resources that brings science to life

Flipbook and Teacher Edition ▶

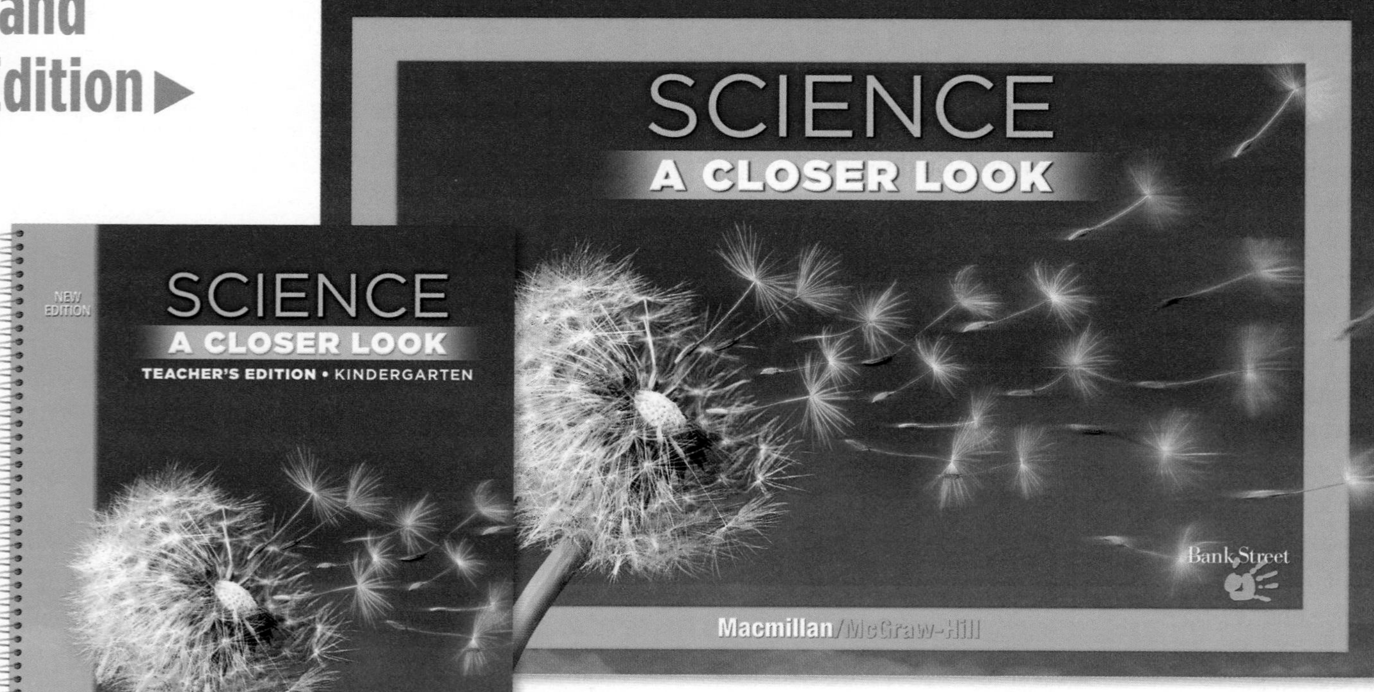

Pre-K also available

Activity Resources ▶

Materials Kits support Science Activities.

Grab 'n Go Activity Bags make preparing for activities quick and easy.

activities & content

Standards-based instruction with activities that work.

access for all

Support for all learners.

ease of use

Center-focused instruction.

Instructional Resources ▼

Reading Resources make science accessible and fun to read.

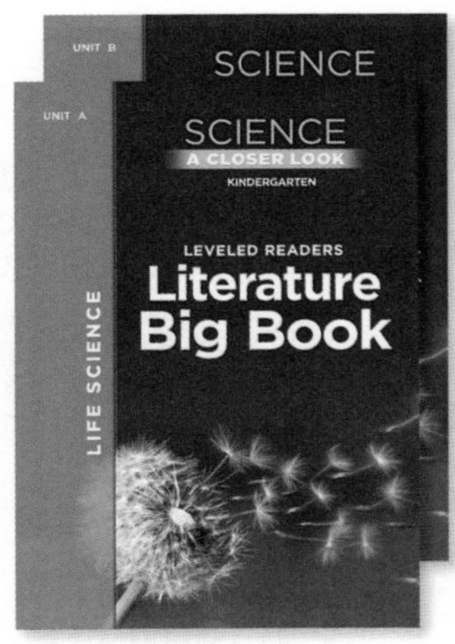

Supporting Resources provide for a wide variety of student needs.

A to Z Activity Book

Floor Puzzles

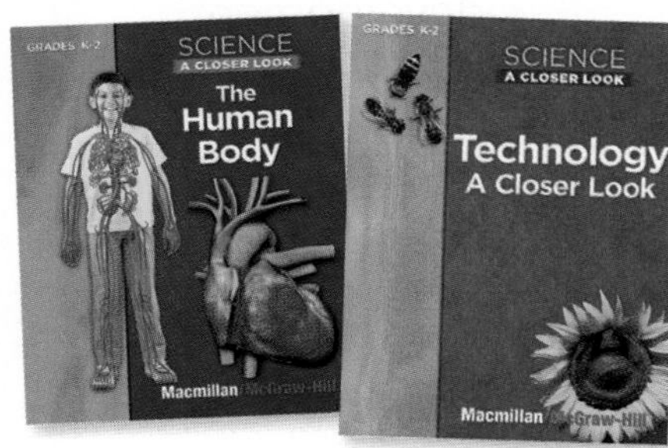

also with technology support

Technology for Kindergarten

Practice and Activities ▼

Science in Motion ▼

Animated key concepts with assessment

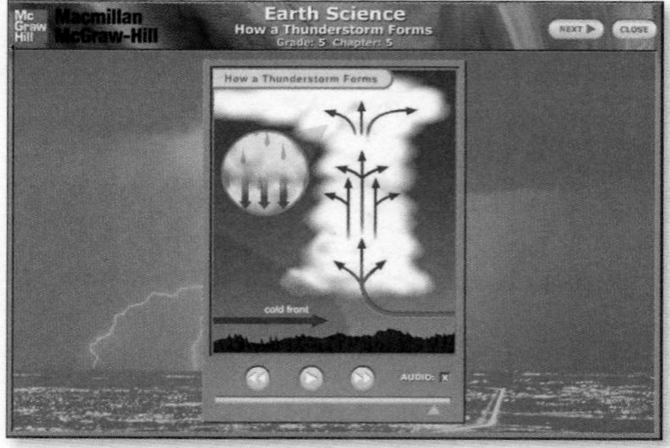

Unit A	Strawberry Plant Life Cycle of a Bean Plant Wheat to Bread
Unit B	The Food Chain Parts of a Fish
Unit C	Beach Rocks
Unit D	The Seasons
Unit E	Water Changes
Unit F	Gravity at Work

Vocabulary Activities ▼

Review vocabulary words and definitions online

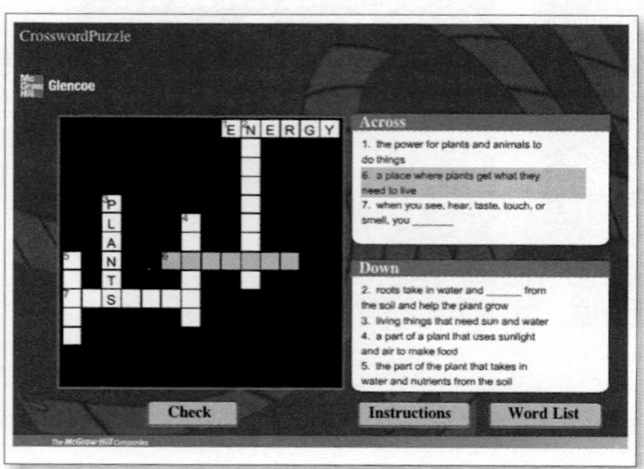

Science Songs Audio CDs ▼

Reinforce science content with songs and music

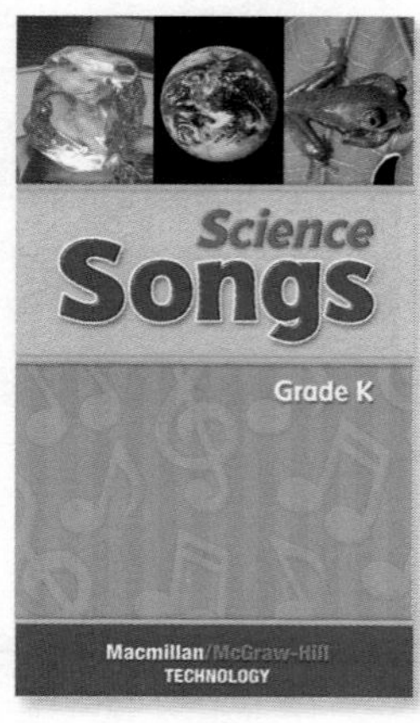

Planning and Instruction ▼

TeacherWorks™ Plus CD-ROMs ▶

Electronic Teacher's Edition with lesson planning and printable resources

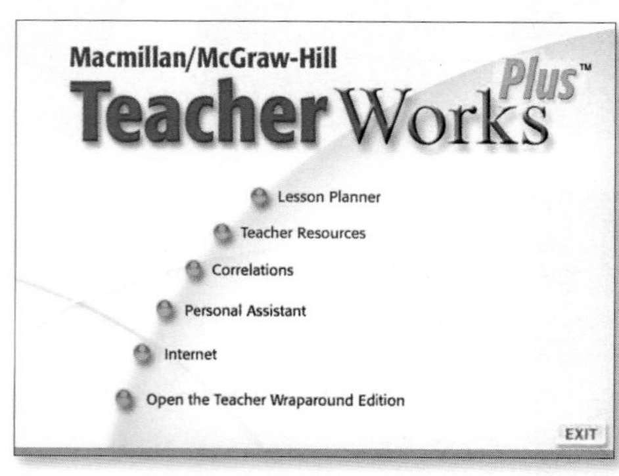

Teacher Resources ▶

Find professional development resources online

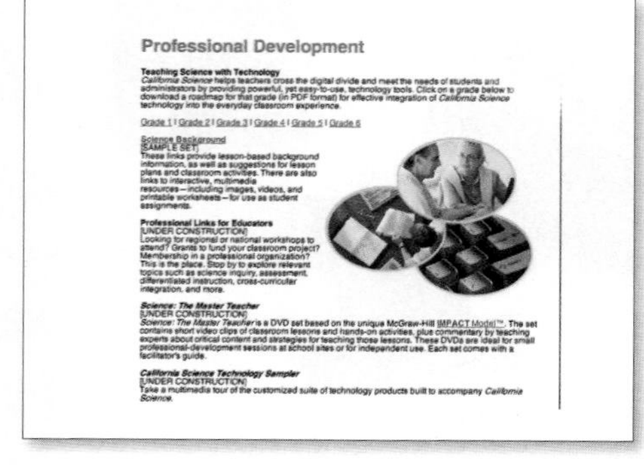

Additional Online Resources ▼

Visit www.macmillanmh.com for online resources.

◀ American Museum of Natural History
The museum offers an extensive collection of resources and programs for students and teachers.

◀ National Science Digital library
Funded by the National Science Foundation, this digital library constitutes an online network of learning environments and resources for science, technology, engineering, and mathematics.

◀ Joint Oceanographic Institute
JOI is a consortium of 29 premier oceanographic research institutions that serves the U.S. scientific community by leading large-scale, global research programs.

◀ Bank Street College
The College offers a rich source of innovative ideas and practices aimed at improving the education of children.

◀ Ranger Rick
A program of the National Wildlife Federation, its mission is to inspire Americans to protect wildlife for our children's future.

◀ TIME for Kids
TIME for Kids brings real news and enlightening information to young people so that they can develop a lifelong interest in, and a connection to, world events.

◀ Sally Ride Science
Through the creation of innovative science programs and publications, Sally Ride Science's mission is to empower girls to explore and actively participate in the fields of science, mathematics and engineering.

program overview

	Pre-K	Kindergarten	Grade 1	Grade 2
Life Science	**Unit A** **Be a Scientist**	**Unit A** **Plants** • Parts of Plants • What Plants Need and How They Grow • Leaves and Flowers and How We Use Plants	**Unit A** **Plants** Chapter 1 **Plants Are Living Things** Chapter 2 **Plants Grow and Change**	**Unit A** **Plants and Animals** Chapter 1 **Plants** Chapter 2 **Animals**
	Unit B **Plants**			
	Unit C **Animals**	**Unit B** **Animals** • Animals Are Everywhere • Animal Needs • How Animals Grow and Change	**Unit B** **Animals and Their Homes** Chapter 3 **All About Animals** Chapter 4 **Places to Live**	**Unit B** **Habitats** Chapter 3 **Looking at Habitats** Chapter 4 **Kinds of Habitats**
Earth and Space Science	**Unit D** **Our Earth**	**Unit C** **Our Earth, Our Home** • Soil and Rocks • Land and Water • Resources and Recycling	**Unit C** **Our Earth** Chapter 5 **Looking at Earth** Chapter 6 **Caring for Earth**	**Unit C** **Our Earth** Chapter 5 **Land and Water** Chapter 6 **Earth's Resources**
	Unit E **Sky and Weather**	**Unit D** **Weather and Sky** • Look at Weather • Seasons • Sun, Moon, Stars	**Unit D** **Weather and Sky** Chapter 7 **Weather and Seasons** Chapter 8 **The Sky**	**Unit D** **Weather and Sky** Chapter 7 **Observing Weather** Chapter 8 **Earth and Space**
Physical Science	**Unit F** **Matter and Motion**	**Unit E** **Exploring Matter** • Paper and Cloth • Wood, Metal, and Clay • Investigate Water	**Unit E** **Matter** Chapter 9 **Matter Everywhere** Chapter 10 **Changes in Matter**	**Unit E** **Matter** Chapter 9 **Looking at Matter** Chapter 10 **Changes in Matter**
		Unit F **Moving Right Along** • Wheels and Motion • Gravity and Sounds • Magnets	**Unit F** **Motion and Energy** Chapter 11 **On the Move** Chapter 12 **Energy Everywhere**	**Unit F** **Motion and Energy** Chapter 11 **How Things Move** Chapter 12 **Using Energy**

	Grade 3	Grade 4	Grade 5	Grade 6
Life Science	**Unit A Living Things** Chapter 1 **A Look at Living Things** Chapter 2 **Living Things Grow and Change**	**Unit A Living Things** Chapter 1 **Kingdoms of Life** Chapter 2 **The Animal Kingdom**	**Unit A Diversity of Life** Chapter 1 **Cells and Kingdoms** Chapter 2 **Parents and Offspring**	**Unit A Diversity of Life** Chapter 1 **Classifying Living Things** Chapter 2 **Cells**
	Unit B Ecosystems Chapter 3 **Living Things in Ecosystems** Chapter 4 **Changes in Ecosystems**	**Unit B Ecosystems** Chapter 3 **Exploring Ecosystems** Chapter 4 **Surviving in Ecosystems**	**Unit B Ecosystems** Chapter 3 **Interactions in Ecosystems** Chapter 4 **Ecosystems and Biomes**	**Unit B Patterns of Life** Chapter 3 **Genetics** Chapter 4 **Ecosystems**
Earth and Space Science	**Unit C Earth and Its Resources** Chapter 5 **Earth Changes** Chapter 6 **Using Earth's Resources**	**Unit C Earth and Its Resources** Chapter 5 **Shaping Earth** Chapter 6 **Saving Earth's Resources**	**Unit C Earth and Its Resources** Chapter 5 **Our Dynamic Earth** Chapter 6 **Protecting Earth's Resources**	**Unit C Earth and Its Resources** Chapter 5 **Changes over Time** Chapter 6 **Conserving Our Resources**
	Unit D Weather and Space Chapter 7 **Changes in Weather** Chapter 8 **Planets, Moons, and Stars**	**Unit D Weather and Space** Chapter 7 **Weather and Climate** Chapter 8 **The Solar System and Beyond**	**Unit D Weather and Space** Chapter 7 **Weather Patterns** Chapter 8 **The Universe**	**Unit D Weather and Space** Chapter 7 **Weather and Climate** Chapter 8 **Astronomy**
Physical Science	**Unit E Matter** Chapter 9 **Observing Matter** Chapter 10 **Changes in Matter**	**Unit E Matter** Chapter 9 **Properties of Matter** Chapter 10 **Matter and Its Changes**	**Unit E Matter** Chapter 9 **Comparing Kinds of Matter** Chapter 10 **Physical and Chemical Changes**	**Unit E Matter** Chapter 9 **Classifying Matter** Chapter 10 **Chemistry**
	Unit F Forces and Energy Chapter 11 **Forces and Motion** Chapter 12 **Forms of Energy**	**Unit F Forces and Energy** Chapter 11 **Forces** Chapter 12 **Energy**	**Unit F Forces and Energy** Chapter 11 **Using Forces** Chapter 12 **Using Energy**	**Unit F Forces and Energy** Chapter 11 **Exploring Forces** Chapter 12 **Exploring Energy**

Focus on Kindergarten

Program Philosophy

Macmillan/McGraw-Hill's science program engages the innate curiosity that Kindergartners bring with them to school.

Each lesson has a variety of whole-class and small-group activities and materials that introduce concepts and provide opportunities for children to work together to investigate and then communicate their ideas and discoveries to teachers and peers.

The introduction to the **science inquiry skills** will provide a firm foundation for children's continued work in science.

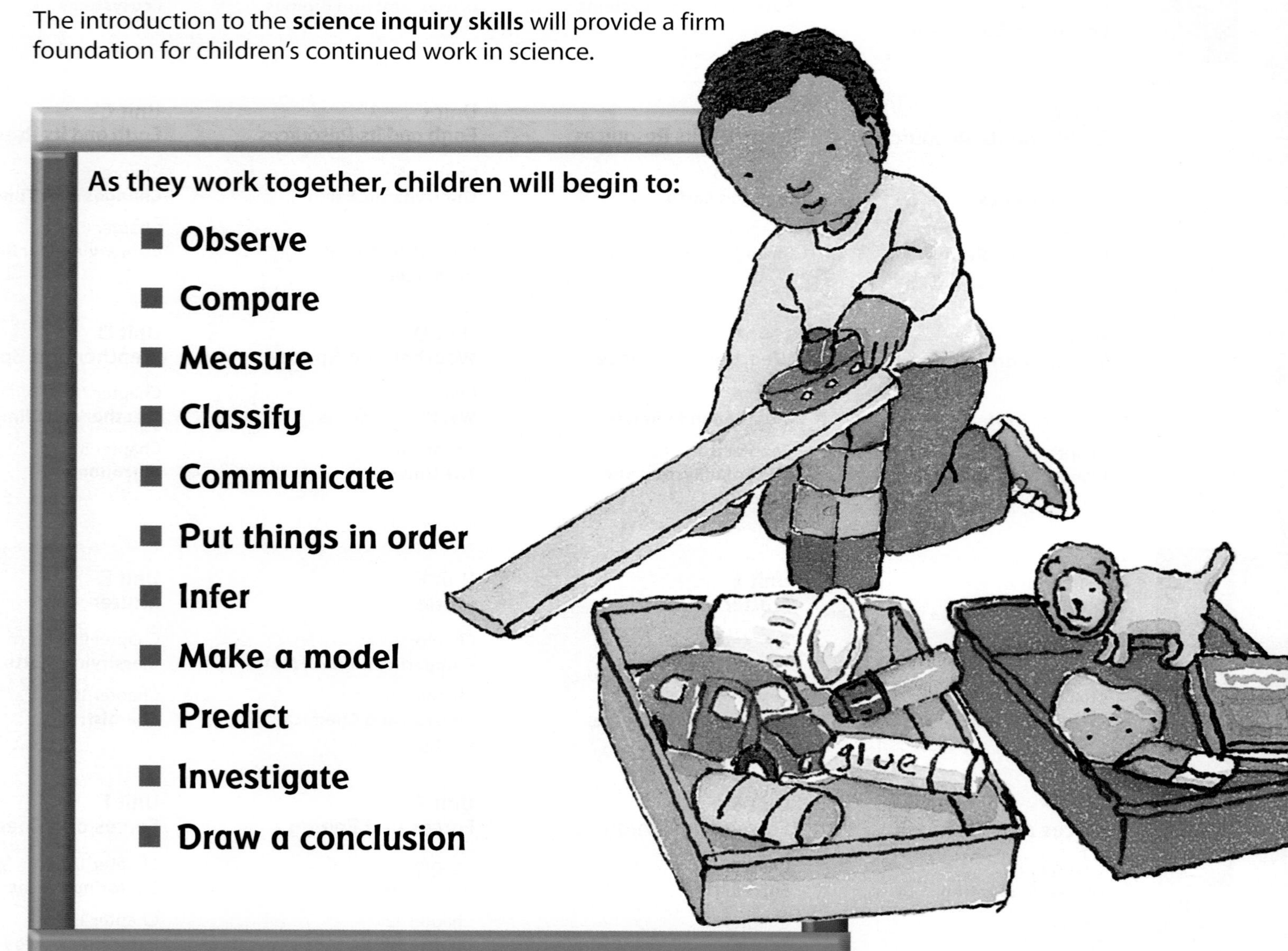

As they work together, children will begin to:

- **Observe**
- **Compare**
- **Measure**
- **Classify**
- **Communicate**
- **Put things in order**
- **Infer**
- **Make a model**
- **Predict**
- **Investigate**
- **Draw a conclusion**

Making It Your Own

What the Research Says . . .

"Inquiry into authentic questions generated from student experiences is the central strategy for teaching science."

(National Science Education Standards, p. 31)

Early childhood teachers must be flexible and respond quickly to "teachable science moments" as they arise. Here is an example:

Willie the Hamster

On a Friday afternoon, a student noticed that the watering can was almost full. When he returned on Monday, he was surprised to see that it was almost empty. At a whole-group meeting the student asked, "Who used the water?" After much denial on the part of the students, the group decided that Willie, the class hamster, must have been drinking the water at night. The teacher then asked, "How can we be sure it's Willie?" The students decided to put Willie's cage in the middle of the Sand Table before they left that day. The next morning—no footprints in the sand! It couldn't be Willie. The teacher then suggested measuring the height of the water each day. The level was decreasing, but not by the same amount each day. In discussing the results of their investigation, a child said, "You know, once my mother put the clothes in the dryer and forgot to turn on the heat. It took a really long time to dry the clothes. Maybe water disappears faster when it is warmer!"

(adapted from the National Science Education Standards, Chapter 6, pp. 124–125)

Rather than provide the answer, this teacher helped her students:

- Investigate the question that came up as part of the daily routine of watering the plants.
- Refine their thinking based on their investigations.
- Engage in a discussion in which they linked what they had observed with what they knew of the outside world.

Macmillan/McGraw-Hill's science program helps you learn how to respond to such "teachable moments" by facilitating lively discussions in each lesson that revolve around questions such as: *How do you know? What could we do to find out?* Then, when someone asks *Who's drinking the water?* you and your students will know what to do. You will be "doing science," too!

Science Content

What the Research Says . . .

The *National Science Education Standards* emphasize that from the earliest grades, students must:

- Develop the ability to ask their own questions.
- Answer them by seeking information from reliable sources and from their own observations and investigations.

(National Science Education Standards, p. 121)

Macmillan/McGraw-Hill Science is designed to help you create an inquiry-based science program.

"Inquiry into authentic questions generated from student experiences is the central strategy for teaching science."

(National Science Education Standards, p. 31)

In Macmillan/McGraw-Hill's science program:

- Each unit begins by asking children to share what they know, what they wonder about, and what they want to find out. This discussion will help you select activities that will be responsive to their interests.
- The images and questions in the Flipbook are designed to encourage broad child participation in science discussions.

"At all stages of inquiry, teachers guide, focus, challenge, and encourage student learning."

(National Science Education Standards, p. 33)

In Macmillan/McGraw-Hill's science program:

- Each lesson includes a variety of small-group hands-on activities that will help you observe, support, and challenge children as they work.
- Formative assessment strategies are offered throughout the unit to help you focus your observations so that you can more effectively meet the needs of individual children.

"Effective science teaching depends on the availability and organization of materials, equipment, media, and technology."

(National Science Education Standards, p. 44)

In Macmillan/McGraw-Hill's science program:

- The "Be a Scientist" section helps you establish daily science routines and effective year-long studies, and provides suggestions on plants, animals, and Science Center activities to incorporate into your class.
- The program components can be used to integrate science into other parts of your day.
- Multiple online resources and bibliographies in the Teacher's Edition will enrich your planning and lesson implementation.

"An important stage of inquiry and of student science learning is the oral and written discourse that focuses the attention of students on how they know what they know and how their knowledge connects to larger ideas, other domains, and the world beyond the classroom."

(National Science Education Standards, p. 36)

In Macmillan/McGraw-Hill Science:

- Each lesson helps you create questions that will promote higher-order thinking and create lively whole-class and small-group discussions.
- Each unit contains activity and center suggestions that help you integrate what you are doing in science throughout your day.
- The collection of Science Leveled Readers and Photo Sorting Cards that accompany each unit provide additional science content to support your literacy instruction.

"Teachers must have theoretical and practical knowledge and abilities about science, learning, and science teaching."

(National Science Education Standards, p. 28)

In Macmillan/McGraw-Hill Science:

- Each lesson provides "Science Facts" that give you additional information about the images in the Flipbook.
- There is a glossary of scientific terms written specifically for the Kindergarten teacher to support his or her science background.

"Teachers who are enthusiastic, interested, and who speak of the power and beauty of scientific understanding instill in their students some of those same attitudes."

(National Science Education Standards, p. 37)

In Macmillan/McGraw-Hill Science:

- The materials and activities are designed to help teachers and children observe and marvel at the beauty of the world around them.

The Developmental Profile of a Five-Year-Old

What the Research Says . . .

Social/Emotional

"Social interaction with peers and adults is essential for children to learn cooperation."

(The Guidelines for Appropriate Curriculum Content and Assessment in Programs Serving Children Ages 3 Through 8, NAEYC, 1990)

Kindergarten is a time when children learn to make choices, share their ideas and feelings, cooperate on a group task, and resolve everyday conflicts. They need time and teacher support to learn essential social skills.

Macmillan/McGraw-Hill's science program provides materials and activity suggestions that enable children to work independently. Children can:

- Initiate and play a matching game with Photo Sorting Cards.
- Gather together a group of friends to work on a Science Floor Puzzle.
- Work in small groups in the Science Center or other classroom centers.

Language

"Scientific literacy means that a person has the ability to describe, explain, and predict natural phenomena . . . it also implies the capacity to pose and evaluate arguments based on evidence and to apply conclusions from such arguments appropriately."

(National Science Education Standards, p. 22)

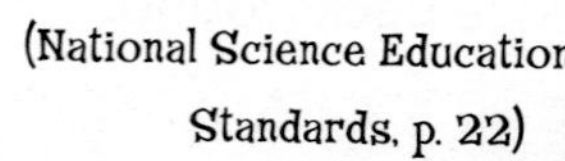

Kindergarten children need many opportunities to communicate their ideas to peers and adults. They also need teacher support to learn words that will accurately describe what they see or experience, and to use evidence to justify their tentative conclusions.

Macmillan/McGraw-Hill's science program provides a variety of ways to help children learn how to talk together.

- Each Flipbook page opens with a question that invites multiple responses.
- Activities are designed to provide teacher support to small groups so that children can be helped to interact more fully with peers and adults.
- Leveled Readers provide images and text to read and discuss.
- Easily memorized song texts teach new vocabulary and provide a safe way for shy children to join in.

Cognitive

"Even at the earliest grade levels, students should learn what constitutes evidence and judge the merits or strength of the data and information that will be used to make explanations."

(National Science Education Standards, p. 122)

Kindergarten children are capable of exploring the idea of cause and effect and drawing conclusions based on their experiences.

Macmillan/McGraw-Hill Science offers experiences that will help children refine and deepen their thinking.

- The "Be a Scientist" activity in each lesson provides opportunities for children to explore and investigate, and then discuss what happened.
- The Science Journal pages provide a record of children's work over time.
- Program materials help children improve their problem-solving abilities, better their visual discrimination skills, and enhance their visual literacy skills as they examine and discuss photographs.

Physical

"Physical development is an integral part of children's well-being. A principle focus is on children's ability to move in ways that demonstrate control, balance and coordination. Fine motor skills are equally important in laying the groundwork for artistic expression, handwriting, and self-care skills."

(Jablon et al. 1994)

Kindergarten children have just begun to develop a sense of control over their bodies and a sense of themselves as competent learners. They need opportunities to strengthen both their large and small muscles.

Macmillan/McGraw-Hill Science integrates music, movement, outdoor trips, and the use of manipulatives.

- Children can move along to the beat as they listen to and sing the Science Songs.
- Building with blocks provides opportunities to develop their fine motor skills as they learn about balance, scale, and the laws of physics.
- Outside, children may climb, stretch, and hike as they travel to nearby parks, city streets, or playgrounds to explore the natural world.

The Kindergarten Classroom

"Even the youngest students can and should participate in discussions and decisions about using time and space for work. The more independently students can access what they need, the more they can take responsibility for their own work."

(National Science Education Standards, p. 45)

Macmillan/McGraw-Hill's science program provides ideas and suggestions for creating a well-organized and well-equipped classroom that invites scientific exploration and discovery.

Art Center

Children can explore properties of matter as they:

- Cut, tear, fold, bend, and arrange paper or cloth when they make collages.
- Combine a variety of materials into sculptures.
- Explore color mixing as they paint and draw with pastels or chalk.
- Shape clay, play dough, and other pliable materials.

Water/Sand Table

Children can explore science as they:

- Pour and measure sand or water.
- Investigate how water flows by using tubes and funnels.
- Create dioramas of animal habitats.
- Explore properties of matter as they investigate what sinks and what floats, and draw tentative conclusions from that investigation.
- Experience the pleasure of working with sensorial materials.

centers

Library Center

Bibliographies of award-winning science books for young children are included so that children can:

- Read pictures and text to gain information about the world around them and explore topics that interest them.
- Reread Leveled Readers independently after hearing them read aloud from the Big Books or discussed in guided-reading groups.

Dramatic Play Center

The dramatic play area fosters cooperative play and expressive and receptive language skills as children:

- Represent what they saw or did on a trip.
- Create their own play script that incorporates information learned in the science curriculum, e.g., what animals do in the winter.

Block Center

When building in the block area, children discover:

- Principles of balance, design, and symmetry.
- How to plan and build structures that please their eye.
- How to work cooperatively with others to share the limited resources of the block area.
- Principles of physics as blocks tumble, cars roll, and things slide over and around ramps.

Organizing Your Day and Year

"Select science content and adapt and design curricula to meet the interests, knowledge, understanding, abilities, and experiences of students."

Macmillan/McGraw-Hill's science program is designed to provide flexibility in planning and implementing a science program. All units suggest using everyday, familiar objects for exploration and discovery. Some studies may require a week or less, some studies might last half a year or more. The program is designed to help you and your students embark upon an exciting voyage of discovery as your students explore the world around them.

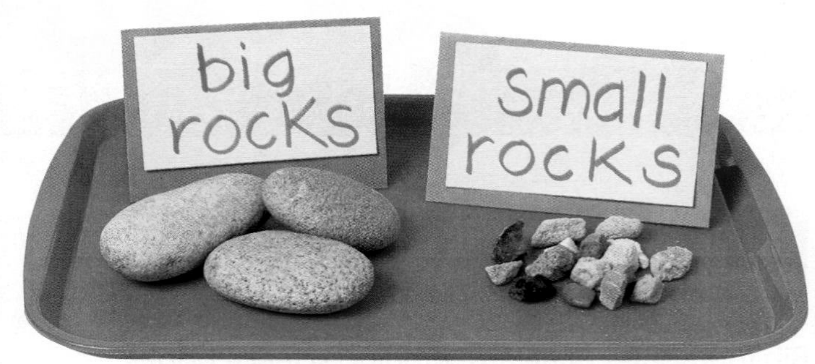

Beginning Your Year

The introductory "Be a Scientist" section is designed to help teachers learn how to develop long-term studies.

- **Investigate Weather** Establish a year-long weather study that develops children's observational skills, helps them learn how to record and save data, and how to organize and graph the data to compare the information over time.
- **The Five Senses** Use apples to explore how we can use the senses to discover more about the properties and characteristics of objects.
- **Living Things** Provide suggestions on how to add habitats for living plants and animals to your classroom Science Center. This allows children to do close observations and experiments with living things in which you and your class are interested.

Science Every Day

Each unit in the program has suggested projects, investigations, and activities that may be sustained for a few weeks, a few months, or more.

- **Unit Projects** provide long-term activities designed to allow children to observe and explore materials over time and to record and display discoveries in a culminating project.
- **Take a Trip** provides suggestions that are designed to help you plan for and implement a science curriculum that takes into consideration the unique physical environment in which you live.
- **"Be a Scientist" Activities** provide in-depth, hands-on explorations related to the lesson content. These activities allow children to take a topic, such as animal homes, and use inquiry skills to observe, investigate, and record what they discover.

"Time, space, and materials are critical components of an effective science learning environment that promotes sustained inquiry and understanding."

Here are ways you can integrate science instruction into your Kindergarten day.

Full-Day Program

8:30–8:50	**Arrival Time** Recording weather
8:50–9:20	**Morning Circle Time** Circle Time activities Flipbook discussion (1–2 a week)
9:20–10:00	**Small-Group Instruction** **Independent Work Time** Be a Scientist activities Photo Sorting Cards Floor Puzzles
10:00–10:05	**Clean-up Time**
10:05–11:00	**Literacy** Literature Big Book read-aloud Leveled Readers Songs Vocabulary activities Be a Reader/Be a Writer activities
11:00–11:30	**Outdoor Play** Take a Trip
11:30–12:30	**Lunch/Quiet Time** Science books read aloud
12:30–1:30	**Math/Reading Instruction** Be a Math Wiz activities Guided reading using Leveled Readers
1:30–2:30	**Center Time** Take a Trip Center activities
2:30–2:45	**End-of-Day Circle Time** Sharing work
2:45–3:00	**Dismissal**

Half-Day Program

8:30–8:45	**Arrival Time** Recording weather
8:45–9:10	**Morning Circle Time** Science Songs Circle Time activities Introducing science activities and materials
9:10–9:55	**Center Time** Be a Scientist activities Center activities Photo Sorting Cards Science Floor Puzzles
9:55–10:00	**Clean-up Time**
10:00–10:30	**Literacy** Flipbook discussions Literature Big Book read-aloud Science Songs Be a Reader/Be a Writer activities
10:30–10:50	**Outdoor Play** Take a Trip
10:50–11:30	**Small-Group Instruction** **Independent Work Time** Be a Math Wiz activities Guided reading using Leveled Readers Be a Scientist (if not done during Center Time)
11:30–11:45	**End-of-Day Circle Time** Sharing work
11:45–12:00	**Dismissal**

Integrating Science

A to Z Activity Book

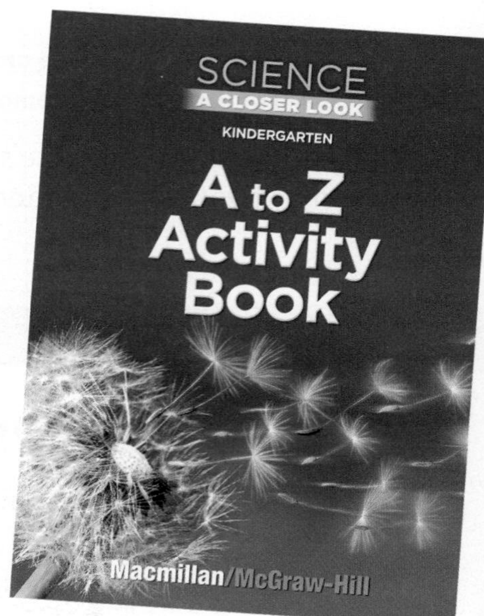

This Macmillan/McGraw-Hill Science teacher resource provides additional science activity ideas for teachers to integrate into their curriculum throughout the year. Kindergarten science themes are aligned with each of the letters of the alphabet; for each theme, four cross-curricular activity suggestions and support materials are provided.

Each activity coincides with a lesson in the Macmillan/McGraw-Hill Science program. Science themes covered include:

- A is for Amazing Ants
- B is for Blowing Bubbles
- C is for Creating Colors
- D is for Digging in Dirt
- E is for Extraordinary Eggs
- F is for Food Chains
- G is for Grow, Grow, Grow
- H is for Hibernation
- I is for Insects
- J is for Jumping Critters
- K is for Kids Like to Touch
- L is for Life Cycles
- M is for Magnets and Much More
- N is for Nature
- O is for Opposites
- P is for Paper Fun
- Q is for Quiet or Not
- R is for Rocks
- S is for Shadows
- T is for Traveling
- U is for Universe
- V is for Vision
- W is for Water
- X is for X-Ray
- Y is for Young Animals
- Z is for Z-z-z at Night

Plants

Animals

Our Earth, Our Home

Weather and Sky

Exploring Matter

Moving Right Along

Teacher Reference Section

Be a Scientist

Establishing Routines

In this first section, children will begin basic classroom routines that will help them to be scientists as they make careful observations of their surroundings. These routines, once established, will help children observe weather patterns, use their five senses to make detailed observations, and learn about living things in the classroom Science Center. Using these skills throughout the year will help turn your students into "scientists."

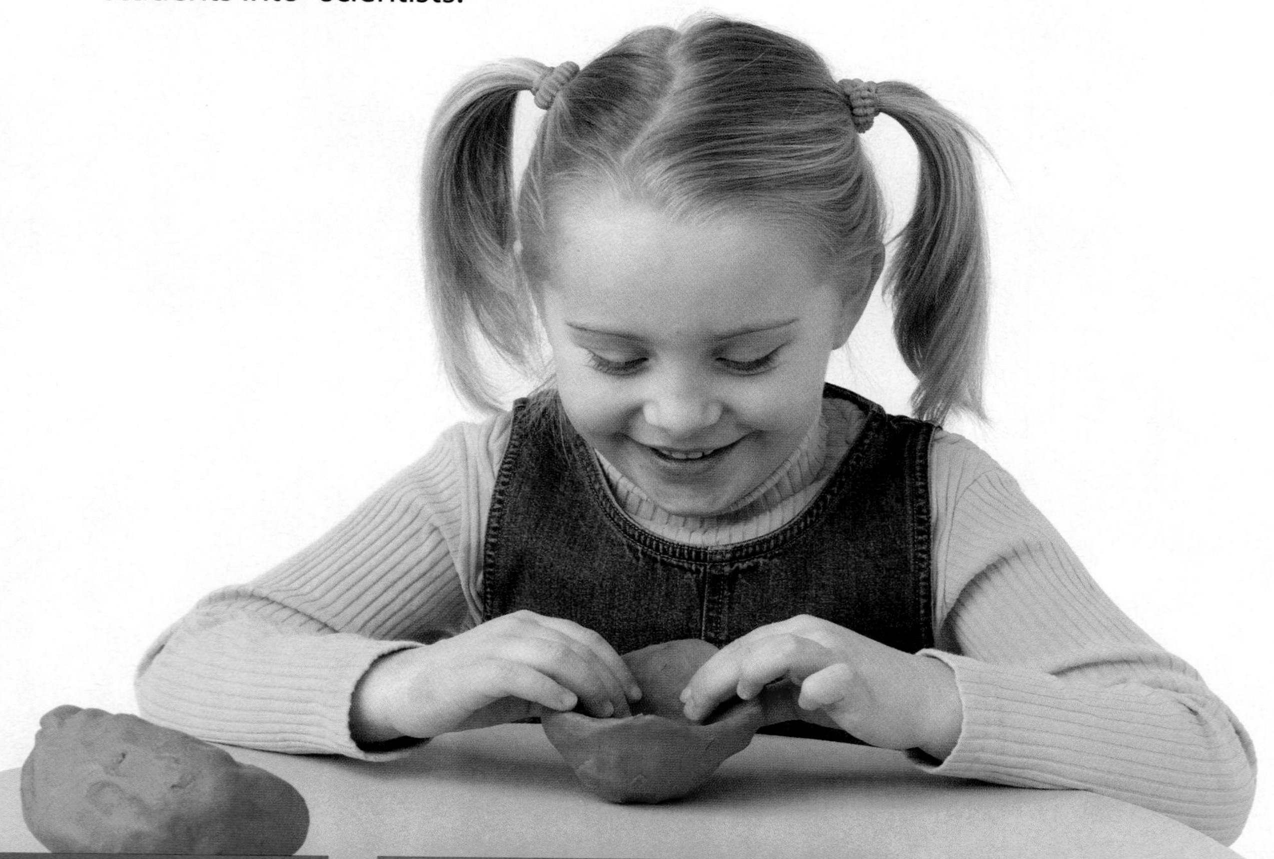

Investigate Weather

As part of a year-long study of weather patterns and seasons, children will begin to observe and record weather by:

- Watching and recording weather.
- Graphing and analyzing monthly weather patterns.
- Identifying and using appropriate weather language.

The Five Senses

Children will begin to use the senses as they carefully observe and investigate their surroundings by:

- Sorting objects by size, shape, and color.
- Identifying characteristics involving touch, sight, and hearing.
- Classifying items using visual and verbal clues.

Living Things

Children will begin to learn about what living things need and how to care for them appropriately by:

- Looking closely at characteristics of living things.
- Revisiting living things to note changes over time.
- Sorting and classifying living and nonliving things based on characteristics.

Professional Development

Establishing science routines in your classroom will enable your students to become scientists as they observe, explore, and investigate their world. Refer to the following pages for strategies to help support your science instruction:

- Talking Science pp. 8–9
  This section provides questioning strategies to develop inquiry and foster children's inquisitive nature.
- Taking Trips pp. 14–15
  This section provides guidance for taking children out into the neighborhood on field trips to explore their environment and to learn to look closely at their surroundings.
- Your Science Center pp. 20–21
  This section provides suggestions and support for setting up your classroom Science Center and introducing living things into the center.

Routine Planner

Lesson	OBJECTIVES	VOCABULARY	RESOURCES and TECHNOLOGY	MATERIALS
Investigate Weather PAGES 4–7 PACING: 3–5 days	■ Observe and record the weather over the course of the year.	sunny windy snowy cloudy rainy	**Flipbook,** p. 2 ■ **Photo Sorting Cards 31–37** ■ **A to Z Activity Book,** pp. 28–29	■ teacher-made calendar, chart paper, crayons, index cards, glue stick
The Five Senses PAGES 10–13 PACING: 3–5 days	■ Use the five senses to sort, classify, and categorize objects.	taste touch smell hear see	**Flipbook,** pp. 3, S1, G1 **Leveled Reader:** *At the Petting Zoo* **Unit A Literature Big Book**	■ apples, paper, pencils, crayons
Living Things PAGES 16–19 PACING: 3–5 days	■ Investigate living things in the classroom.	living nonliving	**Flipbook,** p. 4 ■ **Photo Sorting Cards 1–10** ■ **A to Z Activity Book,** p. 30	■ sweet potatoes, carrots, potted plants, chart paper

PACING Assumes a day is a 20–25 minute session.

LOG ON www.macmillanmh.com for more planning resources and http://nsdl.org/refreshers/science for science resources from NSDL

Investigate Weather

Read Together and Learn

▶ Build on Prior Knowledge

Ask children to share words they know that are used to describe weather. Write them on a chart labeled *Weather Words*.

▶ Use the Visuals

- **Communicate** Read the question, tracking the print with your finger as you read. Encourage children to respond freely to the question.
- After discussing the images, read the list of words on the left side of the page and have children decide which word best describe each picture. When a volunteer responds, ask: **What in the picture makes you think that? Why?**
- After someone responds, ask: **Does anyone disagree?** If they do, have them explain their thinking.

Flipbook

BE A SCIENTIST • INVESTIGATE WEATHER

What do you see?

sunny

cloudy

snowy

rainy

windy

Reading Strategy

Phonological Awareness Ask children to list words that rhyme with *do* (you, boo, who, moo, to, blue, zoo, clue, sue).

Print Awareness Point to the question and ask: **Is this an asking sentence or a telling sentence? How do you know?** Point out the question mark at the end of the sentence and explain that this means the sentence is a question.

Science Facts

Questioning Strategies The Flipbook is designed to help you and your students engage in open-ended discussions. The questions invite multiple responses and will help you learn what your students already know and what they may want to learn more about. These discussions will encourage children to communicate their own ideas and understandings. For more information on questioning techniques, see pages 8–9.

Weather Terms The Science Facts in Unit D provide information that may be useful for you to read as you begin your year-long weather study.

Professional Development For more Science Facts and resources from NSDL visit http://nsdl.org/refreshers/science

▶ Ask Questions

- **How would you describe today's weather? Which picture looks like the weather today? What makes you think that? Does anyone have another idea?**
- Point to the photo of the beach. Ask: **How do we know the wind is blowing? If you are looking out a window, how do you know if the wind is blowing?**
- Point to the photo of the flower: **What is happening to this flower? When the rain stops, what do you think will happen to the water on its leaves and petals? How do you know? Does anyone have another idea?**

Write on It

Have children use a dry-erase marker to draw a line from each term to the picture that represents that type of weather.

Think and Talk

Remind children that we can use all of our senses to learn more about the weather. Ask them to discuss different ways to observe weather (how hot or cold our skin feels; what the rain, wind, or snow feels like; what thunder sounds like; what we see moving in the wind).

More to Read

Like a Windy Day,
by Frank and Devin Asch
(Harcourt, 2002)

Activity As you read this book, have children look for clues in the illustrations that indicate that it is a windy day.

Weather Activities

Observing and recording the weather gives children the opportunity to analyze weather patterns throughout the year.

Watching Weather

You need
- teacher-made calendar on large paper
- crayons
- pencils

Objective: Learn how to observe, record, and discuss weather findings with others.

Inquiry Skills: observe, make a model, communicate, predict, infer

1. **Observe** Introduce the weather calendar during Circle Time and explain that each day a child will draw a picture on the calendar to record the weather. Have children describe that day's weather. Ask: **Is it cold? Warm? Windy?** You can also suggest looking out the windows to help them recall the weather.
2. **Make a model** Have children brainstorm what they could draw that would indicate weather patterns such as sunny, cloudy, rainy, snowy, windy. Make simple line drawings that reflect their ideas and post them on the wall near the calendar. Then model drawing the appropriate symbol(s). Add a word underneath to describe the sky and/or temperature.
3. **Communicate** As children come to Circle Time, have one child describe and record the day's weather and write down a word they use to describe the temperature.
4. **Predict** Ask the class to predict what the weather will be like the next day. Record their prediction so they can discuss the accuracy of their prediction on the following day.

Investigate More

Infer Ask children questions about the weather such as: **Will we be able to go outside today? Will we need to wear coats when we go out today?**

Graph the Weather

You need
- monthly weather calendar
- index cards
- chart paper
- glue sticks
- crayons

Objective: Categorize and graph weather.

Inquiry Skills: classify, communicate, compare, draw a conclusion

- Have the class help you use the information you have gathered all month to graph the weather. Children can color boxes on the graph to indicate the number of days of each type of weather.
- Discuss the data collected on the graph with the class. Invite children to make comparisons between different types of weather. For example, ask: **Were there more sunny days or cloudy days? Were there less warm days or cold days?**
- Have children help you hang the graph in the hall near the classroom in order to share their information with other members of the school community.

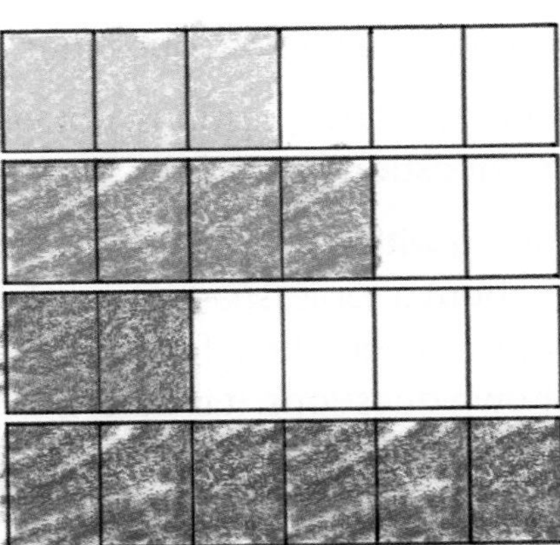

Weather Words

You need
- Photo Sorting Cards 31–37

Objective: Identify and match weather images and words.

Inquiry Skills: observe, compare

- Have pairs of children place both sets of weather cards facedown on the table.
- Have children take turns flipping over two cards looking for a match.
- When they find a match, have them identify the beginning letter of the word at the bottom of the card. Play until all matches have been found.

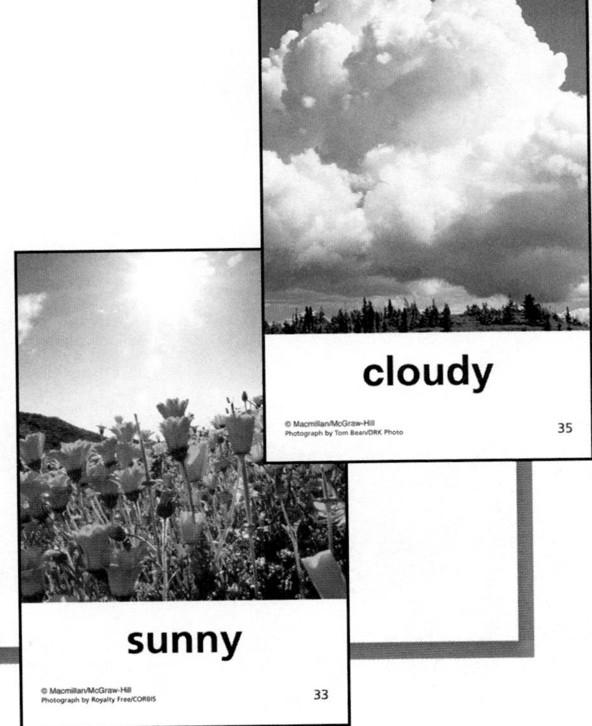

Talking Science

Beginning Your Science Work

Support children's natural curiosity by creating meaningful, hands-on science experiences. You can also help them learn how to investigate their own questions.

In order to meet such goals, you must find time to:

- Listen to what they say.
- Encourage them to ask questions.
- Respond to their questions, not with answers, but with a question instead.

Responding with a question helps children think deeper about a topic, share more of what they know, and may even help them formulate tentative answers to their own questions. It is hard, but worth practicing.

Developing Questioning Strategies

When we ask children questions, it is often because we want to find out if they know something.

- You might begin your science work by asking: **What do you think scientists do?**
- If a child responds by saying: "They look at things," you might be tempted to say: "That's right," and record their response on a chart. Then move on to another child.
- Here is an alternative way to respond: **What kind of things do they look at?** This question gives children an opportunity to share more of their thinking. You might find out that they know scientists look at stars or dinosaur bones. Their response to this question gives you information about what they know, and what they might be curious to know more about.
- You could continue to ask questions to probe further: **Why do you think they look at stars or dinosaur bones?**

Sometimes you will have to answer children's questions with answers. They may insist on an answer from you, or you may not have time to engage in this kind of in-depth discussion at that moment. Remember you can also respond by saying:

- **That is a good question. We do not have time to talk about it now, but I will write it down so we can discuss it later.**
- **I am not sure, what do you think?**
- **I do not know. How could we find out?**

Addressing Misconceptions

As teachers, we are anxious to cover all the standards, teach all the necessary vocabulary, and make sure children know certain "science facts." In fact, we also know that children's behavior and thinking develop over time. They are continually revising their understanding of the world, and no matter what we tell them, they might not be able to understand it yet. We must be careful not to correct children's misconceptions too soon.

Doing What Scientists Do

One Kindergarten teacher, who had guinea pigs in her classroom, was introducing a rabbit to the group. She asked the children: "What do you know about rabbits?" One child said that guinea pigs grew up to be rabbits.

The teacher responded by asking: "Why do you think that?" The child said that as guinea pigs grew, their ears would become longer and they would turn into rabbits.

Why this response? The child was using what she knew to try and figure something out.

- She knew her own body parts were getting bigger over time.
- She knew that rabbit ears were long and that rabbits had fur and four legs.
- She also knew guinea pigs had fur and four legs.

It is not correct, but she was putting together the information she had gathered about the way things work in the world and was trying to make sense of it. She was doing what a scientist would do.

Rather than just telling the child that was incorrect, the teacher asked: "Since we have both guinea pigs and a rabbit in our classroom, how could we find out if guinea pigs turn into rabbits?"

- Some children said they should watch the guinea pigs to see what happened.
- Another child shared that she had guinea pigs at home and one got really old and died, but it never turned into a rabbit.
- Now the children were able to weigh this new evidence and draw their own conclusions.

In this meeting, children talked and listened to one another. The teacher facilitated this discussion by asking questions that helped children share their thinking. She did not just tell children the answers. As a result, they discovered that they could work together to draw conclusions and develop plans to investigate a question that was important to them.

Talking seriously about science and "doing" science together can be an exciting, rewarding experience for you and your class. This type of science teaching helps children craft their own questions, develop hypotheses, and work together to investigate possible answers to their questions.

For more information, read:

Doing What Scientists Do: Children Learn to Investigate Their World,
by Ellen Doris
(Heinemann, 1991)

The Five Senses

Read Together and Learn

▶ Build on Prior Knowledge

Begin by asking children to share what they know about apples. Label a chart *What We Know About Apples* and record their responses. Start a second chart labeled *What We Want to Learn About Apples* and encourage children to generate questions.

▶ Use the Visuals

- **Communicate** Read the question, tracking the print with your finger as you read. Encourage children to respond freely to the question.
- If children describe something that does not appear on the chart *What We Know About Apples*, add their comments to it.
- **Observe** Discuss how the children in the pictures are using their five senses to explore apples. Point out that the girl is using her nose to smell the pie, the girl is using her fingers to touch the apples, the boy is using his mouth to taste the apple and his ears to hear it crunch as he bites into it. Ask: **What other body parts do we use when we use our senses?**

Flipbook

BE A SCIENTIST • THE FIVE SENSES

What can you do with apples?

hear

taste

touch

smell

see

Reading Strategy

Phonemic Awareness Say the word *do* while emphasizing the beginning consonant sound. Have children think of other words that begin with that sound. Record their responses.

Print Awareness Have children help you count the number of words in the sentence. Turn to the previous page and ask: **What do you notice about the question on this page?**

Science Facts

Apples There are over 2,500 varieties of apples grown in the United States. These apples come in a variety of colors, shapes, and sizes. One of the most popular varieties, the Red Delicious, can be easily identified by the five bumps on the blossom end of the fruit. The core of every apple contains five seeds. Most trees begin bearing fruit after four to five years. Because 25% of an apple is air, apples float. Some apple trees have lived more than 100 years.

Apple Pie It takes two pounds of apples to make one 9-inch apple pie.

Professional Development For more Science Facts and resources from NSDL visit http://nsdl.org/refreshers/science

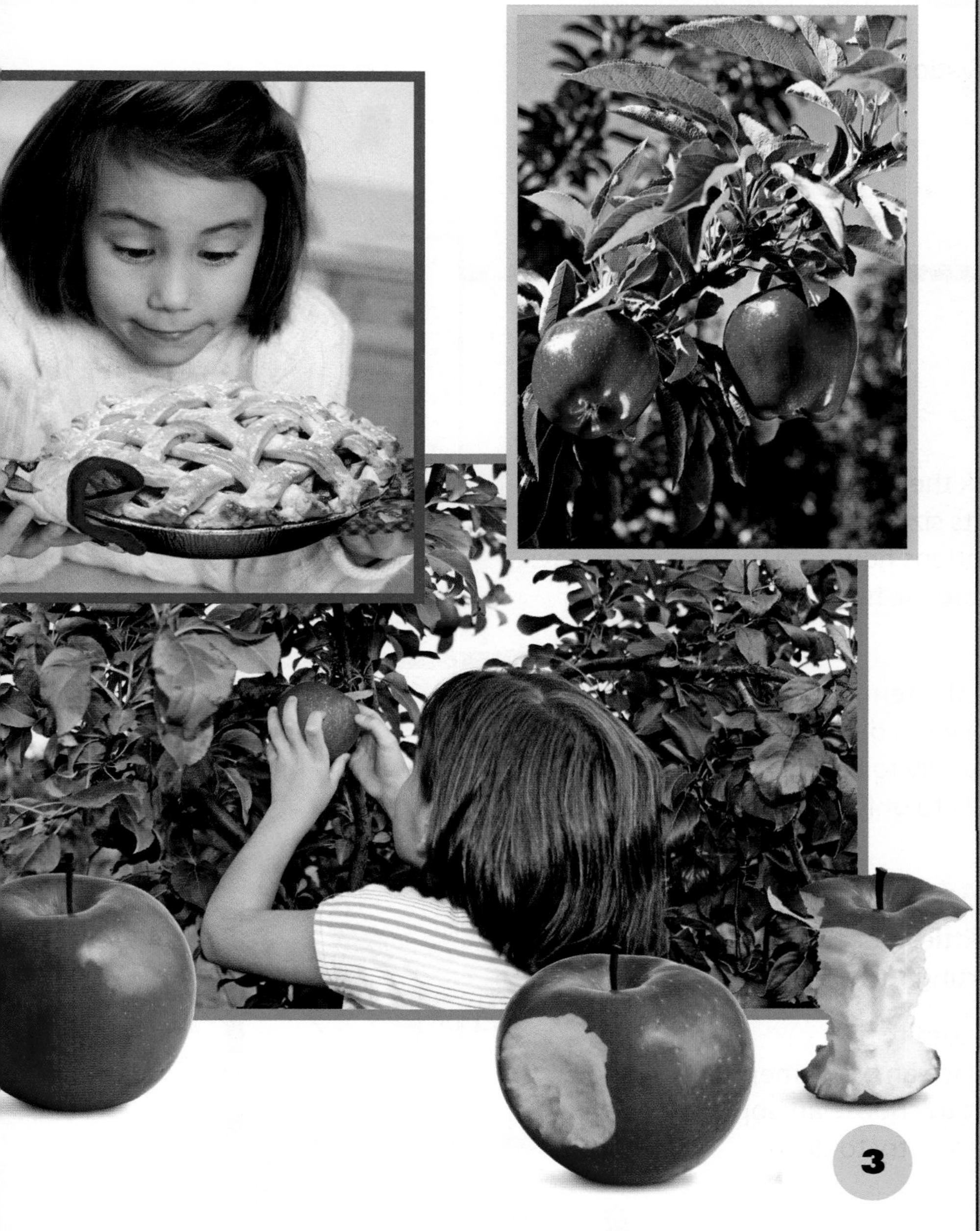

▶ Ask Questions

- **How do the apples on this page look?** Read the list of senses and ask: **What sense did you use to answer this question?** sight, my eyes
- **How might the apples on this page feel? What sense would you use to know for sure?** touch, my fingers
- **How might the apples on this page sound if we bit into them? What sense would you use to know for sure?** hearing, my ears
- **How might the apples on this page taste? What sense would you use to know for sure?** taste, my mouth and tongue
- **How might the apples on this page smell? What sense would you use to know for sure?** smell, my nose

Write on It

Have children use a dry-erase marker to draw a line from each term to the image of the sense(s) each child is using.

Think and Talk

Remind children that scientists observe things carefully in order to learn about them. By using our senses, we can learn more about the characteristics of various objects. Ask: **What other ways do you use your senses?**

More to Read

My Five Senses,
by Margaret Miller
(Simon and Schuster, 1994)

Activity After reading this book, take photographs of objects in your classroom environment and have children help you create a book using the example in *My Five Senses.*

The Five Senses Activities

Using the senses to make thorough and careful observations is a skill that children will use throughout the year.

Sorting Apples

You need
- apples
- paper
- pencils
- crayons
- Flipbook p. G1

Objective: Use the senses to sort apples.

Inquiry Skills: observe, compare, classify, communicate, investigate

1. **Observe** Show the class an apple and ask them to describe what they see. Help them generate a list of terms about its size, color, and texture. Repeat this with each type of apple. If a child mentions how it tastes, you might say: **We have not tasted this apple. What would we have to do to know for sure?**
2. **Compare** Choose two apples and, using the terms the children generated, help them complete a Venn diagram. Label the circles and write the terms that apply to both apples in the intersection and the terms that apply to only one type of apple in the appropriate circle.
3. **Classify** Have a child describe how he or she might sort the apples into two groups (red, not red; big, little; green, not green). Repeat this activity using other sorting rules.
4. **Communicate** Explain that scientists record their observations by drawing what they see. Place an assortment of apples on each table and invite children to choose an apple to draw. Encourage them to look carefully and try to draw what they see.

Investigate More

Investigate Ask: **What other ways can we use our senses to compare the apples?** Possible answers: We can smell them; we can taste them; we can listen while people bite into them; we can feel their skins and stems.

Rough or Smooth?

You need

- Unit A Literature Big Book
- Leveled Reader: *At the Petting Zoo*

Objective: Read a story and discuss how animals might look or feel.

Inquiry Skill: communicate

- Throughout the year, read at least two science books each week. By integrating science into your literacy curriculum, you will help children learn how to communicate their ideas.
- Show the Unit A Leveled Readers Literature Big Book and read the title of the story *At the Petting Zoo*. On each page, discuss what children know about the animal. Then read the word and discuss what sense was used to describe the animal.
- **Guided Reading** Use the Leveled Reader version of *At the Petting Zoo* for small-group, guided-reading instruction.

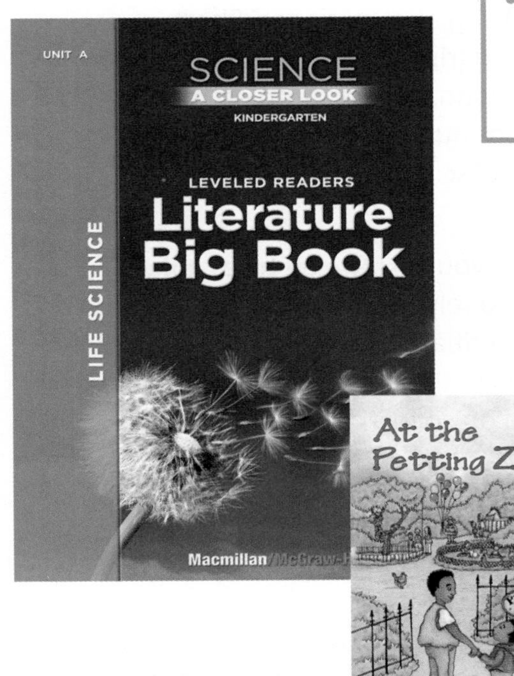

Sing About Apples

Objective: Classify apples according to their color, shape, and size.

Inquiry Skills: classify, compare

- Use music from the **Science Songs Audio CD** all year to support science instruction in your classroom.
- **Listen and Find** Invite children to listen to the song "Apples." Afterward, read each sentence on the Flipbook page and ask volunteers to come up and find apples according to their color, shape, and size. Encourage them to share how the apples are alike and different.
- **Echo and Sing** Say each line slowly, encouraging children to echo you as you track the print from left to right using a pointer or your finger. Then replay the song, inviting children to sing with you.
- See page TR22 for printed music to the song "Apples."

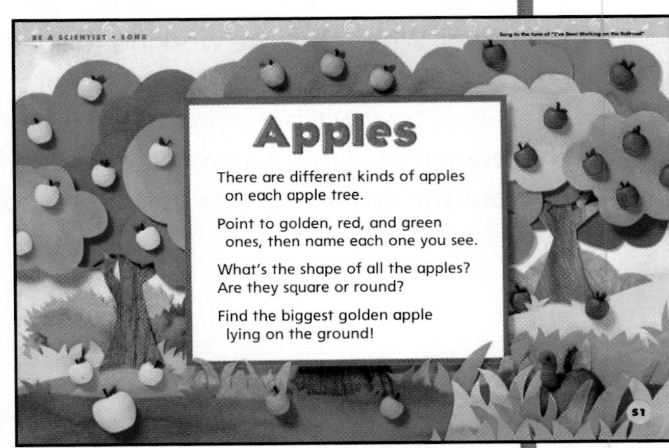

Flipbook p. S1
CD tracks 1–2

Taking Trips

Why Take Field Trips?

We know that young children learn best by "doing" science, as well as talking about science. Short neighborhood walking trips allow children to observe the natural world, collect samples, record observations, and communicate their findings to others, becoming child scientists working directly in the field.

- If, in the classroom, you ask: **Are all trees alike?** children will have to rely on their previous, imprecise, and unverifiable knowledge. If you ask that same question in the field, children can draw conclusions based on solid, first-hand evidence.
- Young naturalists can use the information they gather to make solid inferences, and collect samples to be sorted, classified, measured, and compared.
- By taking numerous trips to the same site, children can begin to predict what they might see. Their predictions will be based on previously observed phenomena, not wild guesses. They can also help make a plan and carry out their own investigations.
- Children will learn how to help articulate the goal of the trip, formulating the questions they hope to answer. When they are working to solve their own problems, and generating some of their own questions, they will become excited to meet the challenge that all good teachers ask: **How can we find out the answer to that good question?**

Where To Go?

- If you have a playground, start there. The first unit in this science program is about plants. Find a plant that children can observe over time.
- If you have only a paved schoolyard, you will still probably find plants. Another alternative is to walk around the block. You will find living things all around.
- After many trips to nearby sites, you may decide to explore a larger space such as a park, a science museum, or a botanical garden. At this point, children will be able to make good use of such a trip. They will not be overwhelmed by the larger space because you will have taught them how to work in the field before traveling far.

What To Look For?

Trips out into your school playground or neighborhood will enable you to create field investigations in the life, earth, and physical sciences.

- While studying the life sciences, revisit the same site for several months, observing natural changes and the interdependence of plants and animals, and perhaps "adopting" one tree to study in depth.
- When you study the earth sciences, children can collect water samples, perhaps observe a landform, and possibly see some evidence of erosion after a storm. By using a nearby site, children will be able to make multiple trips on a single day to observe changing weather conditions, to measure the rainfall, or to watch the movement of clouds on a succession of days.
- As you study the physical sciences, your class can easily take magnets outdoors to see where they stick, search for objects made of wood or metal, or investigate what happens to a puddle over a period of time.

For more help learning how to create your own meaningful field trips, read:

Ten-Minute Field Trips: A Teacher's Guide to Using the School Grounds for Environmental Studies,
by Helen Ross Russell
(National Science Teachers' Association, 1997)

Living Things

▶ Build on Prior Knowledge

Ask: **Has anyone ever worked in a Science Center before? What did you do in the Science Center? What do you know about what scientists do? Where do you think they do their work?**

▶ Use the Visuals

- **Communicate** Read the question, tracking the print with your finger as you read.
- **Observe** Have children discuss what they see in this Science Center. Ask: **Where is the rabbit?** inside the cage **Where is the guinea pig?** on the boy's lap next to the cage **Where are the fish and snail tanks?** on top of the counter
- **Compare** Ask: **How is it like the Science Center in our classroom? How is it different?**

Reading Strategy

Phonics Ask children to listen as you say the words *see* and *me*. Ask: **What do you notice about these words?** Ask children if they can think of other words that rhyme with *see* (tree, flea, tea, bee, he, key, knee, we).

Print Awareness Have children help you count the number of letters in the word *see*. Ask: **Do any words in the sentence have less than three letters?**

Science Facts

Guinea Pigs are easy to hold and cuddle and are usually sweet-natured. They love to grunt and whistle and generally stay awake during the day.

Rabbits require more care than guinea pigs and can bite or scratch if provoked, but they are wonderful to pet.

Fish and water snails require little maintenance and are inexpensive to feed. They do not need to go home during vacations if you use long-lasting pellets.

Land snails reproduce quickly, come out of their shells when they feel safe, and are easy for children to handle. For more on class plants and animals, see pp. TR3–TR6.

Professional Development For more Science Facts and resources from  visit http://nsdl.org/refreshers/science

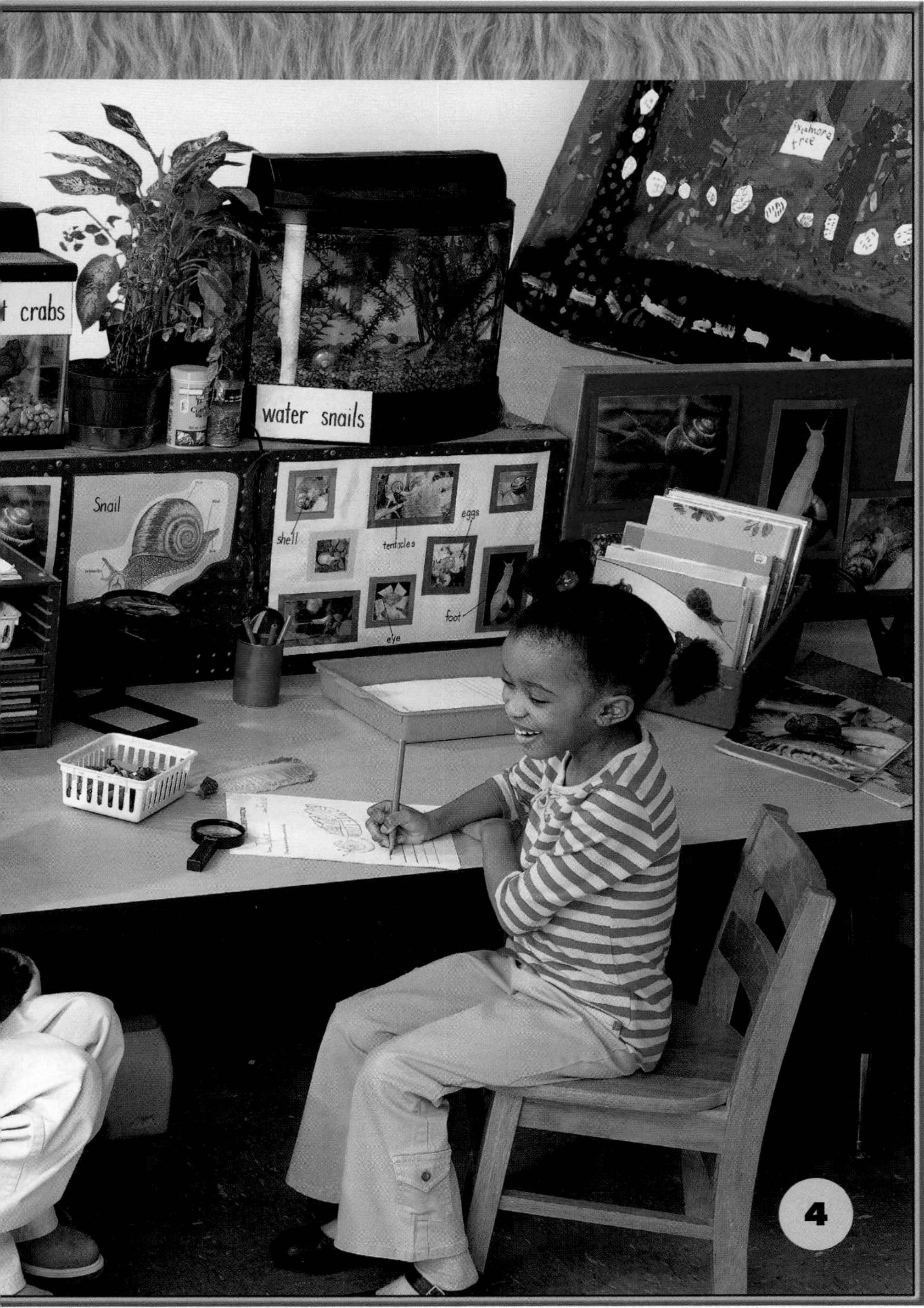

▶ Ask Questions

- **What living things are in this Science Center?** fish, snails, guinea pig, rabbit, hermit crabs, plants, worms, children
- **What nonliving things do you see?** Possible answers: books, table, chairs, tanks, cages, paper, pencils
- **What do you think you might discover about the animals and plants?** Possible answers: We will learn what they need to live; we might find out that they are friendly.
- **Why do you think there are books in the center?** to learn more about science topics
- **How would the work you do here be like the work scientists do?** Possible answer: We can observe, explore, and record things.

Write on It

Have children name what they see in the center, and help them use a dry-erase marker to label the materials, animals, and plants.

Think and Talk

Review how this Science Center is similar to the one in your classroom and how it is different. Discuss how children will find materials, work, and clean up in your Science Center.

More to Read

Living Things,
by Adrienne Mason
(Kids Can Press, 1998)

Activity After reading, discuss whether any of the animals in this book could sleep comfortably in the classroom.

Living Things Activities

Closely observing and revisiting living and nonliving things in your classroom will support children's ability to make detailed scientific drawings and observations.

Living and Nonliving

You need

- sweet potatoes
- carrots
- other vegetables
- potted plants from the Science Center
- chart paper

Objective: Make plant observations and begin to define living and nonliving.

Inquiry Skills: observe, classify, compare, make a model, predict

1. **Observe** Show the vegetables one at a time and have the group describe each one.
2. **Classify** Choose two of the vegetables and record children's observations on chart paper.
3. **Compare** Choose one cut vegetable and one potted plant and ask children how they are alike and different. Then ask: **Do you think these are living things? How do you know?** We know that cut vegetables are no longer living, but children may not. At this point, resist the impulse to correct misunderstandings and define living and nonliving for them. Instead, encourage them to share how they define the terms.
4. **Make a model** Explain that you will help children do an observational drawing of one of the vegetables at the Science Center. As you work with children, you will be able to show them how to find materials in this center, where finished work will go, and how to treat the plants and other materials with respect. You will also be able to support children as they try to draw what they see.

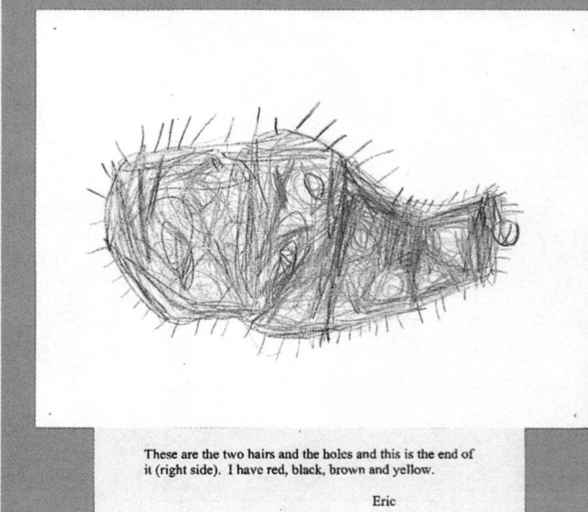

Investigate More

Predict If you plan to cook the vegetables, ask children: **How do you think the vegetables will change when we cook them? How do you know?** Possible answers: They will get mushy; they will fall apart.

Look Closely

You need

- *On Market Street* by Arnold Lobel

Objective: Practice observation and communication skills.

Inquiry Skills: observe, communicate

- Read *On Market Street,* a book by Arnold Lobel, or any picture book with little text but detailed illustrations.
- Before you read each page, ask children to describe what they see. Then ask: **Does anyone notice something else about this page?** This invitation to look again will help children learn to examine illustrations more closely.
- Point out that when they carefully observe the pictures in this book, children are learning to work as scientists do.

Sorting Plants

You need

- Photo Sorting Cards 1–10

Objective: Begin to categorize objects into two or more groups.

Inquiry Skills: classify, put things in order

- Because classifying and putting things in order will be an integral part of their ongoing science work, have children practice sorting using the Photo Sorting Cards.
- As they work together, encourage children to describe the images with as much detail as possible. After sorting the cards using two to three different sorting rules, give children the second set of cards in order to play a concentration game.
- As you work with children, observe how they approach this task. Your notes will help you plan further activities to support their learning.

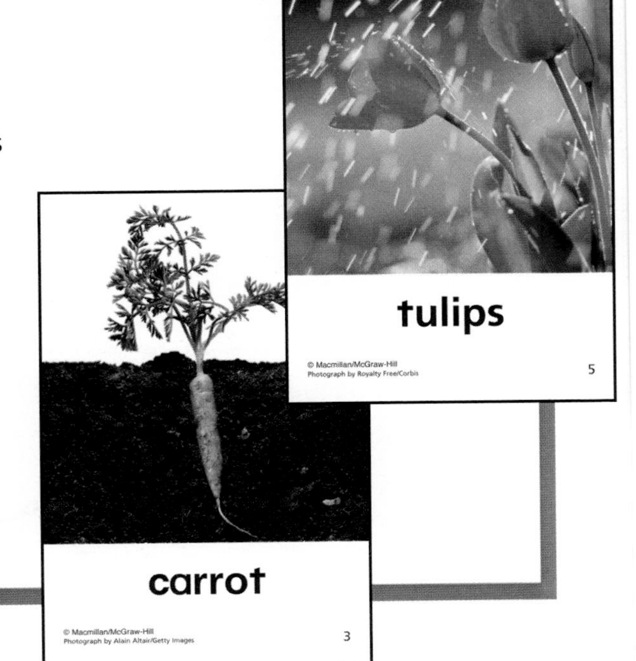

carrot

3

Your Science Center

Early Materials

Do not be afraid to place only a few carefully chosen materials in the Science Center, especially at the beginning of the year. Too many materials can overwhelm children and discourage focused observation.

- Begin by placing some healthy potted plants in the center. Show children how to check the dirt before adding water. When you begin your formal study of plants in Unit A, children will have had some practice caring for and informally observing the plants in the center.
- Add some science observation sheets in a folder or basket and show children where to write their names and add the date stamp so you can file the observational drawings in their portfolios. A small basket of colored pencils and/or crayons and some black pencils will invite drawing.
- A small collection of books that extend the work you are doing in each unit will invite browsing.
- As you begin Unit A, you will want to add some of the class projects (germinating seeds, seedlings) to the center for safe storage where children can observe change over time.

Sand/Water Table

If you have a Sand/Water Table, you may want to begin the year with sand in the table. The clean-up is easier and requires less teacher supervision than water.

- Be sure to include a small hand brush and dustpan near the Sand Table so children can clean up any sand that falls to the floor.
- If some of your children went to the beach in the summer, add some small stones and shovels. The memories of family beach activities that such materials evoke can be comforting for children who may be feeling some separation anxiety.
- As the year progresses, you will add other materials. For now, give children time to explore the sand, learn the clean-up routines, and enjoy "messing about" in a material that will later teach them much about volume and weight, and be used to create models of landforms.

Adding Class Pets

Once children are comfortable and secure in their new classroom environment, you may want to add animals to your Science Center. They can be very important components of your year-long science work.

- By carefully introducing pets to the class and teaching children how to care for and treat them gently, you will foster lifelong skills that every child needs.
- If you choose pet(s) that have short growth cycles, children may be able to observe the complete life cycle of one or more species.
- The investment in a fish tank, filter, and cover will last many years and create an environment for cold-water fish, water snails, and plants.
- Before adding any furry pets, you will need to make sure no child has an allergy to the pet you are considering.

Be prepared to help children answer many of their own questions about the new pets.

- Collect a variety of nonfiction books that discuss the animal(s) you have added. Do not worry if some of them are written for older children. Your children will learn a great deal as they look at the illustrations in the books.
- In Unit B, you will find that many of the suggested activities and investigations can be modified so that children can use the animal(s) in the Science Center to learn more about animal motion, coverings, adaptations, etc.

By the end of the year, children will know a great deal about the animals that you have included in the Science Center, and will have used the science inquiry skills needed to learn more about many other kinds of animals and plants.

In subsequent units, you will want to add materials to support the study of earth and physical sciences, but keep the plants and animals in the center all year. You will be able to collect and analyze the children's drawings and observations of these livings things over the course of the year.

For more information on setting up and using centers, read:

Classroom Routines That Really Work for Pre-K and K, by Kathleen Hayes and Renee Creange (Scholastic, 2001)

Animals in the Classroom: Selection, Care, and Observation, by David C. Kramer (Addison-Wesley Longman, 1989)

Teacher's Notes

Life Science

National Standards

The following K-4 National Science Education Standards are covered in Units A and B:

- Organisms have basic needs. For example, animals need air, water, and food; plants require air, water, nutrients, and light. Organisms can survive only in environments in which their needs can be met. The world has many different environments, and distinct environments support the life of different types of organisms.
- Each plant or animal has different structures that serve different functions in growth, survival, and reproduction. For example, humans have distinct body structures for walking, holding, seeing, and talking.
- Plants and animals have life cycles that include being born, developing into adults, reproducing, and eventually dying. The details of this life cycle are different for different organisms.
- Plants and animals closely resemble their parents.
- Many characteristics of an organism are inherited from the parents of the organism, but other characteristics result from an individual's interactions with the environment. Inherited characteristics include the color of flowers and the number of limbs of an animal. Other features, such as the ability to ride a bicycle, are learned through interactions with the environment and cannot be passed on to the next generation.
- All animals depend on plants. Some animals eat plants for food. Other animals eat animals that eat the plants.
- An organism's patterns of behavior are related to the nature of that organism's environment, including the kinds and numbers of other organisms present, the availability of food and resources, and the physical characteristics of the environment. When the environment changes, some plants and animals survive and reproduce, and others die or move to new locations.
- All organisms cause changes in the environment where they live. Some of these changes are detrimental to the organism or other organisms, whereas others are beneficial.
- Humans depend on their natural and constructed environments. Humans change environments in ways that can be either beneficial or detrimental for themselves and other organisms.

Floor Puzzle

Pond Puzzle Activity

- Introduce your study of Life Science by showing Flipbook page 5. Discuss the image and have children share what living and nonliving things they can find.
- Place the floor puzzle pieces out for children and invite them to construct the puzzle, using the Flipbook image as a guide. Ask questions such as: **What lives in this pond? What are the people doing? What plants do you see? What animals do you see?**
- **Revisit** Encourage children to build, rebuild, and explore the floor puzzle over the course of your study of plants and animals in Units A and B.

Plants

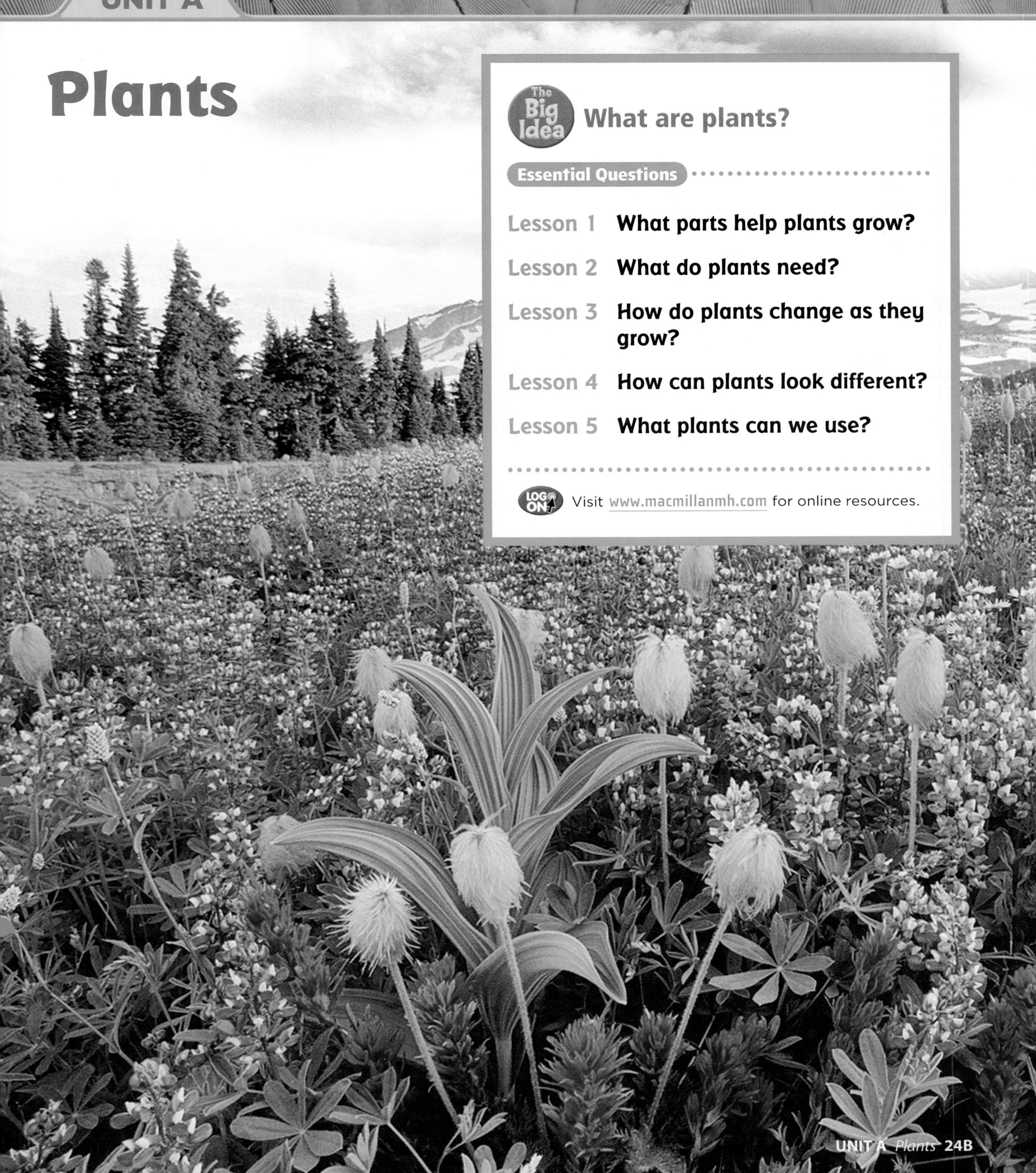

What are plants?

Essential Questions

Lesson 1 **What parts help plants grow?**

Lesson 2 **What do plants need?**

Lesson 3 **How do plants change as they grow?**

Lesson 4 **How can plants look different?**

Lesson 5 **What plants can we use?**

Visit www.macmillanmh.com for online resources.

Science Leveled Readers

All included in the Leveled Readers Literature Big Book.

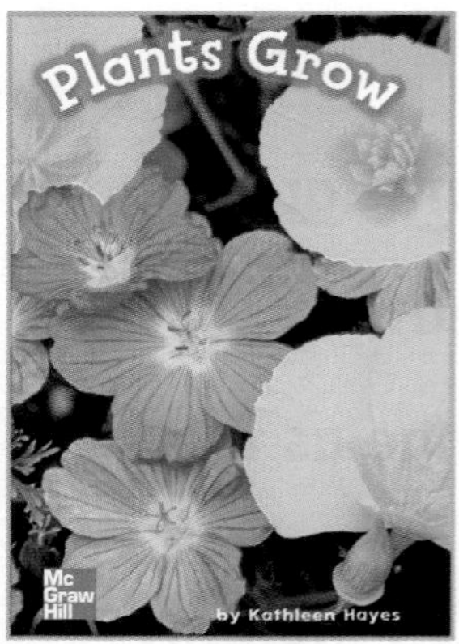

APPROACHING

Plants Grow
Plants have certain needs that must be met in order for them to grow.

ISBN: 978-0-02-281078-8

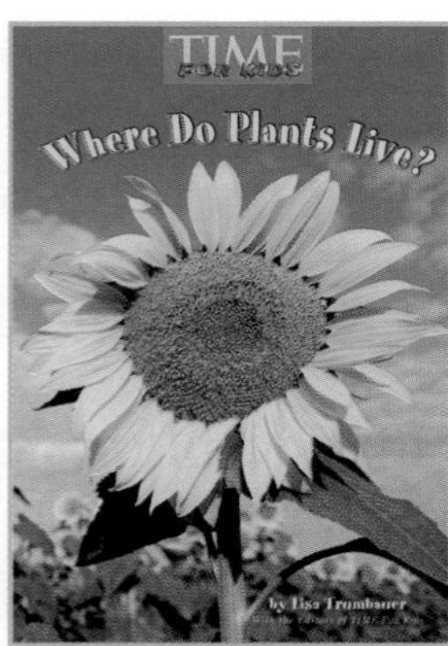

ON LEVEL

Where Do Plants Live?
Plants can live in a variety of places. Plants adapt to the places they live.

ISBN: 978-0-02-284586-5

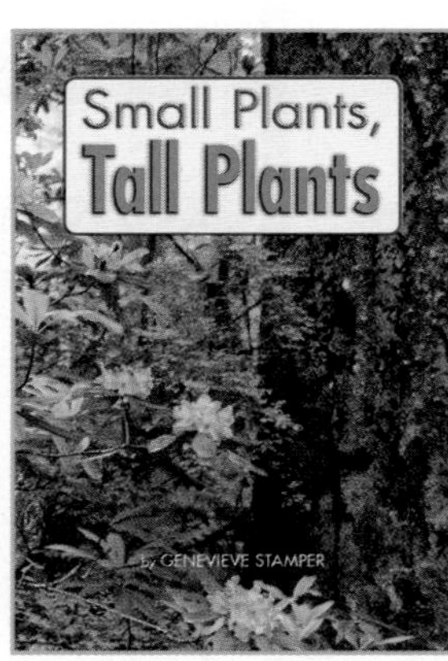

BEYOND

Small Plants, Tall Plants
Certain types of plants that are fully grown stay small while others grow very large!

ISBN: 978-0-02-284585-8

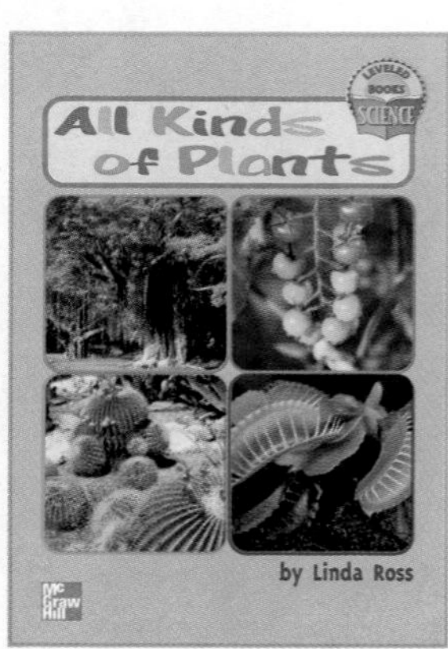

APPROACHING

All Kinds of Plants
There are many different types of plants that come in all shapes, colors, textures, and sizes.

ISBN: 978-0-02-278966-4

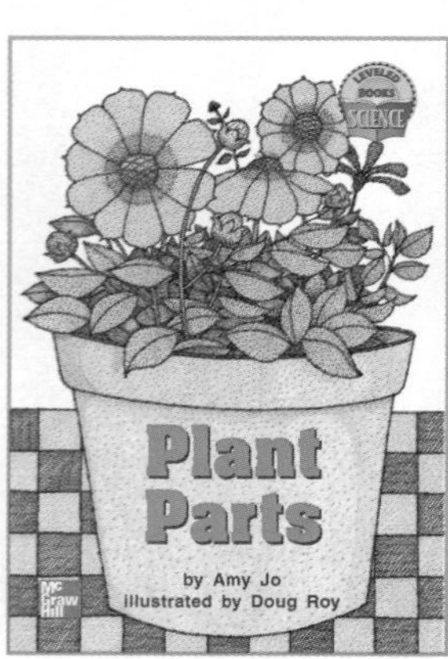

ON LEVEL

Plant Parts
Plants are made up of many parts. Each part is illustrated and explained thoroughly.

ISBN: 978-0-02-278967-1

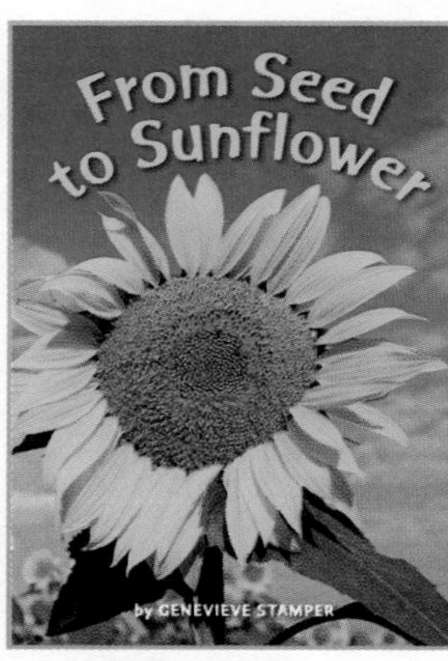

BEYOND

From Seed to Sunflower
See the life cycle of a sunflower plant from seed to flower and back to seed again!

ISBN: 978-0-02-284587-2

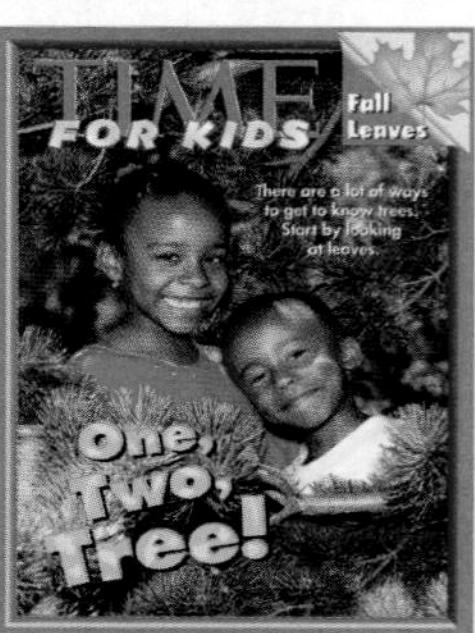

Includes *Time for Kids* magazine about trees and their leaves.

Bibliography

Lesson 1 Parts of Plants

A Fruit Is a Suitcase for Seeds, by Jean Richards (Millbrook, 2002)

Oh Say Can You Seed, by Bonnie Worth (Random House, 2001)

The Pumpkin Book, by Gail Gibbons (Holiday House, 1999)

A Tree Is a Plant, by Clyde Robert Bulla (HarperCollins, 2001)

Lesson 2 What Plants Need

Carrot Seed, by Ruth Krauss (HarperCollins, 2005)

Compost! Growing Gardens From Your Garbage, by Linda Glaser (Millbrook, 1996)

A Handful of Sunshine: Growing a Sunflower, by Melanie Eclasre (Ragged Bears, 2001)

Lesson 3 How Plants Grow

Dandelions: Stars in the Grass, by Mia Posada (Carolrhoda Books, 2000)

Pumpkin Circle: The Story of a Garden, by George Levenson (Tricycle Press, 1999)

Red Leaf, Yellow Leaf, by Lois Ehlert (Harcourt, 1991)

Ten Seeds, by Ruth Brown (Knopf, 2001)

Lesson 4 Look at Leaves and Flowers

Leaves! Leaves! Leaves!, by Nancy Elizabeth Wallace (Marshall Cavendish, 2003)

National Audubon Society First Field Guide: Trees, by Brian Cassie (Scholastic, 1999)

Zinnia's Flower Garden, by Monica Wellington (Dutton Books, 2005)

Lesson 5 Plants We Use

Fruits and Vegetables/Frutas y Vegetales, by Gladys Rosa-Mendoza (Me+Mi Publishing, 2002)

Growing Vegetable Soup, by Lois Ehlert (Harcourt, 1990)

Market Day, by Lois Ehlert (Harcourt, 2002)

Bank Street

For more information on these or other books, visit **www.bankstreetbooks.com** or your local library.

Vocabulary Activities

Science Vocabulary

Vocabulary

root
stem
leaf
air
light
water
soil
seed
seedling
fruit
flower
vegetable

Plant Parts

Explore what children know about plants and their parts by creating a word web.

- Have children name the parts of plants they know. Record their ideas on the word web.
- If children have difficulty identifying some of the plant parts, show them a plant or a picture of a plant. Ask them to look closely for any other parts they may have missed. Revisit and add to the word web throughout the unit.

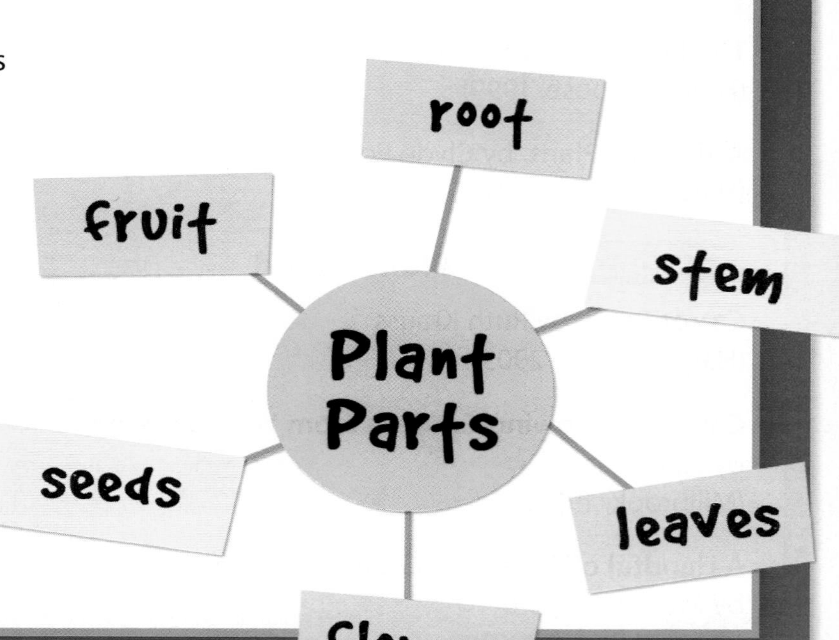

ELL Support

Plants

Write the vocabulary words on chart paper. Draw simple pictures to illustrate each word. Ask children to repeat each word after you as you point to its picture.

- **Beginning** Show children a plant. Point to the plant and say the word *plant*. Invite children to repeat the action after you as they say the word. Repeat with other vocabulary words.
- **Intermediate** Prompt children to point to different parts of the plant. Say: **Point to the flower.** Encourage them to say the name as they point to the part. Repeat with other plant parts.
- **Advanced** Ask children to point to and name parts of the plant. Model using sentences like *This is a flower.*

A to Z Activity Book

Look for additional cross-curricular activities about plants in the A to Z Activity Book.

D **is for Digging in Dirt**, pp. 8–9
Reading, Math, Cooking, Science

G **is for Grow, Grow, Grow**, pp. 14–15
Music, Art, Reading, Bulletin Board

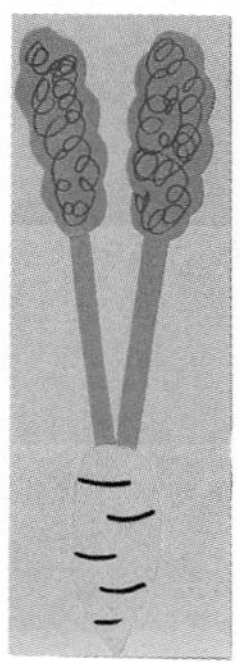

Unit Project

Ask the Experts

Objective: Interview an adult to learn about plants.

You need
- chart paper
- marker
- drawing paper
- pencils
- crayons
- colored pencils

Step 1 Ask parents or school staff who are gardeners or workers in a greenhouse or florist shop to come to your classroom and share their knowledge of plants with the children. You can also arrange a visit to their workplace and interview them there.

Step 2 Explain to the class that they will interview an expert to learn more about plants. Help children think of questions they want to ask. Write the questions on chart paper.

Step 3 On the day of the interview, help children ask the questions on the list and take notes. Following the interview, ask each child to draw a picture and write or dictate information about what they learned. Compile the pages into a class book and place it in the class library.

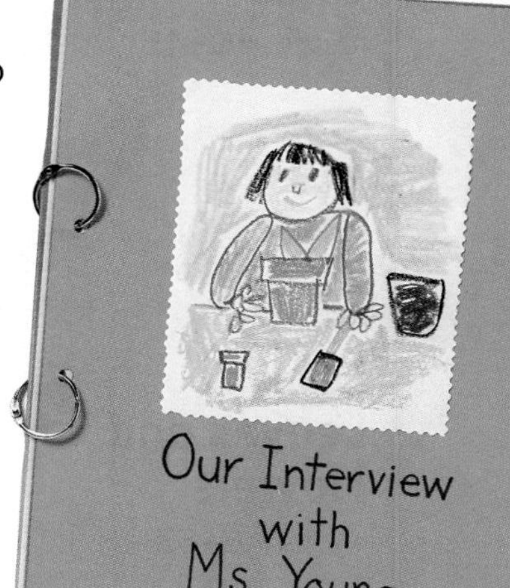

Technology

IWB INTERACTIVE WHITEBOARD READY

LOG ON www.macmillanmh.com

Visit Macmillan/McGraw-Hill Science online for projects and activities for students, teachers, and parents.

www.bankstreet.edu

Visit Bank Street online for teacher resources and more activities.

www.timeforkids.com

Visit Time for Kids online to read about current science news and explore additional science activities.

UNIT A Planner

TeacherWorks™ Plus CD-ROM has interactive unit planner, Teacher's Edition, and worksheets.

Lesson	OBJECTIVES	VOCABULARY	RESOURCES and TECHNOLOGY	
1 Parts of Plants PAGES 26–31 **PACING**: 3–5 days	■ Understand that plants are organisms with parts that help them get what they need to grow and mature.	**root** **stem** **leaf** **seed**	**Flipbook,** p. 7 **Leveled Reader:** *Plant Parts* **Unit A Literature Big Book** ■ **Science Resource Book,** p. 27	■ **Photo Sorting Cards 1–10** ■ **A to Z Activity Book,** pp. 8–9, 14–15 ■ **Science on the Go,** pp. 1–4
2 What Plants Need PAGES 32–39 **PACING**: 3–5 days	■ Recognize that a plant is an organism that needs air, water, light, and soil to survive.	**air** **light** **water** **soil**	**Flipbook,** pp. 8–9 **Leveled Readers:** *Where Do Plants Live?* and *Plants Grow* **Unit A Literature Big Book** ■ **Science Resource Book,** p. 28	■ **Photo Sorting Cards 1–10** ■ **A to Z Activity Book,** pp. 8–9, 14–15 ■ **Science on the Go,** pp. 5–8
3 How Plants Grow PAGES 40–45 **PACING**: 3–5 days	■ Recognize that plants are organisms that grow and change.	**seed** **seedling** **fruit** **flower**	**Flipbook,** pp. 10, S2 **Leveled Reader:** *Small Plants, Tall Plants* **Unit A Literature Big Book** ■ **Science Resource Book,** p. 29	**Science Songs CD** ■ **A to Z Activity Book,** pp. 8–9 ■ **Science on the Go,** pp. 9–12
4 Look at Leaves and Flowers PAGES 46–53 **PACING**: 3–5 days	■ Recognize that plants can be identified by their parts.	**leaf** **leaves** **flower** **flowers**	**Flipbook,** pp. 11–12, S3 **Leveled Reader:** *From Seed to Sunflower* **Unit A Literature Big Book** *Time for Kids* ■ **Science Resource Book,** p. 30	**Science Songs CD** ■ **Photo Sorting Cards 1–10** ■ **Floor Puzzle:** *Pond Life* ■ **Science on the Go,** pp. 1–4
5 Plants We Use PAGES 54–59 **PACING**: 3–5 days	■ Identify and explore plants that we eat and the foods that come from different plants.	**fruits** **vegetables**	**Flipbook,** p. 13 **Leveled Reader:** *All Kinds of Plants* **Unit A Literature Big Book** ■ **Science Resource Book,** p. 31	■ **A to Z Activity Book,** pp. 8–9, 14–15 ■ **Science on the Go,** pp. 15–18

PACING Assumes a day is a 20–25 minute session.

www.macmillanmh.com for more planning resources and http://nsdl.org/refreshers/science for science resources from NSDL

BE A SCIENTIST Activities

Observing Stems *p. 30*

PACING: 20 minutes 

Skills observe, communicate, draw a conclusion, infer

Materials celery stalk with leaves, large jars, food coloring

Window Box Wonder *p. 38*

PACING: 30 minutes 

Skills observe, infer, communicate

Materials small window box or flower pots, eight small plants with roots, soil

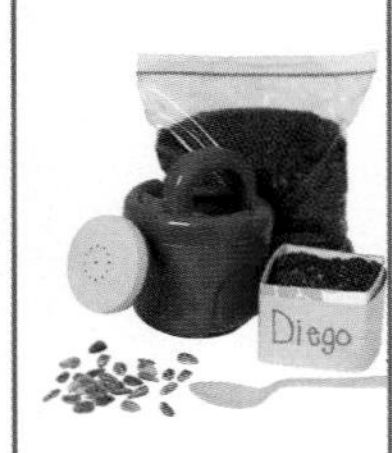

Planting Seeds *p. 44*

PACING: 20 minutes 

Skills observe, communicate, infer

Materials milk containers, masking tape, newspaper, potting soil, seeds, paper cups

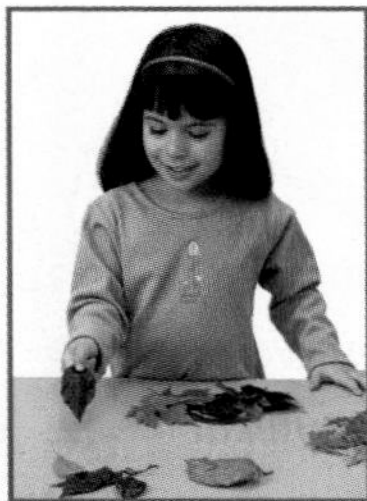

Matching Leaves *p. 52*

PACING: 30 minutes 

Skills observe, classify, investigate, communicate, infer

Materials leaves (or an assortment of pictures of leaves), aluminum trays, chart paper, markers, fresh flowers

Plant Part Soup *p. 58*

PACING: 30 minutes 

Skills classify, predict, observe, draw a conclusion, communicate

Materials produce, large pot, hot plate, knife, cutting board, broth

UNIT ASSESSMENT

Formative Assessment, pp. 29, 37, 43, 51, 57
Performance Assessment, p. 60
Summative Assessment, p. 61
Portfolio Assessment, p. 61

SCIENCE KIT MATERIALS

- seeds
- large jars with lids
- small aluminum trays
- string
- watering can
- modeling clay

COLLECTIBLES

Begin collecting these items for Unit A:

- milk containers
- newspaper
- leaves (or an assortment of pictures of leaves)
- craft sticks
- chenille stems
- pieces of cardboard

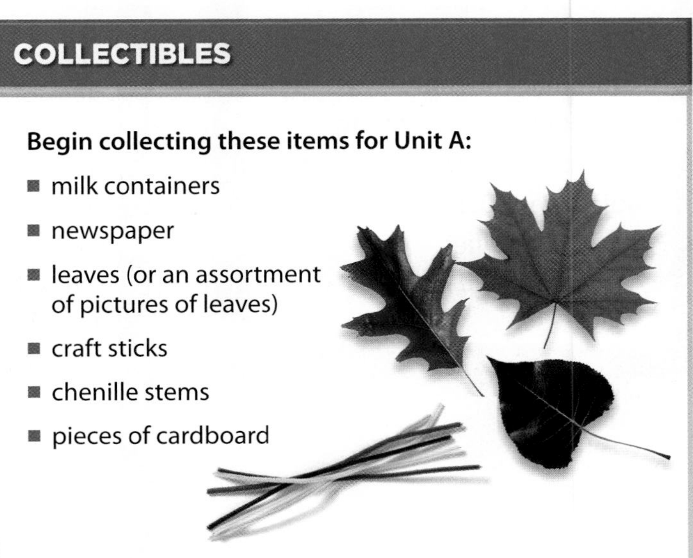

UNIT A

English Language Learner Support

Academic Language

English language learners need help in building their understanding of the academic language used in daily instruction and science activities. The following strategies will help to increase children's language proficiency and comprehension of content and instruction words.

Strategies to Reinforce Academic Language

- **Use Context** Academic Language should be explained in the context of the task. Use gestures, expressions, and visuals to support meaning.
- **Use Visuals** Use charts, transparencies, and graphic organizers to explain key labels that help children understand classroom language.
- **Model** Use academic language as you demonstrate the task to help children understand instruction.

Technology

For additional language support and oral vocabulary development, go to www.macmillanmh.com

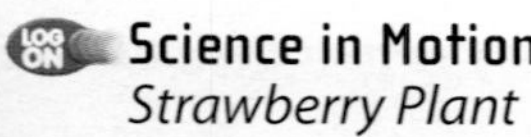

Science in Motion *Strawberry Plant*

Vocabulary Games

Language Transfers

Phonics Transfer
Native speakers of Cantonese may have difficulty with the long vowel sound /ē/ as in *leaf*.

Grammar Transfer
In Haitian Creole, Hmong, Spanish, and Vietnamese, an adjective usually follows the noun it modifies.

I have a leaf red.

Academic Language Vocabulary Chart

The following chart shows unit vocabulary and inquiry skills as well as some Spanish cognates. **Vocabulary** words help children comprehend the main ideas. **Inquiry Skills** help children develop questions and perform investigations. **Cognates** are words that are similar in English and Spanish.

Vocabulary		Inquiry Skills	Cognates	
			English	Spanish
root, p. 26	real, p. 36	observe, p. 30	observe, p. 30	*observar*
stem, p. 26	fantasy, p. 36	communicate, p. 30	communicate, p. 30	*comunicar*
leaf, p. 26	fruit, p. 42	draw a conclusion, p. 30	conclusion, p. 30	*conclusión*
seed, p. 28	seedling, p. 42	infer, p. 30	infer, p. 30	*inferir*
light, p. 32	flower, p. 42	classify, p. 52	real, p. 36	*real*
water, p. 32	leaves, p. 48	investigate, p. 52	fantasy, p. 36	*fantasía*
soil, p. 32	vegetables, p. 54	predict, p. 58	fruit, p. 41	*fruta*
air, p. 34			classify, p. 52	*clasificar*
			investigate, p. 52	*investigar*
			vegetables, p. 54	*vegetales*

Vocabulary Routine

Use the routine below to discuss the meaning of each word on the vocabulary list. Use gestures and visuals to model all words.

Define A *leaf* is the part of a plant that makes food.

Example This *leaf* is from a tree.

Ask Why do you think this *leaf* is green?

Children may respond to questions according to proficiency level with gestures, one-word answers, or phrases.

Vocabulary Activities

Help children understand that a leaf is part of a plant.

BEGINNING Show a real plant or draw one on chart paper. Say: *This is a plant,* as you make a circle around it with your finger. Write, say, and have children repeat *plant*. Continue: *A plant has parts. The leaf is one part of a plant.* Point to a leaf. Have children point to other leaves.

INTERMEDIATE Show Sorting Card 9. Have children describe the leaves. Say: *Point to the leaves. What color are they? Where else can you find leaves? What do they look like?*

ADVANCED Show Flipbook page 7. Say: *The sunflower and the carrot are plants. Plants have parts. One of them is the leaf.* As you point to a leaf on each plant, ask: *How is this leaf like this other leaf? How is it different?*

Objective: Children will discover plant characteristics by observing and comparing plants and their parts.

Before Reading

- Have children share what they know about the plants that appear on the Flipbook page.
- Explain that in this unit, children will be scientists as they observe, compare, and classify plants and plant parts.

During Reading

- Read the poem, pointing to the words as you read.
- As you read again, pause and invite children to point to images on the page that correspond to the lines in the poem.
- Reread the poem two to three times, encouraging children to chime in as you read.

Flipbook

UNIT A

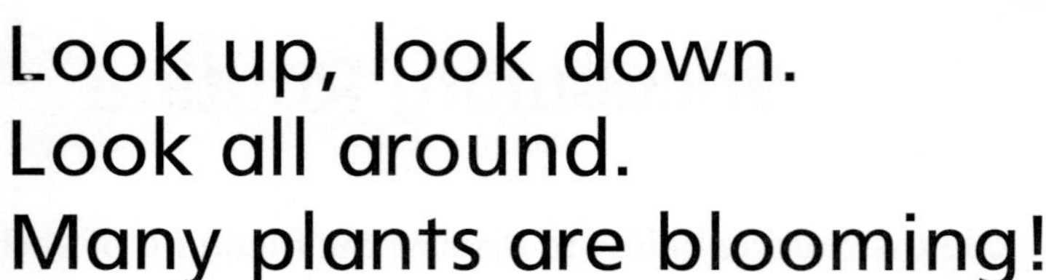

Plants

Look up, look down.
Look all around.
Many plants are blooming!

Some reach high, some are round.
Some spread out along the ground
Some are very, very tall.
Some have flowers very small.

Look up, look down.
Look all around.
Many plants are blooming!

by Susan Greenebaum

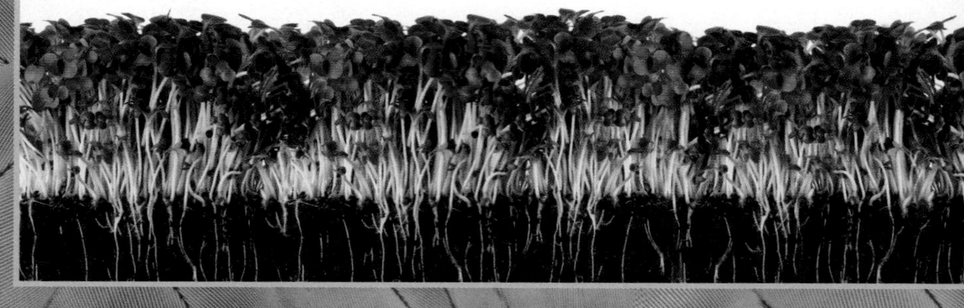

Reading Strategy

Conventions of Print Ask children to help you count how many times the word *look* appears in the first stanza. Then have them help you count the number of letters in the word *look*.

Print Awareness Have children follow along as you read the second stanza so that they notice the first word of each line. As you read, point to the word *some* to help them identify the pattern.

Science Facts

Barrel Cactus *Ferocactus*, meaning "wild cactus," is barrel-shaped and among the largest cacti of the North American deserts. Its flowers grow at the top of the plant. Ripe barrel cactus fruit can be juicy, but is not usually edible.

Beech trees can grow up to 40 meters tall and can live up to 300 years. They usually flower in the spring and drop their nuts in the fall. Oil found in beech nuts can contain around 20% protein, making them quite nutritious.

6

After Reading

- Begin a class chart with the title *Plants We Know.* Ask children to name any plants they recognize on the Flipbook page.
- Invite children to name other plants they know. If children say "flower," ask if they know the names of any flowers.

Write on It

Using a dry-erase marker, have volunteers draw an *x* on all the different varieties of plants they see on the page.

▶ Revisit

Before rereading the poem, ask children if they can make their bodies into a round plant, a tall plant, and a plant that spreads along the ground. Ask for volunteers to demonstrate and encourage them to all move together as you read the poem.

Science Background For more information on plants go to www.macmillanmh.com.

Sprouts are young, edible plants grown from seeds. Sprouts come in several varieties. They can grow from vegetables, beans, and grains. Alfalfa sprouts can contain as much Vitamin C as an orange. Bean sprouts are high in protein and other valuable vitamins and minerals.

Pansy These flowering plants are very fragrant. The most familiar species is known for its bright colors and darker center, often referred to as its "face." Pansies bloom in cooler weather, and cannot survive in hot temperatures.

Professional Development For more Science Facts and resources from NSDL visit http://nsdl.org/refreshers/science

School to Home

At the beginning of Unit A, give children copies of the Home Letter on page 1 in the **Science Resource Book**. Read the letter aloud and help children write their names. Have children bring the letter home to share with their families.

At the end of the unit, send children home with a copy of the Home Letter on page 3.

See pages 2 and 4 in the Science Resource Book for Spanish versions of these letters.

Parts of Plants

▶ **Essential Question**
What parts help plants grow?

▶ **Objective**
Understand that plants are organisms with parts that help them get what they need to grow and mature.

▶ **Vocabulary**

root	leaf
stem	seed

Resources

Flipbook, p. 7

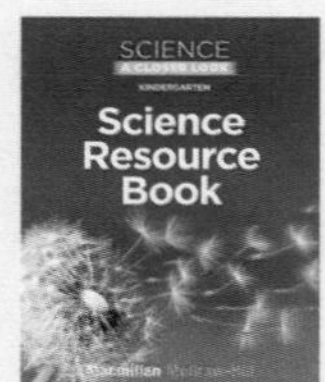

Science Resource Book, p. 27

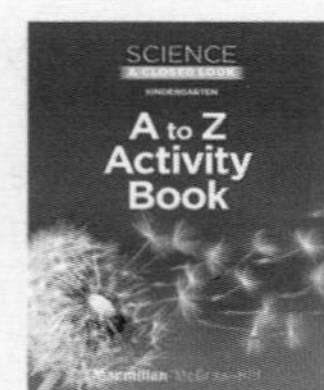

A to Z Activity Book, pp. 8–9, 14–15

Leveled Reader: *Plant Parts*

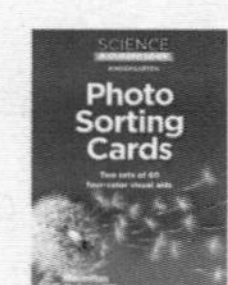

Photo Sorting Cards 1–10

Technology

Science in Motion *Strawberry Plant*

www.macmillanmh.com

NSDL

Circle Time

WHOLE CLASS

Make a Plant

You need
- index cards
- markers

Inquiry Skills: classify, make a model, communicate

- Make a set of five root cards, five stem cards, and enough leaf cards so that each child will have a card.
- Show children the three types of cards. Hold up and read each card and have children repeat after you.
- Distribute the cards and have children find the partners they need to make a plant with all three parts.

Communicate Ask: **What are some other parts of plants?** flower, fruit

Time to Move!

Invite children to move like they were taking care of a plant. Encourage them to act out digging a hole in dirt with a shovel, placing a seed in the ground, and covering it with dirt. They can also act like the Sun and rain.

Be a Reader Activity

You need
- Unit A Literature Big Book
- Leveled Reader: *Plant Parts*
- index cards with plant part words

Read the Big Book

Objective: Identify and match words that describe plant parts.

Inquiry Skills: communicate, compare

- Have children discuss what they know about the parts of plants. Then read aloud *Plant Parts* in the Unit A Literature Big Book.
- Give each child a plant-part word card and help them find the same word in the text. Repeat until each plant part has been identified. Read the story again, inviting children to chime in as you read.
- **Guided Reading:** See the inside back cover of the Leveled Reader version of *Plant Parts* for activities.

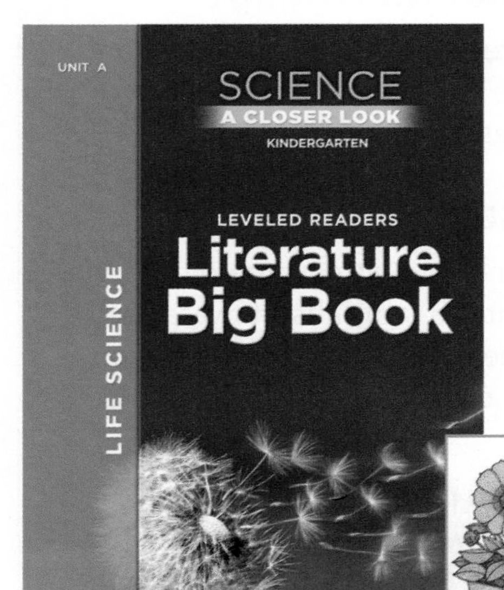

Be a Math Wiz Activity

You need
- Photo Sorting Cards 1–10

Find a Match

Objective: Match sets of cards using one-to-one correspondence.

Inquiry Skills: observe, compare

- Have children place both sets of cards facedown on the table.
- Have children take turns flipping over two cards looking for a match. When they find a match, have them describe the stem, leaves, and any other features of the plant.
- Continue play until all matches have been found.

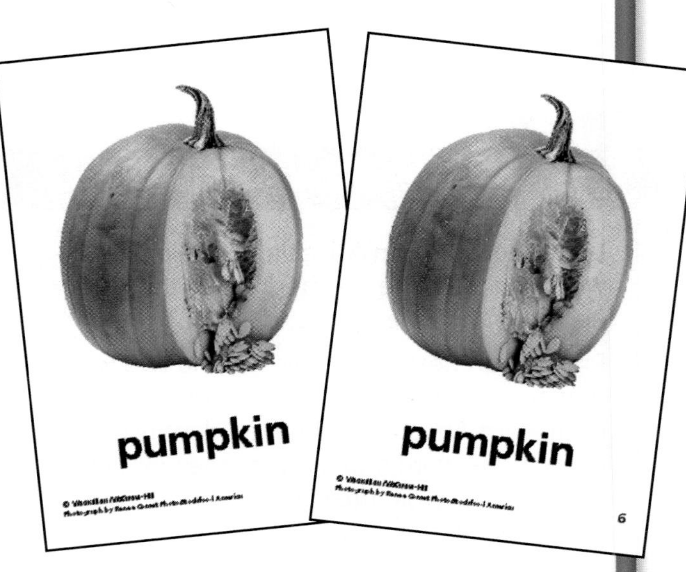

Read Together and Learn

What are the parts of plants?

▶ Build on Prior Knowledge

Ask children to share what they know and what questions they have about the parts of plants. Write their responses on chart paper.

▶ Use the Visuals

- **Communicate** Invite volunteers to point to the roots. Explain that the roots absorb water and nutrients from the soil. Point to the stems and explain that they carry water and food to the leaves.
- Invite volunteers to point to the leaves. Explain that the leaves use water that is carried up the stem, light from the Sun, and nutrients from the soil to make food.
- Point to the fruits and flowers and discuss that they are also parts of many plants.

▶ Develop Vocabulary

roots, **stem**, **leaf, seed** Revisit the cards used in the Circle Time Activity on page 26. Have children choose a plant-part card and match it to the corresponding part shown on the Flipbook page. Discuss that plants grow from seeds.

Flipbook

UNIT A • LESSON 1

What are the parts of plants

Reading Strategy

Letter Recognition Ask children to locate the words on the page that begin with the letter *s*. Repeat with the letters *r* and *l*.

Print Awareness Have children find two words that are the same on the page and point to each word. Encourage them to identify each letter in the word and then say the word aloud. Help them use picture clues if they are having difficulty identifying a word.

Science Facts

Roots can look like a carrot (taproot) or branching such as grass (fibrous). All roots anchor plants and absorb water and minerals from soil.

Stems can be thin or very thick, but they all carry water and minerals to the leaves.

Leaves vary in shape, color, and size, but they all use water, minerals, and light from the Sun to make food for the plant.

Flowers are the bloom or blossom of a plant. It is the part of a plant that produces seeds.

Fruits protect the seeds of a plant and can be eaten.

Seeds contain an embryonic new plant, complete with its first shoot and root and a store of food.

Professional Development For more Science Facts and resources from  visit http://nsdl.org/refreshers/science

▶ Ask Questions

- **What happens when you suck water through a straw?** It goes up the straw to my mouth.
- **Why can the water move up the straw?** Possible answer: Because the straw is hollow and the water can fit inside. Help children make the connection between the tubes in the stem of a plant and a straw.
- **How are the roots of a flower and roots of a carrot different? How are they alike?** Help children notice and describe the differences and similarities among the plant parts on the page.

Write on It

Invite volunteers to use a dry-erase marker to draw a line from each plant part label on the Flipbook page to its corresponding illustration.

Think and Talk

Review the parts of a plant. Have children share what they learned about plants, and the name of each part. Encourage children to explain the function of each part and how they work together to help plants grow.

Formative ASSESSMENT

Observe children as you explore the Flipbook page to help assess what they already know about plant parts and what they might be confused about.

ELL Support

Draw a plant. Point to and name the plant parts. Invite children to repeat after you. Label the parts.

Beginning Name various plant parts and ask children to point to them.

Intermediate Point to various plant parts and ask children to name them.

More to Read

A Tree Is a Plant, by Clyde Robert Bulla (HarperCollins, 2001)

Activity Have children help identify the parts of the tree on each page. After reading, take children outside to find a tree to explore and have them identify each part.

Be a Scientist Inquiry Investigation  

Observing Stems

Objective: Demonstrate the function of a plant stem and record observations.

Inquiry Skills: observe, communicate, draw a conclusion, infer

You need **celery stalk with leaves, large jars, pencil, food coloring, Science Resource Book p. 27**

1. Show the group a stalk of celery with leaves at the top. Explain that it is the stem of the celery plant. Place the celery stalk in a jar with colored water.
2. **Observe** Ask the group to observe the celery throughout the week. Have them discuss their observations with a partner.
3. **Communicate** Have children use their Science Journal page to draw and record what they observe happening over time.
4. **Draw a Conclusion** Ask: **Is there more or less water in the container after 4 days?** less **Why?** It moved up the celery stalk.

Teacher Tip

You can use vegetables and fruits to help children identify what plant parts we eat. For example, we eat the roots of potatoes and carrots, the stems of celery and asparagus, the flowers of broccoli and cauliflower, and various fruits.

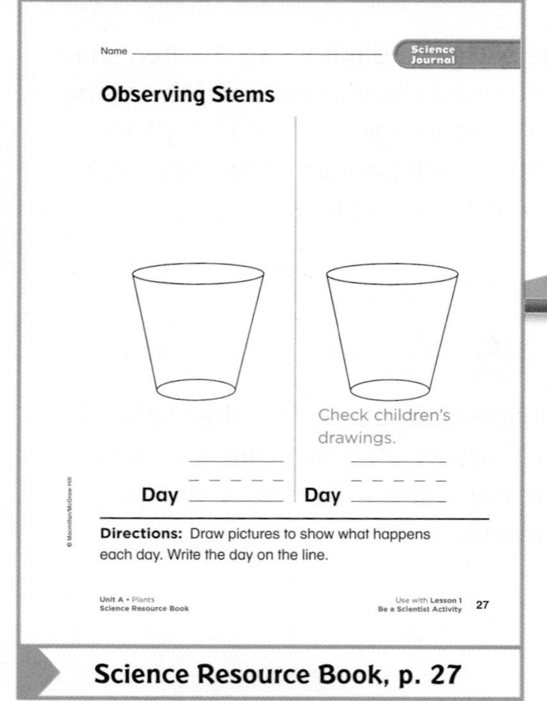

Name ______________ Science Journal

Observing Stems

Check children's drawings.

Day ______ Day ______

Directions: Draw pictures to show what happens each day. Write the day on the line.

Unit A • Plants Science Resource Book — Use with Lesson 1 Be a Scientist Activity 27

Science Resource Book, p. 27

Investigate More

Infer Ask: **Why do you think the celery changed color?** Possible answers: The colored water traveled up the stem. When the plant drinks the colored water, it changes color. **Why do you think a plant needs a stem?** Possible answer: So the water can go from the roots to the leaves.

centers

Art

Root Prints

Objective: Develop small motor skills.

Inquiry Skills: observe, compare, communicate

You need
- potatoes
- carrots
- tempera paint
- painting paper
- newspaper

- Cut potatoes and carrots into pieces that children can grasp. Cover a table with newspaper and arrange small paint containers and paper at the table.
- Show children how to dip the bottom of the vegetable into the paint and then make a print on the paper. Encourage children to make a design that fills the entire piece of paper.
- Ask children: **Do the marks made with the carrot look the same as the ones made with the potato? Why or why not?**

Technology

More about Plants

- For games and activities about plants, help children log onto www.macmillanmh.com.
- Children can use these fun games throughout this unit to reinforce what they learn about plants. Encourage children to revisit these games throughout the school year to review the concepts.

LOG ON www.macmillanmh.com for more science online.

Cooking

Mashed Potatoes

Objective: Recognize that some roots are edible.

Inquiry Skills: observe, predict, infer

You need
- large pot
- hot plate or stove
- knife
- potatoes
- potato masher
- butter
- milk
- salt
- peelers

- Have children help peel the potatoes. Help children describe the skin texture of the potato before cooking. Cut the potatoes into small pieces, cover with water, and heat. Remind children not to touch the hot plate. Ask: **How will the potatoes change when we heat them?**
- When the potatoes are soft, drain the liquid. Ask: **What made the water change?**
- Have children help mash the potatoes and add milk, salt, and butter to taste.

Movement

Trees

Objective: Develop new vocabulary.

Inquiry Skill: communicate

You need
- chart paper
- markers
- pictures of trees

- Show children pictures of a variety of trees. Help them describe the different parts of the trees (trunk, limbs, branches, needles, and leaves).
- Teach children this rhyme to reinforce this new vocabulary. Invite children to use their bodies to act out the poem.
- You may want to copy this poem onto chart paper and use it for shared reading once the children know the poem.

Trees

Elm trees stretch and stretch so wide.
Their limbs reach out on every side.
Pine trees stretch and stretch so high.
They very nearly reach the sky.
Willows droop and droop so low.
Their branches sweep the ground below.

For more on plant parts, use pp. 7–8 in the **Activity Book**.

What Plants Need

▶ **Essential Question**

What do plants need?

▶ **Objective**

Recognize that a plant is an organism that needs air, water, light, and soil to survive.

▶ **Vocabulary**

air water
light soil

Resources

Flipbook, pp. 8–9

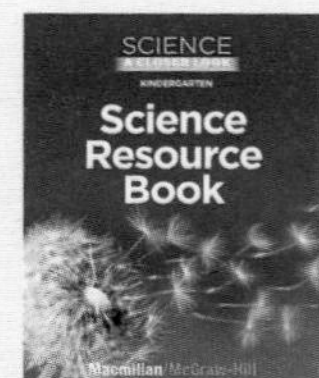

Science Resource Book, p. 28

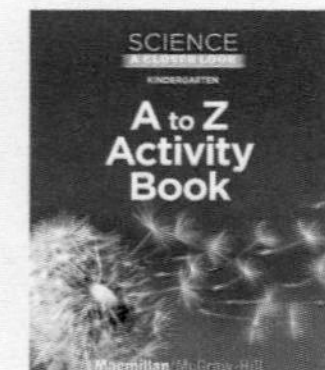

A to Z Activity Book, pp. 14–15

Leveled Reader: *Where Do Plants Live?*

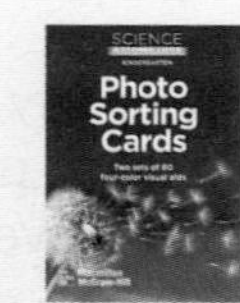

Photo Sorting Cards 1–10

Technology

Circle Time

WHOLE CLASS

Observe a Plant

You need

- a healthy plant
- chart paper
- markers

Inquiry Skills: communicate, predict, infer

- Show children a plant and ask them what they think it will need to stay alive. Record their responses. Help children come to the idea that the plant needs water and light.
- Have children decide what jobs will need to be added to the class job chart to care for the plant. Then have children decide where to put the plant.

Infer Ask children: **Why would a closet not be a good place for the plant?** Possible answer: There is no light.

Time to Move!

Invite children to reach for the sky when you say the words *air* and *light*. Have them reach for their toes when you say the words *water* and *soil*. Randomly call out the words to get children moving.

Be a Reader Activity

Read the Big Book

Objective: Observe and describe the similarities and differences of plants.

Inquiry Skills: communicate, infer

- Read the title *Where Do Plants Live?* and invite children to answer the question. Read the story.
- Cover words that tell where the plants live with sticky notes and reread the story. When you come to a covered word, ask: **What word do you think this is?** Encourage children to use picture clues.
- **Guided Reading:** See the inside back cover of the Leveled Reader version of *Where Do Plants Live?* for activities.
- For more to read on what plants need, see *Plants Grow* found in the Unit A Big Book and as a Leveled Reader.

You need
- Unit A Big Book
- Leveled Reader: *Where Do Plants Live?*
- sticky notes

Be a Math Wiz Activity

How Tall?

Objective: Use nonstandard units to measure plants.

Inquiry Skills: predict, measure, compare

- Hold up a connecting cube and ask: **How many cubes do you think you will need to measure the plant?** Hold up a large paper clip and ask: **How many paper clips do you think you will need to measure the plant?**
- Show children how to cut a string the length of the plant. Have children use connecting cubes to measure the string. Then have them use paper clips. Compare the results.
- Revisit this activity at the end of the unit. Ask: **Has the plant grown? How can we find out?**

You need
- connecting cubes
- large paper clips
- string
- scissors
- plant

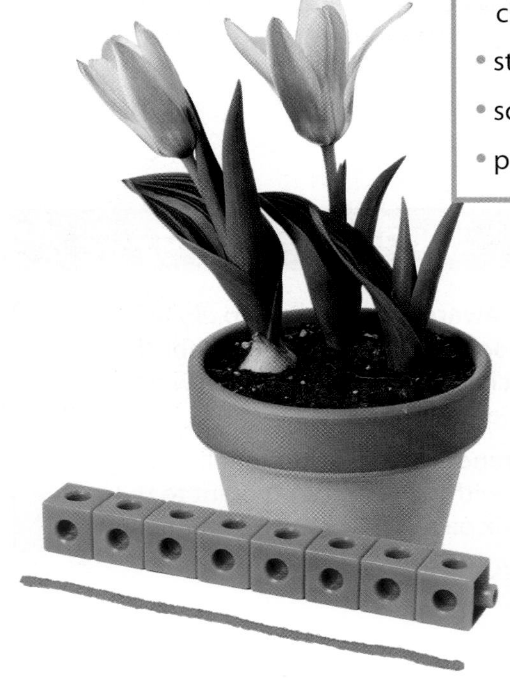

Read Together and Learn

What do plants need to live?

▶ Build on Prior Knowledge

Ask children if they have taken care of a plant. Have them describe what they did to help the plant live.

▶ Use the Visuals

- **Observe** Have volunteers describe what they see in each of the pictures. Record their ideas.
- Help children determine if these pictures answer the question *What do plants need to live?*
- **Communicate** Have volunteers point to a plant on the Flipbook page and name something they see in the picture that helps it stay alive.

▶ Develop Vocabulary

air, **light**, **water**, **soil** Make labels with the vocabulary words on sticky notes or index cards. Read each label and have volunteers stick or tape it on the page where they see the corresponding plant need. Help them to notice that even though we may not be able to see air, it is all around the plants in the picture.

Flipbook

UNIT A • LESSON 2

What do plants need to live?

Reading Strategy

Phonemic Awareness Read aloud *air, light, water,* and *soil* clapping for each syllable in each word. Ask children to identify which word has more than one syllable.

Print Awareness Invite children to find words that begin with the letter *l* and point to them on the Flipbook page. live, Lesson Continue using the letter *p*.

Science Facts

Soil provides minerals, support for the plant, and an anchor for the roots to grow. Decaying plants and animals leave behind minerals in the soil that help plants grow.

Space Plants need adequate space for roots to grow. Roots packed tightly in a pot do not take up nutrients efficiently.

Light Leaves take in energy from the Sun and change it to glucose (sugar) in a process called photosynthesis. Plants use this energy to grow and to reproduce.

Water Plants soak up water through their roots.

Air Green plants take in carbon dioxide from air and use it during photosynthesis to make food.

Professional Development For more Science Facts and resources from NSDL visit http://nsdl.org/refreshers/science

8

▶ Ask Questions

- **Where do plants get their food?** from soil and their leaves
- **What might happen to a plant that does not get what it needs? Why?** It will not grow. It may die.
- **Do all plants need the same amount of water, light, air, and soil in order to survive?** No **Why or why not?** Plants adapt to their environment. For example, plants that live in a rain forest may thrive on lots of water; plants that live in the desert adapt to the dryness and store what little water they receive.

Using a dry-erase marker, invite children to circle where they see light, soil, and water on the Flipbook page. Have them draw an *x* where they think the air might be.

Think and Talk

Review the elements that plants need to live. Draw pictures on index cards to symbolize air, water, light, and soil to use with children throughout the unit as a form of review.

Differentiated Instruction

Enrichment Help children set up an experiment to prove that plants need water in order to survive. Use two plants, giving one water regularly, and the other no water. Repeat this experiment with light and air.

More to Read

Carrot Seed,
by Ruth Krauss
(HarperCollins, 1973)

As you read this book to the class, point out what the boy is doing to help his plant grow. Read it again and ask for volunteers to act out the story.

Read Together and Learn

How are they alike? How are they different?

▶ Build on Prior Knowledge

Ask children to tell about plants they have seen or read about in stories that do things that real plants cannot.

▶ Use the Visuals

- **Observe** Point to the picture of the grapes. Ask children to describe what they see. Repeat with the picture of the toy grapes.
- **Compare** Encourage children to discuss what is the same and different about the two pictures. Explain that one is a living plant and one is a nonliving toy. Repeat this routine with the cactus images.

▶ Additional Vocabulary

real, fantasy Explain to children that the *real* plants are living things that grow whereas the toys are real but nonliving. Tell a story about a cactus that sleeps, eats, and talks. Explain to children that this story is *fantasy,* which means it is not real.

Flipbook

UNIT A • LESSON 2

How are they alike? How are they different?

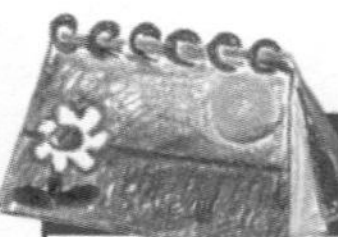

Reading Strategy

Comprehension Point to the word *alike.* Ask: **What does it mean when two things are alike?** Encourage children to give examples of things that are alike, and explain why. For example, a flower and a tree are alike because they are both living plants. Repeat the process with *different*.

Phonemic Awareness Invite children to find words that begin with the *h* sound (/h/) and point to them on the page. Repeat using the *a* sound (/a/).

Science Facts

Grapes are small, smooth-skinned berries that grow in clusters on a vine. Grapes can be eaten fresh or dried (as raisins) or used in wine making, jams, jellies, juices, or vinegars. Green, red, and black seedless grapes are popular for eating. The United States is one of the leading grape-producing countries in the world.

Cactus species, native to North and South America, grow in hot, dry environments. Their large root systems can stretch through rocky and sandy soil. Cacti are leafless, tough-skinned plants, that can hold enough water to last an entire year. Armed with sharp spines, cactus skin protects the plant from animals in search of water.

Professional Development For more Science Facts and resources from NSDL visit http://nsdl.org/refreshers/science

▶ Ask Questions

- Point to the grapes on the plant. Ask: **Do these grapes grow aboveground or underground?** aboveground **Do they need light to grow?** Yes **How do they get it?** Leaves of the grape plant absorb light.
- Point to the toy cactus. Ask: **Does the toy cactus grow? Why or why not? Does it need the same things as the real cactus?**
- **What do you think would happen if we planted a toy grape in soil? What would happen if we watered a toy cactus?**

Using a dry-erase marker, invite volunteers to write *R* on images of real living things. Then have volunteers write *F* on images of fantasy toys that are nonliving.

Think and Talk

Review the similarities and differences between the real living plants and the toys. Prompt children to discuss what real plants need to live.

Formative ASSESSMENT

If children say that grapes do not need light to grow and stay alive, then revisit Flipbook page 8. Discuss the idea that plants need light, water, and air to stay alive.

ELL Support

Bring in pictures of real plants and fantasy plants.

Beginning Point to each picture, saying whether it is real or fantasy. Have children repeat after you.

Intermediate Point to each picture. Have children tell if it is real or fantasy. Model using complete sentences, such as *This plant is real.*

More to Read

The Giving Tree,
by Shel Silverstein
(HarperCollins, 1986)

After reading, discuss what the "Giving Tree" could do that a real tree could not. Have the class come up with their own fantasy story about a plant. Record their story and have children illustrate it.

Window Box Wonder

Objective: To begin to understand the function of roots.

Inquiry Skills: observe, infer, communicate

You need **small window box or flower pots, eight small plants with roots, soil, Science Journal p. 28**

1. **Observe** Remove the soil from the roots of a plant. Ask children to describe the roots.
2. **Infer** Ask: **Why are the roots spread out? What would happen to roots if the seedlings were planted close together? Should the soil be tightly or loosely packed? Why?** The roots need space to grow and find water.
3. Have the group plant a seedling. Ask: **What will the plant need to live?** water, air, light, soil Allow everyone to have an opportunity to help plant a seedling in the window box. Throughout the unit, have children give their plant what it needs to grow and stay alive. Remind children to wash their hands after touching soil.
4. **Communicate** Have children use their Science Journal page to identify what a plant needs to stay alive. Later in the unit, have children examine the plant and draw a picture to show how the plant has changed.

Teacher Tip

How will your class know when their plants need water? You could make two sample pots of soil, one sufficiently moist and another in need of water. Then ask children to carefully feel the soil.

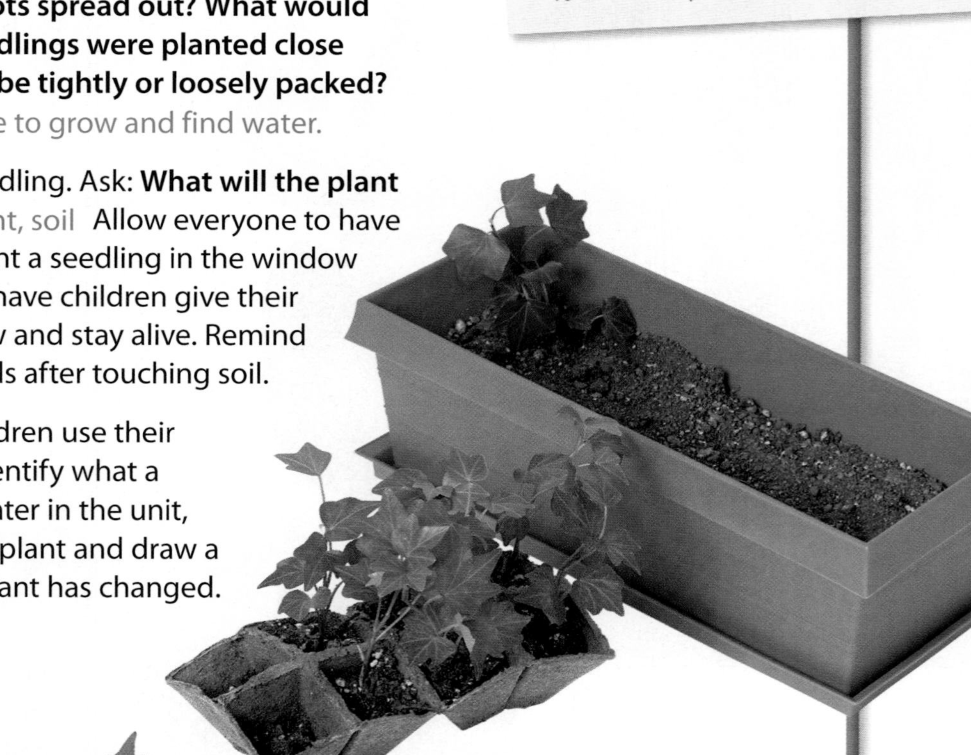

Science Journal Name ____________

Window Box Wonder

Soil

MILK

Candy

Directions: Circle pictures that show what plants need. Draw an X on pictures that show what plants do not need.

28 Unit A • Plants Science Resource Book | Use with Lesson 2 Be a Scientist Activity

© Macmillan/McGraw-Hill

Science Resource Book, p. 28

Investigate More

Infer Do you think roots grow? Why? Possible answer: The plant has used up all the water near the plant and it needs to get water from farther away. **What might happen if the roots grew longer than the pot?** Possible answers: The plant would die; we would need a bigger pot.

centers

Art

Plant Collage

Objective: Develop small motor skills.

Inquiry Skills: communicate, make a model

- Have children cut or tear construction paper into small pieces and place the pieces in the Art Center with additional paper and glue.
- Ask children to think about how they might arrange small pieces of paper on a background to make a design. Ask: **Will you make a real plant or a fantasy plant? A plant with only a few parts? A plant with many parts? If your plant is fantasy, what can it do? Can it sing, dance, or walk?**
- Hang children's plant collages on a bulletin board display. Arrange them by *real* and *fantasy*.

You need
- construction paper
- scissors
- glue

Dramatic Play

Be a Gardener

Objective: Act out the process of taking care of plants.

Inquiry Skills: infer, communicate

- Invite children to act out being gardeners. Tell a story about how a gardener planted vegetables, cared for them, and waited for them to grow.
- At the end of the story, have the gardener take home the vegetables and cook them.
- Encourage children to make up their own story about taking care of plants and act it out.

Sand Table

Stick Garden

Objective: Develop fine motor skills.

Inquiry Skill: make a model

- Ask children how they might use craft sticks to create plants for the Sand Table.
- Provide craft sticks, chenille stems, paper, and glue, and encourage children to design their gardens.
- Invite children to place their plants in containers in the Sand Table. Help them add labels to their sand garden.

You need
- craft sticks
- chenille stems
- construction paper
- glue
- shovels
- watering can
- container

For more on plant needs, use pp. 9–10 in the **Activity Book.**

How Plants Grow

▶ **Essential Question**
How do plants change as they grow?

▶ **Objective**
Recognize that plants are organisms that grow and change.

▶ **Vocabulary**
seed **fruit**
seedling **flower**

Resources

Flipbook, pp. 10, S2

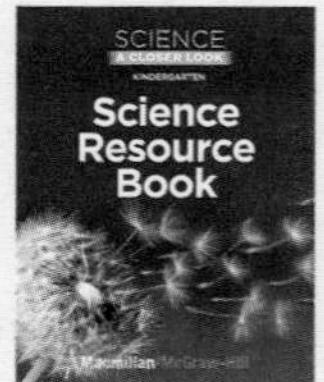

Science Resource Book, p. 29

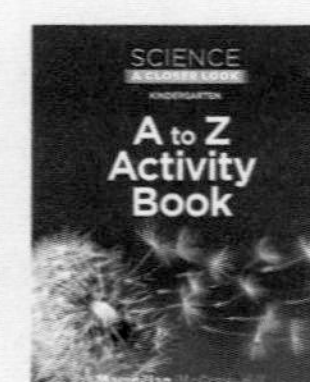

A to Z Activity Book, pp. 8–9

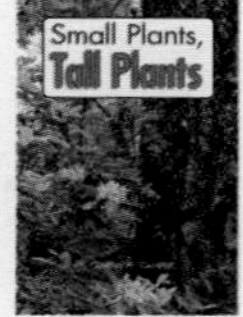

Leveled Reader: *Small Plants, Tall Plants*

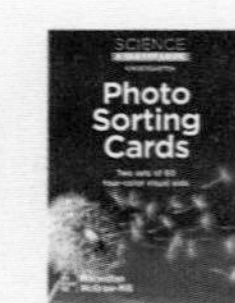

Photo Sorting Cards 1–10

Technology

Science Songs CD Tracks 3–4

Science in Motion
Life Cycle of a Bean Plant

www.macmillanmh.com

NSDL

Circle Time

WHOLE CLASS

Sprouting Seeds

You need
- lima bean seeds
- resealable bags
- moist paper towel

Inquiry Skills: observe, predict, infer

- Show children some lima bean seeds. Tell children what they are. Explain that plants grow from seeds and when these seeds grow, they will all become lima bean plants.
- Place some beans in a resealable bag. Place others on a moist paper towel and put them into another resealable bag. Have children predict what might happen to the seeds. Have children observe their seeds for three days.

Infer **Why did some seeds sprout, or grow, but not others?** Some had water and others did not.

Time to Move!

Invite children to move like a growing plant. Have children begin as a seed in the ground, curled up in a ball. Next have them uncurl and come up to their knees, then stand and stretch their arms, legs, and bodies.

Be a Reader

You need
- Unit A Big Book
- Leveled Reader: *Small Plants, Tall Plants*
- chart paper
- markers

Read the Big Book

Objective: Observe and describe the similarities and differences of plants.

Inquiry Skill: communicate

- Read the title *Small Plants, Tall Plants*. Point to each word as you read. Ask: **What do you think this book is about?**
- Read the sentence on page 2. Invite volunteers to point to the words. Tell children that a sentence is made up of separate words. Repeat this process as you continue to read.
- **Guided Reading:** See the inside back cover of the Leveled Reader version of *Small Plants, Tall Plants* for activities.

Be a Writer

You need
- paper
- crayons
- markers
- Photo Sorting Cards 1–10

Plant Sequence

Objective: Show the sequence of plant growth.

Inquiry Skills: communicate, make a model

- Look at the lima bean seeds that you sprouted in a wet paper towel earlier and invite children to share what they notice.
- Fold a piece of paper into thirds and give one to each child. Have children illustrate what happened first, next, and last when sprouting the lima bean seed.
- Help children write words to describe each drawing. Children can also use the Photo Sorting Cards for help with words.

Be a Math Wiz Activity

You need
- apple and orange with seeds
- red and orange connecting cubes

Predict How Many

Objective: Estimate, count, and compare how many seeds in fruits.

Inquiry Skills: predict, compare

- Have children recall a time when they ate an apple and an orange and to think about how many seeds were in each fruit.
- Show children an apple and orange and have them predict how many seeds are in each. Ask: **How can we find out for sure?** Cut open the fruit and look inside.
- Cut the apple in half, and as you remove a seed have each child take one red cube. Continue until you have found all the apple seeds. Repeat the process with the orange, using orange connecting cubes. Ask: **Which fruit had more seeds? How are the seeds alike and different?**

Read Together and Learn

How do plants grow?

▶ Build on Prior Knowledge

Ask children if anyone has helped a plant grow. Have them describe what they did to make sure the plant(s) grew.

▶ Use the Visuals

- **Communicate** Have volunteers describe the sequence of growth and how the plant changes over time. Read the words together as a class as you point to each stage of growth.
- Explain that the seeds came from a parent plant. The seeds will produce a new plant that looks similar to the parent plant.
- **Infer** Have volunteers point to the roots on the page. Ask: **How are the roots changing in the soil as the plant is growing?** The roots are growing down into the soil and getting longer. They are spreading out.

▶ Develop Vocabulary

seed, **seedling**, **fruit**, **flower** Cover the labels on the Flipbook page with sticky notes. Have children guess the name of each image and then lift the sticky note to see if they were correct.

Flipbook

UNIT A • LESSON 3

How do plants grow?

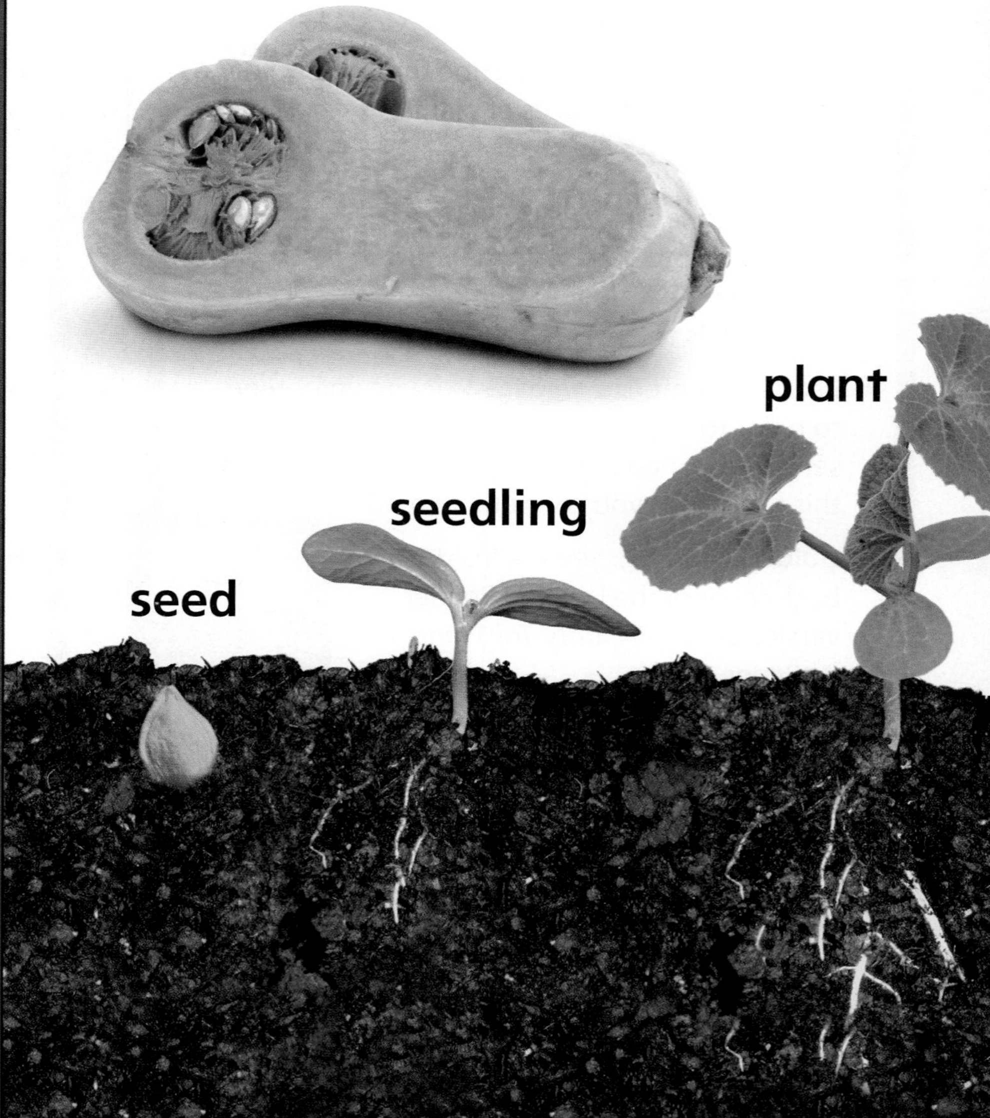

Reading Strategy

Sequence Events Ask children to help you count how many stages are in the plant life cycle. As you count, number the stages 1–5. Then go back and ask what happens first, second, and so on.

Print Awareness Invite children to find words that begin with the letter *s* and point to them on the page. Continue with the letters *p* and *f*.

Science Facts

Seeds, Seedlings Plants come from seeds, which grow into seedlings and then adult plants.

Flower, Fruit Adult plants produce seeds in their fruit, which develop from flowers.

Survival Plant seeds can be carried by wind and water or by animals and insects and deposited in a new location to grow.

Life Cycle Plants have different life cycles, varying from one season to thousands of years. Bristlecone pines, for example, live more than 5,000 years.

Squash This fruit varies in color from white to orange to green. Squashes grow to different sizes. Familiar types of squash include zucchinis and pumpkins.

Professional Development For more Science Facts and resources from  visit http://nsdl.org/refreshers/science

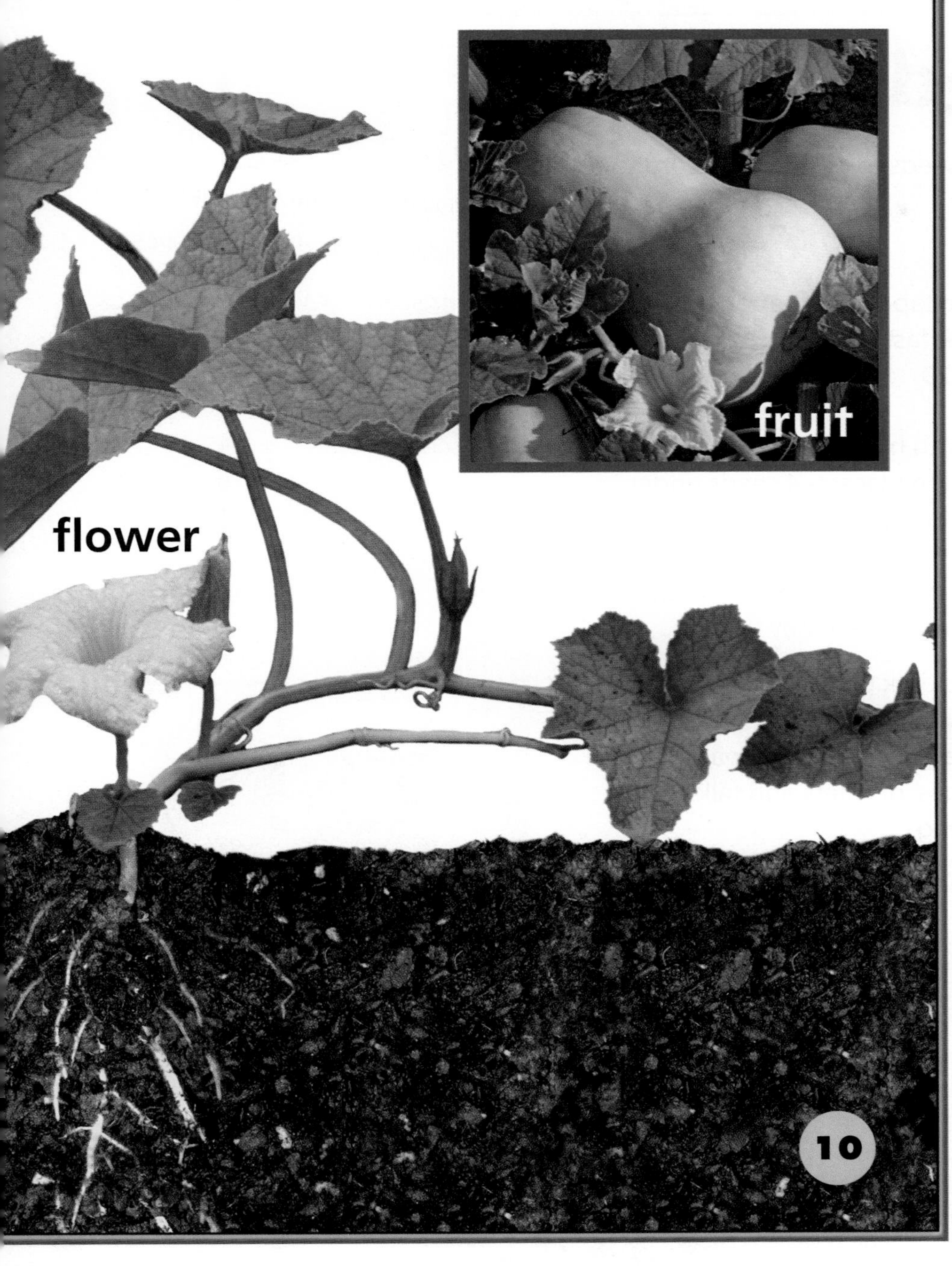

▶ Ask Questions

- **What does the squash seed need in order to grow?** water, space, sunlight, soil **What do you notice about the roots as the plant grows?** The roots grow too. They grow longer. There are more roots as the plant gets bigger.
- **What part of the plant is a squash?** the fruit **What would happen if we planted all the squash seeds in one pot?** Some would grow, others would not because there is not enough space.
- **What happens to a plant if we eat its fruit?** It continues to grow and may grow new fruit.

Write on It

Invite children to circle the parts of the plants using a dry-erase marker. Write labels for the parts to help children revisit these terms while reviewing the life cycle.

Think and Talk

Review the stages of plant growth with children. Have volunteers share what they know about each stage. Make a chart listing the stages that occur and invite children to illustrate each stage.

Differentiated Instruction

Extra Support Give children a blank piece of paper divided into five parts. Tell a brief story of the life cycle of a squash. Have children draw a picture of what the squash plant looks like at each stage of growth. Refer to the Flipbook page as you tell the story.

More to Read

Dandelions: Stars in the Grass, by Mia Posada (Carolrhoda Books, 2000)

As you read, point out the stages of the plant life cycle. After reading, place this book in the Writing Center so children can use it as a reference when making their own plant growth books.

Observe if children describe the stages of plant growth correctly or incorrectly. Encourage children to draw a picture of their sprouting lima bean seeds every three days to record what happens over time.

Planting Seeds

Objective: Explore what is needed for seeds to grow and record the change that occurs over time.

Inquiry Skills: observe, communicate, infer

You need **milk containers, masking tape, newspaper, potting soil, radish seeds or other fast-growing seeds, paper cups, Science Journal p. 29**

1. Give each child a milk container with his or her name on the side. Have children add soil and then place 3–4 seeds under the surface of the soil. Remind children to give each seed some space to grow. Have children use a paper cup to add water to the container. Remind children not to put seeds or soil in their mouths and make sure they wash their hands when they finish.
2. **Observe** Place the containers in a dark place. Once the seeds have sprouted, move them to a sunny place. Have children observe their plants daily. Encourage children to measure their plants with string, connecting cubes, or rulers and talk about how they are growing and changing.
3. **Communicate** Have children record what they observe on their Science Journal page.

Teacher Tip

You may want to have groups of children plant different types of seeds. As the plants grow, the class can compare the different plants.

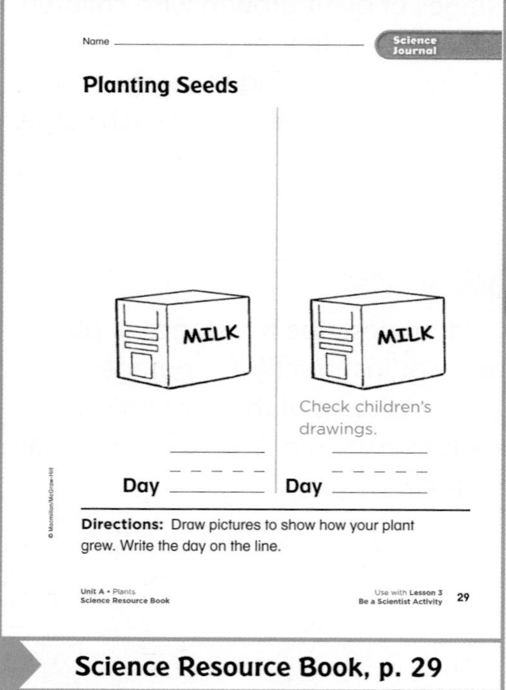
Name ______ Science Journal

Planting Seeds

MILK MILK

Check children's drawings.

Day ______ Day ______

Directions: Draw pictures to show how your plant grew. Write the day on the line.

Unit A • Plants Science Resource Book — Use with Lesson 3 Be a Scientist Activity 29

Science Resource Book, p. 29

Investigate More

Infer Ask: **After the seeds started to grow, why did you move them to a sunny place?** Young plants need light to grow.

Music

In My Garden

Objective: Sing about and discuss how flowers grow.

Inquiry Skills: communicate, put in order

- **Sing and Talk** Play the recording of "In My Garden," tracking the print with your finger as you sing. Play the song again, inviting children to chime in. Afterward, ask children to describe the steps it takes to grow a plant, using the text and pictures to help them recall what happens first, next, and last.
- **Act It Out** Encourage children to make up actions to the song, such as planting and watering the seeds, drawing the shape of the Sun in the air, pointing to themselves on the words *my* and *I*, and so on.
- See page TR23 for the printed music for "In My Garden."

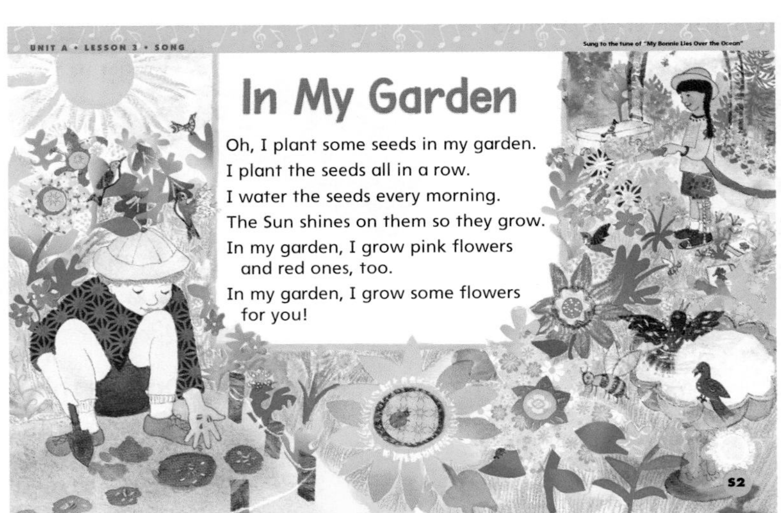

Flipbook p. S2
CD Tracks 3–4

Art

Bean Collages

You need
- dry beans of different shapes, sizes, and colors
- glue
- small pieces of cardboard
- containers

Objective: Sort and classify beans.

Inquiry Skill: classify

- Place a collection of beans in a bin. Ask children to sort the beans into containers. Help them locate the lima beans.
- Have children glue beans onto the cardboard to make their own design.

Blocks

Block Garden

You need
- colored paper
- tape
- scissors

Objective: Begin to understand what plants need to thrive.

Inquiry Skill: create a model

- At Circle Time, ask children: **Why do you think some gardens have fences and gates?** Possible answer: To keep out animals that would eat the plants.
- Provide children with paper, scissors, and tape to make flowers for their gardens. Invite them to bring their flowers to the Block Area.
- Encourage children to work together to create one or more gardens in the Block Area.

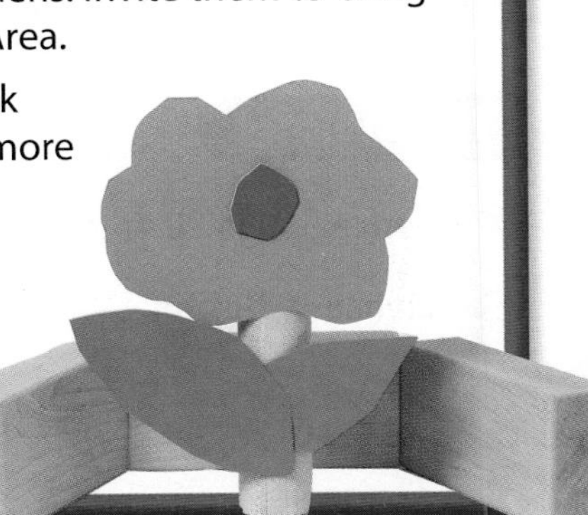

For more on how plants grow, use pp. 11–12 in the **Activity Book.**

▶ **Essential Question**
How can plants look different?

▶ **Objective**
Recognize that plants can be identified by their parts.

▶ **Vocabulary**
leaf **flower**
leaves **flowers**

Resources

Flipbook, pp. 11–12, S3

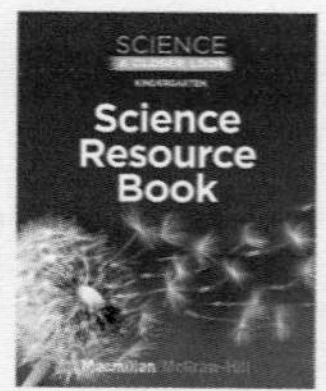

Science Resource Book, p. 30

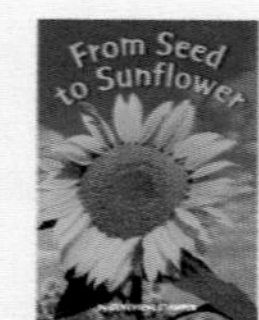

Leveled Reader: *From Seed to Sunflower*

Floor Puzzle: *Pond Life*

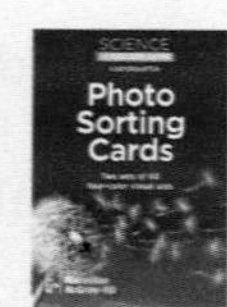

Photo Sorting Cards 1–10

Technology

Science Songs CD Tracks 5–6

Science in Motion
Strawberry Plant

www.macmillanmh.com

Look at Leaves and Flowers

Circle Time WHOLE CLASS

Sorting Plants

You need
- Photo Sorting Cards 1–10
- string or yarn

Inquiry Skills: communicate, classify, infer

- Ask: **Do you think every plant has flowers?** Explain that some plants have flowers and some do not.
- Use string or yarn to make two circles on the floor. Explain that one circle will be for plants with flowers and the other for plants without flowers.
- Hold up a card and ask a volunteer to place it in the correct circle. Continue until all cards have been sorted.

Infer Ask: **Why do you think some plants have flowers and others do not?**

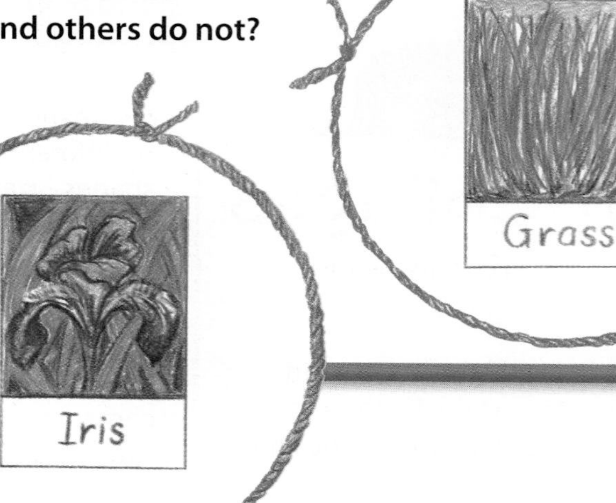

Time to Move!

Play soft, slow music. Invite children to move like leaves on a tree gently blowing in the breeze. Then play faster music and encourage children to move like leaves swishing around on a very windy day.

Be a Reader

Read the Big Book

Objective: Observe and describe seed bearing plants.

Inquiry Skills: communicate, predict

You need
- Unit A Literature Big Book
- Leveled Reader: *From Seed to Sunflower*
- sunflower seeds
- chart paper
- markers

- Read the title *From Seed to Sunflower*. Ask children if they have ever seen a sunflower or a sunflower seed. Give each child a sunflower seed. Ask: **What might happen if we planted these seeds?** Record children's predictions.
- Read the book. Then invite volunteers to retell what happened to the seed. Revisit children's predictions.
- **Guided Reading:** See the inside back cover of the Leveled Reader version of *From Seed to Sunflower* for activities.

Be a Writer

Favorite Flower

Objective: Use a model to draw and write about a flower.

Inquiry Skills: observe, communicate

You need
- fresh flower or flowering plant
- paper
- crayons
- colored pencils

- Place a single flower or a flowering plant in the Drawing/Writing Center. Encourage children to look carefully at the flower and draw what they see.
- Support children by asking: **Where will you start your drawing? What will you draw first? What shapes and colors will you use?**
- Have children write a word to describe their drawing or offer to take dictation.

Be a Math Wiz

Pattern Flowers

Objective: Identify shapes of pattern blocks.

Inquiry Skills: communicate, make a model

You need
- pattern blocks
- chart paper
- marker
- paper
- pencils

- Prepare a recording sheet to help children record their work. Ask: **What shapes would you use to make a flower?** Record the names of the pattern blocks on chart paper.
- Show the recording sheet and explain that children will use pattern blocks to make flower designs. Children will count and record how many of each shape they used on the recording sheet.
- Have children work in pairs to create and record their own pattern block flowers. Invite children to trace their designs.

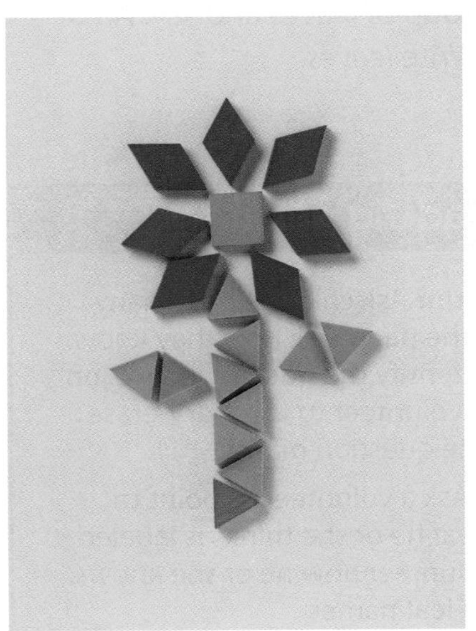

Read Together and Learn

How are these leaves alike? How are they different?

Build on Prior Knowledge

Ask if anyone has seen a leaf. Ask volunteers to describe what a leaf looks like and record their responses.

Use the Visuals

- **Communicate** Have volunteers point to leaves on the page that they have seen before. Invite them to repeat the name of the leaf.
- Ask children if they think any of the leaves on the page are from the same tree. Have children explain their thinking.
- **Compare** Point to a leaf and ask a volunteer to describe it. Then point to another leaf and ask: **How are they alike/different?** Repeat with different volunteers and other combinations of leaves.

Develop Vocabulary

leaf, **leaves** Write on the board: *This is a* _____, and *All of these are* _____. Point to the maple leaf and say: **This is a leaf.** Write the word *leaf* in the blank. Point to a group of leaves and say: **All of these are leaves.** Write *leaves*.

Flipbook

UNIT A • LESSON 4

How are these leaves alike? How are they different?

maple

ginkgo

cottonwood

Reading Strategy

Conventions of Print Ask children how many questions are on the page. Ask how they know. Help children to identify the question marks on the page. Invite a volunteer to use a dry-erase marker to circle the question marks.

Print Awareness Ask a volunteer to point to the name of the leaf he or she thinks is labeled *willow*. Ask the volunteer how he or she knows. Repeat with other leaf names.

Science Facts

Leaf Types A broad leaf has a blade (lamina) which is the flattened part. This type of leaf is deciduous and drops during the fall (maple, cottonwood). Evergreens, or the gymnosperms, are plants which do not lose all their leaves (fir).

Simple and Compound Leaves A simple leaf has a single blade on a stalk (petiole) attached to a woody stem. A compound leaf has a blade divided into leaflets attached to a single petiole.

Leaf Identification First determine if the leaves have opposite (ash) or alternate (cottonwood) arrangement on the branch. Next determine if it is simple or compound. Then determine if it has rounded, pointy, or no lobes.

LOG ON **Professional Development** For more Science Facts and resources from  NSDL visit http://nsdl.org/refreshers/science

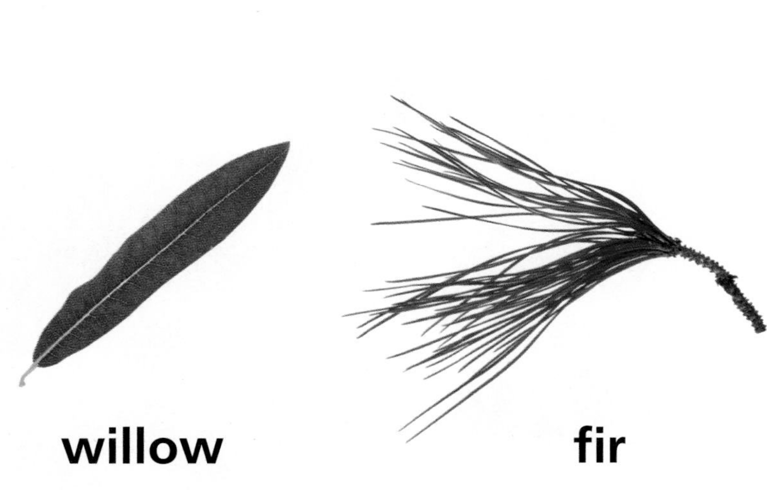

willow

fir

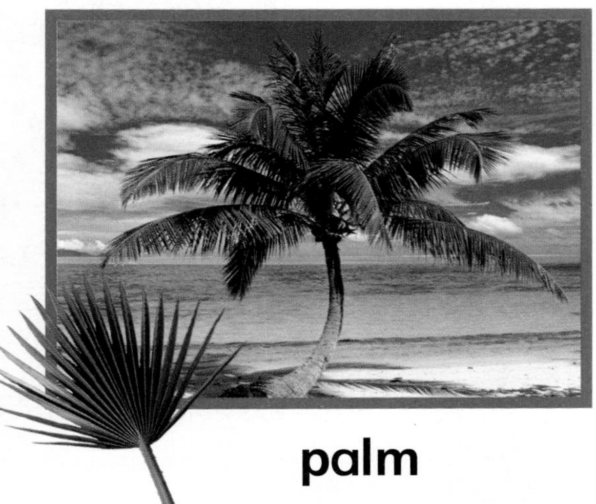

niper

palm

11

▶ Ask Questions

- **Do you think that all the leaves on this maple tree will look alike? Why or why not?** They look alike because they come from the same plant. They can look different, such as in size.
- **If you take a walk in a forest, how would you know what kind of trees you see?** Look at the shape of the leaves.
- **Why do you think trees have leaves?** for food, protection, shelter **Have you ever seen a tree without leaves? Why does it not have leaves?** They fell to the ground; it's winter.

Write on It

Invite children to trace around the edges of the leaves on the Flipbook page using a dry-erase marker. Ask children what they notice about the different edges.

Think and Talk

Ask children what kind of tree a maple leaf comes from. Repeat the question using each type of leaf shown on page 11.

Take a Trip

Nature Walk Bring children on a walk to collect leaves. Remind them to only collect leaves that they find on the ground. Give each child a collection bag and ask them to find as many different leaves as possible. Remind children to be safe when collecting leaves. If they see a leaf they do not know or recognize, remind them to ask an adult before they touch it.

Back in the Classroom Children can use these leaves in the Science Center, for sorting exercises, or for leaf rubbings in the Art Center.

Read Together and Learn

How are these flowers alike? How are they different?

▶ Build on Prior Knowledge

Ask volunteers to name flowers they already know. Encourage them to describe how they look.

▶ Use the Visuals

- **Communicate** Have a volunteer count how many different kinds of flowers he or she sees on the page. Ask: **How do you know the flowers are different?**
- Ask children if they think any of the flowers are from the same plant. Have children explain their thinking.
- **Compare** Point to a particular flower and ask a volunteer to describe it. Record his or her response. Then point to another flower. Ask: **How are they alike/different?** Repeat with different combinations of flowers.

▶ Develop Vocabulary

flower, **flowers** Write *flower* and *flowers* on sticky notes. Invite children to place them on pictures of single or multiple flowers. Explain that adding the letter *s* at the end of a word changes its meaning.

Flipbook

UNIT A • LESSON 4

How are these flowers alike? How are they different?

Reading Strategy

Print Awareness Have children count how many times the letter *w* appears on the page. Have them describe the position of the letter within the word. Ask: **Is it at the beginning, middle, or end of the word?**

Phonemic Awareness Write the word *flower* on a sticky note and have children find the word on the page. Then ask what sound they hear in the middle of the word *flower*. (*ow* sound) Ask: **What other words have this sound?** (*now, how, owl, cow, fowl, brown*)

Science Facts

Flower Parts of a flower include petals, sepals, carpels (female reproductive organs), and stamens (male reproductive organs).

Flower clusters Some flowers are actually a cluster of many flowers on one stem. Some flowers feature a globe-shaped cluster on a long, leafless stem.

Single bud Most flowers grow from a single bud that opens into a single flower. Flowers come in a variety of colors and bloom in different seasons.

Tree flowers Some trees have flowers that bloom in spring. In the case of the apple blossom, these flowers develop into fruit (apples) whereas the dogwood flowers do not.

Professional Development For more Science Facts and resources from NSDL visit http://nsdl.org/refreshers/science

12

▶ Ask Questions

- **What colors do you see on the page? How do these flowers look different?**
- Explain that children who come from the same parents tend to look a little, but not exactly, alike. Ask: **Could any of these flowers be in the same family? Why or why not?**
- **Do all plants have flowers? Do trees have flowers? How do you know?** Not all plants have flowers; not all trees have flowers.

Have children use a dry-erase marker to label the colors they see in the flowers. Provide a list of colors and ask children to write the first letter of the color with the corresponding color dry-erase marker.

Think and Talk

Ask a volunteer to point to a flower on page 12 and describe it. Repeat for all the flowers shown.

Formative ASSESSMENT

If children are unable to identify the names of some of the leaves and flowers on the pages, have them join you in a concentration game using Photo Sorting Cards 1–10. Reinforce the leaf and flower names as you play the game.

TIME FOR KIDS

Shared Reading Read *One, Two, Tree!* with the class. Use this magazine, found in the **Unit A Literature Big Book,** as a photographic review of trees and their leaves.

More to Read

Zinnia's Flower Garden, by Monica Wellington (Dutton Books, 2005)

Activity After reading the story, invite children to write and illustrate their own story about a garden they would like to grow. Provide gardening magazines so children can cut out flowers to use in their illustrations.

Matching Leaves

Objective: Identify and classify leaves.

Inquiry Skills: observe, classify, investigate, communicate, infer

You need **leaves (or an assortment of pictures of leaves), aluminum trays, chart paper, markers, Science Journal p. 30, fresh flowers**

1. **Observe** Bring in leaves or have children collect leaves. Hold up two different leaves and ask children if they are from the same tree. Ask why or why not and record their responses. Explain that they will become researchers to find out the names of the leaves.
2. **Classify** Give children a collection of leaves. Have them sort the leaves into piles that they think came from the same tree.
3. **Investigate** Have children use Flipbook p. 11, and any field guides you have, to help them identify the leaves.
4. **Communicate** Have children choose their favorite leaf to tape and label on their Science Journal page.

Teacher Tip

You will want to have a few tree and flower field guides that children can use as research tools. You can also refer to the Internet for more tree identification help.

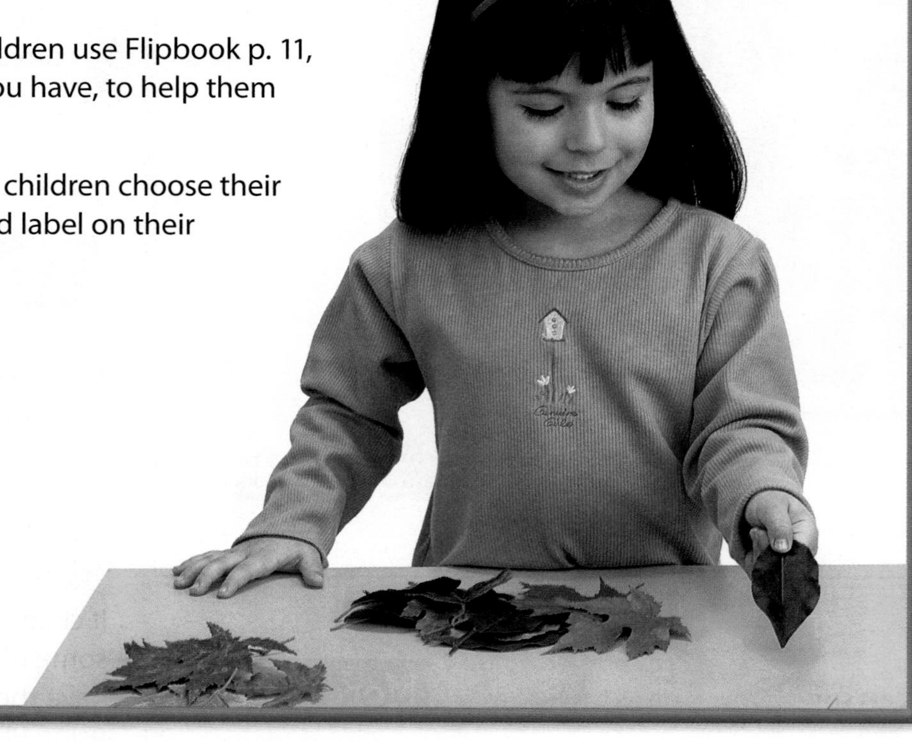

Science Journal Name ______

Matching Leaves

This is a ______ leaf.

Check children's work.

Directions: Tape a leaf, or your leaf rubbing, to the page. Write the name of your leaf below.

30 Unit A • Plants Science Resource Book — Use with Lesson 4 Be a Scientist Activity

© Macmillan/McGraw-Hill

Science Resource Book, p. 30

Investigate More

Infer Hold up two flowers of the same kind and ask: **Are these flowers from the same plant? How do you think we could find out what kind of plant these flowers come from?**

Music

Take a Look

Objective: Compare how leaves are alike and different.

Inquiry Skills: observe, compare

- **Listen and Find** Have children listen to "Take a Look" and look at the Flipbook images carefully. Afterward, read each line slowly, inviting volunteers to point to the brown, yellow, pointed, and round leaves. Have children find the two biggest leaves. Then have them compare the leaves by observing their color, shape, and size. Play the recording again, and encourage the class to sing along.
- **Innovate** Challenge children to think of other ways to describe leaves, such as red, green, fuzzy, or small. Record their new lyrics on chart paper and have them sing their new song using the instrumental track.
- See page TR24 for the printed music for "Take a Look."

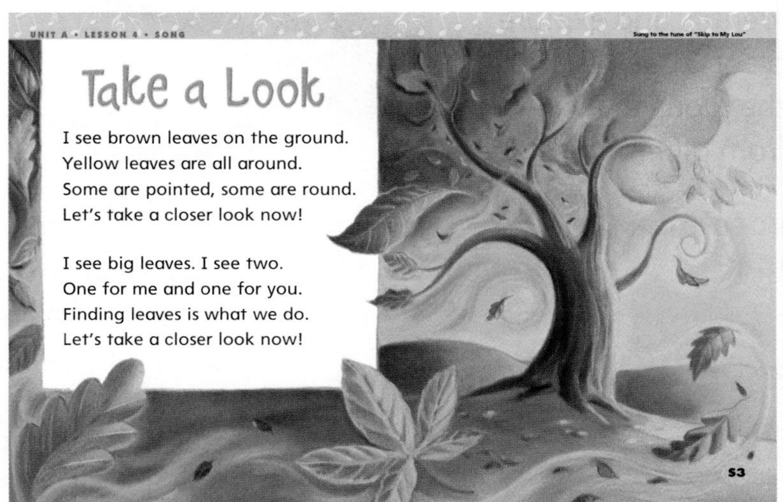

Flipbook p. S3
CD Tracks 5–6

Blocks

Make a Park

You need
- colored construction paper
- scissors
- tape

Objective: Create park structures.

Inquiry Skills: make a model, communicate

- Explain to children that this week the Block Area will be a park. Help them brainstorm structures they could build for the park.
- Ask: **How can we make trees, flowers, and bushes for our park?**
- If possible, keep up the park structure until everyone has had a turn to add to it.

Art

Leaf Rubbings

You need
- leaves
- paper
- crayons
- index cards

Objective: Notice leaf patterns.

Inquiry Skills: observe, communicate, compare

- Place 2–3 different leaves in the Art Center. Make a label for each leaf using index cards. Show children how to do a leaf rubbing.
- Have children do rubbings of each leaf and label them, copying the names from the cards you made.
- Help children describe how the patterns in their rubbings are alike and different.

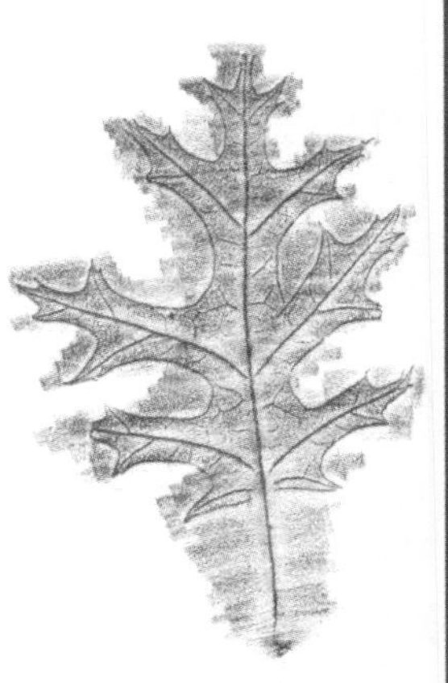

For more on leaves and flowers, use pp. 13–14 in the **Activity Book.**

Plants We Use

▶ **Essential Question**
What plants can we use?

▶ **Objective**
Identify and explore plants that we eat and the foods that come from different plants.

▶ **Vocabulary**
fruits vegetables

Resources

Flipbook, p. 13

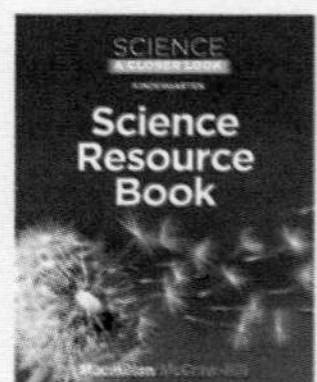
Science Resource Book, p. 31

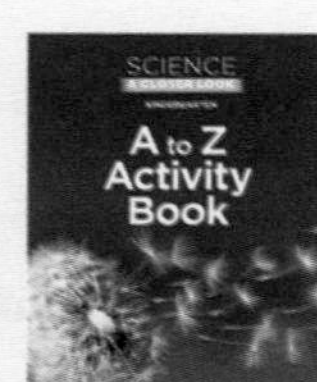
A to Z Activity Book, pp. 14–15

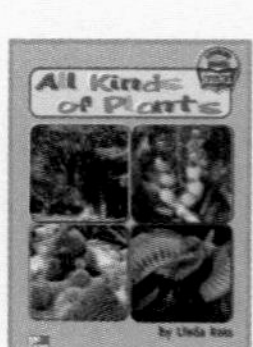
Leveled Reader: *All Kinds of Plants*

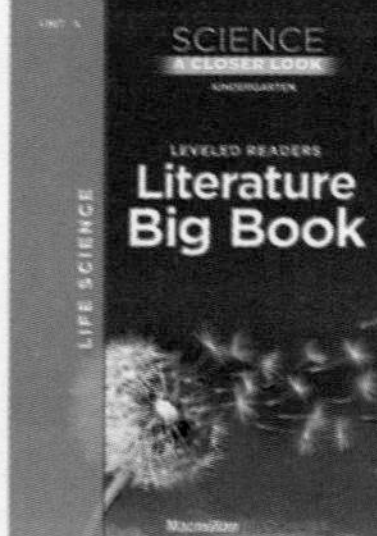
Unit A Literature Big Book

Technology

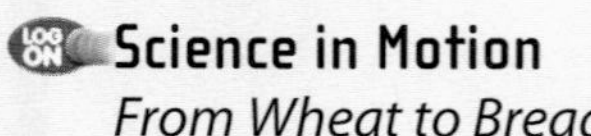
From Wheat to Bread

Circle Time — WHOLE CLASS

Feely Box

You need
- 5 fruits and vegetables
- small box with a hole in the top

Inquiry Skills: observe, infer, communicate

- Show children the fruits and vegetables and have them name and describe what they notice about each item.
- Show children the "Feely Box" and then place the fruits and vegetables into the box. Have children reach into the box, feel one piece, describe what they feel, and say what they think it is. Have them remove the item to see if they were correct. Repeat until everyone has had a turn.

Infer Ask: **Which of these plants are vegetables? Fruits? How can we find out?** Find the seeds.

Time to Move!

Help children to form a circle. Invite volunteers to come to the center of the circle and name their favorite fruit. Invite children in the circle to stand on one foot if they like the same fruit. Repeat with vegetables.

Be a Reader Activity

Read the Big Book

Objective: Use picture clues to read and write about plants.

Inquiry Skills: communicate, infer

- Read *All Kinds of Plants* to the class. After reading, ask children what other plants we eat. Model writing as you record their responses. Ask children to help you listen for beginning sounds as you write. Reread the story, asking children to chime in.
- Have children write and illustrate their own plant stories. Remind children that they can use the Big Book to help them spell words.
- **Guided Reading:** Use the Leveled Reader version of *All Kinds of Plants* for small-group guided-reading instruction.

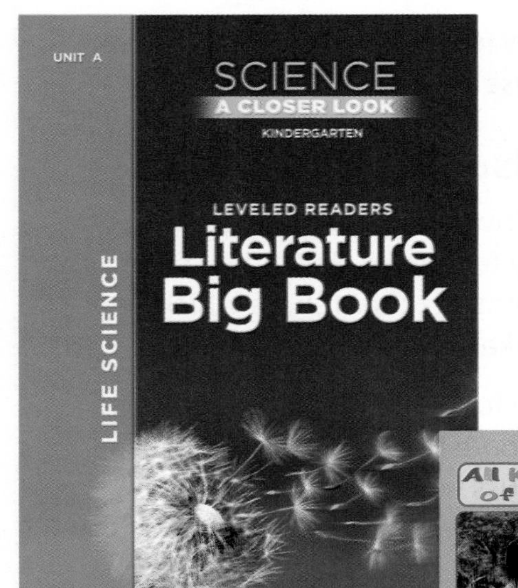

You need
- Unit A Big Book
- Leveled Reader: *All Kinds of Plants*
- lined paper
- pencils
- crayons

Be a Math Wiz Activity

Counting Peas

Objective: Count sets using one-to-one correspondence and graphing skills.

Inquiry Skills: observe, compare, predict, communicate

- Make a blank bar graph with six columns and the numbers 3–8 along the bottom.
- Open a snow pea pod and show the seeds. Ask: **How many seeds do you think there are?** Take the seeds out one at a time, count them, and record the results on the graph.
- Then have pairs open each of their pods, count the number of seeds in each, and record their findings on the class graph. Encourage them to predict which number will be most common.

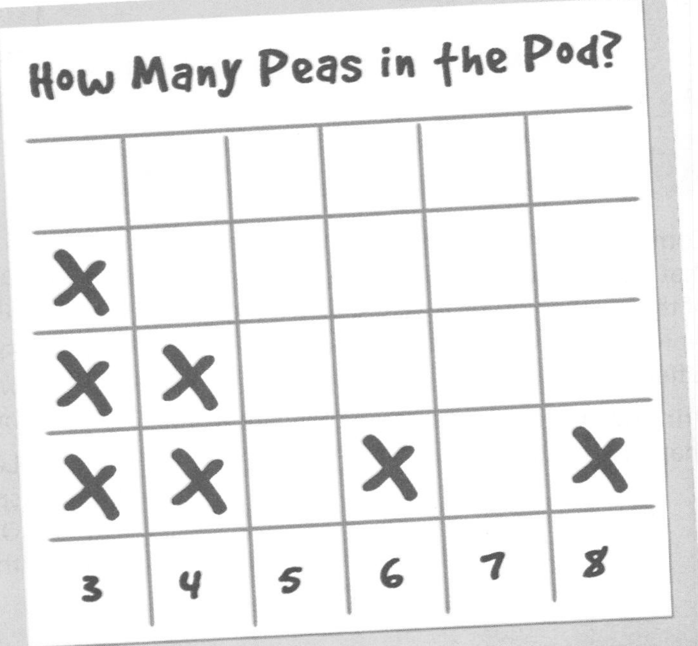

You need
- 10 snow pea pods per pair
- chart paper
- markers

Read Together and Learn

What plant parts do we eat?

▶ Build on Prior Knowledge

Ask children if they might eat any plants today. Have them list these plants.

▶ Use the Visuals

- **Communicate** Invite volunteers to point to any vegetables they recognize. Ask if they have eaten any of them. Let them share which ones they like or dislike.
- Ask if any of the foods on the page are plants. Have children share how they know. Invite children to discuss where the fruits and vegetables on the page come from.
- Encourage children to discover which fruits and vegetables around the center picture are also shown in the supermarket scene.

▶ Develop Vocabulary

fruit, **vegetable** Make a two-column chart labeled *fruits* and *vegetables*. Have children share ideas for each column. Explain that a fruit is the part of the plant that you can eat that contains the seeds. A vegetable is the stems, leaves, and roots of the plant that you can eat.

Flipbook

UNIT A • LESSON 5

What plant parts do we eat?

broccoli

yams

blueberries

peppers

Reading Strategy

Comprehension Ask children to locate words that end in *s*. Ask what happens to the meaning of the word if you take off the final *s*. Say the words aloud both ways so children can hear the difference.

Print Awareness Ask children to locate words that have double letters. Say the words out loud.

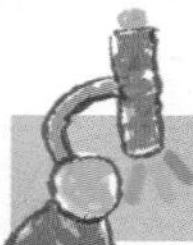

Science Facts

Bulbs are thin, tightly folded layers of leaves that grow underground.

Flower Buds Broccoli is a plant with dense clusters of flower buds.

Leaves Some leaf plants like spinach have loose leaves. Others, like cabbage, have leaves that are packed very tightly.

Roots become fat because the plant takes in more food than it can use; it stores the extra food in its roots.

Seeds can be eaten dried or fresh.

Stems and Stalks of a plant support its leaves, flowers, and fruits.

Professional Development For more Science Facts and resources from NSDL visit http://nsdl.org/refreshers/science

tomatoes

iwi

peas

pineapple

13

▶ Ask Questions

- **What parts of the plants do you see on the broccoli?** stem, flowers **Peas?** seeds **Tomatoes?** fruit
- **Which of these foods have you eaten? Did someone cook them or did you eat them uncooked?**
- **What other plants do you like to eat?**

Write on It

After discussing fruits and vegetables, ask volunteers to use dry-erase markers to write *V* for vegetable or *F* for fruit next to the appropriate picture on the Flipbook page.

Think and Talk

Review the fruits and vegetables on the Flipbook page. Have children share what they learned about plant parts that we eat. Encourage children to share what they know about how plants get to the market.

Formative ASSESSMENT

Observe children as you explore the Flipbook page to help you assess what they already know about the parts of plants we eat.

ELL Support

Hold up six vegetables. Write their names on the board. Ask children to say the plant names after you.

Beginning Ask children to hold up vegetables as you name them.

Intermediate Put vegetables in a bag and have children pick one and name it.

More to Read

Growing Vegetable Soup, by Lois Ehlert (Harcourt, 1990)

Activity Have children sort the vegetables used in the book into vegetables that grow above ground and vegetables that grow below ground.

Be a Scientist | Inquiry Investigation

Plant Part Soup

Objective: Understand what parts of plants we eat.

Inquiry Skills: classify, predict, observe, draw a conclusion, communicate

You need **produce, large pot, hot plate, knife, cutting board, broth, Science Journal p. 31**

1. **Classify** Help children discuss which part of the plant each item represents. Have them wash their hands, then prepare and add a specific type of vegetable to the soup. Remind children that the hot plate is hot and they should not to touch it.
2. **Predict** Ask: **What will happen to these vegetables when we cook them in a soup?** They will get soft and mushy because they soak in water.
3. **Observe** As the soup cooks, ask children to observe what happens to the vegetables. Ask: **Did they change color? Get soft?** Have children record their observations on their Science Journal page.
4. **Draw a Conclusion** Ask: **Why did the vegetables change form?** Possible answers: They soaked up the water. They were cooked.

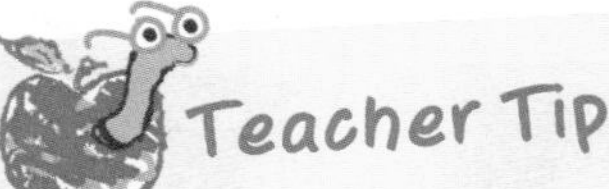

Teacher Tip

In this activity, use vegetables children are likely to know such as carrots and potatoes (roots), broccoli and cauliflower (flowers), spinach and cabbage (leaves), celery and green onions (stems), tomatoes and lemons (fruits), corn and peas (seeds).

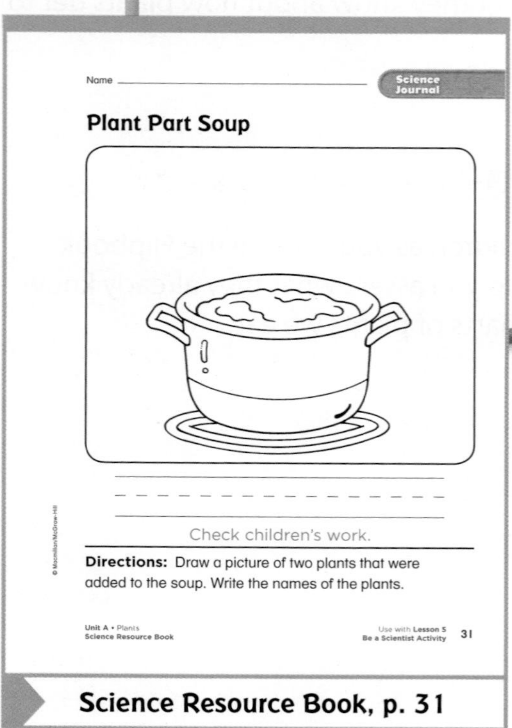

Name ______ Science Journal

Plant Part Soup

Check children's work.

Directions: Draw a picture of two plants that were added to the soup. Write the names of the plants.

Unit A • Plants
Science Resource Book

Use with Lesson 5
Be a Scientist Activity 31

Science Resource Book, p. 31

Investigate More

Communicate Ask children: **What plants do you eat at breakfast, lunch, and dinner?**

centers

Drawing and Writing

Our Produce Store

You need
- paper
- crayons
- colored pencils
- markers

Objective: Draw and write signs.

Inquiry Skill: communicate

- Ask children to make signs to advertise the produce market that they will make in the Block Area.
- Have children brainstorm what words they might need to use. Make a list for future reference.
- Have children draw pictures and write words to advertise the market. Encourage children to use the words from the list to help them.

Technology

More about Plants

- For games and activities about plants, help children log onto www.macmillanmh.com.
- Children can use these fun games to review and reinforce what they have learned about plants.

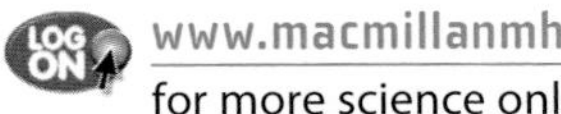

for more science online.

Art

Play Dough Plants

You need
- play dough or other air-drying clay
- paint
- paintbrushes
- produce

Objective: Develop small motor skills.

Inquiry Skills: observe, make a model

- Have children make clay fruits and vegetables for their produce market.
- Bring in some real fruits and vegetables for children to use as models as they work with clay.
- When dry, have children paint their clay fruits and vegetables.

Blocks

Produce Market

You need
- clay produce
- blocks
- child-made signs

Objective: Recognize and experience different aspects of a produce market.

Inquiry Skills: infer, make a model

- Explain that pairs will build their own produce stand in the Block Area. Discuss what they might build, such as shelves or a counter.
- When everyone has had a turn to build, have children add the clay props and signs.
- Encourage children to buy and sell produce to one another as they play.

For more on plants we use, see pp. 15–16 in the **Activity Book.**

ASSESSMENT

DOK 2

Plant Parts and Needs

Objective: Identify the main parts of a plant; identify what plants need to live.

- Place a potted plant on the table. Ask the child to point to and name as many of the parts as he or she can.
- Ask the child to describe the part of the plant that is hidden in the soil. Encourage him or her to draw the parts.
- Ask the child to tell you what helps a plant grow.

You need
- potted plant
- paper
- crayons

Scoring Rubric

Use the Scoring Rubric to evaluate each child's completion of the Performance Assessment task.

points = Child identifies three parts of the plant (roots, stem, leaves) and knows at least two things that plants need to live (water, sunlight, air, soil).

points = Child identifies at least two parts of the plant (roots, stem, leaves) and knows at least one thing that plants need to live (water, sunlight, air, soil).

point = Child is unable to identify more than one part of the plant (roots, stem, leaves) or any of the things that plants need to live (water, sunlight, air, soil).

Summative ASSESSMENT

DOK I

Use the Unit A Assessment booklet on pp. 63–64 of the **Science Resource Book** to assess children's understanding of the parts of plants and what plants need to live.

- Help children fold their page in half to create a book and write their name. On the first page, read the words, then read the sentence and have children circle the word that best completes the sentence.
- Have children copy the word on the line below the drawing. Repeat until each page has been completed.

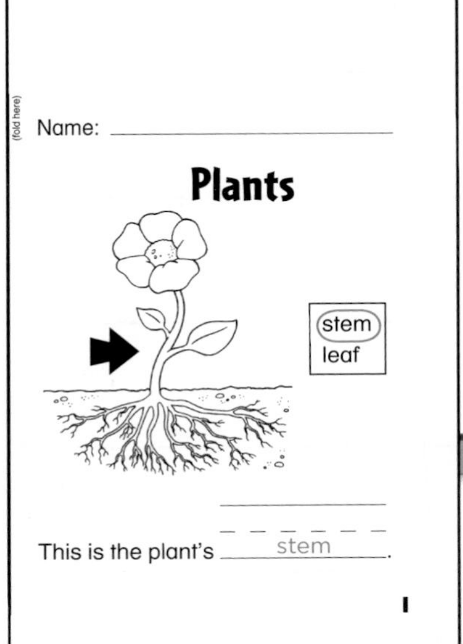

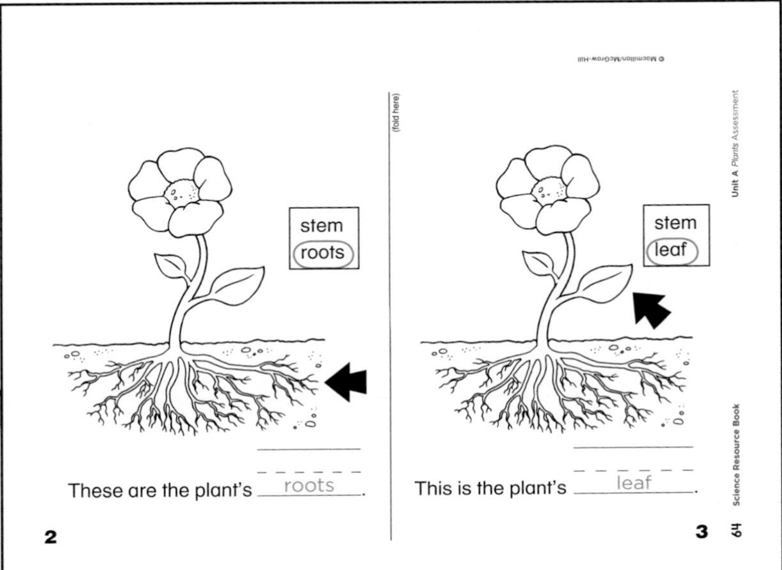

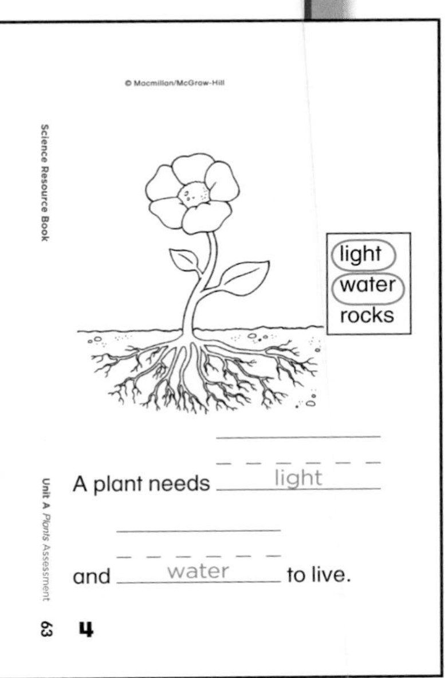

Science Resource Book, pp. 63-64

Portfolio ASSESSMENT

- Have children choose one piece of work from their Science Journal to place in their portfolio. Have children describe what they did on the chosen page and why they chose that piece of work to be included.
- You can use the Unit A Checklist from the **Science Resource Book** page 77 to record children's progress. Place checklists and notes in each child's portfolio.

Teacher's Notes

Animals

What are animals?

The Big Idea

Essential Questions

Lesson 1 Where do animals live?

Lesson 2 What do animals need?

Lesson 3 How can bugs look alike?

Lesson 4 Where do reptiles live?

Lesson 5 How can animals move?

Lesson 6 How do animals stay safe?

Lesson 7 How do animals change as they grow?

Lesson 8 What do we get from animals?

LOG ON Visit www.macmillanmh.com for online resources.

More to Read

Science Leveled Readers

All included in the Leveled Readers Literature Big Book.

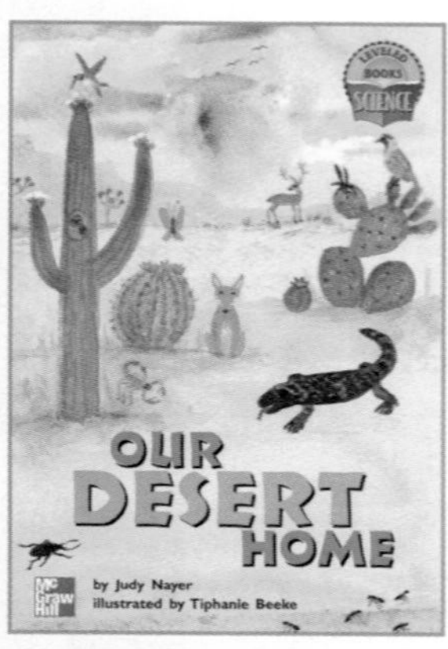

APPROACHING

Our Desert Home The desert is home to many plants and animals.

ISBN: 978-0-02-278972-5

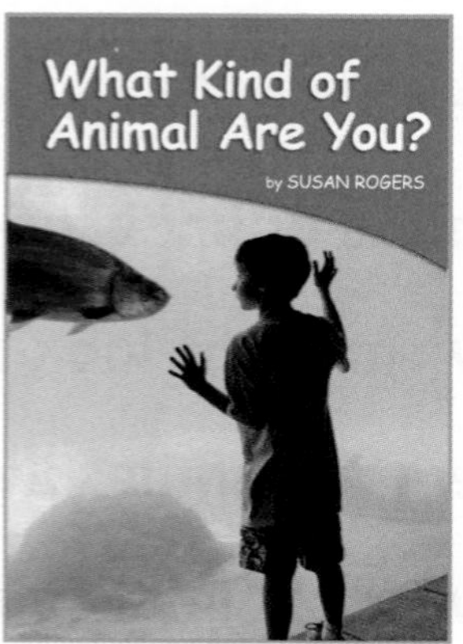

ON LEVEL

What Kind of Animal Are You? Different animals have special parts that help them.

ISBN: 978-0-02-284588-9

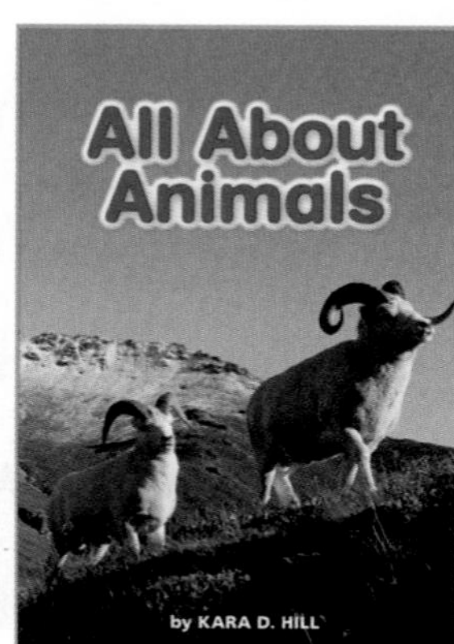

BEYOND

All About Animals All animals have unique body parts to help them adapt to their environments.

ISBN: 978-0-02-284591-9

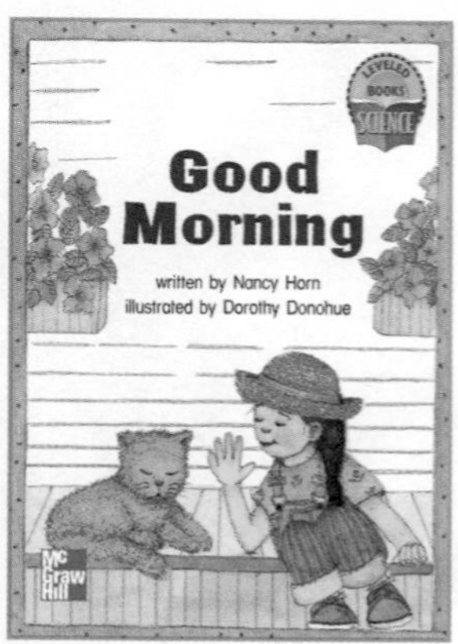

APPROACHING

Good Morning A young girl starts her day greeting a variety of animals.

ISBN: 978-0-02-278459-1

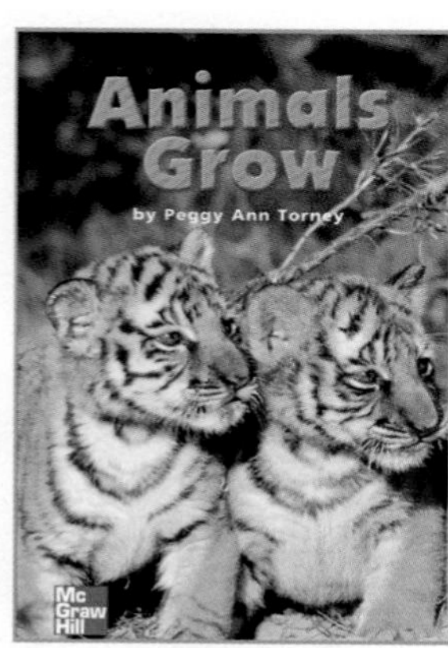

ON LEVEL

Animals Grow Animals grow and change.

ISBN: 978-0-02-281079-5

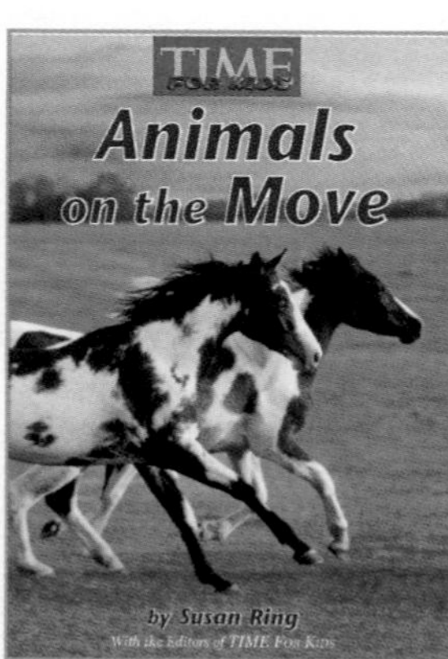

BEYOND

Animals on the Move Animals get from place to place in different ways.

ISBN: 978-0-02-284589-6

Includes *Time for Kids* magazine about how animals prepare themselves for winter.

Bibliography

Lesson 1 Animals Are Everywhere

Central Park Serenade, by Laura Godwin (HarperCollins, 2002)

Do You Know the Difference?, by Andrea and Michael Bischhoff-Miersch (North-South, 1995)

Leaving Home, by Sneed B. Collard III (Houghton Mifflin, 2002)

Lesson 2 What Animals Need

Birds Build Nests, by Yvonne Winer (Charlesbridge, 2002)

Duck on a Bike, by David Shannon (Blue Sky Press, 2002)

Giant Pandas, by Gail Gibbons (Holiday House, 2002)

Lesson 3 Bugs and More Bugs

Bugs Are Insects, by Anne Rockwell (HarperCollins, 2001)

Butterfly: Watch Me Grow, by Lisa Magloff (DK Publishing, 2004)

Insects, by Robin Bernard (National Geographic Society, 2002)

Lesson 4 Reptiles

Dig, Wait, Listen: A Desert Toad's Tale, by April Sayre (Greenwillow, 2001)

Slinky Scaly Slithery Snakes, by Dorothy Patent (Walker, 2000)

The Yucky Reptile Alphabet Book, by Jerry Pallotta (Library Binding, 1999)

Lesson 5 Up Above and Down Under

Jellies: The Life of Jellyfish, by Twig George (Millbrook, 2000)

The Mountain That Loved a Bird, by Alice McLerran (Simon & Schuster, 2000)

Over in the Ocean: In a Coral Reef, by Marianne Berkes (Dawn Publications, 2004)

Lesson 6 Staying Safe

How Animal Babies Stay Safe, by Mary Ann Fraser (HarperCollins, 2001)

I See Animals Hiding, by Jim Arnosky (Scholastic, 2000)

Red Eyes or Blue Feathers: A Book About Animal Colors, by Patricia M. Stockland (Picture Window Books, 2005)

Lesson 7 Grow and Change

Make Way for Ducklings, by Robert McClosky (Viking Press, 2001)

A Mother's Journey, by Sandra Markle (Charlesbridge Publishing, 2005)

Lesson 8 People and Animals

Farm Animals, by Tammy Schlepp (Copper Beech Books, 2000)

The Year at Maple Hill Farm, by Alice and Martin Provensen (Simon & Schuster, 2001)

Units A and B

Pond, by Gordon Morrison (Houghton Mifflin, 2002)

For more information on these or other books, visit **www.bankstreetbooks.com** or your local library.

Vocabulary Activities

Science Vocabulary

Vocabulary

animal
food
water
air
shelter
mammal
bird
fish
insect
reptile
habitat
fur
feathers
scales

Animal Homes

Explore what children know about animals and create a chart.

- Have children brainstorm different animals that they know. List them in the left-hand column of the chart, under *Animal*.
- Invite children to share where each animal might make its home. Then list their ideas in the middle column, under *Home*.
- Have children share what they know about each animal's characteristics (such as, *has fur, eats seeds, has four legs*). Write their ideas in the third column, under *We Know*.

What We Know About Animals

Animal	Home	We Know
bear	cave	has fur hibernates
bird	nest	makes nests flies

ELL Support

Animals

Invite children to repeat the vocabulary words as you read them. Ask children to identify what is the same about the following words: *food, fish, fly, fur, feathers,* and *fins.*

- **Beginning** Name objects in the classroom. Ask children to stand up when they hear a word that begins with /f/.
- **Intermediate** Ask: **Which word begins with /f/, the first or the second?** *fly/guy, buy/fly, hi/fly, why/fly*
- **Advanced** Ask children to name more words that begin with /f/. Encourage children to use these words in a sentence.

Ff

food
fish
fly
fur
feathers
fins

A to Z Activity Book

Look for additional cross-curricular activities about animals in the A to Z Activity Book.

A is for Amazing Ants, pp. 2–3
Reading, Writing, Math, Art

E is for Extraordinary Eggs, pp. 10–11
Reading, Writing, Math, Art

F is for Food Chains, pp. 12–13
Music, Reading, Art, Dramatic Play

H is for Hibernation, pp. 16–17
Music, Science, Reading, Math

I is for Insects, pp. 18–19
Science, Math, Art

J is for Jumping Critters, pp. 20–21
Math, Reading

L is for Life Cycles, pp. 24–25
Reading, Math, Art, Social Studies

Y is for Young Animals, pp. 50–51
Reading, Writing

Z is for ZZZ at Night, pp. 52–53
Science, Music, Reading, Math

Unit Project

Make a Worm Habitat

Objective: Make a habitat for worms that can be observed over time.

Step 1 In order to create a suitable classroom home for worms, help children observe the worms' natural environment as they collect the worms.

Step 2 In the classroom, help children set up a tank with soil, grass, leaves, and/or small twigs. Add the worms and a little water to the tank. Cover the tank securely with mesh and dark construction paper and put the tank in a cool place. Remind children to wash their hands after handling the worms and the soil.

Step 3 Each day, have children add a little water to the tank and observe and draw what they see. Help them write or have them dictate their observations to you. Ask questions to help children form hypotheses about worm behavior. Compile children's observations in a class book.

You need

- mesh (for a lid)
- clear plastic container
- soil
- grass
- dead leaves
- dark construction paper
- tape
- worm food (apple peelings, dry oatmeal, cornmeal)
- spoon

Technology

IWB INTERACTIVE WHITEBOARD READY

LOG ON www.macmillanmh.com

Visit Macmillan/McGraw-Hill Science online for projects and activities for students, teachers, and parents.

Bank Street www.bankstreet.edu

Visit Bank Street online for teacher resources and more activities.

TIME FOR KIDS www.timeforkids.com

Visit Time for Kids online to read about current science news and explore additional science activities.

UNIT B Planner

TeacherWorks™ Plus CD-ROM has interactive unit planner, Teacher's Edition, and worksheets.

Lesson	OBJECTIVES	VOCABULARY	RESOURCES and TECHNOLOGY
1 Animals Are Everywhere PAGES 64–69 **PACING**: 3-5 days	Understand the basic definition of an animal and explore animals in your neighborhood.	**animal** **habitat**	**Flipbook**, p. 15 **Leveled Reader:** *What Kind of Animal Are You?* **Unit B Literature Big Book** **Science Resource Book** p. 32 **Photo Sorting Cards 11–20** **A to Z Activity Book** pp. 20–21, 52–53
2 What Animals Need PAGES 70–77 **PACING**: 3-5 days	Recognize that animals are organisms that need air, water, food and shelter to stay alive.	**air** **water** **shelter** **food** **space**	**Flipbook** pp. 16–17, S4 **Science Resource Book,** p. 33 **Science Songs CD** **Photo Sorting Cards 11–20** **A to Z Activity Book,** pp. 12–13, 16–17
3 Bugs and More Bugs PAGES 78–83 **PACING**: 3-5 days	Learn about bugs, their attributes, and where they live.	**insect**	**Flipbook** p. 18 **Science Resource Book,** p. 34 **Photo Sorting Cards** 11–20 **A to Z Activity Book,** pp. 2–3, 18–19
4 Reptiles PAGES 84–89 **PACING**: 3-5 days	Understand the basic definition of a reptile, its attributes, and where it lives.	**reptile** **snake** **lizard** **mammal**	**Flipbook**, p. 19 **Leveled Reader:** *Our Desert Home* **Unit B Literature Big Book** **Science Resource Book,** p. 35 **Photo Sorting Cards 11–20**
5 Up Above and Down Under PAGES 90–97 **PACING**: 3-5 days	Learn about birds, fish, and other water animals.	**wings** **fins** **reef**	**Flipbook**, pp. 20–21, S5 **Leveled Reader:** *Animals on the Move* **Unit B Literature Big Book** **Science Resource Book,** p. 36 **Science Songs CD** **Photo Sorting Cards 11–20**
6 Staying Safe PAGES 98–103 **PACING**: 3-5 days	Explore how animals have adapted to their environments.	**skin** **fur** **scales** **shell** **feathers**	**Flipbook**, p. 22 **Leveled Reader:** *All About Animals* **Unit B Literature Big Book:** *Time for Kids* **Science Resource Book,** p. 37 **Photo Sorting Cards 11–20** **A to Z Activity Book** pp. 16–17
7 Grow and Change PAGES 104–111 **PACING**: 3-5 days	Understand how animals grow and change as they mature.	**grow** **change**	**Flipbook**, pp. 23–24 **Leveled Reader:** *Animals Grow* **Unit B Literature Big Book** **Science Resource Book:** p. 38 **Photo Sorting Cards 11–20** **A to Z Activity Book** pp. 10–11, 24–25, 50–51
8 People and Animals PAGES 112–117 **PACING**: 3-5 days	Explore relationships between people and animals.	**beekeeper** **farmer**	**Flipbook**, p. 25 **Leveled Reader:** *Good Morning* **Unit B Literature Big Book** **Science Resource Book,** p. 39 **Photo Sorting Cards 11–20**

PACING Assumes a day is a 20–25 minute session.

www.macmillanmh.com for more planning resources and http://nsdl.org/refreshers/science for science resources from NSDL

BE A SCIENTIST Activities

Animal Habitat p. 68

PACING: 20 minutes 

Skills make a model, observe, communicate, infer

Materials small clear tank, pebbles, sand, cold-water fish, fish food

Animal Homes p. 76

PACING: 30 minutes 

Skills observe, infer, draw a conclusion, communicate

Materials hand lenses, paper, pencils

Bug Collection p. 82

PACING: 20 minutes 

Skills investigate, predict, observe

Materials small clear containers, eyedroppers, hand lenses, bug net

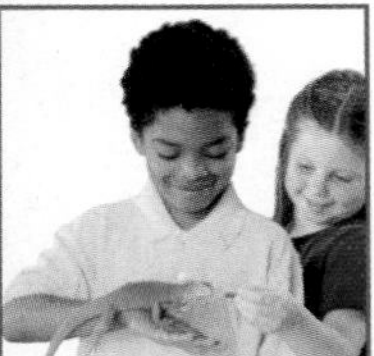

Reptile Guest p. 88

PACING: 20 minutes 

Skills communicate, observe, infer

Materials drawing paper, pencils or crayons, chart paper, markers

Bird Feeder p. 96

PACING: 20 minutes 

Skills investigate, observe, classify, compare

Materials empty pint milk cartons, string, lard, oatmeal, flour, wild birdseed, bowl, spoons, pencils, markers, scissors

Wormy Behavior p. 102

PACING: 20 minutes 

Skills investigate, communicate, compare, infer

Materials earthworms, paper, eyedroppers

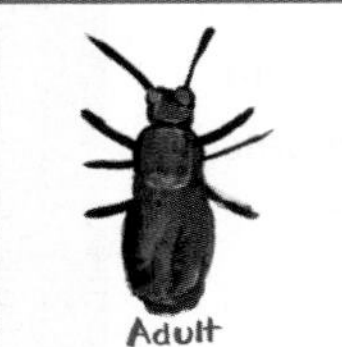

Growing Animals p. 110

PACING: 20 minutes 

Skills observe, make a model, communicate, put in order, infer

Materials container, oatmeal, mealworms, apple, hand lenses

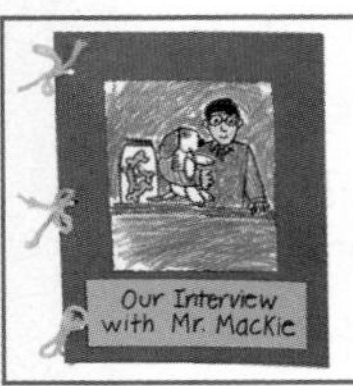

Ask an Expert p. 116

PACING: 20 minutes 

Skills communicate, investigate, infer

Materials chart paper, drawing paper, pencils, crayons

UNIT ASSESSMENT

Formative Assessment, pp. 67, 75, 81, 87, 95, 101, 109, 115
Performance Assessment, p. 118
Summative Assessment, p. 119
Portfolio Assessment, p. 119

SCIENCE KIT MATERIALS

- colored modeling clay
- small clear tank
- clear containers
- aluminum pans
- attribute shapes
- hand lenses
- string
- eyedroppers
- blocks

COLLECTIBLES

Begin collecting these items for Unit B:

- empty pint milk cartons
- craft sticks
- chenille stems
- sunglasses
- supermarket flyers
- paper bags
- small empty boxes
- egg cartons

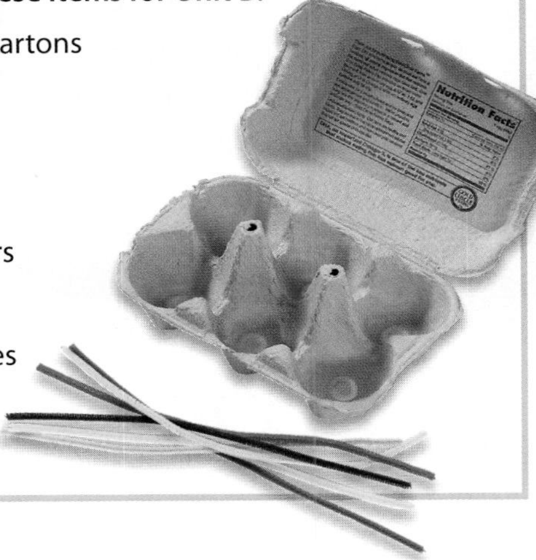

English Language Learner Support

Academic Language

English language learners need help in building their understanding of the academic language used in daily instruction and science activities. The following strategies will help to increase children's language proficiency and comprehension of content and instruction words.

Strategies to Reinforce Academic Language

- **Use Context** Academic Language should be explained in the context of the task. Use gestures, expressions, and visuals to support meaning.
- **Use Visuals** Use charts, transparencies, and graphic organizers to explain key labels that help children understand classroom language.
- **Model** Use academic language as you demonstrate the task to help children understand instruction.

Academic Language Vocabulary Chart

The following chart shows unit vocabulary and inquiry skills as well as some Spanish cognates. **Vocabulary** words help children comprehend the main ideas. **Inquiry Skills** help children develop questions and perform investigations. **Cognates** are words that are similar in English and Spanish.

Vocabulary		Inquiry Skills	Cognates	
			English	Spanish
animal, p. 64	fins, p. 90	make a model, p. 68	animal, p. 64	*animal*
habitat, p. 66	reef, p. 90	observe, p. 68	habitat, p. 66	*hábitat*
shelter, p. 70	skin, p. 98	communicate, p. 68	model, p. 68	*modelo*
food, p. 70	fur, p. 98	infer, p. 68	observe, p. 68	*observar*
space, p. 70	scales, p. 98	draw a conclusion, p. 76	communicate, p. 68	*comunicar*
insect, p. 78	shell, p. 98	investigate, p. 82	infer, p. 68	*inferir*
reptile, p. 84	feathers, p. 98	predict, p. 82	space, p. 70	*espacio*
snake, p. 84	grow, p. 104	classify, p. 96	conclusion, p. 76	*conclusión*
lizard, p. 84	change, p. 104	compare, p. 96	insect, p. 78	*insecto*
mammal, p. 84	farm, p. 112	put in order, p. 110	investigate, p. 82	*investigar*
wings, p. 90	beekeeper, p. 114		predict, p. 82	*predecir*
			reptile, p. 84	*reptil*
			classify, p. 96	*clasificar*
			compare, p. 96	*comparar*

Technology

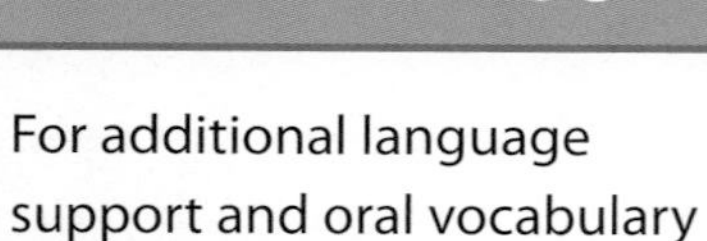

For additional language support and oral vocabulary development, go to www.macmillanmh.com

Science in Motion
The Food Chain
Parts of a Fish

Vocabulary Games

Language Transfers

Phonics Transfer
Haitian Creole, Hmong, Korean, and Vietnamese do not have the *r*-controlled vowel sound /ûr/.

Grammar Transfer
In Haitian Creole, Hmong, Spanish, and Vietnamese, possession is often expressed with a prepositional phrase.

The cat of Tom has brown fur.

Vocabulary Routine

Use the routine below to discuss the meaning of each word on the vocabulary list. Use gestures and visuals to model all words.

Define *Fur* is the hair that covers the body of some animals.

Example The cat has *fur* on its body.

Ask What other animals have *fur*?

Children may respond to questions according to proficiency level with gestures, one-word answers, or phrases.

Vocabulary Activities

Help children identify animals with fur.

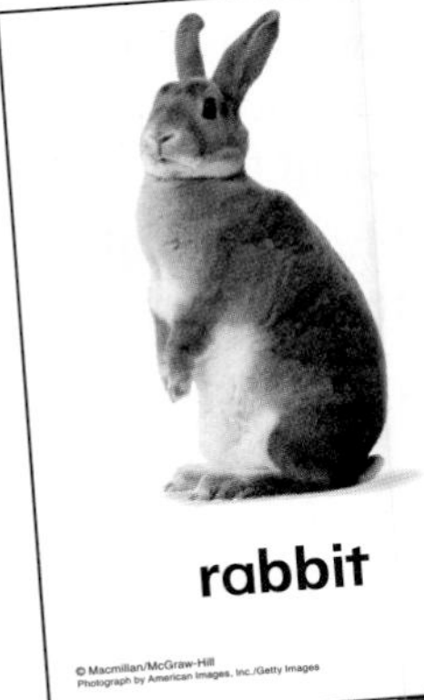

BEGINNING Show page 16 of the Flipbook. As you point to the animals, ask: *Which animal do you like? The turtle? The bird? The dog? The cat? I like animals with fur on their bodies, like this cat.* (Pretend to pet the cat's fur.) Write, say, and have children repeat, *fur*. Continue: *What other animal on this page has fur?*

INTERMEDIATE Show page 14 of the Flipbook. Say: *Some of these animals have fur on their bodies. Which animals have fur? Which animals do not have fur?* Fill in a Venn diagram as children respond.

ADVANCED Form three groups. Take out Sorting Cards 12, 14, and 15. Give one to each group, facedown. Tell the groups: *Your cards have pictures of animals, but only one of the animals has fur. Turn your cards over and tell me what animals you see. Which animal has fur?*

UNIT B

Objective: Children will classify animals by discussing similarities and differences.

Before Reading

- Have children share what they know about the animals that appear on Flipbook, page 14.
- Explain that in this unit, children will be scientists as they observe, compare, and classify animals.

During Reading

- Read the first verse of the poem, pointing to the words as you read.
- Invite children to say which animals on the page have no legs, two legs, four legs, or more. Repeat this process with the next verse.
- Reread the poem two to three times, encouraging children to chime in as you read.

Flipbook

UNIT B

Animals

I see...
Animals with no legs.
Animals with two.
Animals with four legs or more!
Do you?

I see...
Animals who climb in trees.
Animals who swim in the sea.
Animals who live right near me!
Do you?

by Kathleen Hayes

beetl

puppy

ant

Reading Strategy

Conventions of Print Have children help you count the number of times the word *animals* appears in the poem. Then have them help you count the number of letters in the word *animals*.

Phonological Awareness Have children identify the words that rhyme. Record their responses.

Science Facts

American Black Bear This North American mammal can grow to weigh as much as 600 pounds.

Chickens are raised throughout the world, and all produce eggs. A chicken's heart beats 350 times a minute, four times as fast as the heart rate of a child.

Tree Boa These brightly-colored snakes live in trees, where they can hide and wait for their prey.

Puppy (Dog) Descendants of the wolf family, dogs are considered to be the first domesticated animal. There are over 300 breeds of dogs.

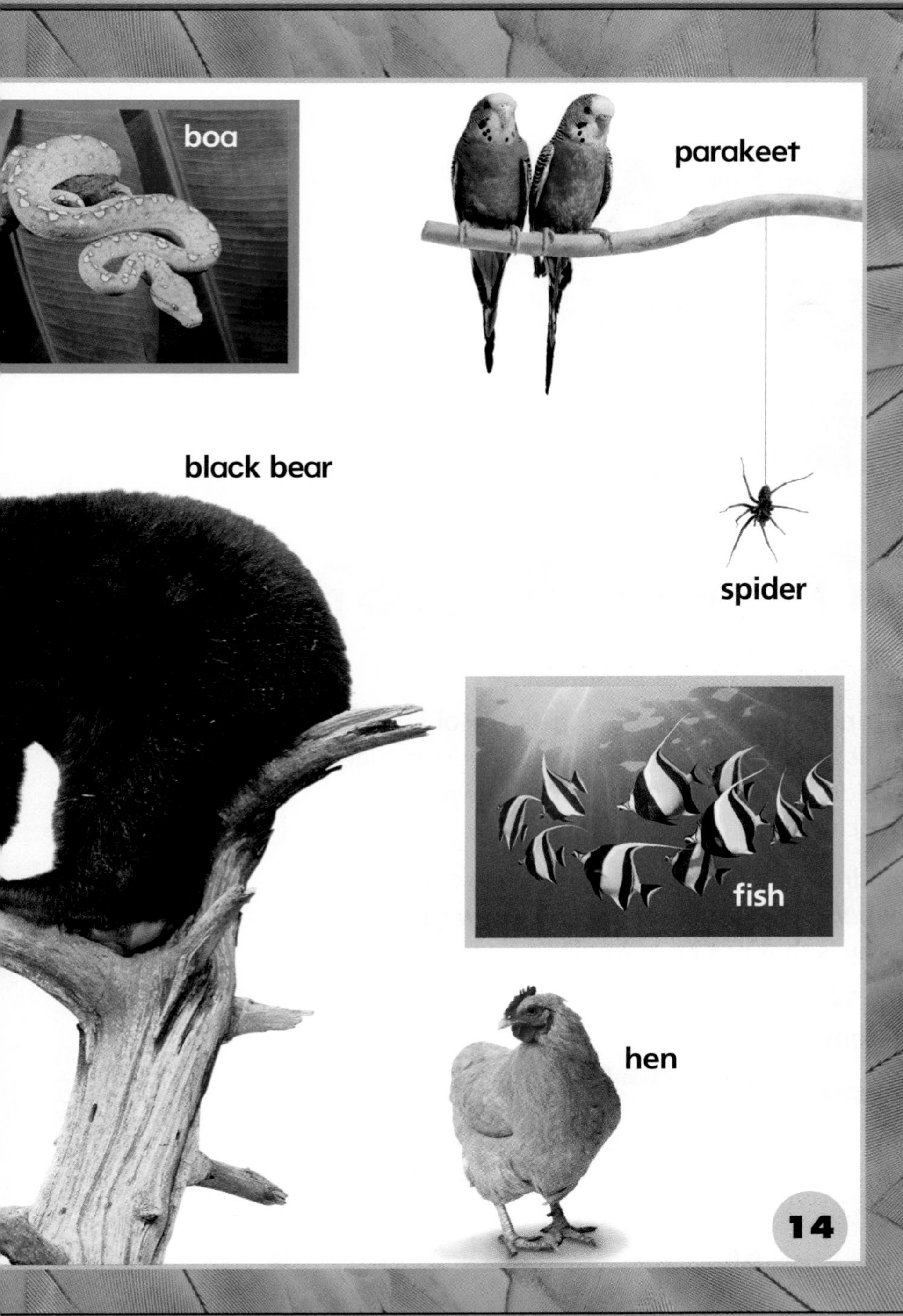

After Reading

- Write *Animals with Legs* and *Animals with No Legs* at the top of a piece of chart paper. Ask children to help you decide in which column to list the animals pictured in the Flipbook.
- Invite each child to name a new animal and say where it belongs.

Write on It

Have volunteers use a dry-erase marker to circle animals with no legs, put an *x* on animals with two legs, draw a line under animals with four legs, and make a check mark next to animals with more than four legs.

▶ Revisit

Ask children to move like a bear with four legs climbing a tree. Then have them act out climbing a tree like a snake with no legs. Finally, have them move like a fish in the sea. Invite children to move together as you reread the poem.

Science Background For more information on animals go to www.macmillanmh.com.

Parakeet Long ago, the black, wavy lines over a parakeet's colorful feathers helped it hide from predators in the grasslands of Australia, its native habitat.

Ants live in big groups called colonies. Every ant has a job—the queen lays eggs, and the workers dig tunnels and get food.

Spider Most spiders use spinnerets at the tip of their abdomen to spin silk webs.

Beetle A beetle has a hard shell that protects it, and hard front wings called elytra, which cover its thinner back wings.

Moorish Idol Fish This fish uses its long snout to find food in small cracks and crevices.

Professional Development For more Science Facts and resources from NSDL visit http://nsdl.org/refreshers/science

School to Home

At the beginning of Unit B, give children copies of the Home Letter on page 5 in the **Science Resource Book.** Read the letter aloud and help children write their names. Have children bring the letter home to share with their families. At the end of the unit, send children home with a copy of the Home Letter on page 7.

See pages 6 and 8 in the Science Resource Book for Spanish versions of these letters.

▶ **Essential Question**
Where do animals live?

▶ **Objective**
Understand the basic definition of an animal and explore animals in your neighborhood.

▶ **Vocabulary**
animal
habitat

Resources

Flipbook, p. 15

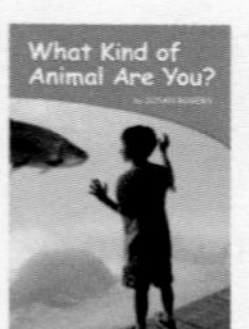
Leveled Reader: *What Kind of Animal Are You?*

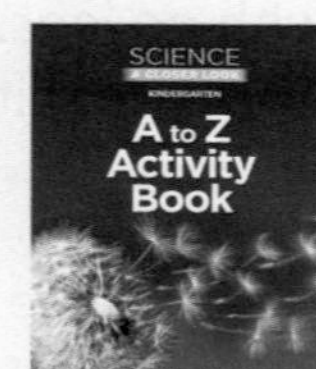
A to Z Activity Book, pp. 20–21, 52–53

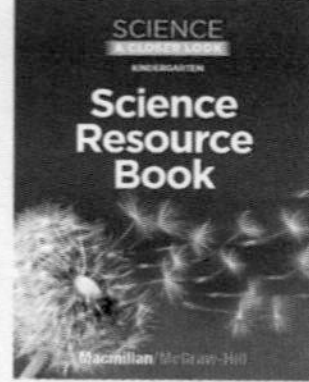
Science Resource Book, p. 32

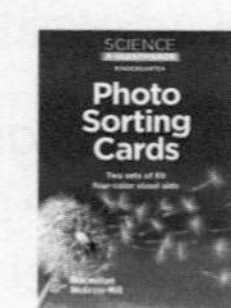
Photo Sorting Cards 11–20

Technology

Science in Motion
The Food Chain

www.macmillanmh.com

Animals Are Everywhere

Are They Animals?

You need
- Photo Sorting Cards 11–20
- yarn
- chart paper

Inquiry Skills: communicate, classify, compare

- Make a sorting ring on the floor with yarn. Hold up the sorting cards one by one and invite volunteers to place the card inside the circle if an animal is shown and outside the circle if it is not an animal.
- Record children's work on chart paper. Help children define *animals* by asking: **How do you know that is an animal? Why is that not an animal?**

Compare Ask: **What is the same about all the animals? What is different?**

Time to Move!

Have children follow simple directions that contain opposites, such as the following: *Hop fast like a rabbit. Crawl slowly on your hands and feet like a spider. Quack very loudly like a duck. Flutter very quietly like a butterfly.*

Be a Reader Activity

You need
- Unit B Big Book
- Leveled Reader: *What Kind of Animal Are You?*
- chart paper
- markers

Read the Big Book

Objective: Use illustrations to help read a story.

Inquiry Skills: classify, infer

- Use large sticky notes to cover the photo of each animal. Keep close-up photos uncovered.
- Read the title. Take a picture walk, inviting children to identify each close-up photo. As children identify feathers, for example, ask: **What kind of animal has feathers?** Repeat for scales and fur.
- Read the book, pausing to let children answer each question. After reading, discuss the idea that people are mammals. Brainstorm other mammals.
- **Guided Reading:** See the inside back cover of the Leveled Reader version of *What Kind of Animal Are You?* for activities.

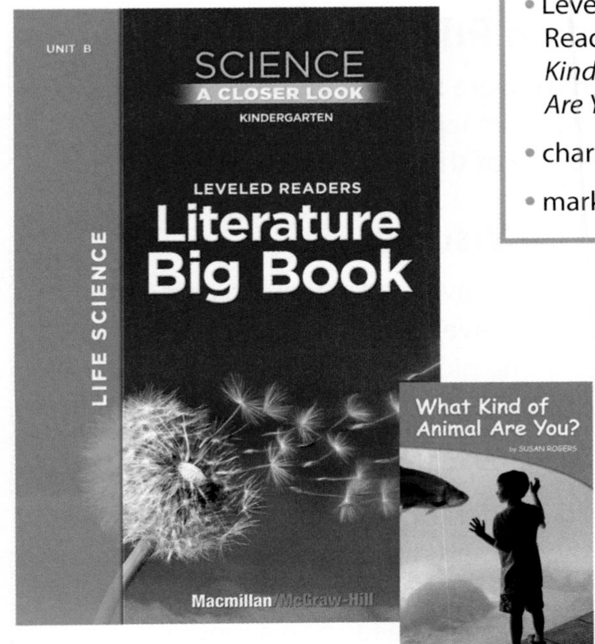

Be a Math Wiz Activity

You need
- chart paper
- marker
- clipboard and paper
- 4-row graph or Flipbook p. G2
- dry-erase marker

Find and Graph Animals

Objective: Make a graph.

Inquiry Skills: investigate, communicate

- Encourage children to brainstorm what animals they might see near the school. List their responses. Take a trip nearby to count the number of each type of animal they see. As children see an animal, make a tally mark on your recording sheet.
- After your trip, discuss what you saw. Have children help graph the results. If children find more than four types of animals, choose three animals and label the fourth row on the graph *other animals.*
- Ask: **Why do you think we saw more ____ than ____?**

Neighborhood Animals

squirrel	X	X				
bird	X	X	X	X		
dog	X					
cat	X					

Read Together and Learn

Which animals could you see in this park?

▶ Build on Prior Knowledge

Say: **When we were looking at pictures of animals with two legs, four legs, and no legs, did we see any of these animals?**

▶ Use the Visuals

- Communicate Invite volunteers to name the animals, and have them talk about which ones they see in the picture of the park.
- Help the group determine which other animals might be found in the park.
- Discuss the different types of places where the rest of the animals might live. Invite children to point to the animals and share their ideas. Possible answers: on a farm, in a jungle, in the woods

▶ Develop Vocabulary

animal, **habitat** Explain that an *animal* is any living thing that takes in food and is able to move about. As children discuss where the animals might live, have them describe each animal's home and what surrounds it, such as walls, trees, or dirt. Explain that where an animal lives is called its *habitat*.

Flipbook

UNIT B • LESSON 1

Which animals could you see in this park?

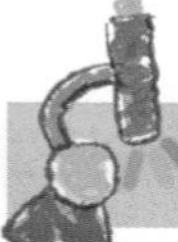

Reading Strategy

Print Awareness Write the animal names from the Flipbook page on sticky notes. Place these notes on the incorrect animals. Have children work together to rearrange each note, matching it with the correct animal.

Phonological Awareness Have children think of words that end in *-ark*. List their responses (*ark, park, bark, dark, hark, lark*).

Science Facts

Butterfly, Spider Insects and arachnids have many predators; one of them is the robin.

Robin In spring, robins migrate.

Squirrels and chipmunks, which eat eggs, are robins' enemies.

Dogs sometimes chase after small birds and mice.

Ducks have oily and waterproof outer feathers that keep cold water away from their skin. Female ducks can lay 5–12 eggs.

Raccoons can eat many different foods, including nuts, fruits, insects, plants, and small rodents.

Rabbit Like birds, rabbits build nests, but the mother stays away to avoid attracting predators, then returns to feed the babies.

Professional Development For more Science Facts and resources from NSDL visit http://nsdl.org/refreshers/science

▶ Ask Questions

- **How do the animals that live in a park find their food?** Possible answers: People feed them; they eat grass; they eat nuts and leaves from trees; they eat other animals.
- **What are some changes that animals living in a park might make to their habitat?** Possible answers: Some animals dig up the dirt; birds build nests in the trees.

Invite children to use a dry-erase marker to circle animals that might live in the park, and draw an *x* on animals that would not. Have volunteers underline animals that they might see on a farm.

Think and Talk

Review the animals that are shown on the Flipbook page. Ask children: **What makes the park a good habitat for some animals? What makes it a bad habitat for other animals?**

Formative ASSESSMENT

If children are having difficulty identifying what makes animals different from each other, review Flipbook pages 14 and 15.

ELL Support

Display and name pictures of common animals. Teach "Old McDonald" and have children use the pictures as props.

Beginning Ask children to hold up pictures as they hear their animal's name.

Intermediate Sing: **Old McDonald had a ____**, and point to a picture.

More to Read

Leaving Home,
by Sneed B. Collard III
(Houghton Mifflin, 2002)

Activity Read this book with children to reinforce their understanding of where animals make their homes. Have children share their favorite animal from the story and then draw it in its respective home.

Animal Habitat

Objective: Make a classroom plant and animal habitat.

Inquiry Skills: make a model, observe, communicate, infer

You need **small clear tank (optional: air filter, aquatic plants, water snails), pebbles, sand, cold-water fish, fish food, Science Journal p. 32**

1. **Make a model** Have children help prepare a freshwater aquarium. Place sand and small pebbles in the tank. Gently put plants in the sand, spreading out their roots. Slowly add tap water and attach an air pump and filter.
2. After 24 hours, add cold-water fish (such as guppies or goldfish). Add water snails and cover the tank. Show children how to feed the fish sparingly.
3. **Observe** Have children observe the aquarium often. Discuss their observations.
4. **Communicate** Have children use their Science Journal page to draw what they observe in the aquarium.

Teacher Tip

Placing snails and live plants together in a habitat enables children to explore the relationship between plants and animals. They can also compare how different types of animals move in water. Adding a filter increases the initial cost, but it helps to ensure a quality environment.

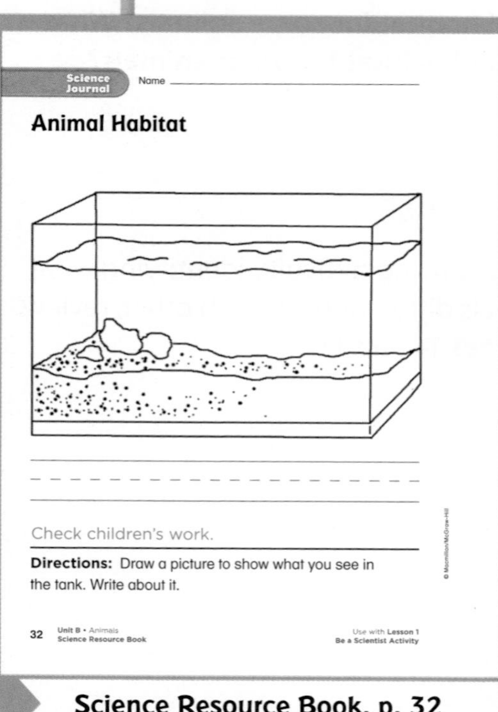
Science Journal Name

Animal Habitat

Check children's work.

Directions: Draw a picture to show what you see in the tank. Write about it.

32 Unit B • Animals Science Resource Book Use with Lesson 1 Be a Scientist Activity

Science Resource Book, p. 32

Investigate More

Infer Ask: **What helps the fish and snails move?** Possible answers: fins, foot **How do you know? What else have you noticed about the animals and plants?**

centers

Art

Animal Mural

Objective: Develop small motor skills.

Inquiry Skills: communicate, make a model

- Show children a large piece of paper. Explain that it will be the background for a mural of animals that live nearby.
- Have volunteers use green paint for grass, gray paint for cement, and blue paint for sky.
- Have children draw, color, and cut out animals they have seen in the neighborhood. After gluing their animals to the background, help them write labels and glue the appropriate label next to each animal.

You need

- paper
- paint
- crayons
- colored pencils
- glue
- scissors

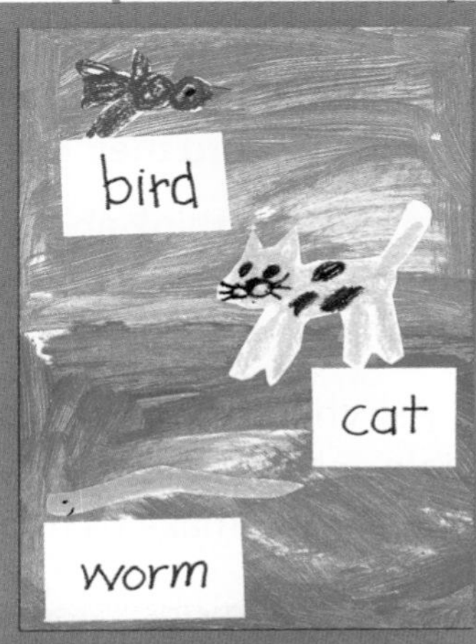

Technology

More About Animals

- For games and activities about animals, help children log onto www.macmillanmh.com.
- Children can use these games throughout this unit to reinforce what they learn about animals. Encourage children to revisit these games throughout the school year to review the concepts.

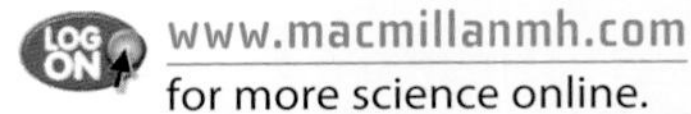

Movement

How They Move

Objective: Sort and categorize animals by how they move.

Inquiry Skills: classify, communicate

- Reread *What Kind of Animal Are You?* in the Unit B Big Book. Invite children to move like each animal.
- Afterward, ask a volunteer to demonstrate how he or she moved as a bird, fish, and horse.
- Make a three-column chart with the following ways to move as titles: *fly, swim,* and *walk*. Write the appropriate animals in each column. Ask volunteers to name other animals and decide where they belong on the chart.

You need

- chart paper
- marker

Blocks

Animals All Around

Objective: Use fine motor skills to build and explore.

Inquiry Skills: make a model, communicate

- At Circle Time, explain that if children build with blocks, they are to build a model of the school and place animals near it.
- Encourage children to draw an animal, cut it out, and tape it to a block. Then they can place it near the block school.
- Take a photograph of the block structure and allow children to use the structure throughout the day.

You need

- paper
- crayons
- colored pencils
- scissors
- tape

For more on animals, use pp. 17–18 in the **Activity Book.**

What Animals Need

Essential Question
What do animals need?

Objective
Recognize that animals are organisms that need air, water, food, and shelter to stay alive.

Vocabulary
air, food, water, space, shelter

Resources

Flipbook, pp. 16–17, S4

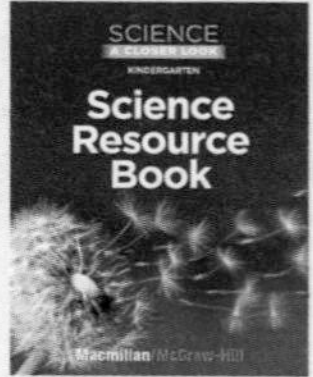
Science Resource Book, p. 33

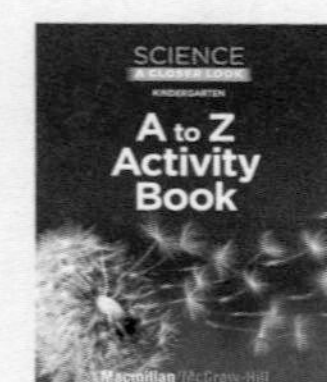
A to Z Activity Book, pp. 12–13, 16–17

Photo Sorting Cards 11–20

Technology

Science Songs CD Tracks 7–8

Science in Motion *The Food Chain*

www.macmillanmh.com

Circle Time

WHOLE CLASS

Pet Care

You need
- chart paper
- markers

Inquiry Skills: communicate, infer

- Ask children how they care for any pets they may have at home. Record their responses on chart paper.
- Help children think about the purpose of pet care by asking: **Why do you give your pets water? Food? A home? What do all pets need to grow up healthy?**

Infer Ask children: **How do pets take care of and help us?** Possible answers: A dog may protect our home. Pets are our friends and make us happy.

Caring for our Pets

Rachel: I give my cat food every day.

Keshawn: I take my dog for a walk.

Time to Move!

Call out a letter of the alphabet. Invite children to move like an animal that begins with that letter. For example, call out the letter *c* and children can move like a cat, a cow, a caterpillar, and so forth.

Be a Writer Activity

You need
- paper folded in half to make a 4-page book
- pencils
- colored pencils
- crayons

Make a Pet Book

Objective: Describe how humans care for domesticated animals.

Inquiry Skills: communicate, put in order

- At Circle Time, show children the 4-page blank books. Invite them to use these books to show three things they do for the pet they have or would like to have.
- As children work, offer to take dictation to write about what they do to care for their pet. Ask questions such as: **What are you doing to help your pet in this picture? What does your pet need to stay alive?**
- Encourage children to share their book with a friend and take it home to read to a family member.

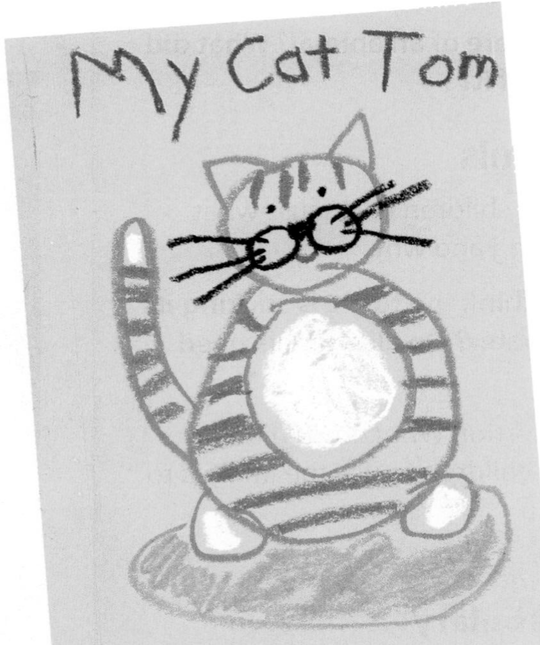

Be a Math Wiz Activity

You need
- chart paper
- markers

Our Pets

Objective: Collect and analyze data.

Inquiry Skills: classify, compare, communicate

- Create a graph. List the kinds of pets children own along the bottom. Make a column for *no pet* in case children do not have a pet at home.
- Have children add their names to the graph.
- Discuss the graph. Ask: **How many pets all together? Which column has more? Less? The same number? Do all the pets have fur? If no, how many do not? Do all of the pets live on land?**

Class Pets

				Jane
				Terry
	Karen			Mary
	Juan			Sara
Tim	Zoe	Josh		Louis
Dena	Emily	Mike		Ben
Lisa	Bill	Sally	Lana	Sue
dog	cat	fish	snake	no pet

Read Together and Learn

What do animals need to live?

▶ Build on Prior Knowledge

Ask: **Who has taken care of an animal? What did you do to take care of it?**

▶ Use the Visuals

- Communicate Ask children to discuss what each animal is doing and why.
- Ask: **Why do you think the turtle is sticking its head out of the water? Do all animals need air? Why?**
- Infer Read the question *What do animals need to live?* and invite children to use the images to try answering it.

▶ Develop Vocabulary

air, **water**, **shelter**, **food**, **space** On chart paper, list children's ideas about what all animals need to live. Help them arrive at a list of most animals' five basic needs: *air, water, shelter, food,* and *space*. Review what plants need to live: soil, air, light, water, and space.

Flipbook

UNIT B • LESSON 2

What do animals need to live

water

Reading Strategy

Comprehension Point to the question mark and ask children to explain what it is and how it is used. Have children think of other questions they could ask about the images on this page.

Phonics Ask children to name words that rhyme with *air*. As you write their answers, help children see that the same long *a* sound can be written differently (*fair, fare, mare, tear, stare, hair, pair, stair, wear*).

Science Facts

Snow Goose Snow geese follow a vegetarian diet—they eat grains, grasses, and water plants. Geese may have entirely white feathers, or may have bluish-gray feathers.

Bluebirds have blue and reddish-brown feathers, and male birds tend to be more brightly colored than females. These birds scan the ground and swoop down to catch their prey.

Turtle Pet water turtles need to be fed in the water because that is the only way they can swallow. When they are hungry, they will swim up to the glass and open their mouths.

Dog When people began keeping dogs as pets thousands of years ago, the dogs had jobs such as herding and hunting.

Professional Development For more Science Facts and resources from NSDL visit http://nsdl.org/refreshers/science

air

food

16

▶ Ask Questions

- **Do these animals need the same kind of food? What might the turtle eat?** leaves **How about the dog?** dog food **What are the goose and bird eating?** berries, grass, water plants
- **What does the turtle need that is different from what the dog, the bird, and the goose need?** a habitat in water **What do people need that pets also need?** food, air, shelter, water
- **Is this goose in the wild or is it a pet? How might it get its food?**

Write on It

Invite children to use a dry-erase marker to find and underline every letter *a* on the page. Discuss the sound that it makes in each word. Ask: **How is it different from the sound of the letter *a* in the word *snake*?**

Think and Talk

Point out the dogs on Flipbook pages 14–16. Have children identify the animals. Then prompt a discussion about how they are the same kind of animal, but are different in some ways.

Take a Trip

Visit a Pet Store Plan a trip to a local pet store. Have children observe the animals to see how they are getting what they need. Encourage children to ask one of the workers questions about how they take care of the animals. If a trip is not possible, invite a pet-store worker to visit your class.

Back in the Classroom Children can draw pictures about what they observed and learned at the pet store. Have children write words or sentences about their pictures. Display their work on a bulletin board.

Read Together and Learn

How are these animals alike? How are they different?

▶ Build on Prior Knowledge

Ask: **Can real animals talk, sing, dance, and play like they do in some of the movies and stories you read? Why or why not?**

▶ Use the Visuals

- Compare Ask children to talk about what is the same and different about the lions shown. Ask: **What do you think the lion on the left needs in order to live? Does the stuffed lion need the same things? Why or why not?**
- Prompt children to discuss what is the same and different about the butterflies. Ask: **Which of these butterflies needs food in order to live? What else do you think they need?**

▶ Additional Vocabulary

real, **fantasy** Create a two-column chart labeled *real* and *fantasy*. Have children brainstorm ideas that fall under each category. Record their ideas.

Flipbook

UNIT B • LESSON 2

How are these animals alike? How are they different?

Reading Strategy

Comprehension Write the words *real* and *fantasy* on separate sticky notes. You may wish to write just the beginning letters of each word. Invite volunteers to place the sticky notes on the appropriate pictures. Encourage them to talk about why the animal pictured is real or fantasy.

Print Awareness Invite children to find words on the Flipbook page that are the same. Then ask if they can find anything else on the page that is the same.

Science Facts

Lions The lion is the second-largest feline predator in the world. They are also the only cat species that live, hunt, and move in a social group—a pride. Lions typically live in savanna and grassland regions, where they prey on large animals, including wildebeests, zebras, wild pigs, and other hoofed mammals.

Butterflies Of all the pollinators, butterflies are the second largest group. They use a straw-like extension of their mouth to drink and transport nectar from plant to plant. The largest butterfly is the Queen Alexandra's Birdwing butterfly, found in Papua New Guinea. Its wingspan can reach up to twelve inches.

LOG ON **Professional Development** For more Science Facts and resources from  visit http://nsdl.org/refreshers/science

▶ Ask Questions

- Ask: **How were you able to tell the difference between the living (real) and the nonliving (fantasy) animals in the pictures?**
- Ask: **Do you think a child could hug a real lion like a stuffed lion? Why or why not?** No, because it might be dangerous.
- **Where is each butterfly? Do you think a real butterfly can hold papers to a refrigerator? Why or why not?**

Write on It

Write *lion* under the lion pictures with a dry-erase marker. Invite a volunteer to underline the *l* in *lion*. Then have children practice writing the letter *l*. Repeat with the letter *b* for *butterflies*.

Think and Talk

Discuss the differences between what plants and animals need in order to live. Ask: **What is different about the way animals and plants get what they need? What is similar about the needs of plants and animals?** Both need air, food, and water

Differentiated Instruction

Enrichment Have children bring a stuffed animal to school. Encourage them to share information about their stuffed animals, and to talk about how each stuffed animal is different from the real animal of the same kind. Have children draw and write about their stuffed animal.

More to Read

Duck on a Bike,
by David Shannon
(Blue Sky Press, 2002)

Activity Read this book with children and invite them to identify what is real and what is fantasy. You may want to ask questions like: **Can a duck ride a bike? Can children ride a bike?**

Formative ASSESSMENT

If children are having difficulty understanding what animals need to live, then revisit page 16 of the Flipbook. Help them use the pictures to summarize animal needs.

Animal Homes

Objective: Discover where animals live in the school neighborhood.

Inquiry Skills: observe, infer, draw a conclusion, communicate

You need **hand lenses, paper, pencils, Science Journal p. 33**

1. Before you take a trip, ask the class to predict where animals might be living in the neighborhood. List their responses.
2. **Observe** Take a small group out to look for animal homes. Help children look under rocks and logs, in piles of dead leaves, in trees, on bark, on stems and leaves, on telephone wires, and on the foundations of buildings. Remind children to be safe.
3. **Infer** Ask children: **Why do you think the animal lives where it does?** Possible answers: It is the same color as its home so other animals cannot see it; its home is close to drinking water; the home is in the tree so other animals cannot reach it.
4. **Draw a Conclusion** Ask: **Where do animals in the school neighborhood make their homes?** Record children's responses.
5. **Communicate** Have children use their Science Journal page to record what they observed.

Teacher Tip

If you are unable to take your class outside to look for animal homes, ask children to look for them on their way home from school, or in their own backyard. Children can use the Science Journal page the next day to record what they observed.

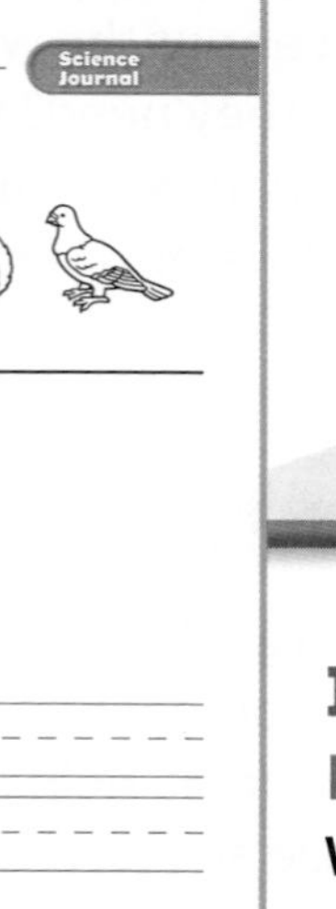

Name ______________________ Science Journal

Animal Homes

Check children's work.

Directions: Circle one animal whose home you found. Draw and write about the animal's home.

Unit B • Animals Science Resource Book | Use with Lesson 2 Be a Scientist Activity 33

Science Resource Book, p. 33

Investigate More

Infer Hold up Photo Sorting Cards of different animals and ask: **What do you think this animal might use to make its home? Where does the animal make its home?** Examples: A duck makes its home in a pond; a spider uses its silk to make a web.

Music

Animals Need

Objective: Discuss and compare what people and animals need to stay alive.

Inquiry Skills: communicate, compare

- **Talk About It** Draw a word web on chart paper, and write *Needs* in the circle. Ask children: **What do people need to live?** Help children to compare and contrast the needs of people and animals.
- **Listen and Find** Play the song "Animals Need." Encourage children to point to the animals that are eating, drinking water, breathing, and living in their homes.
- **Sing and Innovate** Have children sing the song. Then have them make up lyrics about people's needs, such as "People like me need food to eat."
- See page TR25 for printed music for "Animals Need."

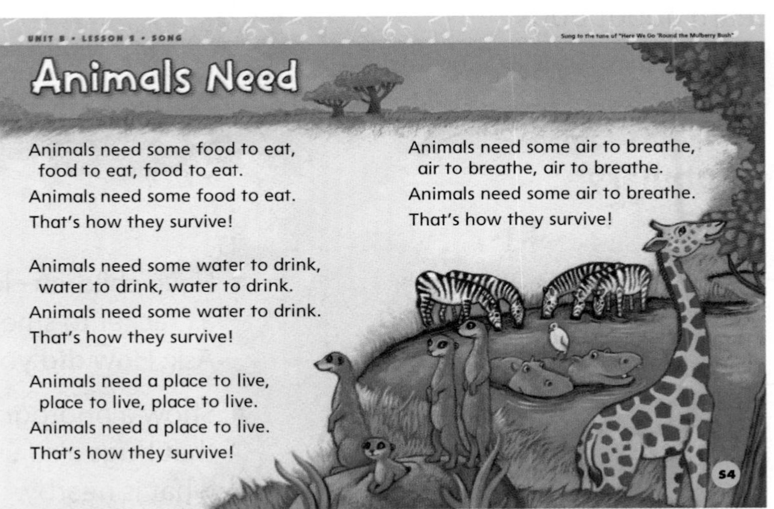

Flipbook p. S4
CD Tracks 7–8

Art

Build a Nest

Objective: Use art materials to create bird nests.

Inquiry Skill: make a model

- Have children make a "nest" out of clay.
- Ask: **What will you add to make your nest strong? Soft?**
- Once the nest is constructed, have children add a clay bird and eggs to the nest.

You need
- clay or play dough
- twigs
- craft sticks
- string
- pieces of felt
- scraps of newsprint
- cotton balls
- leaves
- grass

Movement

Guess My Pet

Objective: Act out common pet behaviors.

Inquiry Skill: communicate

- Ask for a volunteer to act like a pet. Explain that the volunteer should not tell what animal he or she is imitating.
- Have the other children observe the "pet behavior" and determine what animal it is. Ask: **How did you know?**
- Have three or four children act like a pet. On subsequent days, repeat this activity until everyone has had a turn.

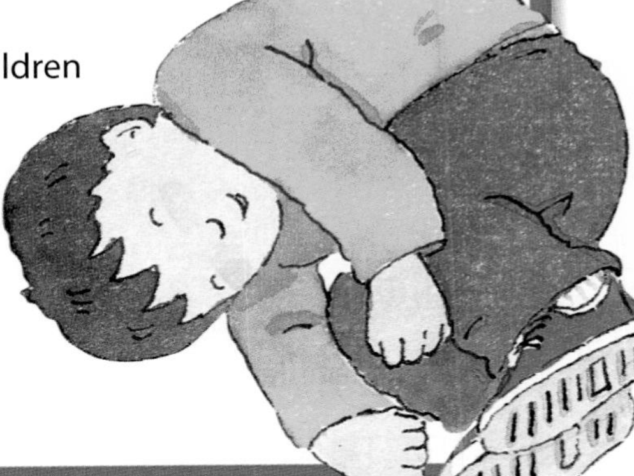

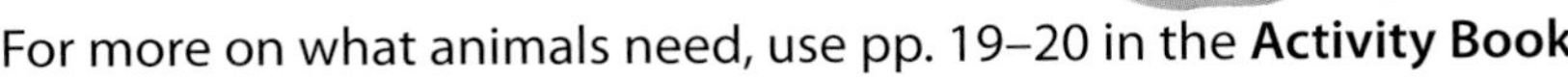
For more on what animals need, use pp. 19–20 in the **Activity Book.**

▶ **Essential Question**
How can bugs look alike?

▶ **Objective**
Learn about bugs, their attributes, and where they live.

▶ **Vocabulary**
insect

Resources

Flipbook, p. 18

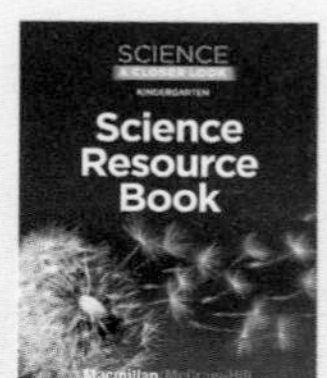
Science Resource Book, p. 34

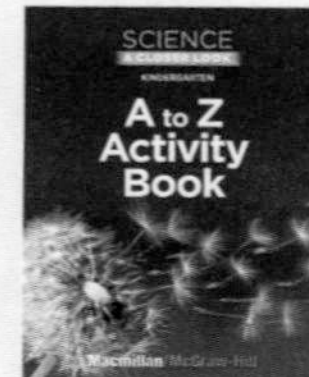
A to Z Activity Book, pp. 2–3, 18–19

Photo Sorting Cards 11–20

Science on the Go Cards pp. 27–30

Technology

www.macmillanmh.com

NSDL

Bugs and More Bugs

Circle Time

WHOLE CLASS

Be a Bug

Inquiry Skills: infer, compare

- Have children close their eyes and give them a block and a cracker to smell and feel. Have them open their eyes. Ask: **How did you know that one object was food?**
- Show a photograph of a bug showing its feelers. Explain that bugs also use touch and smell to help them know what is nearby.

Compare Ask: **How do we use our senses of touch and smell? How is it different from bugs? How is it the same?**

You need

- square attribute blocks
- square crackers
- picture of a bug
- chart paper
- marker

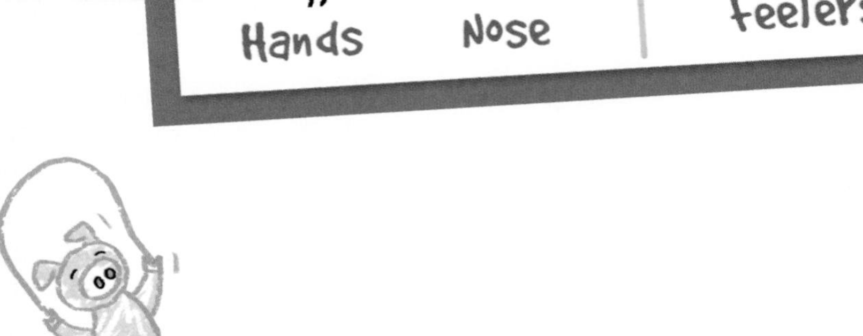

Time to Move!

Have children take turns moving and making sounds like different types of insects (such as a bee, ladybug, or firefly) while the rest of the class tries to guess the type of insect.

Be a Reader Activity

You need
- Photo Sorting Cards 11–20

Animal Match

Objective: Match cards that have the same bug names.

Inquiry Skill: communicate

- Have children identify the first letter of the name of each animal on the Photo Sorting Cards.
- Mix up the cards and place them facedown. Have children take turns flipping two cards over to read the labels and see if they match. Allow children to keep the matching cards they find.
- At the end of the game, have children count how many matches they each found. Ask if anyone found a match of bug cards. snail, spider, or butterfly Have them count how many bugs they have.

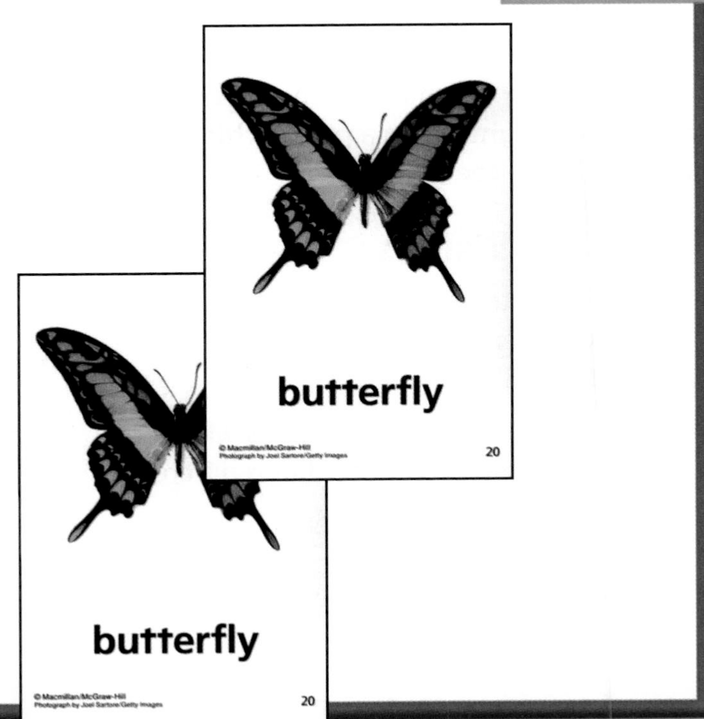

Be a Math Wiz Activity

You need
- paper
- pencils
- chart paper

Favorite Bugs

Objective: Survey people about bugs and graph their responses.

Inquiry Skills: communicate, compare, draw a conclusion

- Have children survey people around school or their classmates to find which of the following bugs they like: mosquito, spider, ant, or ladybug.
- When children have completed their survey, help them transfer the information they collected onto a four-column graph.
- Discuss the results. Ask: **Which was the favorite bug? The least favorite? Why do you think some bugs were more popular than others?**

Favorite Bugs

mosquito	spider	ant	ladybug
			X
			X
	X		X
	X		X
	X	X	X

Read Together and Learn

How are these bugs alike? How are they different?

▶ Build on Prior Knowledge

Say: **Remember when we discussed which animals we might see in a park? Which of these bugs might we see?**

▶ Use the Visuals

- Ask children to identify any bugs they can, and then help them read the labels for those they cannot identify.
- Compare Have the group discuss what is the same about the bugs and what is different.
- Communicate Invite children to point to bugs they have seen and to share what they know about them.

▶ Develop Vocabulary

insect Have children find two beetles on the page and discuss their similarities (six legs, hard body, actually have wings but you cannot see them). Tell children that a ladybug is a kind of beetle. Explain that a beetle is an insect and that all insects have three body parts and six legs. Ask: **Is the tarantula an insect?** No, it has eight legs.

Flipbook

UNIT B • LESSON 3

How are these bugs alike? How are they different?

Reading Strategy

Comprehension Point to the word *fly* and explain that this word is hiding in two other words. Have children find them. Then ask: **Why do you think *fly* is part of *butterfly* and *dragonfly*?**

Phonemic Awareness Write the word *bug* on a sticky note and have children find it in a word on the page. (*ladybug*) Then ask what sound they hear in the word *bug*. (short *u* sound) Ask: **What other words have this sound?**

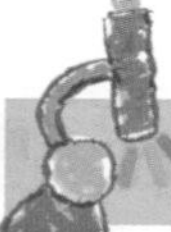

Science Facts

Bee A bee's "buzz" is the sound of its fanning wings.

Ladybugs are a kind of beetle. They eat insects that attack plants, so farmers use them to protect plants.

Butterfly A butterfly's feet have taste sensors, so it stands on its food to taste it.

Beetles have two pairs of wings. One pair, called *elytra,* feels hard, and is used only for protection.

Dragonfly Although the dragonfly has six legs, it does not walk. It has four wings and flies.

Tarantula Tarantulas are arachnids, not insects, and have eight legs.

Grasshopper These insects can leap about 20 times their own body length.

Professional Development For more Science Facts and resources from NSDL visit http://nsdl.org/refreshers/science

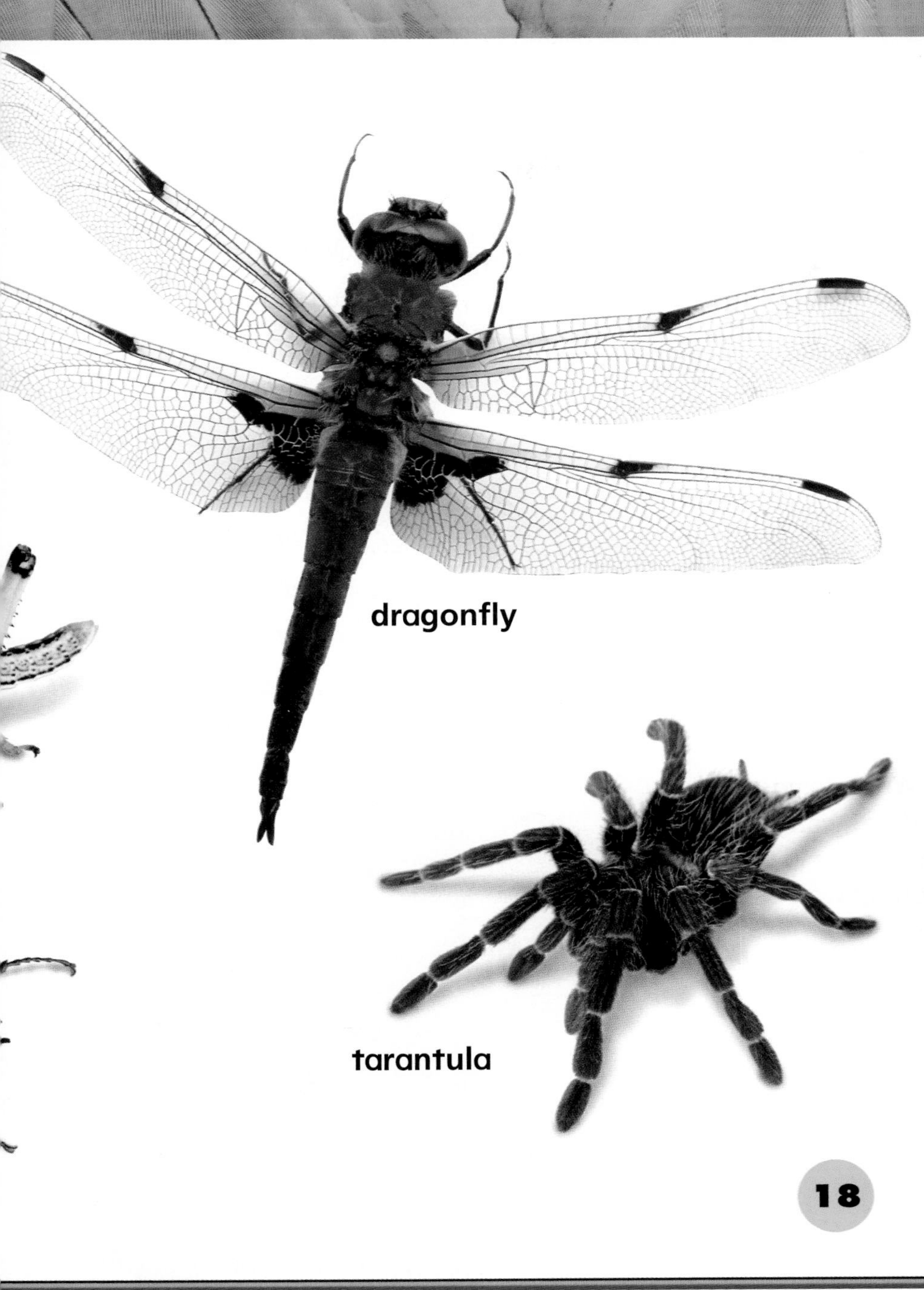

▶ Ask Questions

- **How do you think the bugs' colors and patterns help them stay alive?** They can blend in with leaves or flowers.
- **Look at the ladybug and the beetle. How are they alike? How are they different?** Help children notice that the same type of animal still has differences.
- **What makes the spider different from the other bugs on the page?** It has eight legs.

Invite children to use a dry-erase marker to circle the bugs with wings and draw a line under the bugs without wings.

Think and Talk

Review the bugs on the Flipbook page. **Ask: Which bugs use wings to move? What other body parts are similar? If you were a bug, what do you think might be good about having wings? What might be good about not having wings?**

Formative ASSESSMENT

As children discuss how bugs use wings or do not use them, you will learn what they know about this topic. In addition, children will learn from each other as they listen to their classmates share information.

Differentiated Instruction

Enrichment Have children sort the Photo Sorting Cards into two categories: *bugs* and *not bugs*. Have them try to count the number of legs on the bugs in their pictures. Invite them to re-sort their group of bugs into two new categories: *wings* and *no wings*.

More to Read

Butterfly: Watch Me Grow, by Lisa Magloff (DK Publishing, 2004)

Read this book with children to reinforce their understanding of how a caterpillar changes into a butterfly. Record the stages of metamorphosis on chart paper. Have children draw a picture that shows one stage.

Bug Collection

Objective: Collect and observe bugs.

Inquiry Skills: investigate, predict, observe

You need **small clear containers, eyedroppers, hand lenses, bug net, chart paper, markers, apples or bananas, Science Journal p. 34**

1. Explain that the class will be going outside to collect bugs to observe and examine. Remind children to be safe and ask a teacher before touching any bug, and to be gentle with the bugs they pick up. Ask: **Where are some places that we might find bugs?** under the dirt, at the base of trees, on branches
2. **Investigate** Bring children outside to collect bugs. Put bugs of the same kind together in the same containers. Remind children to wash their hands after handling the bugs.
3. **Predict** Back in the classroom, ask: **What do you think these bugs will need to stay alive?** water, food, air, shelter Give children bits of food to place in their bug containers. Help them place drops of water onto leaves to place in the bug containers.
4. **Observe** Have children observe the bugs and record their observations on their Science Journal page.

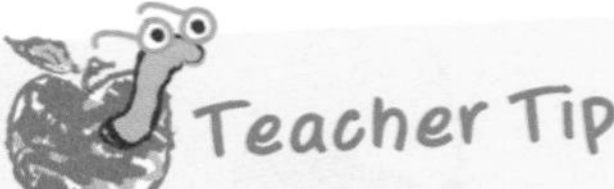

If you are unable to collect bugs, you may want to buy crickets, mealworms, or other bugs that are sold at pet stores. Different bugs eat different foods. Ants can eat honey, grasshoppers can eat vegetables, ladybugs can eat boiled potatoes, and crickets can eat vegetables and crackers.

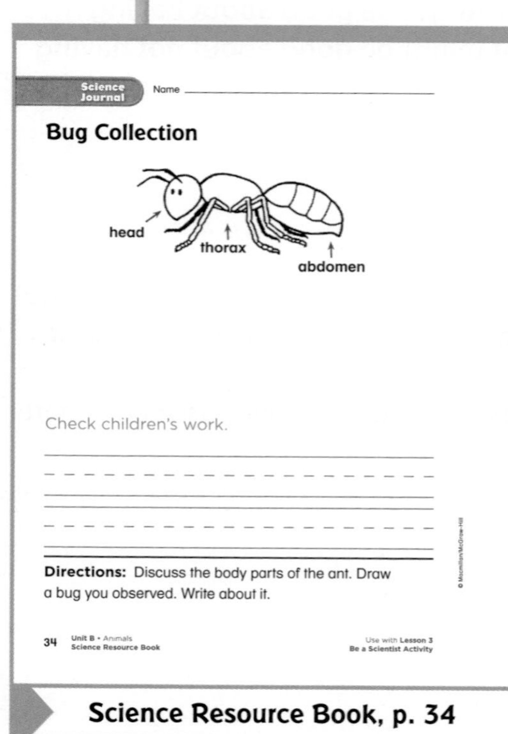
Science Journal Name ____

Bug Collection

head
thorax
abdomen

Check children's work.

Directions: Discuss the body parts of the ant. Draw a bug you observed. Write about it.

34 Unit B • Animals Science Resource Book — Use with Lesson 3 Be a Scientist Activity

© Macmillan/McGraw-Hill

Science Resource Book, p. 34

Investigate More

Investigate **Ask: What other plants do you think the bugs might like to eat?** If they are eating apples or bananas, maybe they will like other kinds of fruit. Children can predict and test which kinds of fruit disappear most quickly.

centers

Blocks

You need
- plastic insects

Bug Homes

Objective: Build a bug habitat.

Inquiry Skill: make a model

- At Circle Time, show children pictures of bug homes.
- Say: **If we build a home for "bugs" in the Block Area, what kinds of structures will you make? How will the bugs stay safe? Where will they get their food?**
- Encourage children to talk about what they made. Keep the bug homes up until everyone has a turn to interact with them.

Drawing and Writing

You need
- paper
- markers or crayons
- pencils

Bug Stories

Objective: Tell a story about a bug.

Inquiry Skill: communicate

- Have children make up a story about a bug.
- Offer to write down the story they tell. If they need help adding details, you can ask: **Where would a bug live? What would a bug want to eat?**
- Have children draw pictures to illustrate their story.

Art

You need
- paper plates
- craft sticks
- paint
- paintbrushes
- markers
- glitter
- chenille stems
- glue

Bug Masks

Objective: Create masks of favorite bugs.

Inquiry Skill: make a model

- Display bug pictures on the board. Explain to children that they will each be making a bug mask. Have them choose a bug.
- Distribute the materials. As children work, come around to help them make eye holes in the masks.
- When masks are dry, attach craft sticks so children can hold their masks in front of their faces. Write the name of each bug on the craft stick.

For more on bugs, use pp. 21–22 in the **Activity Book.**

▶ **Essential Question**
Where do reptiles live?

▶ **Objective**
Understand the basic definition of a reptile, its attributes, and where it lives.

▶ **Vocabulary**
reptile lizard
snake mammal

Resources

Flipbook, p. 19

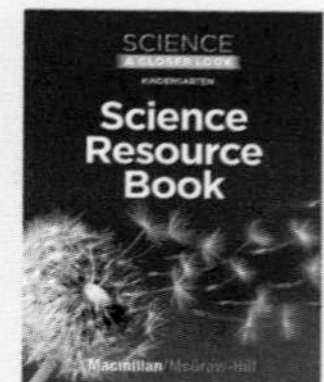
Science Resource Book, p. 35

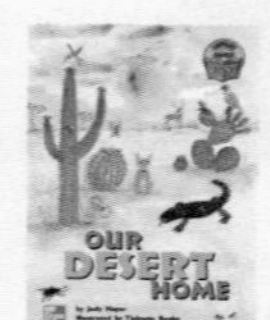
Leveled Reader: *Our Desert Home*

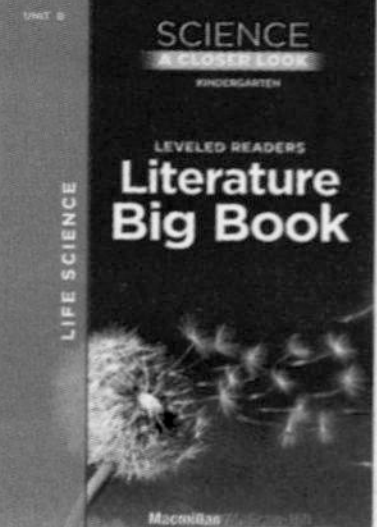
Unit B Literature Big Book

Photo Sorting Cards 11–20

Technology

www.macmillanmh.com

NSDL

Reptiles

Circle Time
WHOLE CLASS

Reptile or Not?

You need
- Photo Sorting Cards 11–20
- chart paper
- markers

Inquiry Skills: compare, infer

- Show pictures of three mammals. Ask: **What is the same about these animals?** Explain that these animals are called mammals.
- Next show pictures of three reptiles. Ask: **What is the same about these animals?** Explain that these animals are called reptiles. Ask: **What is different about each group? What is the same?** Record children's responses.

Compare Ask: **Do reptiles and mammals live together? How?**

Time to Move!

Invite children to slither quickly around the classroom like a snake. Then have them move slowly around the classroom like a turtle.

Be a Reader Activity

You need

- Unit B Literature Big Book
- Leveled Reader: *Our Desert Home*

Read the Big Book

Objective: Identify beginning letter names of animals.

Inquiry Skills: classify, infer

- Show children the cover of *Our Desert Home* in the Unit B Literature Big Book. Ask children: **What do you think it is like in a desert?** Explain that it is hot and dry most of the time. Then read the story, tracking the print as you read.
- After reading, turn to page 5 and have children find something that is not a reptile. Encourage them to discuss how they know. Ask: **Why is the rattlesnake partly underground? Why is the gecko looking at the ants? What plants do you see?**
- **Guided Reading:** Use the Leveled Reader version of *Our Desert Home* for small-group guided-reading instruction.

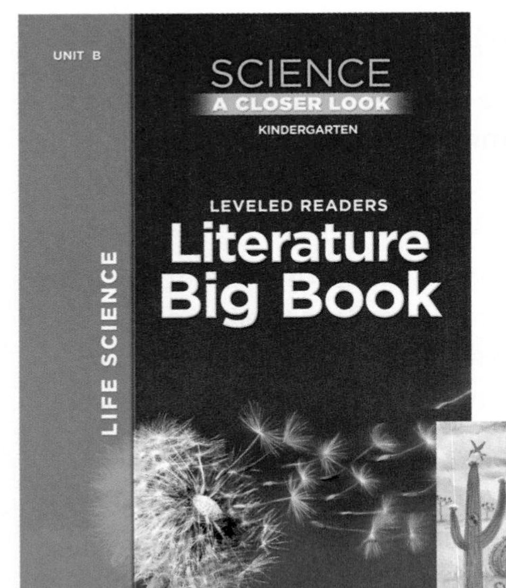

Be a Math Wiz Activity

You need

- connecting cubes
- paper with outline of a connecting-cube train

Snake Patterns

Objective: Create and identify patterns.

Inquiry Skills: put in order, make a model, predict

- Make a connecting-cube "snake" in a two-color pattern.
- Show the snake to the children and say: **My snake is growing. What color could come next in this pattern?** Have children help to extend the pattern in your snake.
- Have children make their own connecting-cube snakes and record their snake pattern on the recording sheet. Encourage children to use up to three colors in their pattern.

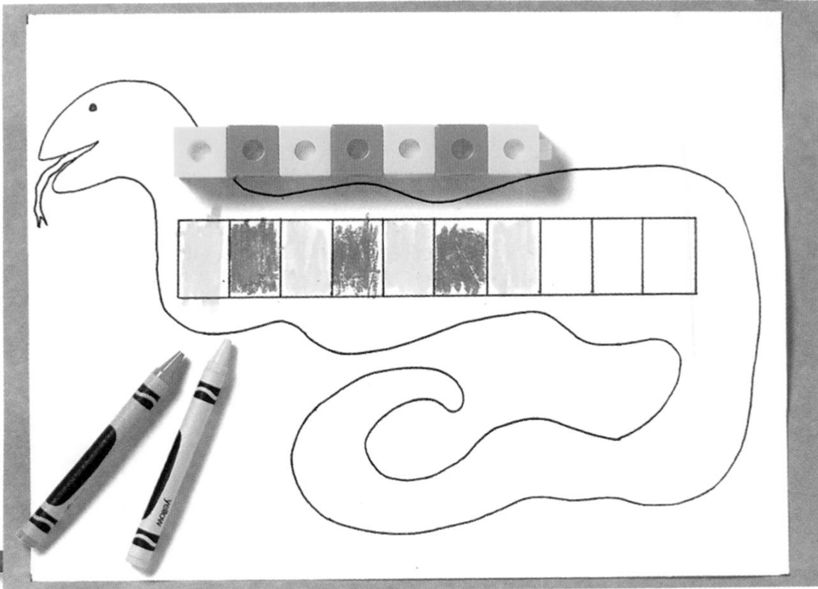

Read Together and Learn

What do you notice about these reptiles?

▶ Build on Prior Knowledge

Ask: **Which of these reptiles have legs? Which do not? What other animals do not have legs?**

▶ Use the Visuals

- **Compare** Ask children to discuss what they notice about the reptiles. Have them look for similarities and differences. Prompt children to discuss how the milk snake and rattlesnake are the same and different.
- **Communicate** Explain that the center picture shows the reptiles' habitat. Ask children to talk about what they see in the habitat: **How does the ground look? What is growing in the ground? Does it look wet or dry?** Explain that this kind of habitat is called a desert.

▶ Develop Vocabulary

mammal, **reptile**, **lizard**, **snake** Explain that a *mammal* is a warm-blooded animal with a backbone and that a *reptile* is a cold-blooded animal with a backbone. Alligators, turtles, *snakes*, and *lizards* are reptiles. Ask: **How does a snake move if it does not have legs?** Invite different children to try moving like a snake.

Flipbook

UNIT B • LESSON 4

What do you notice about these reptiles?

milk snake

desert tortoise

Reading Strategy

Phonics Write the word *snake* on chart paper. Have children brainstorm words that rhyme with it. Ask: **What letters and sounds do the words have in common? What letter is at the end of each word?**

Phonemic Awareness Give each child enough play dough to make a capital *S* and lowercase *s*. Have children suppose that each *s* is a snake, and have them make a snake's hissing sound.

Science Facts

Reptiles are cold-blooded, which means their body temperature changes with the air around them. Reptiles' bodies have adapted to help them warm up or cool down. Their dry, scaly skin keeps them from losing moisture in hot environments. Some lizards have skin that becomes darker to soak up sunlight when they need warmth. Snakes also use their scales for gripping the ground as they move, or for climbing trees. Other reptiles climb using their long toes.

Professional Development For more Science Facts and resources from NSDL visit http://nsdl.org/refreshers/science

rattlesnake

desert

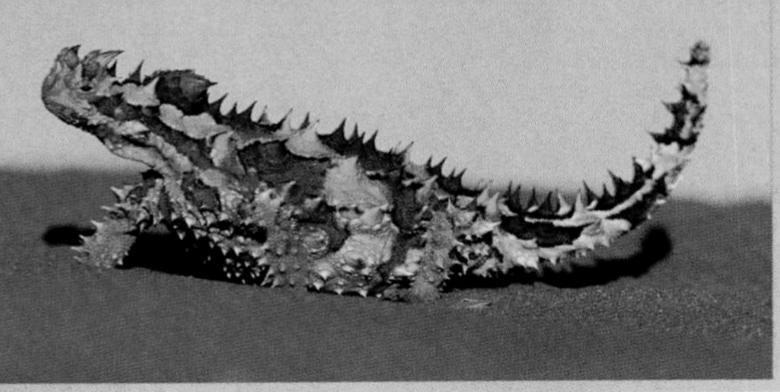

thorny devil

19

Ask Questions

- **What do you notice about the reptiles' skin? How might their skin help them survive in the desert?**
- **What do you notice about their feet? What would be helpful about having long claws in this environment?** for digging into hard ground; to move quickly
- **Where are the reptiles eyes?** on either side of their head **How might this help them survive?** They can see enemies on both sides.

Write on It

Invite children to use a dry-erase marker to draw an arrow showing where they think each reptile's nose might be.

Think and Talk

Review the reptiles that are shown on the Flipbook page. Ask children: **How would you describe the reptiles' environment? How do the reptiles' bodies help them survive in this environment?**

Formative ASSESSMENT

Observe children as you explore the Flipbook page to assess what they understand about reptiles and their environment.

Differentiated Instruction

Enrichment Have children spend some time looking through books on animals and reptiles. Help them write the names of one reptile that they find and draw a picture. Then, bring them together and invite them to talk about how they knew each was a reptile.

More to Read

The Yucky Reptile Alphabet Book, by Jerry Pallotta (Library Binding, 1999)

Activity Have children choose one letter of the alphabet and draw a picture of an animal whose name begins with that letter. Help children write the letter and the name of the animal.

Reptile Guest

Objective: Observe and ask questions about reptiles.

Inquiry Skills: communicate, observe, infer

You need **drawing paper, pencils or crayons, chart paper, marker, Science Journal p. 35**

1. On the day before your visitor comes, tell children about the visitor and the reptile he or she is bringing in to share. Help children generate questions they want to ask. On the day of the visit, remind children to be gentle and quiet so they do not scare the animal. Remind children to wash their hands after touching the reptile.
2. **Communicate** Introduce your guest and help children ask their questions.
3. **Observe** Give children time to do an observational drawing of the reptile. If you use an online visit, have children choose a photograph to draw. Give children the Science Journal page to record their observations.
4. The following day, collect children's drawings and take dictation as they describe what they learned. Bind their drawings into a class book.

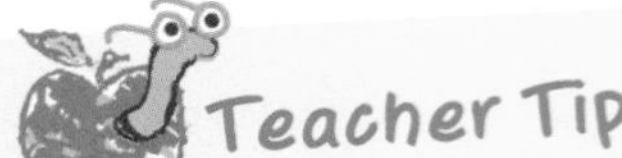

Teacher Tip

Ask children if any of them has a reptile and could bring it in. If not, you can arrange for a local zoo or pet store to bring in a lizard, snake, or turtle to your classroom. You can also try a "virtual field trip," exploring reptile photographs online.

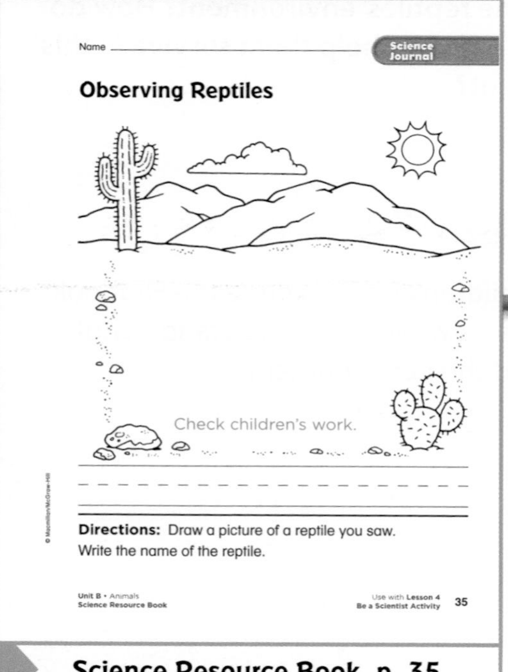

Name ________ Science Journal

Observing Reptiles

Check children's work.

Directions: Draw a picture of a reptile you saw. Write the name of the reptile.

Unit B • Animals Science Resource Book — Use with Lesson 4 Be a Scientist Activity 35

Science Resource Book, p. 35

Investigate More

Infer Ask: **Why do reptiles have scales?** To protect them from predators. To help them slither or climb.

centers

Water Table

Air Holes

Objective: Communicate the idea that reptiles breathe air even in the egg.

Inquiry Skill: make a model

You need
- 1–2 uncooked organic eggs
- clear plastic containers
- straws

- Have two to three children use straws to blow bubbles. Ask: **What makes the bubbles?** Hold up an organic egg, and ask: **When reptiles are in eggs waiting to hatch, how do you think the baby reptile gets air?**
- Have children gently put an egg in a cup of water. Ask: **What do you notice? What do you think makes the bubbles? Why do you think it is air?**
- Explain that an egg's shell has tiny air holes. Have these children help others do the same experiment.

Blocks

Staying Cool

Objective: Build block shelters for desert animals.

Inquiry Skills: investigate, make a model

You need
- crayons
- paper
- scissors
- toy reptiles

- Suggest children build a desert environment with blocks. Brainstorm what they might build (such as tall plants, water holes, and dens).
- They may want to use paper, crayons, and scissors to add details to their block desert.
- Have children discuss what they built and where animals would hide from the Sun in this desert.

Sand Table

Reptile Shelters

Objective: Make shelters for plastic reptiles.

Inquiry Skills: investigate, make a model

You need
- toy reptiles

- Tell children that some reptiles go underground to warm up or cool down. Have them make a hole where a snake could go to stay warm in the winter.
- Add a small amount of water to the sand. Ask: **How will you make a place for the snake to get in? How will you make sure that other animals will stay out?**

For more on reptiles, use pp. 23–24 in the **Activity Book**.

▶ **Essential Question**
How can animals move?

▶ **Objective**
Learn about birds, fish, and other water animals.

▶ **Vocabulary**
wings
fins
reef

Resources

Flipbook, pp. 20–21, S5

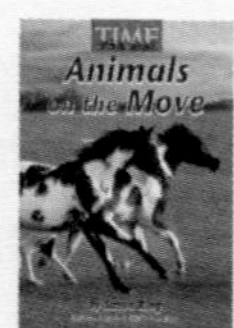

Leveled Reader: *Animals on the Move*

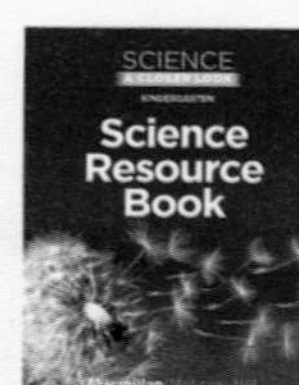

Science Resource Book, p. 36

Floor Puzzle: *Pond Life*

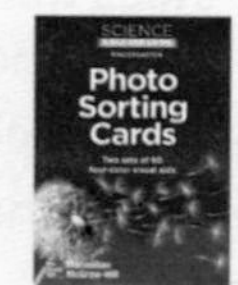

Photo Sorting Cards 11–20

Technology

Songs CD Tracks 9–10

Science in Motion
Parts of a Fish

Up Above and Down Under

Circle Time

WHOLE CLASS

Be a Bird

You need
- chart paper
- marker

Inquiry Skills: communicate, infer

- As children come to the circle, invite them to act like a flying animal. Give each child a chance to tell what animal he or she acted like.
- Ask children: **What other animals fly?** List their responses on chart paper. **What helps them fly?** wings

Infer Ask: **How do other animals move? What helps them move?** Possible answers: Tigers use their legs to run. Frogs use their legs to hop.

Time to Move!

Have children move like a bird flying through the sky. Then invite them to move like a fish swimming through water.

Be a Reader

You need
- Unit B Literature Big Book
- Leveled Reader: *Animals on the Move*
- chart paper
- markers

Read the Big Book

Objective: Understand that animals move in different ways.

Inquiry Skills: communicate, compare

- Before reading, take a picture walk through *Animals on the Move*. Invite children to talk about how the animal on each page is moving.
- Read the book aloud. After reading, encourage children to think about different ways people move. Talk about how they are the same or different than the animals in the book. Record their ideas.
- **Guided Reading:** See the inside back cover of the Leveled Reader: *Animals on the Move* for activities.

Be a Writer

You need
- paper
- pencils
- marker or crayons

Animals in Motion

Objective: Create a class book on animal movement.

Inquiry Skill: communicate

- Explain to children that they are going to use what they know about how animals fly and/or swim to help create a class book.
- Have children draw an animal that swims or flies. Ask questions such as: **Where does your animal live? How does it move?** Offer to write what they tell you about how the animal moves.
- If children are ready to write down the letter sounds they hear, encourage them to do so. Children may also want to use the Flipbook pages or Photo Sorting Cards to help them draw and write.

Be a Math Wiz Activity

You need
- marker
- chart paper

How Many Birds?

Objective: Count and graph the number of birds you see.

Inquiry Skills: observe, compare, put in order

- If possible, bring children outside in small groups to observe birds. If this is not possible, groups can observe birds from the window at separate times.
- Help each group record how many birds they saw on a pictograph by drawing pictures of birds.
- Discuss the pictograph with the whole group. Ask: **Which group saw the most birds? Where were most of the birds? Flying? On telephone wires? On the ground? In bushes?**

Read Together and Learn

What helps this bird fly?

▶ Build on Prior Knowledge

Say: **We have been talking about what different animals need in order to live. What do you think this bird needs to live?**

▶ Use the Visuals

- Communicate Ask children to discuss where this bird is. Invite them to talk about birds they have seen, whether it was in the woods, out in the open, or near water.
- Ask: **Which image shows the bird landing? Why do you think it is landing? Which image shows the bird taking off? How do you know? What helps birds fly, take off, and land?**
- Compare Have children discuss how images of the bird in flight are similar and how are different.

▶ Develop Vocabulary

wings, **fins** Ask children to discuss how fins and wings are alike and different. Use the Venn diagram on page G1 of the Flipbook to record their responses with a dry-erase marker.

Flipbook

UNIT B • LESSON 5

What helps this bird fly?

Reading Strategy

Comprehension Point to the word *what* and help children read it. Have them brainstorm other question words. Have children practice using a question word to ask a question.

Phonics Write the words *what, which, why,* and *when* and ask children what they notice about the words. They all begin with the letters *wh*. Have children practice making the *wh* sound.

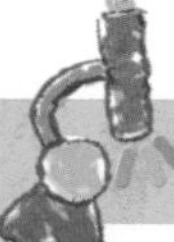

Science Facts

Starling During the 1890s, Eugene Schieffelin released between 60–100 starlings into New York City. These birds have since populated nearly all North American regions. Male and female starlings have similar coloring. Their beaks are short, and are used mainly for eating berries, seeds, and insects. Starling beaks are yellow during the spring and turn dark by fall. Whether it is a chirping bird or human whistle, these birds can imitate many of the noises they hear.

Professional Development For more Science Facts and resources from NSDL visit http://nsdl.org/refreshers/science

20

▶ Ask Questions

- Point to the photos in the center of the page. Ask: **Why do you think the bird is holding its feet like this? What do you notice about the wings?**
- Point to the bird on the far right. Ask: **What is it doing?** landing **What do you notice about its feet? Its wings?**
- **Why might this bird need such a pointy beak?** Possible answer: to eat insects, seeds, and fruits

Write on It

Have children use a dry-erase marker to practice writing *bird* next to each image on the page.

Think and Talk

Prompt children to discuss how birds are the same. Then revisit the images of the birds on Flipbook pages 14, 15, 16, and 20. Invite children to tell how birds can be different.

Differentiated Instruction

Enrichment Have children place a paper strip above their upper lips and blow. Ask: **What happens to the paper?** It lifts up. **What made the paper lift up?** air moving underneath it **What helps a bird stay in the air?** air moving underneath its wings

More to Read

The Mountain That Loved a Bird, by Alice McLerran (Simon & Schuster, 2000)

Activity Read this book with children to review the survival needs of birds. Ask: **How are birds' needs the same as or different from the needs of other animals?** Invite children to draw a picture that shows one thing a bird needs.

Read Together and Learn

What helps these animals move?

▶ Build on Prior Knowledge

Ask: **What is similar about this habitat and the bird habitat? What is different?**

▶ Use the Visuals

- Ask children whether they have been to the ocean. **What did you see there?**
- Observe Have children point to and try to identify all of the animals that they see in the picture.
- Communicate Have children talk about anything they see that is not an animal. Ask children to describe what the fish are swimming around.

▶ Develop Vocabulary

reef Write the word *reef* on chart paper, and have children work together to come to a definition. Ask: **Where is a reef?** in the ocean **What is it made of?** It can be rock, sand or coral—which is actually made of tiny animals called coral polyps. **What grows on a reef?** ocean plants **What animals might you find near a reef?** fish

Flipbook

UNIT B • LESSON 5

What helps these animals m

Reading Strategy

Phonics Write the word *fish* and have children read it with you. Write the word *shark* and read it together. Ask: **What letters do these words have in common? What sound does *sh* make?**

Phonological Awareness Have children brainstorm a list of words that begin with *sh*. Record their ideas on chart paper.

Science Facts

Coral Reef Over 4,000 different kinds of ocean fish live in coral reefs.

Coral is made of animals called coral polyps and of the skeletons they leave behind when they die.

Ocean Fish Like all animals, fish breathe oxygen. Their gills are thin membranes that allow oxygen to pass from the water into their bloodstream.

Sea Turtles have long flippers that help them swim. They usually swim slowly, but can swim more quickly when threatened by sharks or whales.

Octopus An octopus has eight tentacles attached to its head, and on each tentacle are rows of suckers. The octopus uses the suckers to catch its prey.

Professional Development For more Science Facts and resources from NSDL visit http://nsdl.org/refreshers/science

Take a Trip

Aquarium Make arrangements to visit a local aquarium. Look together for ocean animals with a hard shell and without a hard shell. Look for an ocean animal with fins, and one without fins. Give each child a pencil and a piece of cardboard with paper stapled to it so that he or she can sketch a fish or ocean animal.

Back in the Classroom Have children color and cut out their fish pictures and make a class coral reef mural.

▶ Ask Questions

- **What do you notice about the fishes' bodies? How might their colors and patterns help them survive?** They could blend in with the colorful reef to hide from predators.
- **How do you think the octopus's body helps it survive?** camouflage
- **When a turtle swims, how is its body similar to a bird's? How is it different? Why might the shell be more useful to the sea turtle than to a bird?** It has more predators nearby as it swims than a bird has when it flies.

Write on It

Invite children to use a dry-erase marker to circle the parts of the animals that help them to move.

Think and Talk

Look again at the animals on Flipbook pages 20 and 21. Ask children: **What is similar about ocean animals and birds? What is different?** Encourage children to share what they have learned.

Formative ASSESSMENT

By having children discuss similarities and differences between fish and birds, you will see what children understand about this topic.

Bird Feeder

Objective: Observe birds and learn about their behaviors and needs.

Inquiry Skills: investigate, observe, classify, compare

You need **empty pint milk cartons, hole punch, string, lard, hot water, oatmeal, flour, wild birdseed, large bowl, mixing spoons, pencils, markers, scissors, Science Journal p. 36**

1. Cut down one side of each carton to make a ledge. Punch hole(s) at the top and tie string through for hanging. Have a small group help mix the "Bird Feed." Fill and hang the feeders outside your classroom windows or in the schoolyard.
2. **Investigate** Over the course of several weeks, ask: **What kinds of birds have been coming to our feeders? At what time of day do we see the most birds?**
3. **Observe** Have children observe and draw the birds that come to the feeders on their Science Journal pages.
4. **Classify** Help children make a graph of the kinds of birds they see. Ask: **Which kinds of birds have you seen the most often? The least often? Why do you think that is?**

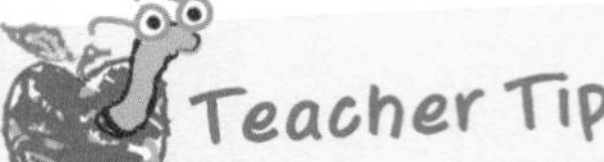

The best way to do this activity is over the course of several weeks, encouraging children to use Science Journal pages often to record what they see, especially when they see a new kind of bird. It is ideal to keep the feeders up for at least the duration of the unit. Be sure to replenish the bird food as needed.

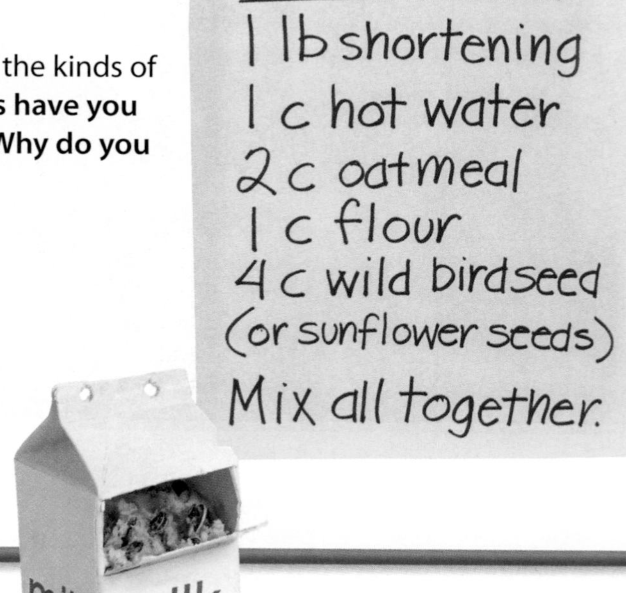

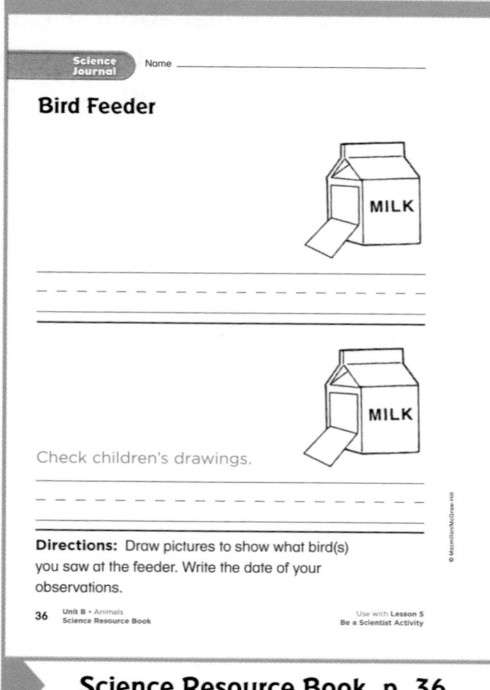
Science Journal Name

Bird Feeder

MILK

MILK

Check children's drawings.

Directions: Draw pictures to show what bird(s) you saw at the feeder. Write the date of your observations.

36 Unit B • Animals Science Resource Book Use with Lesson 5 Be a Scientist Activity

Science Resource Book, p. 36

Investigate More

Compare Ask: **What is similar about the birds you have seen?** Possible answers: Some are the same colors. Some have the same wing shapes. **What is different?** different colors, sizes

Music

Animals in Motion!

Objective: Classify animals according to their movement.

Inquiry Skills: communicate, classify, compare

- **Look and Listen** Ask children to look carefully at the pictures on the page and identify each animal. Invite them to name how each one moves. Discuss that a bee, fly, and bird all have wings that help them to fly, whereas a duck has webbed feet, a fish has fins, a seal has flippers, and a whale has a tail to help swim.
- **Sing and Act It Out** Sing "Animals in Motion!" Then ask children to brainstorm animals that move in other ways. Invite volunteers to become "animals in motion," acting out how each animal moves.
- See page TR25 for printed music to "Animals in Motion!"

Flipbook p. S5
CD Tracks 9–10

Art

Fish Watercolors

You need
- paper
- watercolor paints
- paintbrushes
- crayons

Objective: Have children make fish drawings.

Inquiry Skill: make a model

- Display pictures of fish at the Art Center. Ask: **What colors and patterns do you see? What shape are the fish's fins? Why do you think they have that shape?**
- Have children draw fish using crayons.
- Then have them use watercolor paints over their drawing to show the water.

Cooking

Tuna Salad

You need
- tuna fish
- celery
- mayonnaise
- crackers
- plastic knives
- plastic forks
- paper plates

Objective: Observe characteristics of fish and make tuna salad.

Inquiry Skill: investigate

- Find a picture of a tuna online or in a book, and show it to children. Ask: **What color is it? What are its fins like?** Tell children that the average tuna weighs 20 to 45 pounds, so it can weigh as much as they do.
- Have children help you make tuna salad. Remind children to wash their hands before touching food. They can cut celery with plastic knives. Let them see the tuna in the can before it is mixed with the other ingredients.
- Put it on crackers and enjoy!

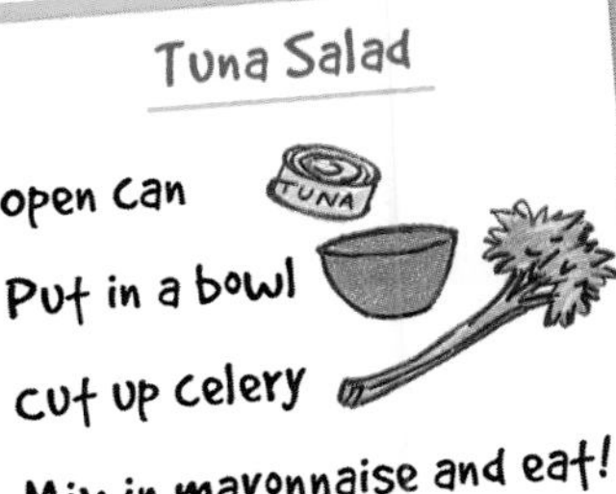

For more on animals in the air and underwater, use pp. 25–28 in the **Activity Book.**

▶ **Essential Question**
How do animals stay safe?

▶ **Objective**
Explore how animals have adapted to their environments.

▶ **Vocabulary**

skin **shell**
fur **feathers**
scales

Resources

Flipbook, p. 22

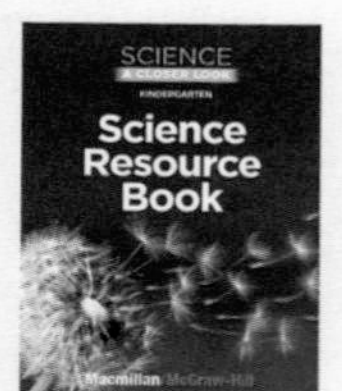

Science Resource Book, p. 37

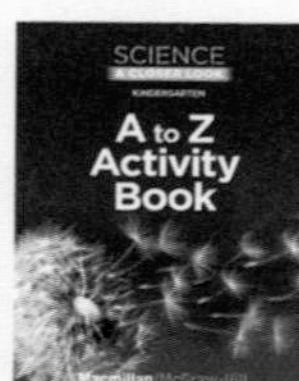

A to Z Activity Book, pp. 16–17

Leveled Reader: *All About Animals*

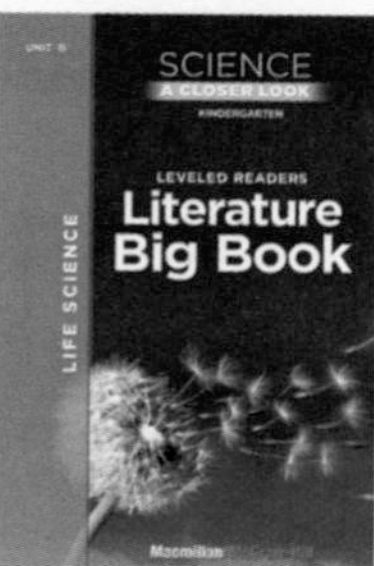

Unit B Big Book: *Time for Kids*

Technology

Staying Safe

Circle Time

People Staying Safe

Inquiry Skill: communicate

- Fill a bag with items that will encourage children to discuss how we get food, stay warm and cool, find shelter, and live together. Invite children to remove an item from the bag and tell how the item helps people. Record children's responses.
- Show children a picture of a polar bear. Ask: **What helps this bear stay warm? Get food?**

Communicate Ask: **What do we do when we are scared or sad?** use words to tell people how we are feeling; get help from adults

You need

- sunglasses
- sunscreen
- mittens
- umbrella
- key
- shoe
- supermarket flyer
- large bag
- chart paper

How do we stay safe and comfortable?

We use umbrellas to stay dry. (Joey)

A key locks our house. (Kathleen)

Time to Move!

Invite children to move like a porcupine with sharp quills, a turtle in its shell, a skunk spraying its stinky spray, and so forth.

Be a Reader

You need
- Unit B Literature Big Book
- Leveled Reader: *All About Animals*

Read the Big Book

Objective: Compare different types of animals.

Inquiry Skills: classify, communicate

- Read the title, pointing to the words, as you read. Before reading the book, take a picture walk. Read the book. On page 3, ask: **What do you think the weather is like here? What might help the polar bear stay warm?** On page 4, ask: **What is the ground like? What might help the bighorn sheep walk on rocky ground?** On page 5, ask: **What is the giraffe eating? What helps it reach the leaves?**
- **Guided Reading:** See the inside back cover of the Leveled Reader *All About Animals* for activities.

Be a Writer

You need
- *I See Animals Hiding,* by Jim Arnosky (or another children's book about animals)
- animal Photo Sorting Cards
- paper
- pencils
- crayons

How Animals Hide

Objective: Write about animals hiding.

Inquiry Skill: communicate

- Show children the cover of the book. Then take a picture walk, asking how each animal's color helps it blend into its environment. Ask: **Why might it be helpful for the animal to blend in?** Possible answer: So it cannot be seen by predators.
- Read the book aloud. Then have children draw and illustrate their own story about animals hiding. Encourage them to use Photo Sorting Cards as a reference when writing.

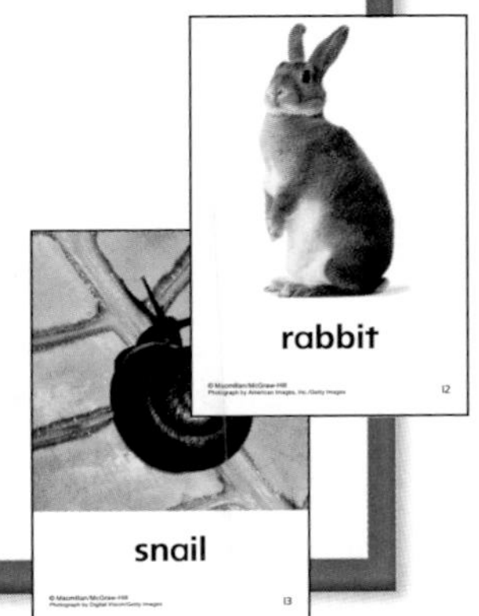

Be a Math Wiz Activity

You need
- connecting cubes
- different-sized plastic cups and containers
- picture of a camel

Fill the Camel's Hump

Objective: Fill and compare capacity.

Inquiry Skills: predict, compare, draw a conclusion

- Show children a picture of a camel. Explain that a camel's hump is not filled with water but with fat and can survive almost two weeks without food. Say: **You are going to fill your "camel's hump" to help it survive.**
- Have children work in pairs to fill cups with cubes. Ask: **Whose cup will hold more?**
- After partners have filled the cups, ask: **Which cup holds more?** Have children take the cubes out of the cups and make trains to compare their results.

Read Together and Learn

What helps these animals stay safe?

▶ Build on Prior Knowledge

Ask if children recognize any of the animals in the pictures. Have them identify each animal, and help them read their names.

▶ Use the Visuals

- **Observe** Invite children to share what they notice about each animal in the pictures.
- **Infer** Read the question at the top of the page and have children tell what they think might help these animals stay safe. Ask: **How do animals stay safe when the seasons change?** Some animals hibernate in the winter.

▶ Develop Vocabulary

skin, **fur**, **scales**, **shell**, **feathers** Point to the picture of the fawn lying in the leaves. Ask children how the color of the fawn's *fur* keeps it safe. Explain that we say an animal is camouflaged when its color helps it blend in with its habitat and surroundings. It would be harder for predators to find the fawn. Explain that the fawn's fur is an animal feature. Animal features are special parts of animals, such as a *skin*, *scales*, *shell*, and *feathers*. Point out that each animal shown has an animal feature that helps to keep it safe.

Flipbook

UNIT B • LESSON 6

What helps these animals stay safe?

fawn

turtle

moth

Reading Strategy

Phonological Awareness Have children think of words that rhyme with *porcupine* (*fine, mine, dine, line*). List their ideas on chart paper. Ask: **What letters do they have in common?**

Phonics Help children think of other words that have a long *i* sound and list them on chart paper (*Mike, bike, fight, like, right*). Invite volunteers to circle words that end in *e* and put an *x* beside words that have *-igh*.

Science Facts

Skunks have glands beneath their tails that spray predators with a bad-smelling liquid.

Fawns (baby deer) have fur with white spots, which makes it different from the plain brown fur of adult deer.

Porcupines move slowly, so they have sharp quills that protect them from predators.

Turtles also move slowly. Their hard shell protects them.

Moth Some moths are colored to blend with their surroundings. This moth camouflages itself as the eyes of an owl.

Kangaroo Baby kangaroos stay safe and warm in their mother's pouch.

Professional Development For more Science Facts and resources from NSDL visit http://nsdl.org/refreshers/science

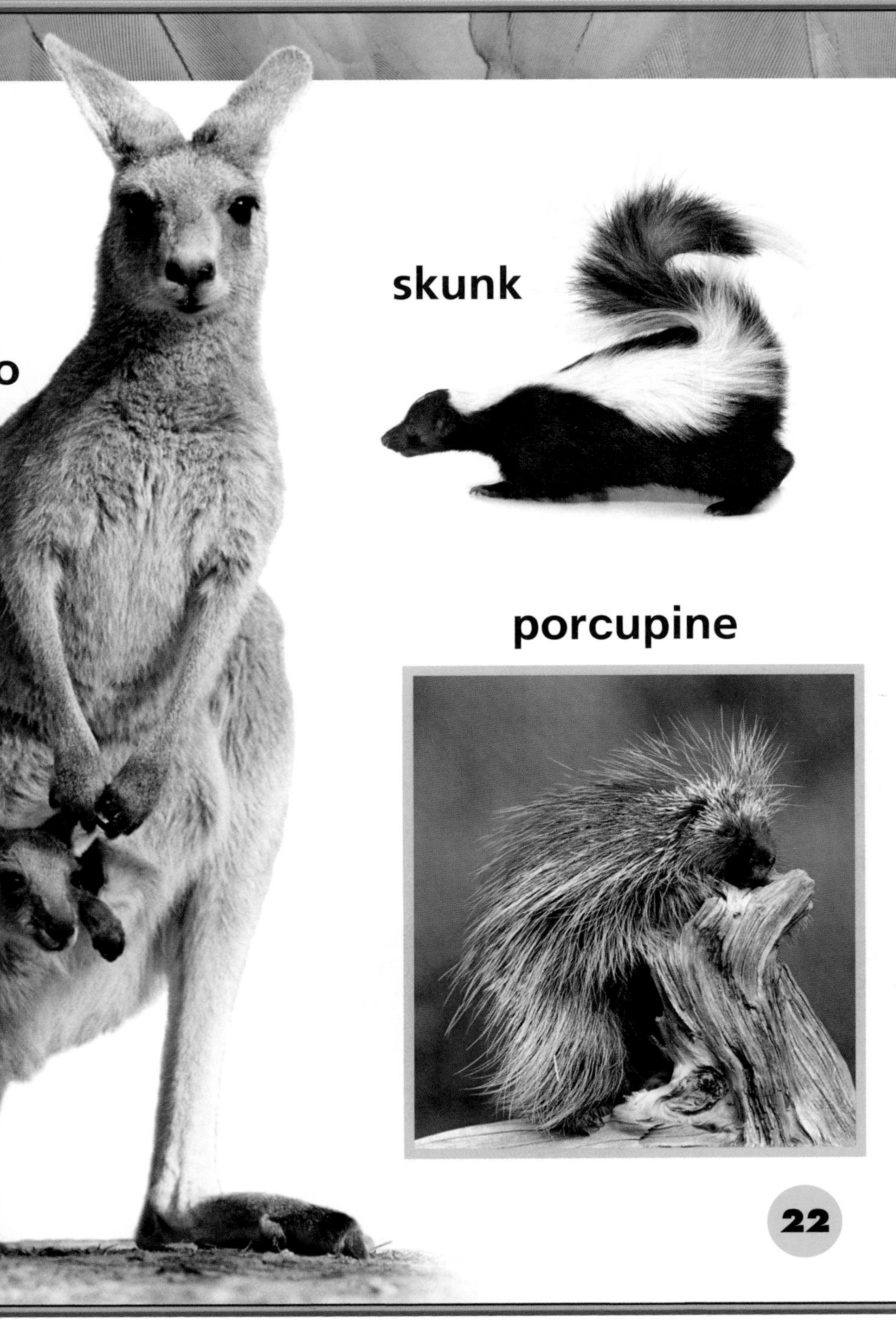

▶ Ask Questions

- **Where is the baby kangaroo hiding?** in its mother's pouch **Why do you think it is hiding there?** to stay safe and warm
- **What would it feel like to touch this porcupine? How do its sharp quills help it survive?** They protect it from predators.
- **What would it feel like to touch the turtle's shell? What would its skin feel like? How does its shell and skin protect the turtle?** It's tough and hard to protect it from enemies.

Write on It

Invite children to use a dry-erase marker to draw arrows pointing to each animal's special feature.

Think and Talk

Ask: **What is similar about the animals on this page?** They all have something special that helps them stay safe. **What is different?** Their features are different. Have children share what they know about these animals' features.

TIME FOR KIDS

Shared Reading Read *Ready for Winter* with the class. Use this magazine, found in the **Unit B Literature Big Book,** as a photographic review of animals and their adaptations.

More to Read

How Animal Babies Stay Safe, by Mary Ann Fraser (HarperCollins, 2001)

Activity After reading this book, have children talk about different ways that people stay safe and draw a picture of one of these ways. Help children write or have them dictate what is happening in their picture.

Observe children as they explore the similarities and differences between the animals to see what they understand about this topic. Children will also learn from each other in the whole-class conversation.

Be a Scientist Inquiry Investigation

Wormy Behavior

Objective: Observe animal adaptations to different environments.

Inquiry Skills: investigate, communicate, compare, infer

You need **earthworms, chart paper, drawing paper, pencils, crayons, colored pencils, eyedropper, Science Journal p. 37**

1. Give pairs of children a worm on a moist paper towel. Say: **We are going to be very gentle with our worms. We are going to change things in their environments and see what they do and how they adapt to the changes.** Remind children to wash their hands after touching the worms.

2. **Investigate** Provide each pair with an eyedropper with water in it. Have them drop a small drop on the worm. Ask: **How does the worm react?** Give each child a small piece of damp sponge to place near the worm. Ask: **How does the worm react?** Then give each pair a piece of paper that is half black and half white. Tell them to put the worm right in the middle. Ask: **Where does it go? Why?** Worms like dark colors better because dark colors absorb heat so they can stay warm.

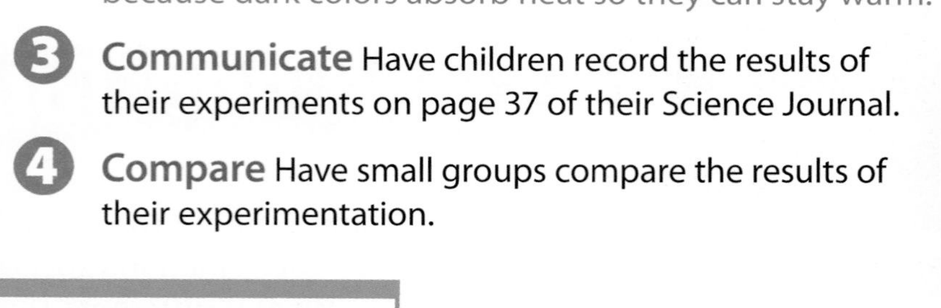

3. **Communicate** Have children record the results of their experiments on page 37 of their Science Journal.

4. **Compare** Have small groups compare the results of their experimentation.

Teacher Tip

If you or the children are not comfortable handling the worms, place a bright light near a worm tank to see what happens. Observe what happens when you clap loudly, reach your hand in, and gently drop water directly on a worm.

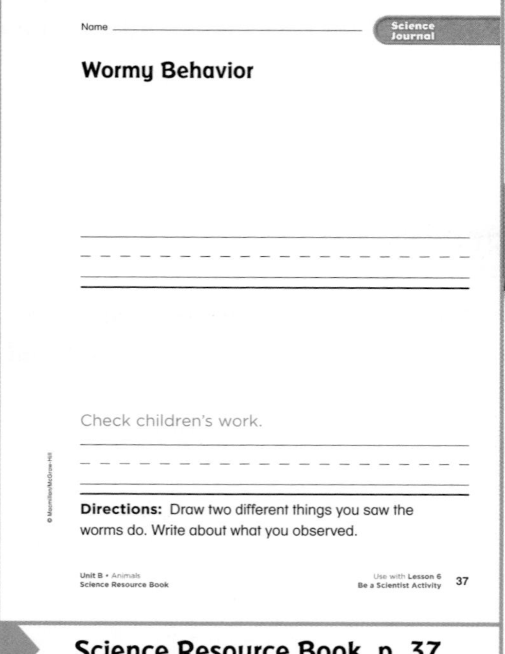

Name ____________________ Science Journal

Wormy Behavior

Check children's work.

Directions: Draw two different things you saw the worms do. Write about what you observed.

Unit B • Animals
Science Resource Book

Use with Lesson 6
Be a Scientist Activity 37

Science Resource Book, p. 37

Investigate More

Infer Ask: **Why do you think your results were different (if they were)? What other things could we use to observe more behavior?** foods, smells, and other colors that may attract worms

centers

Art

Camouflage Collage

Objective: Explore how animals use camouflage to blend in with their environments.

Inquiry Skills: make a model, communicate

- Show children pictures of animals that are camouflaged in their environments. Discuss the purposes of camouflage.
- Have each child draw a habitat. On a separate sheet of paper, have each child draw an animal that would blend into that habitat. Help children cut out their animals and glue them to their "habitats."
- Invite children to write or dictate information on how their animal stays camouflaged.

You need
- paper
- markers
- crayons
- scissors
- glue

Sand Table

Make a Desert

Objective: Make a model of a desert.

Inquiry Skills: classify, make a model

- Invite children to make a desert with a watering hole in the middle.
- If children need help adding detail to their desert, ask: **What kinds of plants will you add? How will you make hollow stems?**
- Encourage children to add animals and ask: **What animals did you add to your desert? What will you make to help them stay safe?**

You need
- straws
- paper
- scissors
- crayons
- paint
- tape
- small plastic desert animals

Cooking

Forest Animals

Objective: Make forest animals.

Inquiry Skills: make a model, communicate

- Have a small group help make play dough or provide children with air-drying clay.
- Children can use the play dough to make forest animals or plants for the Sand Table. Leave children's work out to dry.
- When dry figures can be painted, make more play dough for children to use for another day.

You need
- air-drying clay or play dough
- hot plate
- pot

Play Dough

1 cup flour
1 cup warm water
1/4 cup salt
2 tablespoons oil
2 teaspoons cream of tartar

Mix and cook over medium heat until smooth.

For more on how animals stay safe, use pp. 29–30 in the **Activity Book.**

Grow and Change

Essential Question
How do animals change as they grow?

Objective
Understand how animals grow and change as they mature.

Vocabulary
grow
change

Resources

Flipbook, pp. 23–24

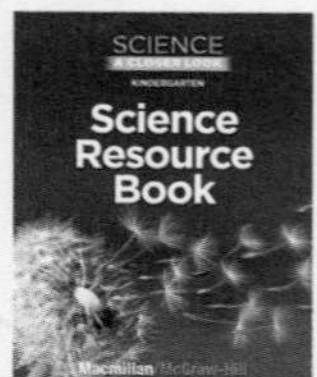
Science Resource Book, p. 38

Leveled Reader: *Animals Grow*

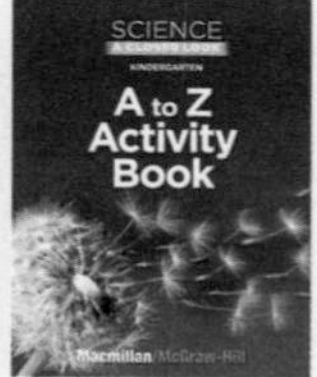
A to Z Activity Book, pp. 10–11, 24–25, 50–51

Photo Sorting Cards 11–20

Technology

Circle Time

WHOLE CLASS

Babies Grow

You need
- picture of a baby from a magazine
- chart paper
- marker

Inquiry Skills: communicate, compare, infer

- Ask: **Who has a baby in their family?** Invite children to share stories about what the baby does or how the baby acts.
- Have children discuss how they have grown and changed since they were babies. Record children's responses.

Infer Ask: **Do you think all babies act the same? Why or why not?**

How I Have Grown and Changed
I can ride my bike. (Matthew)
I can run. (Ana)
I have long hair. (Jenna)

Time to Move!

Invite children to jump up and down when they hear you say the word *baby* in a sentence. Have them put their hands above their heads when they hear the word *adult* in a sentence.

Be a Reader Activity

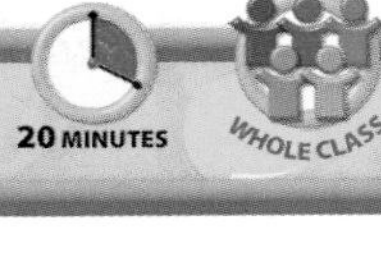

You need

- Unit B Literature Big Book
- Leveled Reader: *Animals Grow*

Read the Big Book

Objective: Use illustrations to explore animal growth.

Inquiry Skill: communicate

- Show the cover of *Animals Grow* in the Unit B Literature Big Book and read the title aloud, tracking the print with your finger.
- Read the story, inviting children to say the predictable text aloud with you as you turn each page. Encourage children to share what they observe about each animal's growth and what changes the animal's experience.
- **Guided Reading:** See the inside back cover of the Leveled Reader version of *Animals Grow* for activities.

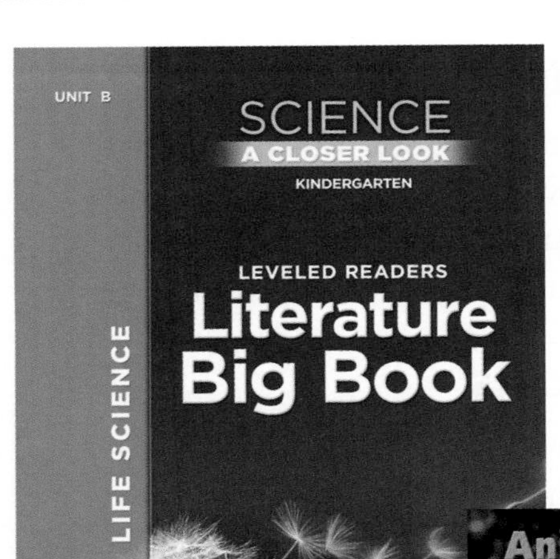

Be a Math Wiz Activity

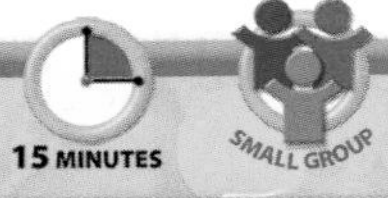

You need

- paper bag
- connecting cubes
- ruler

Measure Up

Objective: Measure length of feet using nonstandard units.

Inquiry Skills: compare, measure

- Have children use connecting cubes to measure their foot and record the results.
- Now that they know the size of their foot, have children decide how long they think their foot would have been when they were a baby, a toddler, and a preschooler.
- As an extension, encourage children to work with a partner to lie on the floor and measure each other's height with connecting cubes or a ruler.

Read Together and Learn

How do these foxes grow and change?

▶ Build on Prior Knowledge

Ask: **What do you notice about the color of the newborn foxes? How might the darker color of the new babies help them survive?** It is closer to the color of the ground so it will help them hide.

▶ Use the Visuals

- **Communicate** Have children discuss how the foxes change as they grow.
- Encourage children to discuss what the foxes in the photograph might be doing.
- **Compare** Invite children to talk about the difference between the adult fox and the image of the mother fox with her kits.

▶ Develop Vocabulary

grow, **change** Ask: **How do you know that the foxes in the pictures are growing?** The foxes are getting bigger. Ask: **How do you know that the foxes in the pictures are changing as they grow?** Their fur changes color and they are getting bigger.

Flipbook

UNIT B • LESSON 7

How do these foxes grow and change?

newborn foxes

baby foxes

Reading Strategy

Use Illustrations Have children use the illustrations to discover what the words say. Point to the labels and ask: **Which of these pairs of words says *adult fox*? How did you know?** Repeat with the other pictures.

Print Awareness Have children circle each letter *h* on the page. Ask: **What is the difference between the two?** Explain that sentences always begin with capital letters.

Science Facts

Foxes Mother foxes may give birth to four to ten kits in a litter. At birth, foxes are deaf, blind, and have short, dark fur. They rely on their mother for food and warmth. Mothers remain close to their babies and depend on the father and other foxes to provide food. After a few weeks, baby foxes can see and begin to explore the world around them. Their dark fur changes into a reddish color, and they begin to eat solid foods. Baby foxes also spend a lot of time playing, like kittens. Young foxes follow adults and learn to find food on their own.

Professional Development For more Science Facts and resources from NSDL visit http://nsdl.org/refreshers/science

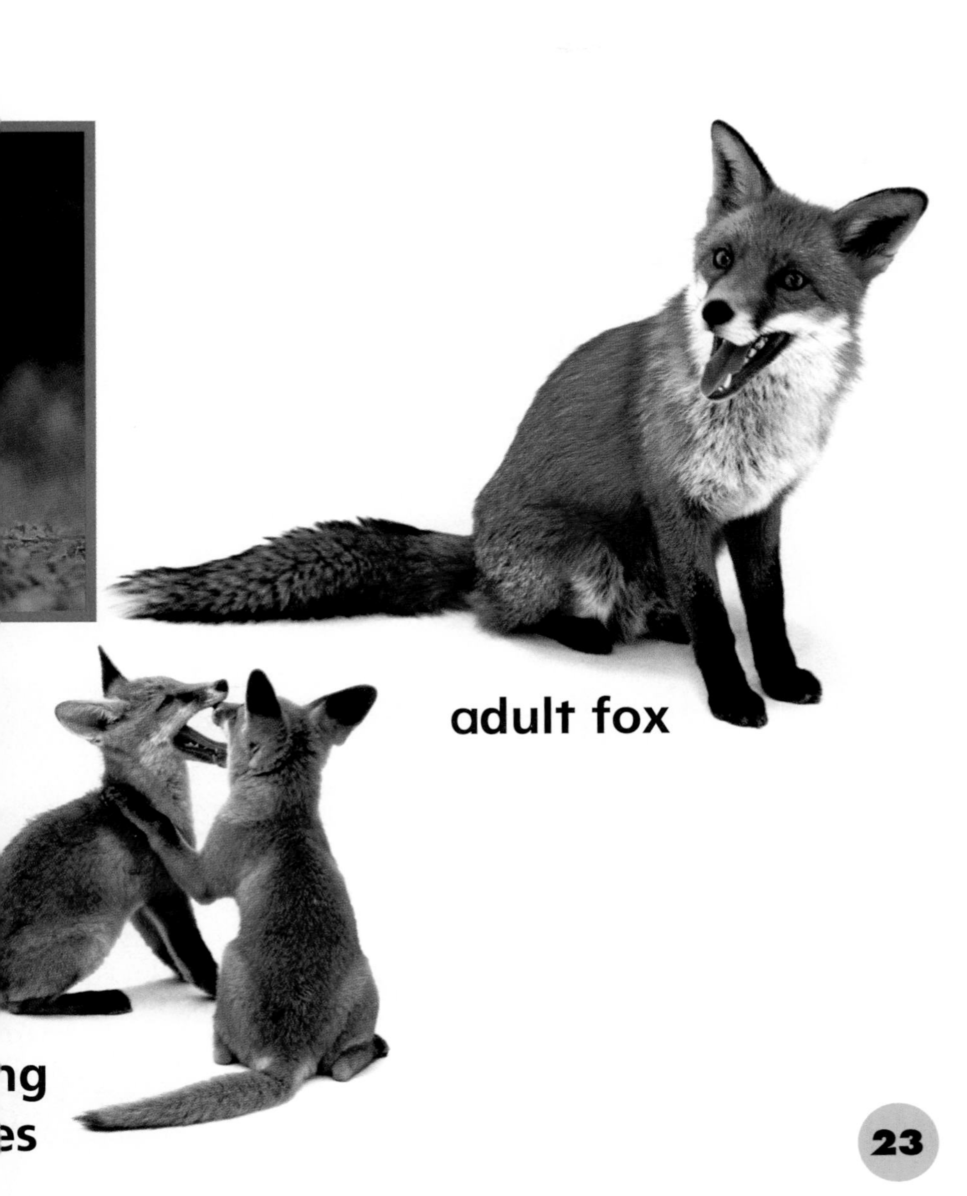

Take a Trip

The Zoo Take a trip to a local zoo or wildlife center. Provide each child with a pencil and a piece of cardboard with a sheet of paper stapled to it. Ask zoo workers to tell the class about how different animals take care of their young. Allow children to observe and draw animals with their young.

Back in the Classroom Have children list what they learned about animals and their babies. Ask: **How close did the baby stay to its mother? What were the baby and the mother doing?**

▶ Ask Questions

- **How might the newborn foxes act?** sleepy, hungry **How might their behavior change when they become older?** They may need their parents less; do more on their own; sleep less.
- **What are the young foxes in the picture doing?** They are following their mother; she might be teaching them to hunt.
- **How might playing help the foxes learn what they need to know as they grow up?** It helps them grow strong and learn fighting strategies to survive.

Write on It

Have children use a dry-erase marker to find and circle the word *fox*. Invite a volunteer to write the word *fox* below the newborn foxes. Repeat with the other pictures.

Think and Talk

Prompt children to discuss how the foxes on page 23 are similar to the animals on pages 16 and 24. How are they different?

Read Together and Learn

How do these baby animals get what they need?

▶ Build on Prior Knowledge

Ask: **How do these baby birds look different from the grown-up bird?** Ask the same question about the other animals.

▶ Use the Visuals

- Communicate Encourage children to talk about any experiences they have had with baby cats or baby birds. Perhaps they have seen grizzly bears, prairie dogs, or penguins at the zoo.
- Have children talk about what the animals in each picture are doing.
- Observe Have them point out what they notice about the habitat of the animals in each picture.

▶ Develop Vocabulary

grow, **change** Have children revisit the words *grow* and *change* by talking about how they think the baby animals in each picture will grow and change.

Flipbook

UNIT B • LESSON 7

How do these baby animals get what they need?

prairie dogs

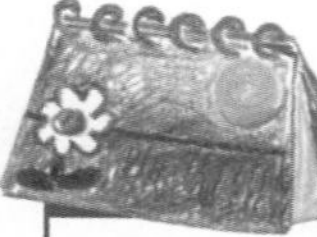

Reading Strategy

Phonological Awareness Write the word *cat* on chart paper. Have children read it with you and ask: **What words rhyme with** *cat?* (*sat, fat, mat, bat, hat, rat*) Have children point out what the words have in common.

Print Awareness Have children point to the letter at the end of each word. Talk about how *s* at the end of a word shows that there is more than one.

Science Facts

Grizzly Bear Female grizzly bears give birth to an average of two cubs every three years. They stay with their young cubs for about two years.

Prairie Dogs are extremely social. They frequently appear to be hugging, kissing, and grooming one another.

Emperor Penguins It is the male penguin who keeps the female's egg, and then the newly hatched chick, safe and warm on his feet.

Cat For up to twelve weeks, kittens rely on their mothers to provide food, warmth, and safety. Throughout this time, mothers teach them valuable lessons.

Robin Mother robins must feed their babies every few minutes, which is about 100 times a day.

Professional Development For more Science Facts and resources from NSDL visit http://nsdl.org/refreshers/science

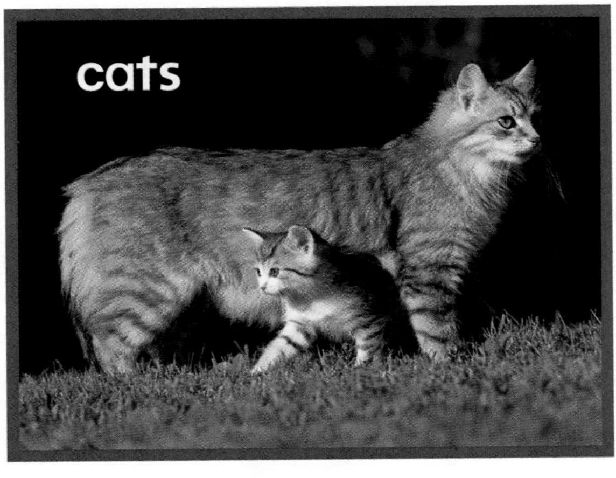

24

▶ Ask Questions

- **Why doesn't the mother bear get a fish for the baby instead of making him do it on his own?** He has to learn how to do it; she is teaching him so he will be able to survive when he is older.
- **Since baby animals cannot talk, how are they showing their parents what they need?** birds opening their mouths in hunger; baby penguins sitting on their parents' feet for warmth
- **What other things do you think baby animals need from their parents?**

Write on It

Have children use a dry-erase marker to circle baby animals that have fur to keep them warm. Ask: **How might the other animals stay warm?**

Think and Talk

Look again at the animals on Flipbook pages 23 and 24. Ask children: **How are the parents helping the baby animals?** Encourage children to share what they have learned about how animals grow and change.

Formative ASSESSMENT

Observe children as you explore the Flipbook pages to see what they do and do not understand about animal growth and care.

Differentiated Instruction

Extra Support Have children draw a picture showing a baby animal and its respective grown-up. Ask: **What does this baby animal need from its grown-up?** Help children write or have them dictate their answer.

More to Read

Make Way for Ducklings, by Robert McCloskey (Viking Press, 2001)

Activity Read this book to children to reinforce their understanding of the idea that parent animals care for their babies. Ask: **How do grown-ups take care of you?** Make a list, and have children draw a picture of one way.

Be a Scientist **Inquiry Investigation**

Growing Animals

Objective: Watch mealworms grow and change.

Inquiry Skills: observe, make a model, communicate, put in order, infer

You need **empty clear plastic box, dried oatmeal or bran cereal, mealworms, construction paper, sliced apple, hand lenses, Science Journal p. 38**

1. **Observe** Have children observe the mealworms on large pieces of paper. Ask: **How does the mealworm move? What does it do when you put a piece of apple in its way? Can you see how many legs it has? Can you find its mouth?**
2. **Make a model** Have children help prepare the mealworms' habitat with dried oatmeal and a piece of apple (for moisture). Then put the mealworms in the container.
3. **Communicate** Over time, have children observe the mealworms. Discuss their observations and ask: **How are the mealworms beginning to change?** Have children use their Science Journal page to record what they observe.
4. **Put in Order** Have children create a large mural that shows the life cycle of the mealworm.

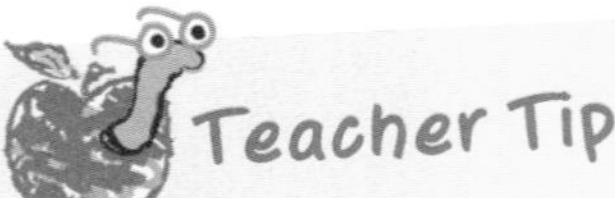

You can find mealworms at the local pet store. You do not need to place a lid on top of their container because they cannot crawl out. If your district has purchased a caterpillar kit for your classroom, you can use it for this activity instead.

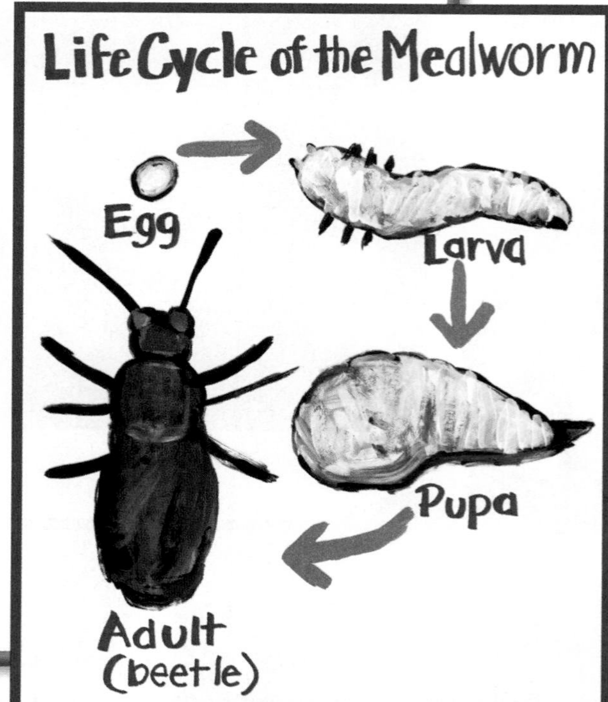

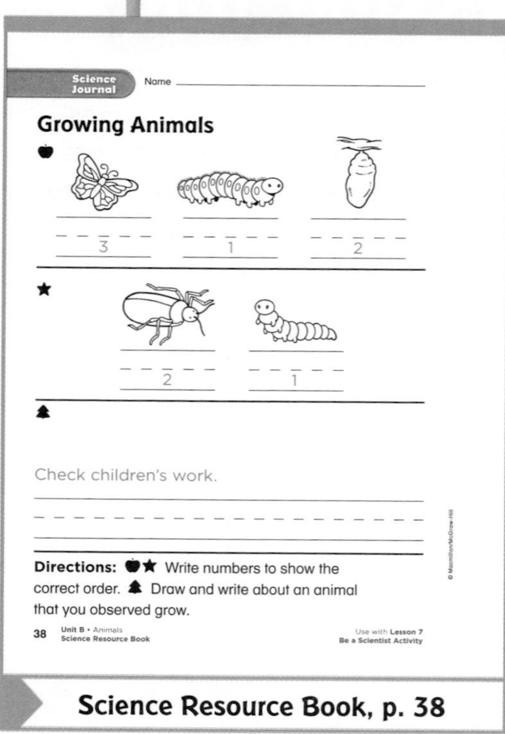

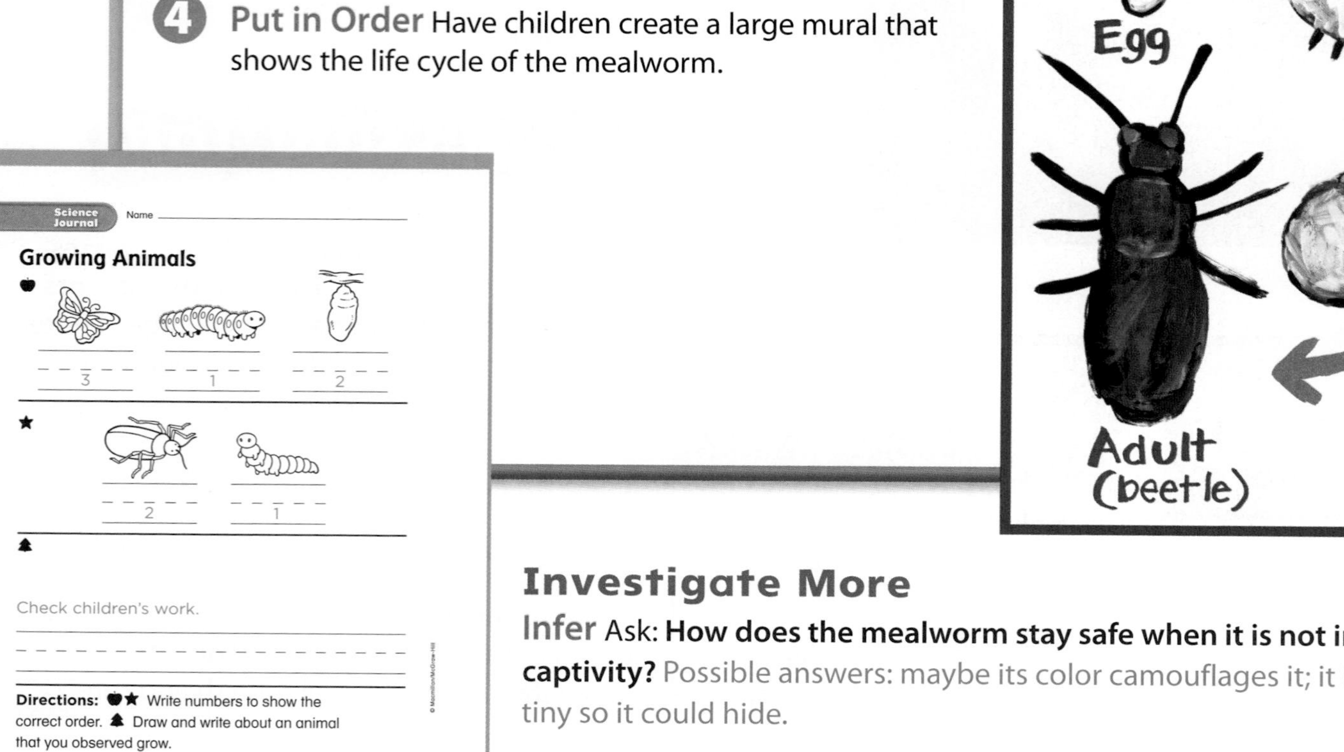

Science Journal Name ____

Growing Animals

3 1 2

2 1

Check children's work.

Directions: Write numbers to show the correct order. Draw and write about an animal that you observed grow.

38 Unit B • Animals Science Resource Book — Use with Lesson 7 Be a Scientist Activity

Science Resource Book, p. 38

Investigate More

Infer Ask: **How does the mealworm stay safe when it is not in captivity?** Possible answers: maybe its color camouflages it; it is very tiny so it could hide.

centers

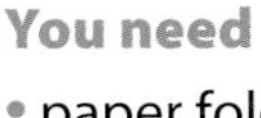

Drawing and Writing

Have You Changed?

Objective: Draw and write about how people grow and change.

Inquiry Skills: put in order, communicate

- At Circle Time, have children discuss what they have learned about how animal babies grow and change. Show children the folded blank books that you are adding to the Writing Center.
- Have children make books that document how they have grown and changed.
- As children work on their books, come around and help them write, or have them dictate their story to you.

You need
- paper folded in half and stapled to make book
- drawing materials

Technology

More about Animals

- For games and activities about animals, help children log onto www.macmillanmh.com.
- Children can use these games to review and reinforce what they have learned about animals. Encourage children to revisit these games and activities throughout the school year.

Art

Hatching Animals

Objective: Make a model of animals hatching from eggs.

Inquiry Skills: make a model, communicate

- At Circle Time, help children list animals that hatch from eggs.
- Have children use clay to make an egg, and then the animal that hatches from the egg.
- Use a shoe box to create a diorama for the hatching animals.

You need
- colored clay
- shoe boxes
- colored paper
- scissors
- cotton balls
- paint
- glue

Dramatic Play

Taking Care

Objective: Act out the parenting behaviors of animals.

Inquiry Skill: communicate

- During Circle Time, show children pictures of parent animals with their babies.
- Ask: **What do you think these animals do to take care of their babies?**
- Invite children to use the Dramatic Play Center to "be the parents" and to "be the babies," to act out how animals care for their young. Ask: **What will you do to help them stay safe?**

You need
- photographs of different parent animals and baby animals

For more on how animals grow and change, use pp. 31-32 in the **Activity Book**.

People and Animals

Essential Question
What do we get from animals?

Objective
Explore relationships between people and animals.

Vocabulary
beekeeper
farmer

Resources

Flipbook, p. 25

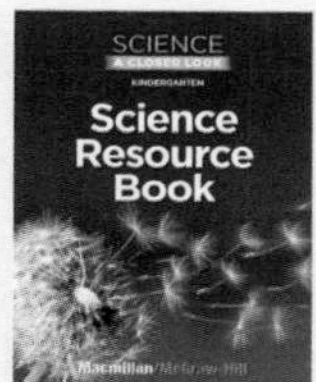

Science Resource Book, p. 39

Leveled Reader: *Good Morning*

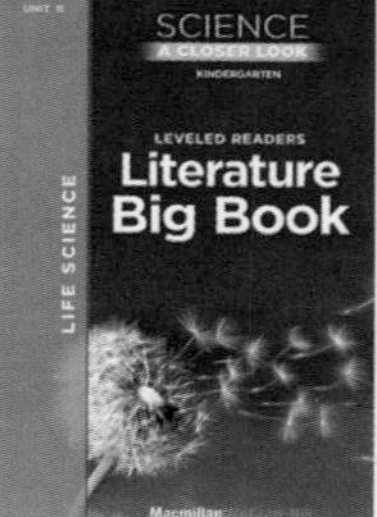

Unit B Literature Big Book

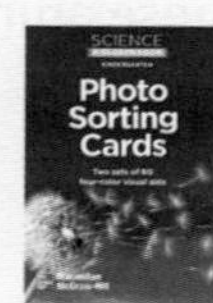

Photo Sorting Cards 11–20

Technology

Circle Time — WHOLE CLASS

Pet Diagram

Inquiry Skills: classify, infer

You need
- Flipbook p. G1
- dry-erase marker

- Have children generate a list of animals that can be kept as pets. Then have them list animals that can be found on a farm. Record this information on the Venn diagram. Ask: **Are there any animals that can live on a farm and be a pet?**
- Finally, ask children to list animals that are not kept as pets or found on farms. Add those animal names outside the circles.

Infer Ask: **What kinds of things does a farmer do to take care of farm animals?**

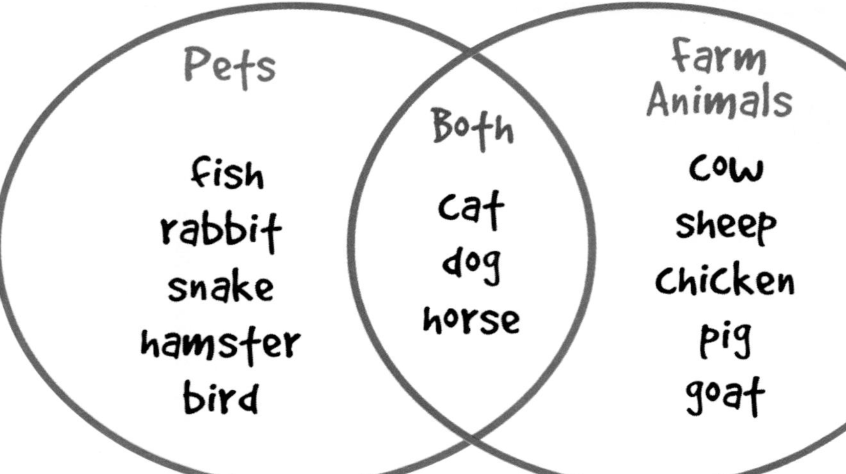

Time to Move!

Invite children to form a circle and sing "Old McDonald Had a Farm." While singing, encourage children to move like each animal (gallop like a horse, waddle like a duck, walk like a chicken, and so forth).

Be a Reader

You need
- Unit B Literature Big Book
- Leveled Reader: *Good Morning*

Read the Big Book

Objective: Use the illustrations to help read a story.

Inquiry Skills: communicate, infer

- Read *Good Morning* in the Unit B Literature Big Book. Reread, inviting children to chime in as you read.
- Go back and have children find something on each page that is not an animal. Ask: **How do you know that it is not an animal?** Ask children what animals they might see if they go for a walk near their house. Record their responses.
- **Guided Reading:** Use the Leveled Reader version of *Good Morning*.

Be a Writer

You need
- paper
- pencils
- coloring materials

Animal Drawings

Objective: Draw and write about animals.

Inquiry Skill: communicate

- Have children draw one of the animals from the Venn diagram.
- As children work, come around and offer to take dictation or to help them write about their drawing. Have them explain whether their animal might be found in a home, on a farm, or in the wild.
- Remind children that they can find the name of the animal they drew on the Venn diagram if they need help spelling the animal's name.

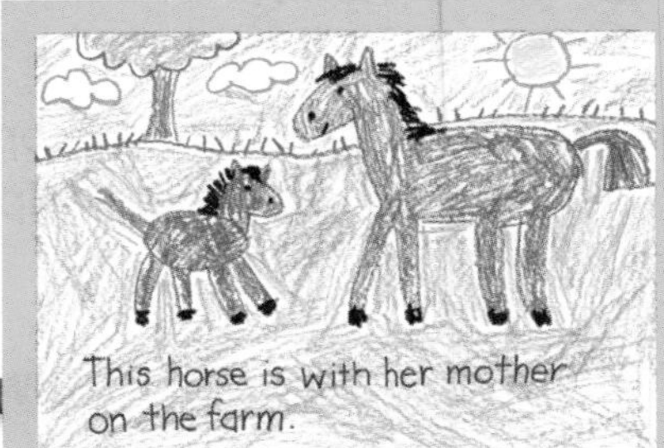

Be a Math Wiz Activity

You need
- 0–5 number cube
- connecting cubes
- paper plates
- paper
- pencils
- egg carton that holds 6 eggs

Collect the Eggs

Objective: Subtract from six.

Inquiry Skills: predict, draw a conclusion

- Explain to children that they will act like farmers collecting eggs before the chicken returns.
- Put 6 connecting cubes on a plate. Have children toss the number cube, take that many "eggs" (cubes) from the "nest" (the plate), and put them in the egg carton. Ask: **How many tosses do you think it will take to fill the carton?**
- Help children use tally marks to record how many tosses it takes to collect all of the eggs.

Read Together and Learn

How do animals help us?

▶ Build on Prior Knowledge

Ask children to name and describe foods on this page that they recognize.

▶ Use the Visuals

- **Observe** Have children describe what they see on the page. Encourage them to share if they have ever seen these products in a grocery store.
- **Communicate** Prompt discussion by asking: **Why do you think the girl is using a basket? Why do you think the man is sitting on a stool? What type of clothing is the beekeeper wearing?**
- Discuss the resources people get from animals as well as how people help animals.

▶ Develop Vocabulary

beekeeper, **farmer** Create a two-column chart. Have children discuss their ideas of what a farmer does. List these under the word *farmer*. Repeat with *beekeeper*. Have children brainstorm other jobs in which a person works with animals (pet store worker, zookeeper, veterinarian, police officer).

Flipbook

UNIT B • LESSON 8

How do animals help us?

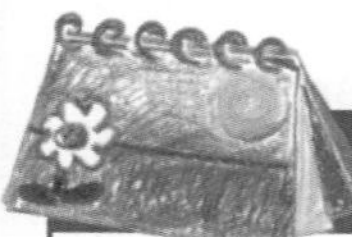

Reading Strategy

Phonological Awareness Write the words *be* and *bee* and help children define each word. Have them think of other word pairs that sound the same but mean different things. Explain that these words are called homophones.

Comprehension On separate index cards, write the words *goat, milk, chicken, egg, bee,* and *honey*. Spread the cards faceup and help children read the words and find matches.

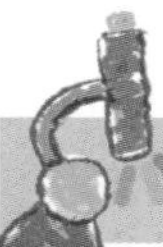

Science Facts

Eggs The eggs that farmers collect for us to eat have not been fertilized by a rooster, which is why there is no chick growing inside.

Honey Bees make and store it in waxy honeycombs, which are attached to wooden frames in the hive. An average hive produces 20 to 30 pounds of honey that the bees do not use, and this is what the beekeeper collects. Beekeepers do get stung, but they minimize the number of stings by wearing protective gear.

Milk Goat's milk is the most widely consumed milk in the world. It is used to make products such as yogurt, cheese, moisturizing lotion, and even soap.

Professional Development For more Science Facts and resources from NSDL visit http://nsdl.org/refreshers/science

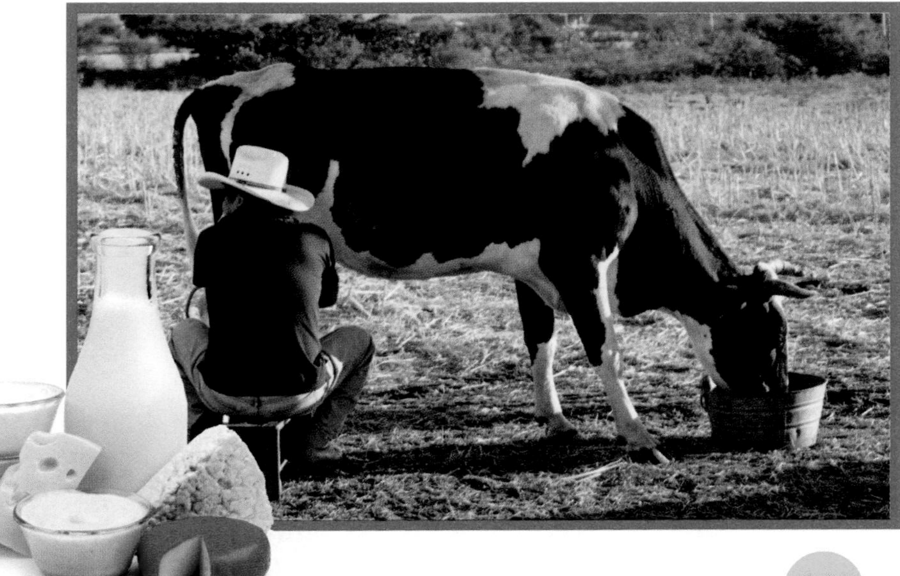

25

▶ Ask Questions

- **What are some other animals that lay eggs?** other birds, alligators, crocodiles, fish, frogs, toads, turtles, snakes
- **Why do cows make milk?** for their babies
- **What other animals have bodies that can make milk for their babies?** dogs, cats, goats, bears, tigers, lions, rabbits Explain that reptiles, amphibians, fish, and birds lay eggs and do not nurse their babies. Mammals do not lay eggs, but they do nurse their babies.

Invite children to use a dry-erase marker to draw an *x* on each food that we get from animals.

Think and Talk

After reviewing the Flipbook page, have children discuss what is the same about all of the pictures. Invite children to share other products they know of that are from animals.

Formative ASSESSMENT

If children are having difficulty making the connection between people and animals, discuss some other examples such as cows and milk or sheep and wool.

ELL Support

Discuss ways people care for animals. Use pictures or realia (honey, eggs, milk) to review how animals help us.

Beginning Invite children to pantomime ways that people care for animals.

Intermediate Pantomime activities like milking a cow and model the question: **Are you ___ ing a cow?**

More to Read

The Year at Maple Hill Farm, by Alice and Martin Provensen (Simon and Schuster, 1981)

Activity Read this book to reinforce children's understanding of how a farm works. Discuss how life on the farm changes throughout the year. Have children draw something that happens on a farm.

20 MINUTES
WHOLE CLASS

Ask an Expert

Objective: Interview an expert to learn about an animal.

Inquiry Skills: communicate, investigate, infer

You need **chart paper, drawing paper, pencils, crayons, colored pencils, Science Journal p. 39**

1. Invite a person whose daily life involves interaction with an animal to visit with your class. Possible visiting "experts" could be a policeman who rides a horse, a person who owns or breeds dogs, a pet store worker, or a local farmer.
2. **Communicate** Before the interview (or trip) help children brainstorm a list of questions that they want to ask. Record their questions on chart paper.
3. **Investigate** On the day of the interview (or trip), help children ask the questions on the list. Take notes as the expert responds to each question.
4. **Communicate** Have children use their Science Journal page to illustrate two ways people and animals interact. Following the interview (or trip), ask children to draw and write or dictate information about what they learned. Compile the pages into a class book.

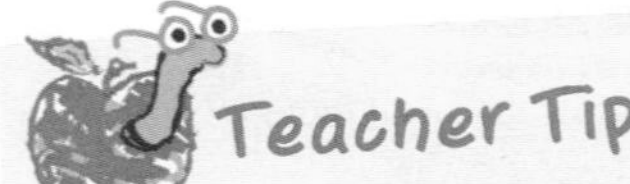

Teacher Tip

During an interview, do not hesitate to interrupt the visitor in order to clarify something for the children, or to help them regain their focus. Remember, you know how to manage a group of five-year-olds and your visitor may not.

Name ______________________ Science Journal

People and Animals

Check children's work.

Directions: Draw and write about one way that animals help people and one way that people take care of animals.

Unit B • Animals
Science Resource Book

Use with Lesson 8
Be a Scientist Activity 39

Science Resource Book, p. 39

Investigate More

Infer Say: **Our visitor talked about caring for his or her animal(s). What other animals need the same kind of care? What are some animals that need to be cared for in a different way?**

centers

Blocks

Build a Farm

You need
- toy farm animals

Objective: Use blocks to create a model of a farm.

Inquiry Skills: make a model, classify, communicate

- During Circle Time, talk to children about building a farm in the Block Area. Ask: **What kinds of structures will you need?** List their ideas on chart paper and post it near the Block Area.
- As children build, ask: **What animals will you have? How will you give them shelter? Where will they exercise? What will they eat?**
- When children are finished, encourage them to act like farmers working on their farm.

Art

Farm Needs

You need
- play dough
- paint
- paper
- crayons
- scissors
- tape

Objective: Make tools, food, and other necessities for their block farm.

Inquiry Skills: classify, make a model, communicate

- Ask children about what kinds of things are needed for the farm they are making in the Block Area. For example, if there is a horse stable, what will the horse need? What kinds of vehicles and tools will the farmers need to do their farming work?
- Have children use art materials to create props for the farm.
- Invite children to place their props in the block farm, and to explain how the farmers or animals will use what they have made.

Cooking

Banana Smoothies

You need
- blender
- cups
- plastic knives
- plain yogurt
- bananas
- honey

Objective: Make smoothies from ingredients we get from animals.

Inquiry Skills: classify, observe

- Hold up a container of yogurt and ask children if they know how we get yogurt. Explain that it comes from milk from a cow. Hold up honey and ask children to discuss how we get honey.
- Have children use plastic knives to cut the bananas into slices. Have children take turns putting yogurt, honey, and banana slices into the blender. Blend until smooth.

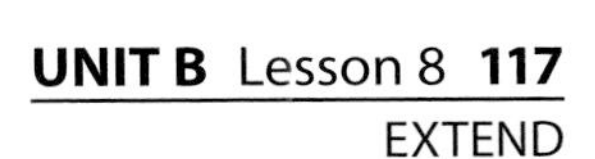

Banana Smoothies

4 cups of plain yogurt

3 bananas

3 teaspoons of honey

Slice bananas. Blend. Drink and enjoy!

Performance ASSESSMENT

DOK 2

Reptiles, Mammals, Birds, and Fish

Objective: Identify and describe reptiles, mammals, birds, and fish.

- Show two pictures, a reptile and a mammal. Ask: **Which is a reptile? Can you point to and talk about something that shows it is a reptile?** Then ask: **Which is a mammal? Can you point to and talk about something that shows it is a mammal?**
- Show the pictures of the bird and fish. Ask: **What do we call these animals? How do they move? Can you point to and say what helps them move?**

You need

- 4 Photo Sorting Cards: reptile, mammal, bird, fish

Scoring Rubric

Use the Scoring Rubric to evaluate each child's completion of the Performance Assessment task.

points = Child is able to identify a reptile, pointing to its scales to show that it is a reptile. Child identifies a mammal, pointing to its fur to show that it is a mammal. Child knows that a bird flies using its wings and that a fish swims using its fins.

points = Child is able to identify the reptile and mammal, but can only point to an identifying characteristic of one of them. Child identifies the bird and the fish, but can only explain what one of them uses to move.

point = Child can identify three or more of the animals, but is unable to talk about any identifying characteristics of these animals.

Summative ASSESSMENT

DOK I

Use the Unit B Assessment booklet on pp. 65–66 of the **Science Resource Book** to assess what children have learned about animals.

- Help children fold their page in half and write their name. On each page, have children circle and write the word to complete the sentence. Encourage children to color each page and add plants and animals to each drawing.
- After reviewing children's work, have them take their books home to share with a family member.

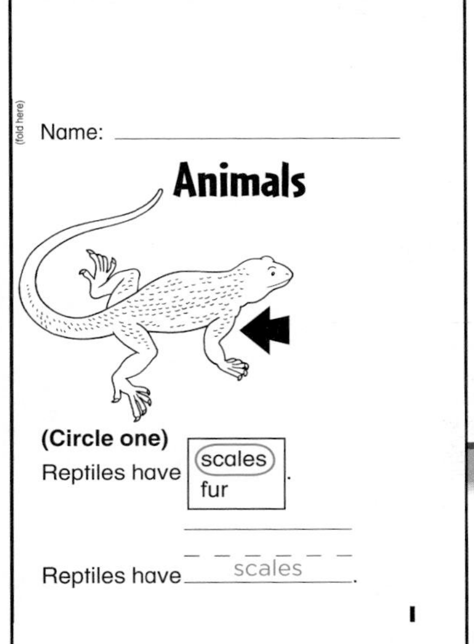

(fold here)

Name: ____________

Animals

(Circle one)

Reptiles have scales / fur.

Reptiles have scales.

1

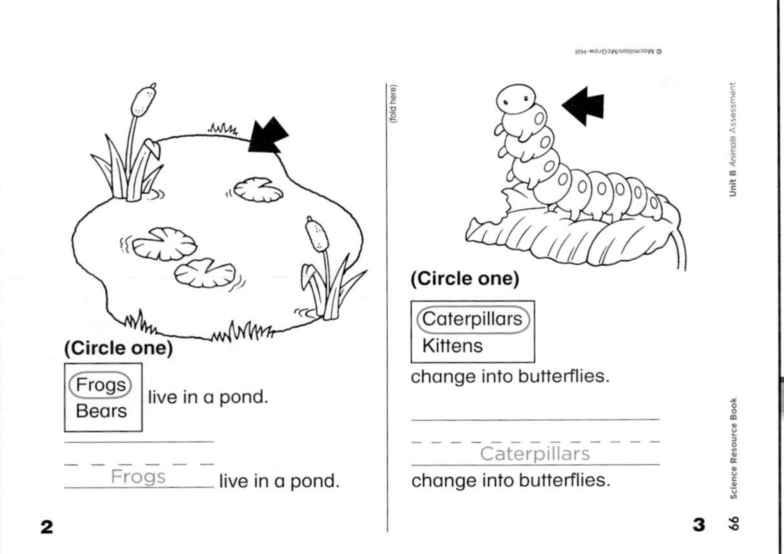

(Circle one)

Frogs / Bears live in a pond.

Frogs live in a pond.

2

(fold here)

(Circle one)

Caterpillars / Kittens change into butterflies.

Caterpillars change into butterflies.

3

Unit B Animals Assessment

Science Resource Book

66

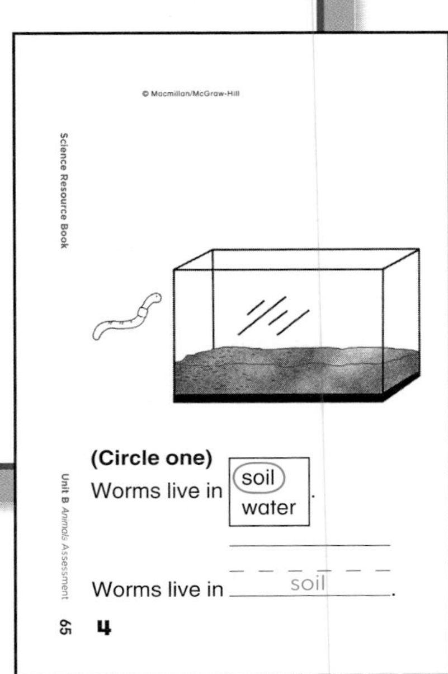

Science Resource Book

(Circle one)

Worms live in soil / water.

Worms live in soil.

Unit B Animals Assessment

65

4

Science Resource Book, pp. 65-66

Portfolio ASSESSMENT

- Encourage children to select work from this unit to include in their portfolio. Discuss their choices and have children share which is their favorite and why.
- Use the Unit B Checklist on page 78 of the **Science Resource Book** to record children's progress. Place checklists and notes in each child's portfolio.

Review TOGETHER

Plants and Animals Working Together

Objective: Explore how animals and plants get what they need to grow and mature together in a particular environment.

What We Learned

- Before showing children the Flipbook page, ask them to describe any of the ponds they have visited or seen in a book.
- Make a list of the plants and animals that they think might live in or near a pond.
- Review what plants and animals need to live: air, water, food, shelter, space, and sunlight.

Flipbook

How do plants and animals use this pond?

duck

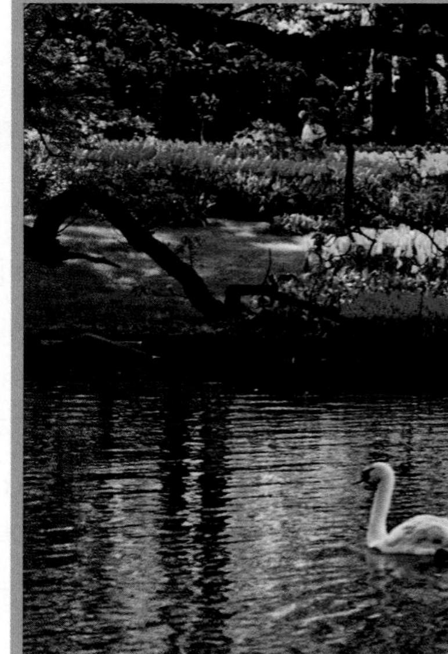

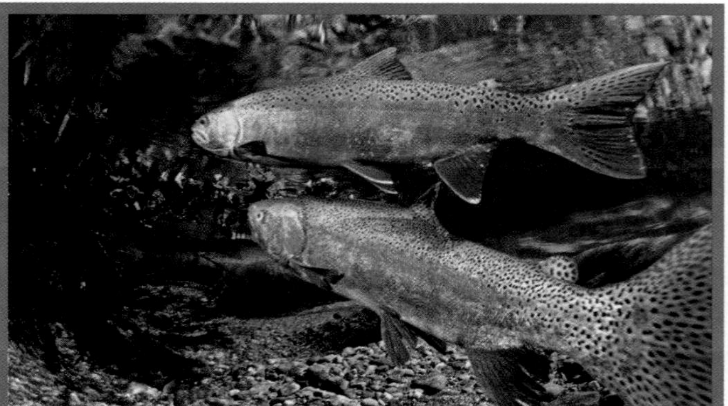
trout

Floor Puzzle

Activity **Pond Life** Revisit the floor puzzle and invite children to recreate the pond image using Flipbook page 5 as a reference. Introduce using the animal and plant pieces and invite children to place them where they may live. You may want to ask: **Where would the heron find its food? Where will the chipmunk get a drink?**

Science Facts

What is happening in this pond?

- The **chipmunk** and other animals drink water from the pond.
- The **duck** hides her nest in **grasses** around the pond.
- **Insects** find food on **plants** in the pond.
- The **frog** gets food by eating insects.
- Some **freshwater fish,** such as the **rainbow trout,** have dark backs to blend in with the murky water.
- The **heron** gets its food by eating **fish** or **frogs.**
- The **water lily's** rounded edges help keep it from tearing when wind makes the water's surface rough.

LOG ON **Professional Development** For more Science Facts and resources from  visit http://nsdl.org/refreshers/science

chipmunk

heron

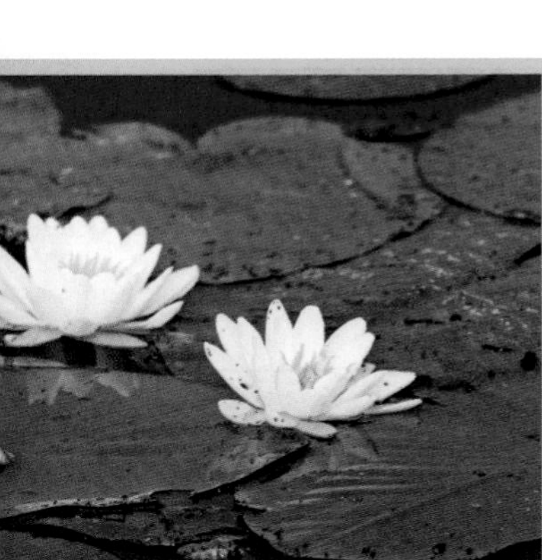

ily and frog

26

Make Connections

How do plants and animals use this pond?

▶ Use the Visuals

- Ask: **How is this pond similar to ponds you have seen? How is it different?**
- Encourage children to discuss each picture so that they better understand the relationship between the plants and animals in and around the pond.
- Point out how these animals and plants work together to survive in their shared habitat.

Write on It

When you discuss the animals' adaptations to the pond's habitat, have children use a dry-erase marker to mark each adaptation with an *x*.

▶ Ask Questions

- **What might happen to the animals and plants if the pond where they live dries up?** Possible answer: They would die; they would have to find a new place to live.
- **Why is it important that animals and plants live together?** They need each other for shelter, food, and air.

Take a Trip

A Pond Habitat Visit a nearby pond so children can investigate the habitat. Give children paper stapled to pieces of cardboard so they can draw and write out in the field. Bring plant and animal field guides to help identify unfamiliar plants and animals. Collect a sample of the pond water.

Back in the Classroom Have children use hand lenses to examine more closely the water sample. Have them create drawings, paintings, a mural, or a book to document what they learned about the pond.

Teacher's Notes

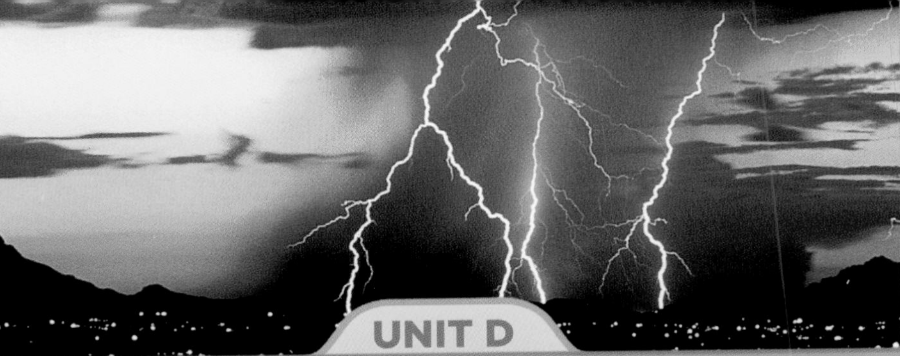

arth and Space Science

Earth and Space Science

National Standards

The following K-4 National Science Education Standards are covered in Units C and D:

- Earth materials are solid rocks and soils, water, and the gases of the atmosphere. The varied materials have different physical and chemical properties, which make them useful in different ways, for example, as building materials, as sources of fuel, or for growing the plants we use as food. Earth materials provide many of the resources that humans use.
- Soils have properties of color and texture, capacity to retain water, and ability to support the growth of many kinds of plants, including those in our food supply.
- The sun, moon, stars, clouds, birds, and airplanes all have properties, locations, and movements that can be observed and described.
- The sun provides the light and heat necessary to maintain the temperature of the earth.
- The surface of the earth changes. Some changes are due to slow processes, such as erosion and weathering, and some changes are due to rapid processes, such as landslides, volcanic eruptions, and earthquakes.
- Weather changes from day to day and over the seasons. Weather can be described by measurable quantities, such as temperature, wind direction and speed, and precipitation.
- Objects in the sky have patterns of movement. The sun, for example, appears to move across the sky in the same way every day, but its path changes slowly over the seasons. The moon moves across the sky on a daily basis much like the sun. The observable shape of the moon changes from day to day in a cycle that lasts about a month.

Landform Puzzle Activity

- Introduce your study of Earth and Space Science by showing Flipbook page 27. Discuss the image and have children point to the landforms that they know or have seen.
- Place the floor puzzle pieces out for children and invite them to construct the puzzle, using the Flipbook image as a guide. Ask questions such as: **Where is the tallest land? What do you see growing where the land is flat? What could people do here? What do you notice about the weather in this image?**
- **Revisit** Encourage children to build, rebuild, and explore the floor puzzle over the course of your study of land and weather in Units C and D.

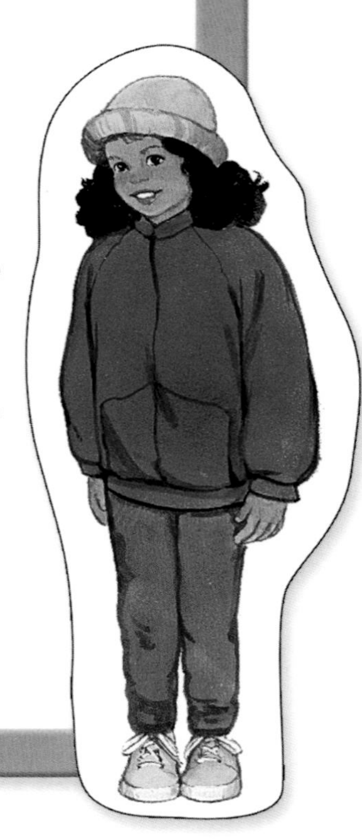

Our Earth, Our Home

What is on Earth?

Essential Questions

Lesson 1 **What uses soil?**

Lesson 2 **How can you describe rocks?**

Lesson 3 **How can land be different?**

Lesson 4 **What are different bodies of water?**

Lesson 5 **What resources from Earth do we use?**

Lesson 6 **How can we take care of Earth?**

Visit www.macmillanmh.com for online resources.

More to Read

Science Leveled Readers

All included in the Leveled Readers Literature Big Book.

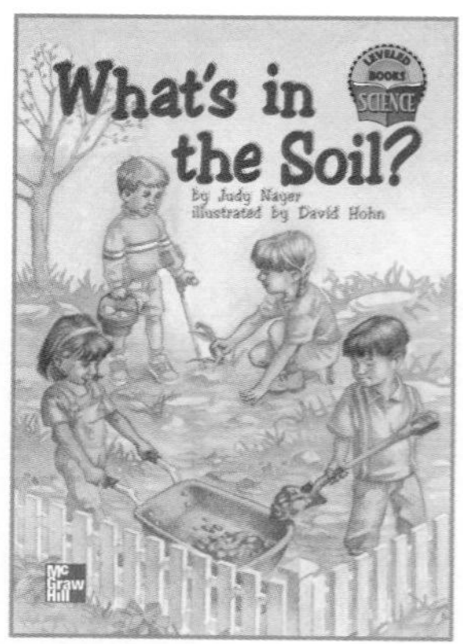

APPROACHING

What's in the Soil?
A group of children have lots of fun making discoveries while working in a garden.

ISBN: 978-0-02-278970-1

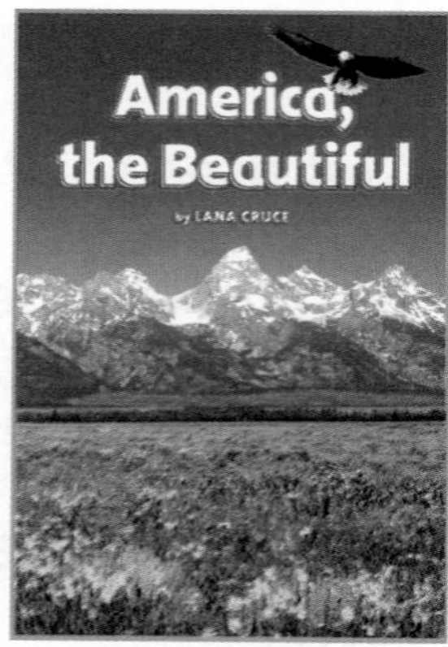

ON LEVEL

America, the Beautiful
Follow an eagle on its journey over various landforms in America.

ISBN: 978-0-02-284592-6

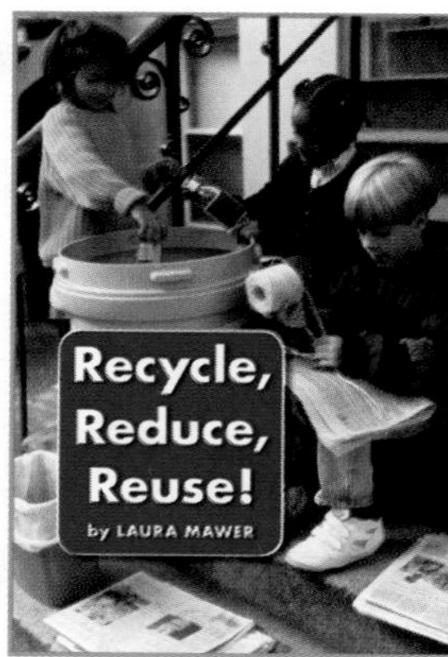

BEYOND

Recycle, Reduce, Reuse!
Children can help take care of our Earth by recycling, reusing, and reducing what we use.

ISBN: 978-0-02-284594-0

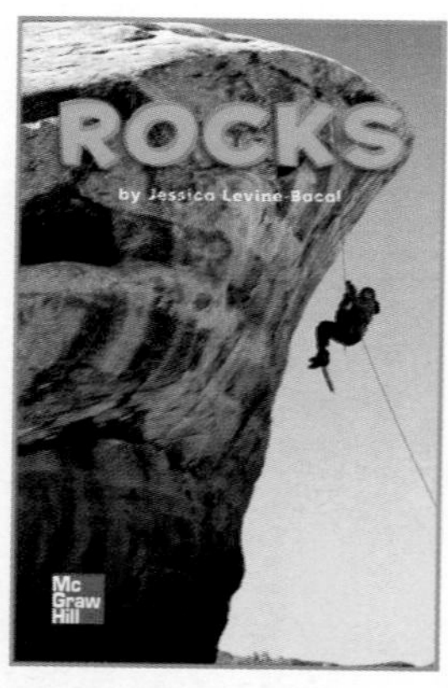

APPROACHING

Rocks
There are many different types of rocks that come in all shapes, colors, textures, and sizes.

ISBN: 978-0-02-281080-1

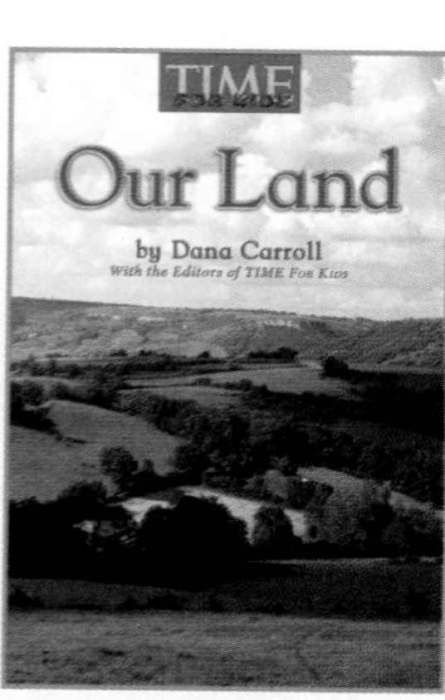

ON LEVEL

Our Land
Different types of landforms and bodies of water have special characteristics.

ISBN: 978-0-02-284593-3

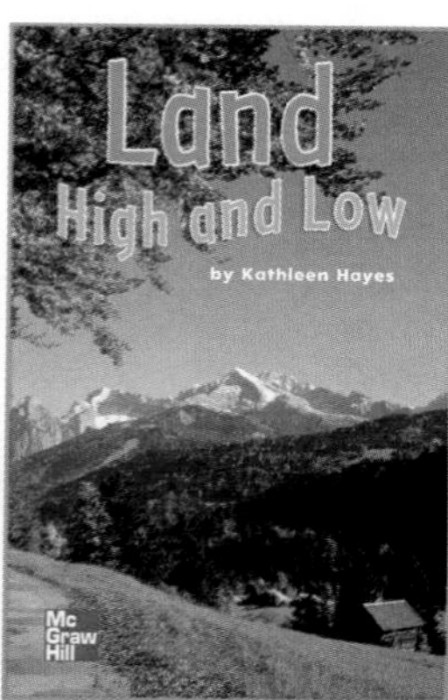

BEYOND

Land High and Low
Earth's landforms provide opportunities for people to work, play, and explore!

ISBN: 978-0-02-281081-8

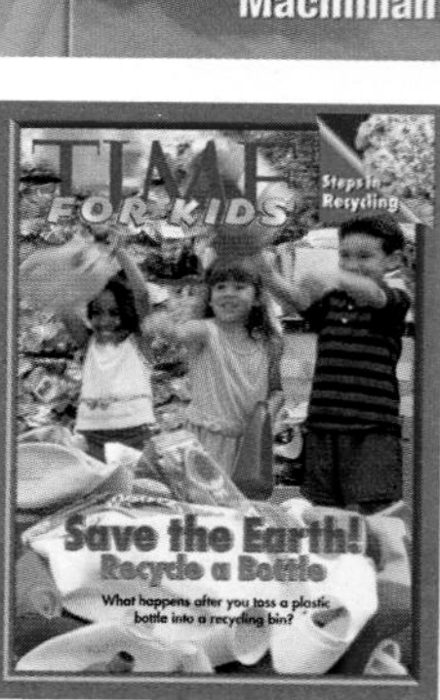

Includes *Time for Kids* magazine about the process of recycling plastic bottles.

Bibliography

Lesson 1 Soil Under Your Feet

Down to Earth, by Melissa Stewart
(Compass Point Books, 2004)

A Handful of Dirt, by Raymond Bial
(Walker & Company, 2001)

Sand, by Ellen J. Prager
(National Geographic, 2006)

Lesson 2 Rocks

If You Find a Rock, by Peggy Christian
(Harcourt Children's Books, 2000)

Rocks: Hard, Soft, Smooth, and Rough,
by Natalie M. Rosinsky
(Picture Window Books, 2003)

Bank Street For more information on these or other books, visit **www.bankstreetbooks.com** or your local library.

Lesson 3 Land High and Low

The Amazing Pop-Up Geography Book,
by Kate Petty
(Dutton Juvenile, 2000)

Danger! Earthquakes, by Seymour Simon
(SeaStar, 2002)

How Mountains Are Made,
by Katherine Weidner Zoehfeld
(HarperCollins Children's Books, 1995)

What's Up, What's Down?, by Lola M. Schaefer
(Greenwillow, 2002)

Lesson 4 Water All Around

The Floating House, by Scott Russell Sanders
(Aladdin, 1999)

Follow the Water from Brook to Ocean,
by Arthur Dorros
(HarperCollins, 1991)

Hello Ocean, by Pam Muñoz Ryan
(Talewinds, 2001)

Lesson 5 Earth's Resources

Dirt: Jump Into Science, by Steve Tomecek
(National Geographic, 2002)

Why Should I Save Water?, by Jen Green
(Barron's Educational Series, 2005)

Lesson 6 Recycle, Reuse

Where Does the Garbage Go?, by Paul Showers
(HarperCollins, 1994)

Why Should I Recycle?, by Jen Green
(Barron's Educational Series, 2005)

Science Vocabulary

Vocabulary

soil
dirt
earth
rocks
sand
mountains
plains
canyons
river
lake
ocean
recycle
reuse

All About Rocks

Encourage children to recognize the similarities and differences between rocks and sand by creating a Venn diagram.

- Have children identify the characteristics of rocks and write them inside the left circle. Invite children to brainstorm characteristics of sand and write their ideas in the right circle.
- Encourage children to think about the characteristics that rocks and sand have in common. Where the circles overlap, write the characteristics children identified. Repeat this routine with mountains/plains and river/lake.
- Use the routine on the **Vocabulary Cards** for further vocabulary development.

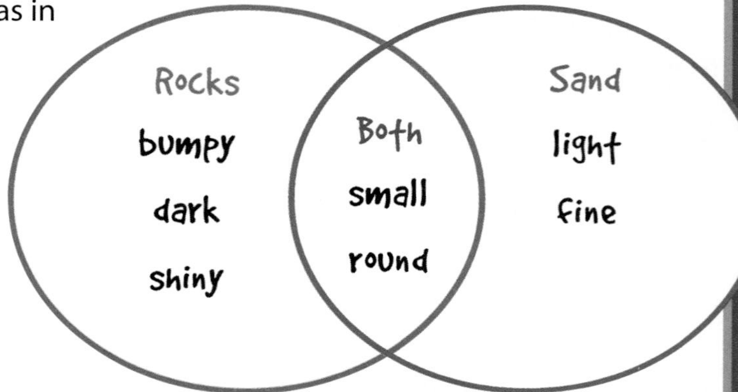

ELL Support

Land and Water

Write the vocabulary words on chart paper and ask children to repeat the words as you read them. Read them again and invite children to clap when they hear words beginning with /r/.

- **Beginning** Use Photo Sorting Cards 21–30 and pictures from magazines to reinforce the vocabulary words. Show children a picture and say the word. Invite children to repeat after you.
- **Intermediate** Show children pictures and ask yes/no questions, such as: **Is this a canyon? Is this an ocean? Is this soil?** Continue using different vocabulary words.
- **Advanced** Show children a picture and say the word. Invite them to say the word and use pantomime to act it out.

A to Z Activity Book

Look for additional cross-curricular activities about our Earth in the A to Z Activity Book.

D is for Digging in Dirt, pp. 8–9
Reading, Math, Cooking, Science

R is for Rocks, pp. 36–37
Science, Math, Art, Cooking

W is for Water, pp. 46–47
Reading, Science, Math

Unit Project

Rock Museum

Objective Make a rock museum.

You need
- rocks
- paper
- crayons
- pencils

Step 1 Send home a letter asking families to help children look for interesting rocks around their homes or in local parks. Bring in a pumice stone (available at drugstores) and other rocks that you can find, or use the rocks from the Science Materials Kit. Make a place to keep the rocks that you and the children collect.

Step 2 Have children work together to make their own categories for the rocks. For example, they might make groups of rocks that are shiny, dark, light, big, small, jagged, or smooth. Help children write or have them dictate how they categorized their rocks.

Step 3 Help children create signs to post around the school, inviting people to their rock museum. When visitors arrive, encourage children to explain how they sorted a particular group of rocks.

Technology

IWB INTERACTIVE WHITEBOARD READY

LOG ON www.macmillanmh.com

Visit Macmillan/McGraw-Hill Science online for projects and activities for students, teachers, and parents.

Bank Street

www.bankstreet.edu

Visit Bank Street online for teacher resources and more activities.

TIME FOR KIDS www.timeforkids.com

Visit Time for Kids online to read about current science news and explore additional science activities.

UNIT C Planner

TeacherWorks™ Plus CD-ROM has an interactive unit planner, Teacher's Edition, and worksheets.

Lesson	OBJECTIVES	VOCABULARY	RESOURCES and TECHNOLOGY	
1 Soil Under Your Feet PAGES 126–131 PACING: 3–5 days	■ Explore the composition and uses of soil.	**soil** **dirt** **earth**	**Flipbook,** p. 29 **Unit C Literature Big Book** **Leveled Reader:** *What's in the Soil?*	■ **Science Resource Book,** p. 40 ■ **A to Z Activity Book,** pp. 8–9 ■ **Science on the Go,** pp. 51–54
2 Rocks PAGES 132–137 PACING: 3–5 days	■ Investigate the characteristics of different rocks.	**rocks** **pebbles** **sand**	**Flipbook,** p. 30 **Unit C Literature Big Book** **Leveled Reader:** *Rocks*	■ **Science Resource Book,** p. 41 ■ **A to Z Activity Book,** pp. 36–37 ■ **Science on the Go,** pp. 55–56
3 Land High and Low PAGES 138–145 PACING: 3–5 days	■ Learn characteristics of geographic features that are high and low.	**mountains** **valleys** **plains** **canyons**	**Flipbook,** pp. 31–32, S6 **Unit C Literature Big Book** **Leveled Readers:** *America, the Beautiful* and *Land High and Low*	■ **Science Resource Book,** p. 42 **Science Songs CD** ■ **Photo Sorting Cards 21–30** ■ **Floor Puzzle:** *Landforms*
4 Water All Around PAGES 146–151 PACING: 3–5 days	■ Learn characteristics of rivers, streams, lakes, and oceans and identify water as a natural resource.	**river** **stream** **lake** **ocean**	**Flipbook,** pp. 33, S7 **Unit C Literature Big Book** **Leveled Reader:** *Our Land*	■ **Science Resource Book,** p. 43 **Science Songs CD** ■ **Photo Sorting Cards 21–30** ■ **A to Z Activity Book** pp. 46–47
5 Earth's Resources PAGES 152–159 PACING: 3–5 days	■ Learn about Earth's natural resources that are used in everyday life and that resources can be conserved.	**resource** **firefighter**	**Flipbook,** pp. 34–35 ■ **Science Resource Book,** p. 44	■ **A to Z Activity Book,** pp. 46–47 ■ **Science on the Go,** pp. 65–68
6 Recycle, Reuse PAGES 160–165 PACING: 3–5 days	■ Learn different reasons for and ways of recycling.	**recycle** **reduce** **reuse**	**Flipbook,** p. 36 **Unit C Literature Big Book:** *Time for Kids* **Leveled Reader:** *Recycle, Reduce, Reuse!*	■ **Science Resource Book,** p. 45 ■ **Science on the Go,** pp. 69–70

PACING Assumes a day is a 20–25 minute session.

LOG ON www.macmillanmh.com for more planning resources and http://nsdl.org/refreshers/science for science resources from NSDL

BE A SCIENTIST Activities

Sampling Soil *p. 130*

PACING: 20 minutes 

Skills observe, investigate, compare, infer

Materials soil, hand lenses, newspaper, large-mesh strainer, fine-mesh strainer

Sorting Rocks *p. 136*

PACING: 25 minutes 

Skills compare, communicate, infer, investigate

Materials classroom rock collection, rocks from outside, hand lenses, crayons

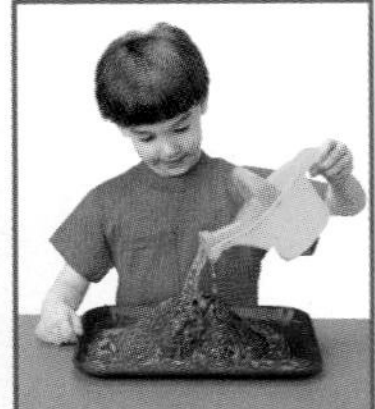

Make It Rain! *p. 144*

PACING: 20 minutes 

Skills make a model, observe, infer, investigate

Materials soil, sand, water, eyedroppers, watering can, cup, container for the mountain (sand table, planter, tray)

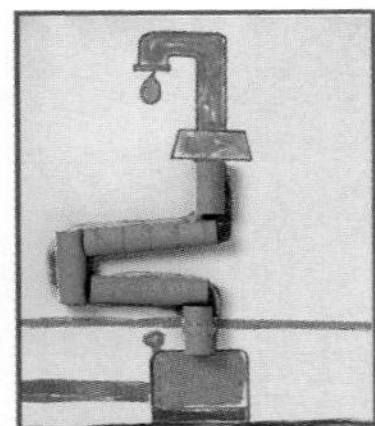

Getting Water *p. 150*

PACING: 30 minutes 

Skills infer, investigate, communicate, make a model

Materials cups, water, chart paper, markers, clipboards, paper, pencils, bulletin board paper, paint, glue, paper towel rolls

Conserve Water *p. 158*

PACING: 30 minutes 

Skills communicate, predict, investigate, compare, draw a conclusion

Materials plastic toy dishes, sink, two large plastic containers, chart paper, markers

Recycling Center *p. 164*

PACING: 20 minutes 

Skills communicate, classify, investigate, infer

Materials yogurt and coffee containers, egg cartons, boxes, newspaper, magazines, paint, scissors, glue, markers, pencils

UNIT ASSESSMENT

Formative Assessment, pp. 129, 135, 143, 149, 157, 163
Performance Assessment, p. 166
Summative Assessment, p. 167
Portfolio Assessment, p. 167

SCIENCE KIT MATERIALS

- hand lenses
- strainers
- rocks
- balance scale
- modeling clay
- string
- eyedroppers
- watering can
- clear containers

COLLECTIBLES

Begin collecting these items for Unit C:

- paper and plastic bags
- newspaper and old magazines
- paper towel rolls
- empty plastic and glass containers with the recycling symbol
- empty soda cans
- pennies
- yogurt containers
- coffee containers
- egg cartons
- boxes

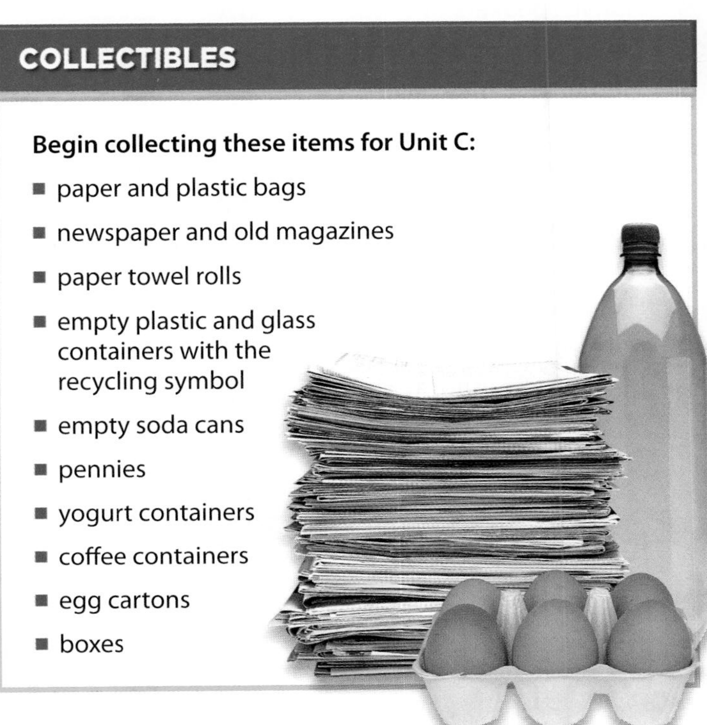

English Language Learner Support

Academic Language

English language learners need help in building their understanding of the academic language used in daily instruction and science activities. The following strategies will help to increase children's language proficiency and comprehension of content and instruction words.

Strategies to Reinforce Academic Language

- **Use Context** Academic Language should be explained in the context of the task. Use gestures, expressions, and visuals to support meaning.
- **Use Visuals** Use charts, transparencies, and graphic organizers to explain key labels that help children understand classroom language.
- **Model** Use academic language as you demonstrate the task to help children understand instruction.

Technology

For additional language support and oral vocabulary development, go to www.macmillanmh.com

LOG ON Vocabulary Games

Academic Language Vocabulary Chart

The following chart shows unit vocabulary and inquiry skills as well as some Spanish cognates. **Vocabulary** words help children comprehend the main ideas. **Inquiry Skills** help children develop questions and perform investigations. **Cognates** are words that are similar in English and Spanish.

Vocabulary		Inquiry Skills	Cognates	
			English	**Spanish**
soil, p. 126	river, p. 148	observe, p. 130	observe, p. 130	*observar*
dirt, p. 126	stream, p. 148	investigate, p. 130	investigate, p. 130	*investigar*
earth, p. 128	lake, p. 148	compare, p. 130	compare, p. 130	*comparar*
rocks, p. 132	ocean, p. 148	infer, p. 130	infer, p. 130	*inferir*
pebbles, p. 134	resource, p. 154	communicate, p. 136	rocks, p. 132	*rocas*
sand, p. 134	firefighter, p. 156	make a model, p. 144	communicate, p.136	*comunicar*
mountains, p. 138	recycle, p. 160	predict, p. 158	mountains, p. 138	*montañas*
valleys, p. 140	reduce, p. 162	draw a conclusion, p. 158	valleys, p. 140	*valles*
plains, p. 140	reuse, p. 162	classify, p. 164	model, p. 144	*modelo*
canyons, p. 140			ocean, p. 148	*océano*
			conclusion, p. 158	*conclusión*
			recycle, p. 160	*reciclar*
			reduce, p. 162	*reducir*
			classify, p. 164	*clasificar*

Language Transfers

Phonics Transfer
Native speakers of Cantonese, Hmong, and Vietnamese may have difficulty with the long vowel sound /ō/.

Grammar Transfer
Native speakers of Cantonese, Haitian Creole, Hmong, Korean, and Vietnamese, may not distinguish between *a/an* and *the*.

Ocean has salt water.

Vocabulary Routine

Use the routine below to discuss the meaning of each word on the vocabulary list. Use gestures and visuals to model all words.

Define An *ocean* is a very large and deep body of salt water.

Example Sharks live in the *ocean*.

Ask What other animals live in the *ocean*?

Children may respond to questions according to proficiency level with gestures, one-word answers, or phrases.

Vocabulary Activities

Help children learn what an ocean is.

BEGINNING Show a globe or world map. Explain that it shows the parts of Earth that are land and the parts that are water. Show that each large body of water is an ocean. Find the word *ocean* and point to it. Read the word and have children repeat.

INTERMEDIATE Show Sorting Card 28. Say: *An ocean is a large part of Earth's water. An ocean can look like this.* Pass the card around. As children look at it, have them describe oceans. Say: *The ocean water moves. It comes and goes.* Have children use gestures to show how ocean water moves.

ADVANCED Show Flipbook page 33. As you talk about the picture of the ocean, elicit from children that rocks are part of Earth's land, and the ocean is part of Earth's water. Point to the photograph of the river. Ask: *Is this an ocean?* Encourage children to describe how oceans are different from rivers.

Objective: Children will discuss different kinds of natural environments.

Before Reading

- Have children share what they know about the places shown on the Flipbook page.
- Explain that in this unit, they will learn about different kinds of environments on our Earth.

During Reading

- Read the first verse of the poem, pointing to the words as you read.
- Invite children to identify which picture shows a mountain, the ocean, the desert, or the plains.
- Read the second verse, and invite children to point to which places on the page show snow, water, warm temperatures, dry-looking ground, or a place to grow food. Reread the poem two to three times, encouraging children to chime in.

Flipbook

UNIT C

Our Earth, Our Home

Would you like to live in
the mountains?
Would you like to live by
the sea?
Would you like to live in
the desert,
Or live on a prairie with me?

We could play in the icy snow,
Or let water tickle our feet.
We could run on the warm,
dry ground,
Or grow food for us all to eat!

by Jessica Bacal

Reading Strategy

Conventions of Print Have children help you count the number of times the phrase *we could* appears in the poem.

Phonemic Awareness Have children find a word in the first stanza that rhymes with *could* (*would*). Have them find a word in the last line that rhymes with *we* (*me*). Encourage them to list other words that rhyme with *sea* and *me* and with *feet* and *eat*.

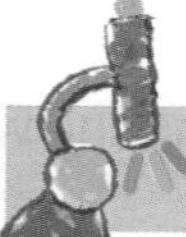

Science Facts

Like plants and animals, people adapt to their environments.

Plains Farmers came to the plains and stayed because of the rich soil that was good for planting food. Indiana ranks fourth in the United States in the production of corn grain. Corn is also used to make many products, such as shoestrings, glue, ketchup, and toothpaste.

Mountains In snowy mountain areas, people build homes with steep-pitched roofs so that snow slides off instead of building up. Glacier National Park in Montana has many types of land, including grasslands, forests, alpine tundra, wildflower fields, and towering mountains. Millions of years of glacial activity carved out the landscape.

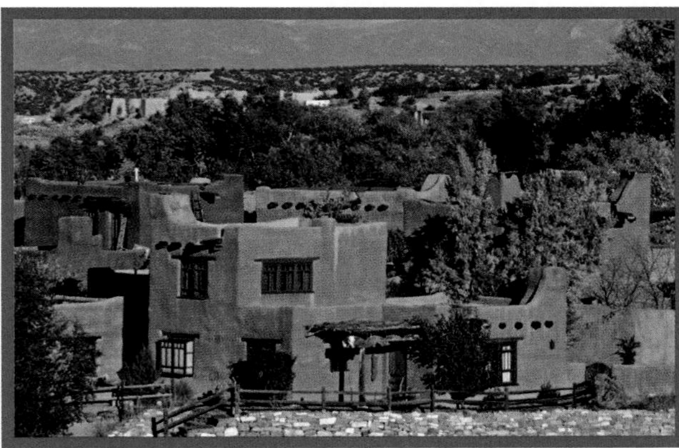

28

After Reading

- Write *Same* and *Different* at the top of a piece of chart paper.
- Have children help you decide what is the same about these homes and what is different. Record their responses.
- Ask: **Which place looks like it would be good for growing food? Why?** The farm land. It is very green and flat.

Write on It

Have volunteers use a dry-erase marker to draw an *x* on a plant or flower growing near each home. Ask: **What kinds of plants might you find at the ocean?**

▶ Revisit

Before rereading the poem, ask children to pantomime each of the activities in the second verse. Encourage children to act out each of these as you read the second verse slowly.

Science Background For more information on our Earth go to www.macmillanmh.com.

Ocean People who live near the ocean often build their homes on stilts so they are safe from flood damage in a storm. South Carolina, for example, is bordered on the east by beaches that spread along the Atlantic Ocean. Many people build homes there.

Desert The first settlers in the southwestern desert built homes with soil, because there were no forests and very few rocks. Adobe (20% clay and 80% sand) is mixed with water, poured into molds, and allowed to dry. Deserts receive less than 25 centimeters of precipitation annually, and as a result, the nutrients in the soil are abundant and the soil is rich in minerals.

Professional Development For more Science Facts and resources from NSDL visit http://nsdl.org/refreshers/science

School to Home

At the beginning of Unit C, give children copies of the Home Activity Letter on page 9 of the **Science Resource Book**. Read the letter aloud and help children write their names. Have children bring the letter home to share with their families.

At the end of the unit, send children home with a copy of the Home Letter on page 11.

See pages 10 and 12 in the Science Resource Book for Spanish versions of these letters.

▶ **Essential Question**
What uses soil?

▶ **Objective**
Explore the composition and uses of soil.

▶ **Vocabulary**
soil
dirt
earth

Resources

Flipbook, p. 29

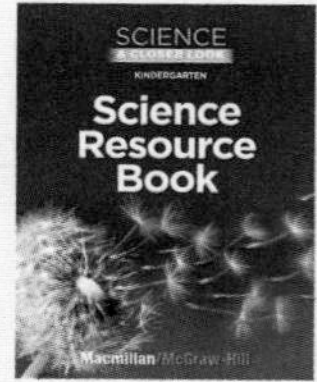
Science Resource Book, p. 40

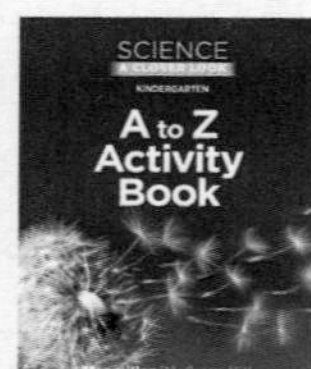
A to Z Activity Book, pp. 8–9

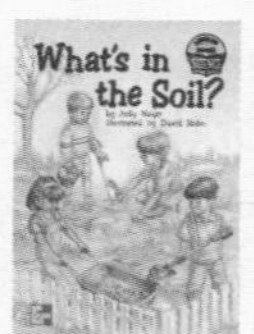
Leveled Reader: *What's in the Soil?*

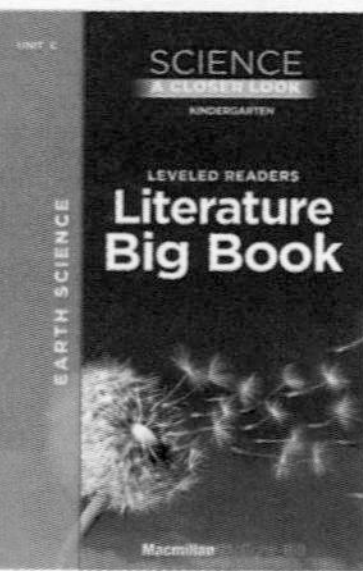
Unit C Literature Big Book

Technology

Soil Under Your Feet

Circle Time
WHOLE CLASS

See Through to Soil

You need
- resealable bags
- soil
- chart paper
- marker

Inquiry Skills: observe, communicate, infer

- Prepare bags of soil. Distribute a bag to each child or to each pair of children. Have them examine the soil without opening the bag.
- Ask children to describe the soil, discuss where we usually find soil, and discuss what it is used for. Record their responses on chart paper.

Infer Ask: **What do you think soil is made of?**
Possible answers: rocks, sand, dirt, twigs, leaves

Time to Move!

Have children move like creatures that live in the soil. Invite children to wiggle like a worm, crawl very slowly like a snail, and march like an ant.

Be a Reader Activity

You need
- Unit C Literature Big Book
- Leveled Reader: *What's in the Soil?*

Read the Big Book

Objective: Use illustrations to get information about a story.

Inquiry Skills: infer, communicate

- Show children the title page of the story. Ask: **What are the children doing? What might they find in the soil?**
- Take a picture walk, inviting children to identify what is found on each page. Read the book aloud. Then read it again, pointing to each word as you read and encouraging children to read with you.
- **Guided Reading:** Use the Leveled Reader version of *What's in the Soil?* for small-group guided-reading instruction.

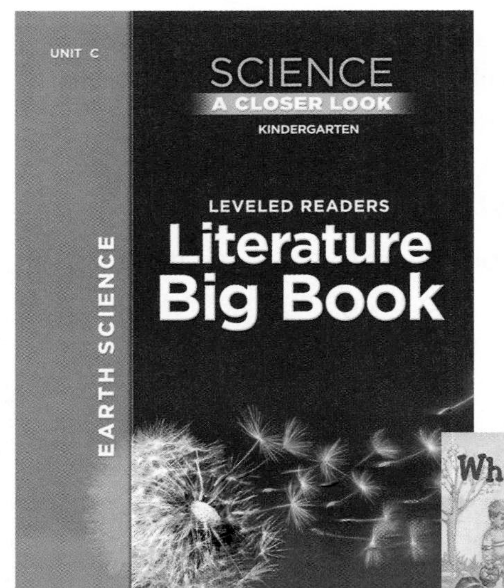

Be a Math Wiz Activity

You need
- small clear containers
- a scoop or small paper cup
- soil
- large bowl
- chart paper

Fill the Jar

Objective: Estimate how many scoops of soil will fill a jar.

Inquiry Skills: predict, investigate, compare

- Have children estimate how many scoops of soil it will take to fill up a small container. Record their estimates.
- Work together to fill and count the number of scoops that go into the container. Record the actual number.
- Ask: **How does the actual number compare to your estimates? If you pack the dirt more tightly, will the jar hold more?** As an extension, pack the dirt more tightly and use smaller or larger containers. Discuss how children's estimates will change.

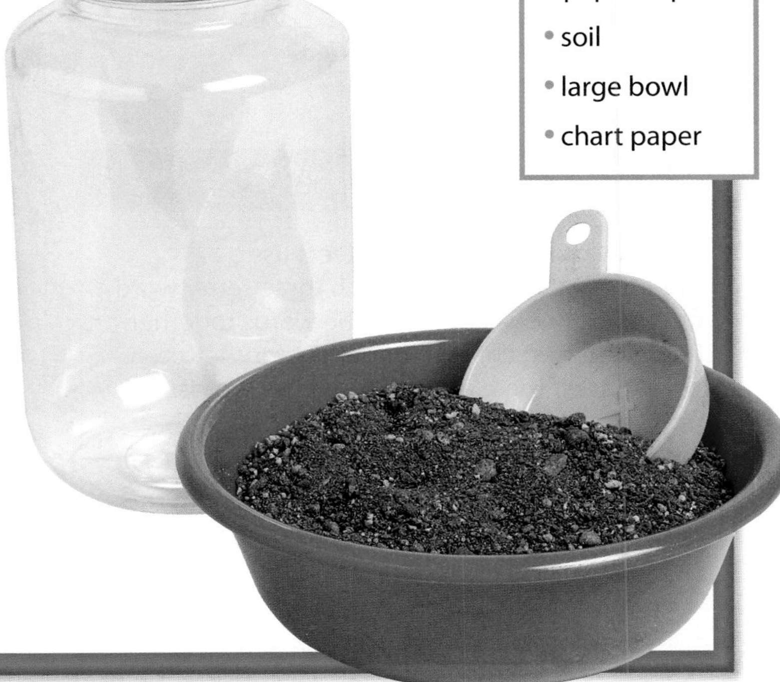

Read Together and Learn

What can you find in soil?

▶ Build on Prior Knowledge

Ask children to talk about times when they have planted things in soil, or dug in soil. Ask: **What did it feel like? Smell like? What did you see?**

▶ Use the Visuals

- **Observe** Invite volunteers to share what they see in the picture. Encourage children to find living and nonliving things in the soil. Ask questions such as: **What is on top of the soil?** grass, pieces of sticks and leaves **What is underneath and in the soil?** rocks, worms
- **Communicate** Help the group determine what else might be found in soil. Possible answers: beetles, spiders, flowers, weeds, carrots

▶ Develop Vocabulary

soil, **dirt**, **earth** As children discuss what might be found in soil, explain that *dirt* and *earth* can both mean *soil*. Use a dry-erase marker to write these words below the word *soil*. Model using these words in sentences. Then invite volunteers to use them in sentences.

Flipbook

UNIT C • LESSON 1

What can you find in soil?

Reading Strategy

Print Awareness Have volunteers use a dry-erase marker to circle each three-letter word on the Flipbook page. Read the words together. Repeat, having children underline four-letter words. Ask if there are any two-letter words.

Phonological Awareness Write the word *in* on chart paper. Help children read it. Help them create new words by adding a consonant to the beginning of the word *(bin, fin, win, tin, pin)*.

Science Facts

Soil Composition Soil is made up of sand, silt, and clay, along with humus, organic material from decomposing vegetation, roots, and animals. Animals like ants and earthworms burrow through the soil, mixing the humus with the sand, silt, and clay. Even smaller animals, called decomposers, eat and digest the material in the humus, helping to break it down so its nutrients dissolve into the soil. Water also helps nutrients dissolve. The soil that is best for growing things is called mollisol. It is thick and dark, and contains an equal mixture of sand, clay, and humus. About half of its volume is air spaces, which sometimes fill with water.

LOG ON **Professional Development** For more Science Facts and resources from  visit http://nsdl.org/refreshers/science

▶ Ask Questions

- **Why might soil be a good place for worms to live?** Possible answers: They can find food; they can hide; they can get water when it rains.
- **How do some living things change the soil?** Worms make tunnels, dogs dig holes, and the roots of plants dig into it and spread out.
- **Plants are growing in this soil. What do you think allows them to grow?** Their roots have space to grow down; the soil does not have too many rocks.
- **What might happen to this soil if it started to rain?** Leaves and other things will wash into the soil and break it into pieces; it will make mud.

Invite children to use a dry-erase marker to draw arrows pointing to any animals in the soil. Have them draw an *x* over rocks and circle roots.

Think and Talk

Review the picture on the Flipbook page. Ask children: **What if we wanted to plant carrots in this soil? What would we find as we begin to dig?** Invite children to share what they have learned about soil.

If children have difficulty understanding what soil is made of, invite them to explore a pile of soil with a hand lens in the Science Center.

ELL Support

Ask children to name items they saw in the book *What's in the Soil?* as you list them. Read the book again.

Beginning Name items found in the soil and have children point to them.

Intermediate Point to various items in the soil and ask children to name them.

More to Read

Down to Earth,
by Melissa Stewart
(Compass Point Books, 2004)

Activity Read this book with children to reinforce their understanding of soil composition. Have children draw a picture of something they might find in soil and write words to describe their drawings.

Sampling Soil

Objective: Explore the contents of different kinds of soil.

Inquiry Skills: observe, investigate, compare, infer

You need **soil, hand lenses, newspaper, large-mesh strainer, fine-mesh strainer, Science Journal p. 40**

1. **Observe** On the first day, spread soil onto newspaper and have children examine it with hand lenses. Ask: **Are there rocks and leaves? How did they get there?** Have children draw what they see on their Science Journal page.
2. **Investigate** On the second day, help children put the soil through a large-mesh strainer so only the large particles are left. Put these on a paper plate. Then put the soil that has fallen through into a fine-mesh strainer. Separate small and medium particles on paper plates. Have children observe, touch, and describe what is on each plate. Remind children to always wash their hands after touching the soil.
3. **Compare** On the third day, give children a chance to compare the contents of their soil with the contents of another group's soil.
4. **Infer** Ask: **Which soil might be better for growing things? Why?**

Teacher Tip

It is ideal to collect soil from different areas around your school. You can also take the group to a local park for soil samples, or have children bring soil from home in resealable bags. If this is not possible, you can purchase soil from a local gardening store.

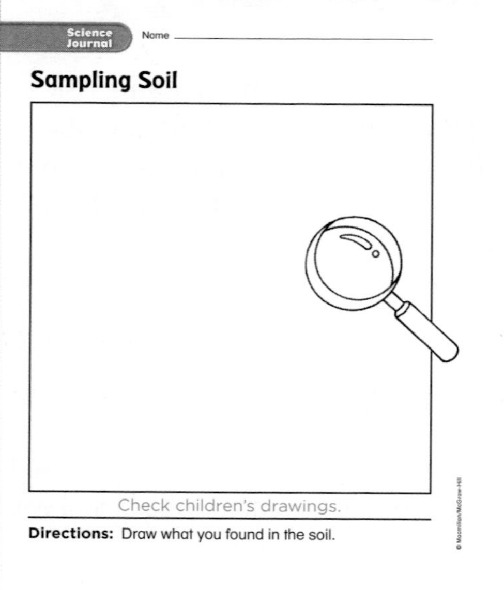
Science Journal Name

Sampling Soil

Check children's drawings.

Directions: Draw what you found in the soil.

40 Unit C • Our Earth, Our Home Science Resource Book — Use with Lesson 1 Be a Scientist Activity

Science Resource Book, p. 40

Investigate More

Investigate Have children plant presoaked lima beans in the different kinds of soil. Ask: **Which is the best kind of soil for planting? Why do you think so?**

centers

Sand Table

You need
- soil
- water

Wet and Dry

Objective: Explore adding water to soil.

Inquiry Skills: predict, investigate

- At Circle Time, show children the soil that you put in the Sand Table (instead of sand). Ask: **What do you think might happen to soil when we add water?**
- Encourage children to explore what happens to the soil when they add a small amount of water.
- Ask: **How does the soil change? What can you do now that you could not do when the soil was dry?**

Technology

More About Our Earth

- For games and activities about our Earth, help children log onto www.macmillanmh.com.
- Children can use these games throughout this unit to reinforce what they learn about Earth. Encourage children to revisit these games throughout the year to review the concepts.

www.macmillanmh.com for more science online.

Art

You need
- root vegetables with some dirt still on them
- paper
- crayons

Beautiful Roots

Objective: Draw root vegetables.

Inquiry Skills: observe, make a model

- Show children the assortment of root vegetables that you are placing in the Art Center. Invite them to describe each vegetable's color and shape.
- Place vegetables in the Art Center, and invite children to use crayons or colored pencils to draw one or more vegetables.

Cooking

You need
- *The Carrot Seed* by Ruth Krauss
- clear container
- soil
- large pot
- blender
- chicken broth
- orange juice
- 5 carrots
- salt and pepper
- cream

Carrot Soup

Objective: Cook using produce grown in the soil.

Inquiry Skill: investigate

- Read the book *The Carrot Seed*. Give children a chance to see how carrots grow under the soil by burying a carrot in a clear container of soil and inviting them to take turns pulling it out.
- Then invite children to help you make carrot soup during Center Time.

Carrot Soup

Peel and slice 5 carrots.
Put in a pot.
Add 1 1/4 cups chicken broth.
Add 1 1/4 cups of orange juice.
Cover and cook until the carrots are soft. Let cool. Blend it in the blender. Add 1/2 cup cream. Reheat, serve, and enjoy!

▶ **Essential Question**
How can you describe rocks?

▶ **Objective**
Investigate the characteristics of different rocks.

▶ **Vocabulary**
rocks sand
pebbles

Resources

Flipbook, p. 30

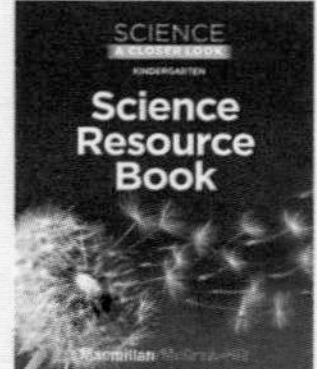
Science Resource Book, p. 41

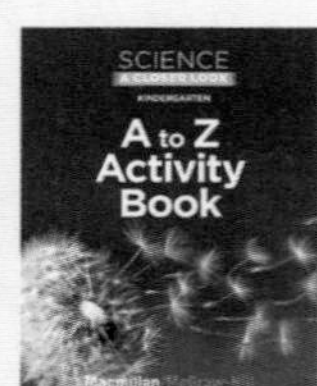
A to Z Activity Book, pp. 36–37

Leveled Reader: *Rocks*

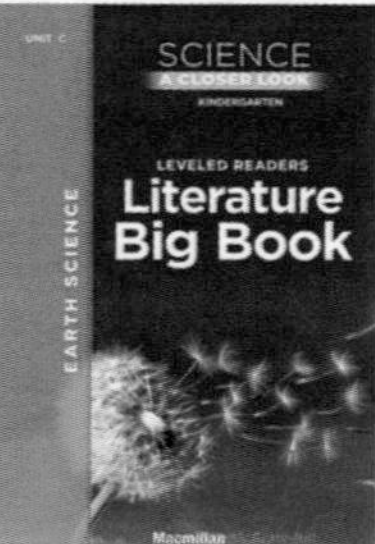
Unit C Literature Big Book

Technology

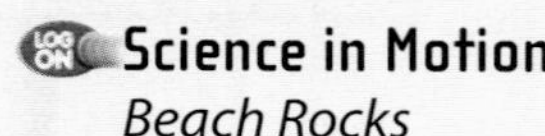
Science in Motion
Beach Rocks

www.macmillanmh.com

Rocks

Circle Time — WHOLE CLASS

My Rock

You need
- different types of rocks
- chart paper
- marker

Inquiry Skills: observe, communicate, compare

- When children gather in the circle, give one rock to each of them. Have them think of at least one word that describes their rock.
- Go around the circle and record what children say on chart paper as they describe their rocks.

Compare Have children compare their rock with a classmate's. Ask: **How are your rocks alike? Different?**

Time to Move!

"Rock and roll" with the children! Fill an empty plastic water bottle with small rocks. Shake the bottle and invite children to move and shake to the sound of the rocks and then freeze in place when the sound stops.

Be a Reader Activity

Read the Big Book

Objective: Read a book about rocks to discover different rock characteristics.

Inquiry Skills: classify, infer, communicate

- Show children the cover of the story and ask them to share what they notice. Take a picture walk, inviting children to chime in with you as you read.
- Read the story a second time, stopping before the last word of each sentence to let children say the ending. On the last page, invite children to share what they notice about the rocks.
- **Guided Reading:** See the inside back cover of the Leveled Reader version of *Rocks* for more activities.

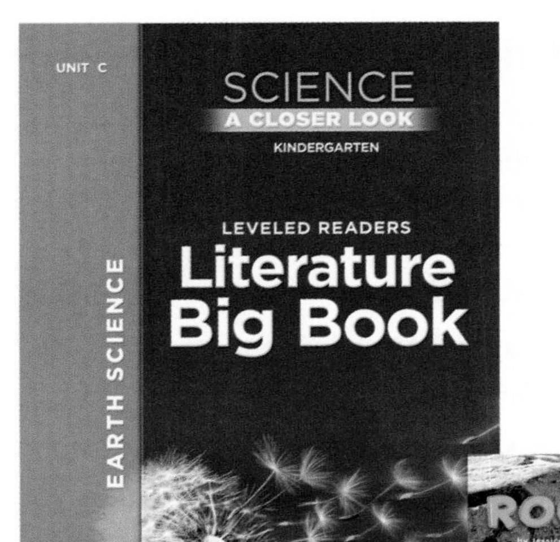

You need

- Unit C Literature Big Book
- Leveled Reader: *Rocks*

Be a Math Wiz Activity

Rock Weight

Objective: Sort rocks by weight.

Inquiry Skills: predict, measure, compare

- Show children an assortment of rocks and a classroom stapler. Have them predict, without touching the rocks, which weighs more or less than the stapler.
- Have children take turns using the scale to test their predictions, sorting rocks into three sorting rings: one for rocks that weigh more, one for rocks that weigh less, and one for rocks that weigh about the same.
- Ask: **Why do you think some rocks weigh more or less?** Possible answer: They are bigger. They are smaller. Have children describe how they used the scale as a tool to make comparisons.

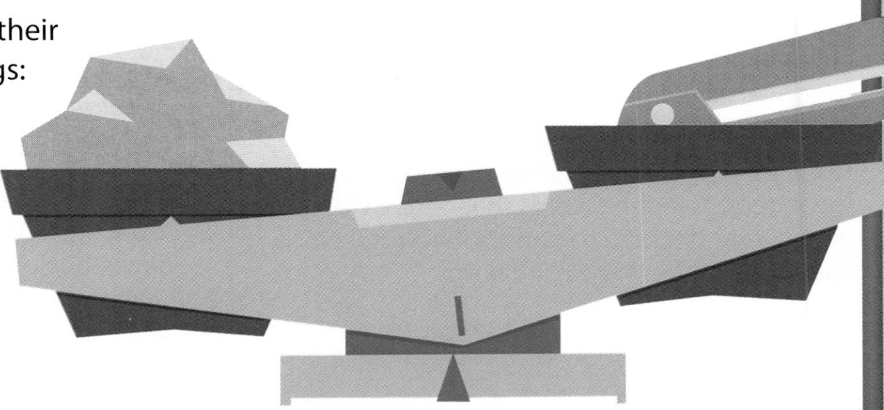

You need

- balance scale
- stapler (or other classroom object)
- rocks
- sorting rings

Read Together and Learn

How are these rocks alike? How are they different?

▶ Build on Prior Knowledge

Ask: **Which of these rocks looks like one you could find if you were digging in soil near the school? Which rock might be hard to find?**

▶ Use the Visuals

- Communicate Encourage children to say whether any of the rocks look familiar.
- Invite volunteers to describe the rocks. Encourage them to use descriptive words such as *shiny, smooth, rough, bumpy, clear, round,* and *square.*
- Help the group read the rock names.

▶ Develop Vocabulary

rocks, **pebbles**, **sand** Write *rocks, pebbles,* and *sand* on chart paper. Ask children to talk about the differences. Draw a large circle to show that we usually mean something larger when we say *rock*. Draw a small circle to show *pebble,* and a dot for *sand.*

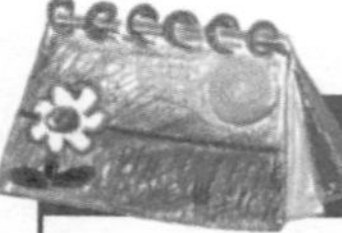

Reading Strategy

Print Awareness Write the word *slate*. Help children find other words in the word *slate* (*slat, late, at, ate*). List the words they find.

Phonics Have children use a dry-erase marker to circle all of the rock names that begin with the letter *g*. Read the words that begin with *g*, and discuss the hard *g* sound. **What letters look like *g* but are not?** The *q* in *quartz*.

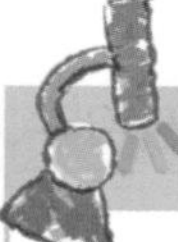

Science Facts

Marble, Slate, and Graphite are examples of metamorphic rocks. They are rocks that have changed in mineral composition after being exposed to heat and pressure. These rocks change from one type to another: limestone to marble, shale to slate, and carbon to graphite.

Granite is an example of an igneous rock—rock made from molten lava from volcanoes. Used for building and in monuments, granite contains several minerals.

Quartz is one of Earth's most common minerals. It is found in several igneous, sedimentary, and metamorphic rocks. This durable mineral is used in jewelry and in home design, and aids in several industrial processes, including glass making and sand blasting.

Professional Development For more Science Facts and resources from NSDL visit http://nsdl.org/refreshers/science

granite

quartz

arble

30

▶ Ask Questions

- **How can we sort these rocks?** Have a volunteer circle rocks that are similar and describe how they are alike. **How are the rocks that are left similar?** Have a volunteer describe how they are alike.
- **Which rocks might be used as part of a building?** marble, granite **Why?** They look solid and hard. **Which rocks do you think would not be used as part of a building?** slate, graphite, quartz **Why?** They do not look as strong.

Write on It

Invite children to use a dry-erase marker to write the first letter of their names next to their favorite rock. Have them talk about why it is their favorite.

Think and Talk

Review the images on the Flipbook page. Ask: **How would you sort these rocks?** Encourage children to share what they have learned about rocks.

Formative ASSESSMENT

By asking children to discuss similarities and differences between rocks, you will learn what they understand. In addition, children will learn as they listen to each other.

Differentiated Instruction

Extra Support Write *smooth* on four sticky notes. Write *rough* on four sticky notes. Help children read the words and talk about what they mean. Have children take turns placing the sticky notes on examples of rough and smooth rocks on the Flipbook page. Repeat with *big, small, shiny,* and *dull.*

More to Read

If You Find a Rock,
by Peggy Christian
(Harcourt, 2000)

Activity Read this book, then have children talk about what they could do with a big rock and a little rock. Have children draw a picture of what they could do. Ask children to write about it or have them dictate their ideas to you.

 Inquiry Investigation

Sorting Rocks

Objective: Sort rocks into two categories.

Inquiry Skills: compare, communicate, infer, investigate

You need **classroom rock collection, rocks from outside, hand lenses, crayons, Science Journal p. 41**

1. **Compare** Have children work in pairs to sort rocks into two categories. Give each pair 6 to 10 rocks. If they need help, you can ask them about the color, shininess, shape, size, or weight of their rocks. Have children draw one of the rocks from each category on their Science Journal page.
2. **Communicate** Invite children to explain how they sorted their rocks. Have them share if their two categories have anything in common.
3. **Infer** Tell children about the categories scientists use for sorting rocks. Show an example of an igneous rock, a sedimentary rock, and a metamorphic rock. Have children find a rock that they think fits one of the categories, and explain why it fits. Possible answer: I think this is igneous because igneous rocks are black and shiny like this one.

Teacher Tip

Provide nonfiction books about rocks for children to use as references. For example, **Let's Go Rock Collecting** by Roma Gans (HarperCollins, 1997), describes the scientific classifications of rocks and how children can recognize them.

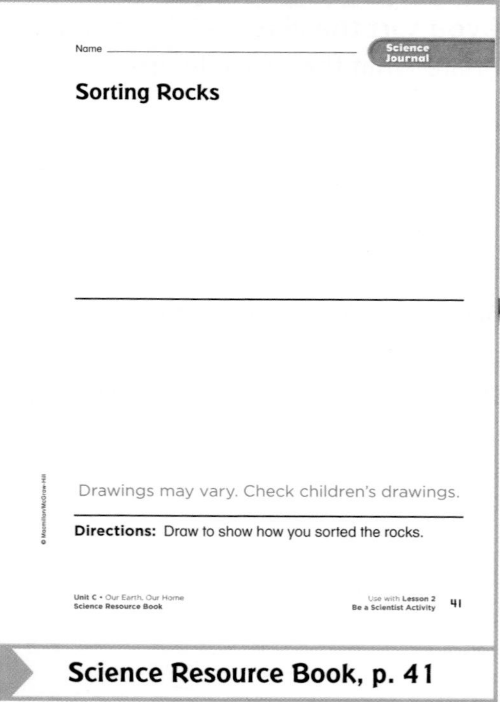
Name ________________ Science Journal

Sorting Rocks

Drawings may vary. Check children's drawings.

Directions: Draw to show how you sorted the rocks.

Unit C • Our Earth, Our Home
Science Resource Book

Use with Lesson 2
Be a Scientist Activity 41

Science Resource Book, p. 41

Investigate More

Investigate Have children try sorting their rocks into three categories instead of just two. Ask: **What about your rocks is the same? Different?**

Games

You need
- Photo Sorting Cards 1–30

Find the Rock

Objective: Match rocks by name and characteristics.

Inquiry Skills: communicate, classify

- Show children Photo Sorting Cards 1–30 (Units A–C). Explain that in this game, children will divide the pile of cards into two sets, one for each player.
- Each player takes a turn flipping over a card from his or her pile and placing the card faceup into a center pile (at a fairly quick pace).
- Each time a "rock" card is flipped onto the center pile, the first child to cover the card with his or her hand gets to take the entire center pile. Play continues until one player has no cards left.

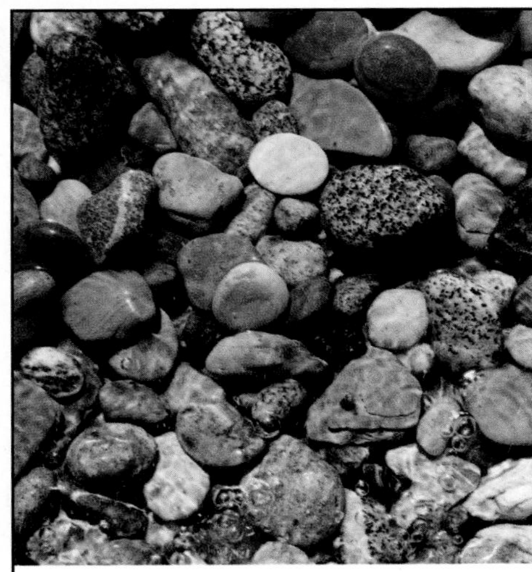

rocks

© Macmillan/McGraw-Hill
Photograph by PhotoDisc/Getty Images 21

Drawing and Writing

You need
- classroom rock collection
- *Sylvester and the Magic Pebble* by William Steig
- paper
- pencils
- crayons

Rock Stories

Objective: Listen to, read, and write stories about rocks.

Inquiry Skill: communicate

- Bring in a piece of jewelry that has a rock in it (such as turquoise) to show the children. Talk about why it is special to you. Ask if anyone else has a story about a special rock.
- Read aloud *Sylvester and the Magic Pebble*. Afterward ask: **Why did Sylvester pick up the pebble? What looks special about it?**
- Invite children to write a story that involves a special classroom rock or any other rock.

Art

You need
- cardboard
- aquarium pebbles
- small rocks
- sand
- glue

Rock Collages

Objective: Make rock collages.

Inquiry Skill: observe

- Invite children to make rock collages by gluing small pebbles and rocks to cardboard.
- Ask: **How will you sort the rocks on your collage? Will you separate rocks by color, or mix them? Will you separate rocks by shape, or mix them?**

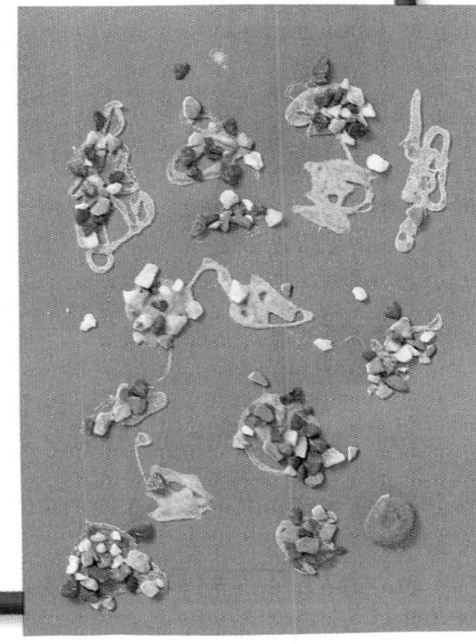

Land High and Low

Essential Question
How can land be different?

Objective
Learn characteristics of geographic features that are high and low.

Vocabulary
mountains plains
valleys canyons

Resources

Flipbook, pp. 31–32, S6

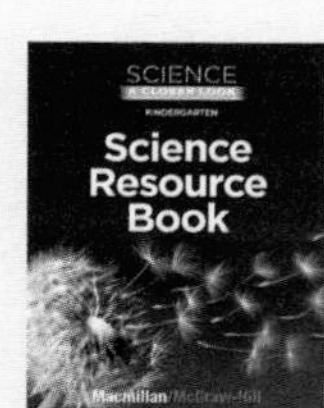
Science Resource Book, p. 42

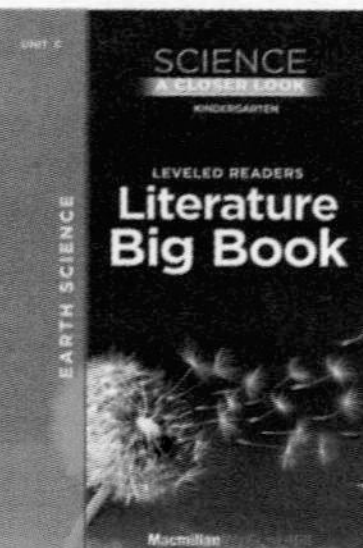
Unit C Literature Big Book

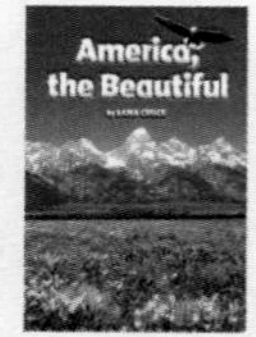
Leveled Reader: *America, the Beautiful*

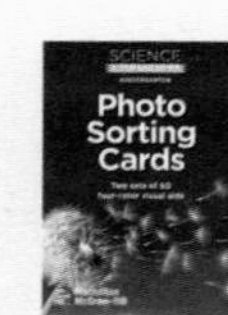
Photo Sorting Cards 21–30

Technology

Science Songs CD Tracks 11–12

www.macmillanmh.com

NSDL

Circle Time — WHOLE CLASS

High Places, Low Places

Inquiry Skills: make a model, communicate

- Explain to children that some mountains are formed when two big pieces of land below Earth's surface push against one another and the pressure causes the land to rise.
- Have children work in pairs, sitting on the floor with their backs touching. Have them push against one another's backs and slowly rise up. Encourage them not to use their hands as they "make a mountain."

Communicate Ask: **How hard did you have to push to make your "mountain?"**

Time to Move!

Have children move in place like they are climbing up a mountain, running through the plains, taking a stroll through a canyon, and skipping through a valley.

Be a Reader Activity

You need
- Unit C Literature Big Book
- Leveled Reader: *America, the Beautiful*

Read the Big Book

Objective: Recognize different types of landforms.

Inquiry Skill: communicate

- Read the title. Have children look at the picture on the cover and ask: **Where is the high land? The low land?**
- Read the book, tracking the print as you read. Reread, inviting children to fill in the last word of each sentence. Ask children if they hear words that rhyme. tree, me
- **Guided Reading:** See the inside back cover of the Leveled Reader version of *America, the Beautiful* for activities.
- For more to read on landforms, read *Land High and Low*, available in the Unit C Literature Big Book and as a Leveled Reader.

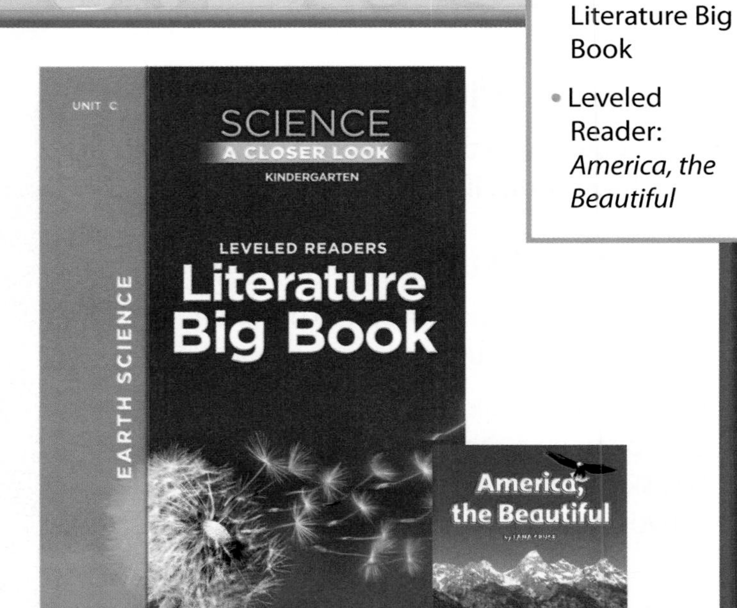

Be a Math Wiz Activity

You need
- modeling clay
- cardboard
- string
- scissors
- connecting cubes
- ruler

'Round the Mountain

Objective: Measure different parts of a model mountain.

Inquiry Skills: predict, make a model, measure

- Have children use clay to make mountains.
- Explain to children that they will use string to measure around their mountains. Ask: **Which part of the mountain will require more string? Which part will require less?**
- Have children measure around the peak, the middle, and the base. Ask: **Which part is the largest? The smallest?** Finally, have children use connecting cubes or a ruler to measure each piece of string.

Read Together and Learn

What makes these places different?

▶ Build on Prior Knowledge

Ask: **We have been talking about rocks. Which of these places looks like it is made of rock?**

▶ Use the Visuals

- Communicate Ask children to describe the places in the pictures.
- Invite them to talk about whether they have ever been on a mountain or a plain. Continue by asking them if they have ever been in a canyon or a valley.
- Observe Ask: **Are there plants growing in each picture? Which has the most plants? Which has the fewest?**

▶ Develop Vocabulary

mountains, **canyons**, **plains**, **valleys** Ask children to use the pictures on the Flipbook page to define each of these words. Write their definitions on chart paper. A *mountain* is land that is high up and is higher than a hill, *plains* are flat land, a *canyon* is rock carved by water and wind, and a *valley* is low land between mountains.

Flipbook

UNIT C • LESSON 3

What makes these places different?

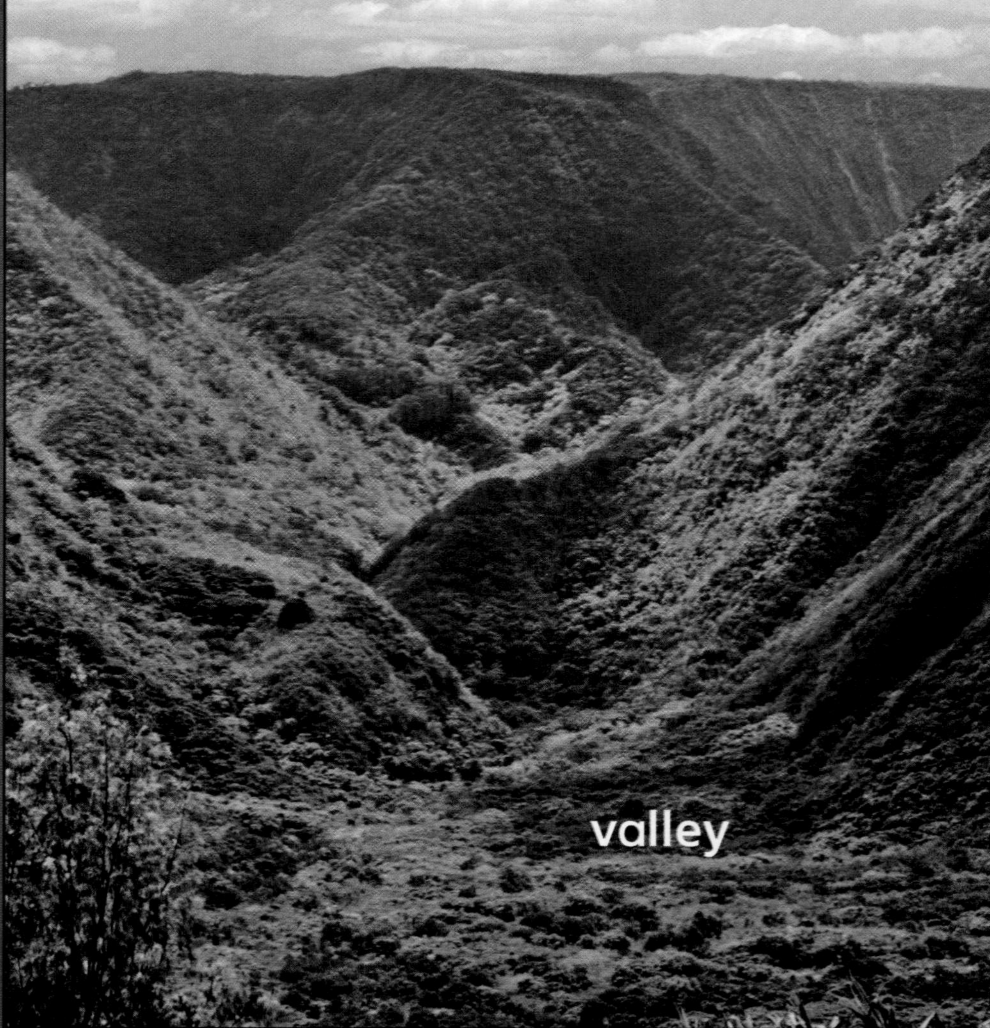

Reading Strategy

Comprehension Write the homonyms *plain* and *plain*. Talk about the definition of *plain* as it applies to land. Ask: **What is the other definition of *plain*?**

One-to-One Correspondence Write *plains, valley, canyon,* and *mountains* on sticky notes and mislabel the pictures. Help children read the words on the notes, and move them so that the pictures are correctly labeled.

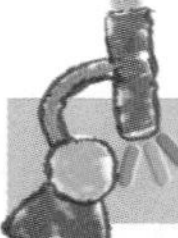

Science Facts

Mountains are formed when Earth's crust is pushed up by volcanic activity or the shifting of tectonic plates. The mid-Atlantic ridge under the Atlantic Ocean is the longest mountain range.

Canyons are made of rock carved by water and wind. The Grand Canyon was carved by the Colorado River as it rushed (for six million years) from the Rocky Mountains to the Gulf of California. The Grand Canyon is a mile deep at its deepest point. The world's deepest canyon is the Colca Canyon in Peru.

Plains are broad areas of land with few or no hills. Plains make up over a third of the United States, and much of their soil is good for farming.

Professional Development For more Science Facts and resources from NSDL visit http://nsdl.org/refreshers/science

plains

untains

canyon

31

▶ Ask Questions

- **Which of these places looks like it would be good for planting vegetables in the ground? Why?** plain; because it is flat
- **Is there a part of the mountain picture that looks like a good place to grow things?** in the valley **Which picture does not look like a good place to grow things?** canyon
- **When you look at the sides of this canyon, do they look smooth or rough?** both **What do you think could have smoothed them out?** wind, water

Write on It

Ask children whether the canyon would be a good home for animals. Tell them that many animals, including lizards, fish, and mountain lions, live in the Grand Canyon. Have children use a dry-erase marker to draw an *x* where they think one of those animals might make its home.

Think and Talk

Help children notice the water in the canyon. Ask: **Where do you think the water came from?** Encourage children to share their ideas with the class.

Mountain, Valley, or Plain? Bring children outside near the school. Have them look around carefully and decide whether they think they are on a mountain, in a valley, or on a plain. Ask: **How do you know? What helps you decide?**

Back in the Classroom Have children draw a picture of what they saw. Help them write about how they knew where they were.

Read Together and Learn

How do volcanoes change the earth?

▶ Build on Prior Knowledge

Ask children: **Does this look like a plain, a valley, a canyon, or a mountain?**

▶ Use the Visuals

- **Observe** Have children describe the *before* picture. Ask: **What do you notice about the top of the mountain? What do you notice about the area around the mountain?** The top is pointed and covered with snow; there are trees around the bottom.
- Have children describe the picture labelled *after*, again describing what they notice about the top of the mountain and the surrounding area.

▶ Additional Vocabulary

lava Write the word *lava* on chart paper, and have children try to read it by sounding out the letter sounds. Help them come to a definition such as, "It is rock that gets so hot that it melts to liquid and comes out of a volcano." Show children a pumice stone, which is made of cooled lava.

Flipbook

UNIT C • LESSON 3

How do volcanoes change the earth?

Reading Strategy

Comprehension Write the word *change,* and explain that there are two meanings of the word: "to make something different" and "coins like pennies, dimes, nickels, and quarters." Explain that words with the same spelling but different meanings are called *homonyms*.

Phonological Awareness Have children brainstorm a list of words that rhyme with *how* (*cow, now, allow, wow, pow, chow*).

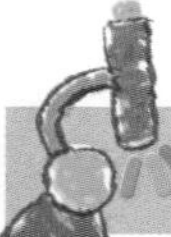

Science Facts

Mount St. Helens Scientists knew that Mount St. Helens would erupt when they began to see steam coming from the top of the mountain. They were able to warn people living in the area. Many animals left because they felt their habitat's temperature rising. Slowly, the forest around Mount St. Helens has begun to recover.

This volcano, in the state of Washington, erupted on May 18, 1980. It was so loud that people living 200 miles away could hear it. The eruption burned 230 square miles of the surrounding forest. However, some plants and trees were protected by a snowy covering. Many underground animals were also protected.

Professional Development For more Science Facts and resources from NSDL visit http://nsdl.org/refreshers/science

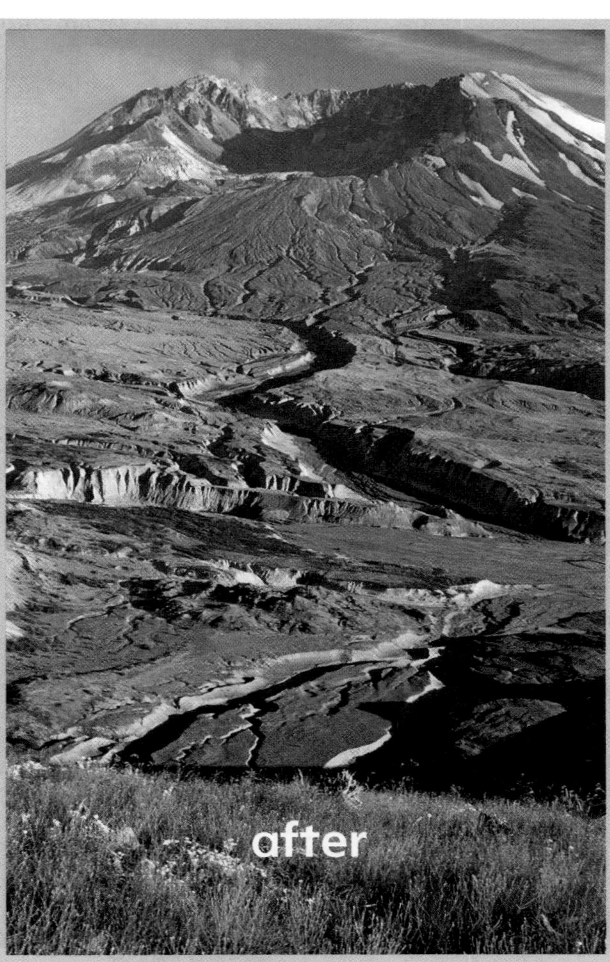

32

▶ Ask Questions

- **What do you know about volcanoes?** Help children understand that deep in the center of Earth, it is so hot that rocks melt to liquid. The heat makes pressure that pushes the liquid rock up, and then it bubbles out of the top. Ask: **Do you think this happens fast or slow? Why?**
- **What do you see coming out of the volcano? What do you think it is?** smoke, ash, lava
- **When the scientists knew that the volcano was going to erupt, they told the people who lived nearby. What do you think those people did?** Possible answer: People probably left. **Why?** to stay safe **Do you think the animals knew it was going to erupt? Why or why not?** Yes, because they left before many people.

Have children use a dry-erase marker to circle something that is the same in the before and the after picture. Have them draw an *x* on something that is different.

Think and Talk

Look again at the images on Flipbook pages 31 and 32. Have children share what they learned about mountains, plains, valleys, and canyons. Show page 32 and ask: **What happened to make the mountain look different in the second picture?**

Have children discuss what they know about different landforms. If they need further support, invite them to use the Photo Sorting Cards.

Differentiated Instruction

Enrichment Show Flipbook page 31 and have children brainstorm words to describe each place. List the words beneath the labels *mountain, valley, canyon,* and *plain*. For example, under *canyon* write *swirling, red, curvy, striped,* and *winding*. Then have children dictate sentences using the listed words.

More to Read

The Amazing Pop-Up Geography Book,
by Kate Petty
(Dutton Juvenile, 2000)

Activity Read this book with children to help them build on their understanding of our Earth. Help children create their own pop-up book about landforms.

Make It Rain!

Objective: Learn about erosion.

Inquiry Skills: make a model, observe, infer, investigate

You need **soil, sand, water, eyedroppers, watering can, cup, container for the mountain (sand table, planter, tray), Science Journal p. 42**

1. **Make a Model** Line the bottom of the Sand Table (or large planter) with rocks. Have children use a mixture of sand, soil, and water to build a mountain in it. Remind children to wash their hands after handling the mixture.
2. **Observe** Have children use different vessels to make it "rain" on the mountain. Ask: **What happens when you use an eyedropper to drip water on one spot? A watering can? A pitcher? What happens if you make it rain on top of the mountain? On one side only? Where does the sand and soil go?** Some of it goes to the bottom. Have children record their observations on their Science Journal pages.
3. **Infer** Ask: **What do you think happens on a real mountain when it rains?** Some soil falls down. Explain that this is called erosion.

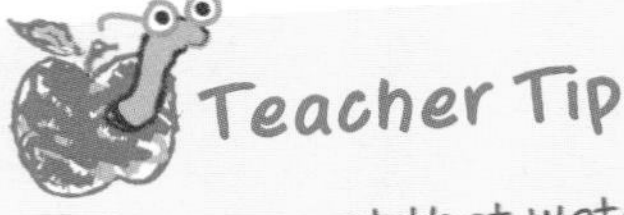

The concept that water always flows down (from high places to low places) is an important one. Give each child time to experiment with making it "rain" on the mountain. You might want to do this investigation over the course of a few days.

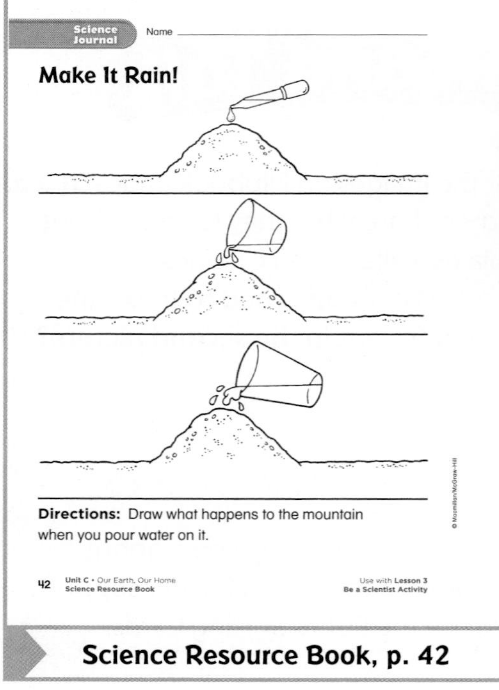
Science Journal Name ____________

Make It Rain!

Directions: Draw what happens to the mountain when you pour water on it.

42 Unit C • Our Earth, Our Home Science Resource Book

Use with Lesson 3 Be a Scientist Activity

© Macmillan/McGraw-Hill

Science Resource Book, p. 42

Investigate More

Investigate Have children cover the mountain with real leaves and make it "rain." Ask: **How did the leaves affect the erosion on the mountain?** they slowed it down; less sand fell down

Music

Let's Look

Objective: Compare and contrast different kinds of landforms.

Inquiry Skills: observe, compare

- **Listen and Talk** Have children listen to the song "Let's Look." Ask volunteers to use the song Flipbook page to answer the following: **Which is Earth's highest land? What kind of land is mostly flat? What kind of land is smaller than a mountain and rhymes with the word *mill*?**
- **Act It Out and Sing** Invite children to make up actions, such as pointing to their eyes on the word *look,* making the shape of a mountain with their arms on the word *mountain*, and so on. Then invite them to sing the song again.
- See page TR27 for printed music for "Let's Look."

UNIT C • LESSON 3 • SONG

Sung to the tune of "On Top of Old Smokey"

Let's Look

Let's look at the mountain.
It's Earth's highest land.
I wish we could climb it.
This mountain is grand!

Let's look at the large plain.
This land is so flat.
Let's look at the corn grow.
Let's eat some of that!

Let's look at the hill now.
This land is not flat.
A hill's not a mountain.
It's smaller than that!

S6

Flipbook p. S6
CD Tracks 11–12

Movement

You need
- Photo Sorting Cards 21–30

Mountain, Canyon, Plain

Objective: Use your body to make the shape of a mountain, canyon, or plain.

Inquiry Skill: make a model

- Review what mountains, canyons, and plains look like by showing children the Photo Sorting Cards.
- Play a game in which you have a leader call out *mountain, canyon*, or *plain*. Then have pairs make the shape of each landform.
- Partners can put their arms up over their heads, hold hands and drop them below waist level, or lie on the floor.

Art

You need
- brown, blue, and green modeling clay
- 6" x 6" pieces of cardboard

Clay Mountains

Objective: Make a model of a mountain.

Inquiry Skill: make a model

- Show children the materials for making a clay mountain. Demonstrate working with the clay.
- Have children work individually or in pairs to build brown or green clay mountains. As they work, ask: **How big will the base of your mountain be? Will your mountain have a high peak or a low peak?**
- Give children blue clay to make a lake or stream.

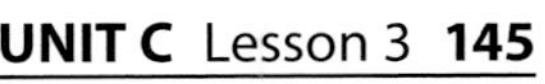

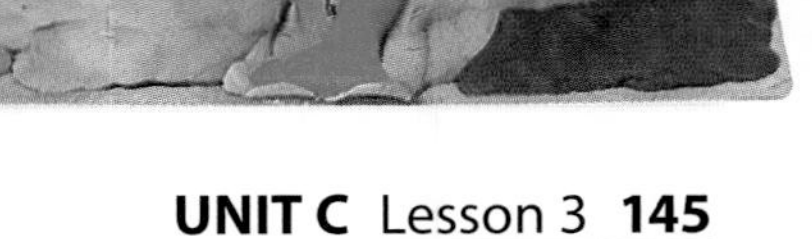

For more on landforms, use pp. 33–34 in the **Activity Book.**

▶ **Essential Question**
What are different bodies of water?

▶ **Objective**
Learn characteristics of rivers, streams, lakes, and oceans and identify water as a natural resource.

▶ **Vocabulary**

river **lake**
stream **ocean**

Resources

Flipbook, pp. 33, S7

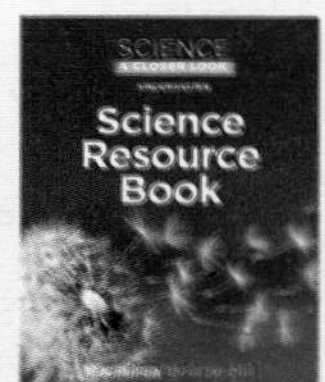

Science Resource Book, p. 43

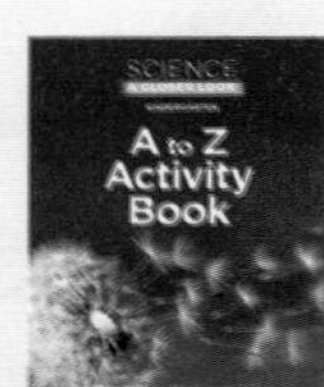

A to Z Activity Book, pp. 46–47

Leveled Reader: *Our Land*

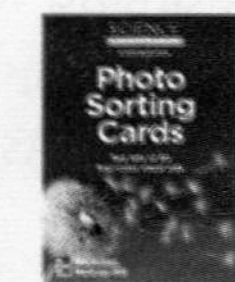

Photo Sorting Cards 21–30

Technology

Science Songs CD Tracks 13–14

Science in Motion *Beach Rocks*

www.macmillanmh.com

Water All Around

Circle Time — WHOLE CLASS

Swim to the Circle

You need
- chart paper
- marker

Inquiry Skills: classify, infer, communicate

- Invite children to pantomime swimming to the circle. Then ask: **Where were you swimming?**
- List the different kinds of bodies of water as children say them, and encourage children to think of other bodies of water. Ask: **Which of these kinds of water is salty? Which kinds are usually not salty?**

Infer Ask: **What animals do you think live in salt water? Fresh water?**

Time to Move

Encourage children to move like they are rowing a boat on a river, swimming in the ocean, fishing in a lake, and tiptoeing through a stream.

Be a Reader

You need
- Unit C Literature Big Book
- Leveled Reader: *Our Land*
- chart paper
- markers

Read the Big Book

Objective: Observe and describe different types of landforms.

Inquiry Skills: communicate, classify

- Read the title, *Our Land*, tracking the print as you read. Take a picture walk. Have children tell what they see.
- Read the book. Then reread it, omitting the last word of the sentence. For example, *Our land has tall mmm*________. mountains Help children fill in the last word. Ask: **How would you describe these mountains?** Record children's descriptions on chart paper. Repeat for each page.
- **Guided Reading:** See the inside back cover of the Leveled Reader version of *Our Land* for activities.

Be a Writer

25 MINUTES
WHOLE CLASS

You need
- paper
- crayons
- markers
- pencils

Water and Me

Objective: Write a story about experiences with water.

Inquiry Skill: communicate

- Have a class discussion about the ways people use water. Encourage children to share experiences they have had with water, such as swimming at the beach, drinking from a water fountain, and brushing their teeth. Introduce the concept that water is a resource that all living things need.
- Invite each child to draw about their experience. Help them to write words about their drawing.
- Bind the pages together into a class book about water. Place the book in your Library Center.

Be a Math Wiz Activity

You need
- Photo Sorting Cards 21–30

How Many Matches?

Objective: Collect and count the number of cards.

Inquiry Skills: compare, classify

- Have children set up the cards facedown in four rows.
- Invite children to take turns turning over the cards two at a time. If they find a match, they keep the cards. Play continues until all matches have been found.
- Have children count and compare the number of matches they found. Ask: **Who had the most? Who found the bodies of water?**

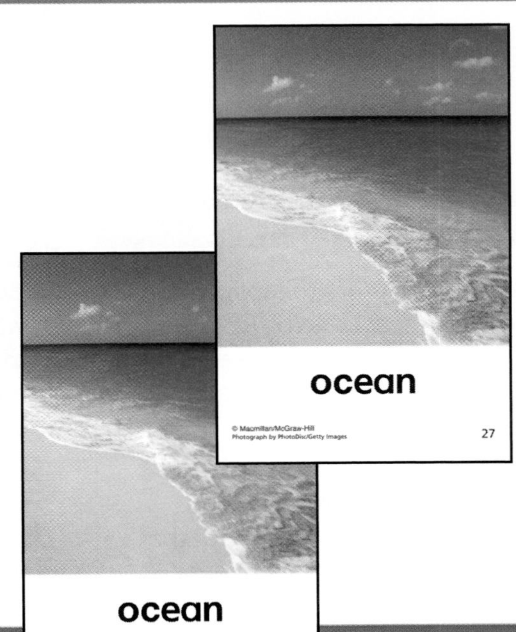

Read Together and Learn

How can water change the earth?

▶ Build on Prior Knowledge

Ask children how these bodies of water look different from the water they have seen before.

▶ Use the Visuals

- **Observe** Have children describe the water in each picture and what it is doing. Invite them to talk about the plants and animals that live by each place.
- **Communicate** Encourage them to talk about whether they have been to an ocean, to a waterfall, or to a river.
- Discuss how flowing water can cut into land, smooth surfaces, and create smooth rocks, sand, and canyons.

▶ Develop Vocabulary

river, stream, lake, ocean Create a four-column chart labeled river, stream, lake and ocean. Record children's ideas of what these bodies of water have in common and what their unique qualities are on the chart.

Flipbook

UNIT C • LESSON 4

How can water change the earth?

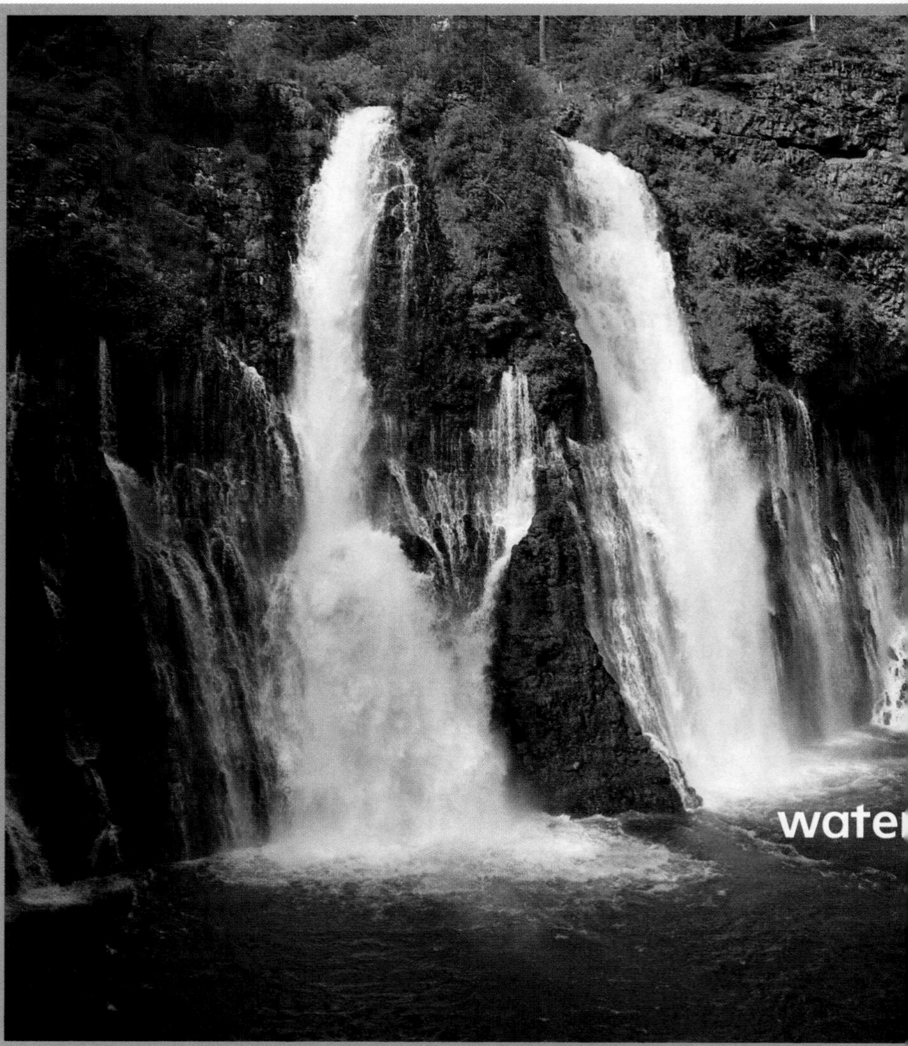

Reading Strategy

Vocabulary Development Write the following words on chart paper: *fish, deer, raccoon, turtle.* Read them aloud and have children use each word in a sentence that tells what these animals do in or around water.

Phonological Awareness Write the word *lake* and ask children to list other words that end with *-ake.* Record their responses. Point out that *lake* has a long vowel sound and a silent *e*.

Science Facts

Waterfalls happen when flowing water makes a steep drop. Over many years, waterfalls erode the rock that they are falling over, and slowly become less steep.

Rivers carry rain and melting snow from the mountains to the ocean. The Mississippi River is the second-longest river in the United States. As rivers flow, they wear away at bedrock, pick up sediment, and carry it out to the ocean.

Oceans Two-thirds of the world is covered with ocean. Gravity from the Sun and the Moon pull the oceans' waters, causing tides. Large waves start out at sea and then travel to beaches. As waves pound a coast, it slowly erodes and changes.

Professional Development For more Science Facts and resources from  NSDL visit http://nsdl.org/refreshers/science

33

▶ Ask Questions

- **What effect do you think the waterfall will have on the rocks it is going over?** wear them away, make them smaller
- **Do you think a flat river flowing over rocks would have the same effect? Why or why not?** Yes, but it would take longer.
- **What will happen to this beach as the waves go out?** The waves will bring some sand with them. **What do you think might happen over many, many years?** The sand might wash away.

Write on It

Invite children to use a dry-erase marker to draw a check mark on the water that seems to be flowing the hardest and fastest.

Think and Talk

Invite children to tell about each image on the Flipbook page and discuss whether they think the water is moving fast or slow.

Formative ASSESSMENT

Observe children as you explore the Flipbook page to see who understands the differences between bodies of water. If children are having difficulty, reinforce the concept by displaying additional labeled pictures at the Science Center.

ELL Support

Collect pictures of a *stream, river, lake,* and *ocean.* Ask children what they see.

Beginning Ask children to point to the picture of a stream, river, lake, or ocean.

Intermediate Give children pictures and ask: **Is it a stream or a river? Is it a lake or an ocean?**

More to Read

The Floating House, by Scott Russell Sanders (MacMillan, 1995)

Activity Read this book to build an understanding of rivers and the different ways that we use them as resources. Have children draw a picture showing something the family saw or did as they traveled.

Getting Water

Objective: Learn how the school building gets water.

Inquiry Skills: infer, investigate, communicate, make a model

You need cups, water, chart paper, markers, clipboards, paper, pencils, bulletin board paper, paint, glue, crayons, paper towel rolls, Science Journal p. 43

1. **Infer** Give each child a glass of water. Ask: **Where does our water come from?** Possible answers: the faucet, water fountain, sink **How does it get there?** List their ideas on chart paper. If they bring up the idea of water pipes, ask: **Where are the school's pipes? How does the water get in the pipes?**
2. **Investigate** Arrange to take your class to see the school's water heater and pipes. Help children generate questions they will ask. Remind children to be safe when visiting this area.
3. **Communicate** Have children use their Science Journal page to show the path of the water through the school.
4. **Make a Model** Have children work together to create a mural that shows how water gets into the classroom.

Teacher Tip

Before children start their mural, have them brainstorm what needs to be shown, and how the paper should be divided. While they watch, use their ideas to divide the paper into a section for the classroom, the school's basement, the pipes outside of the school, and the original water source.

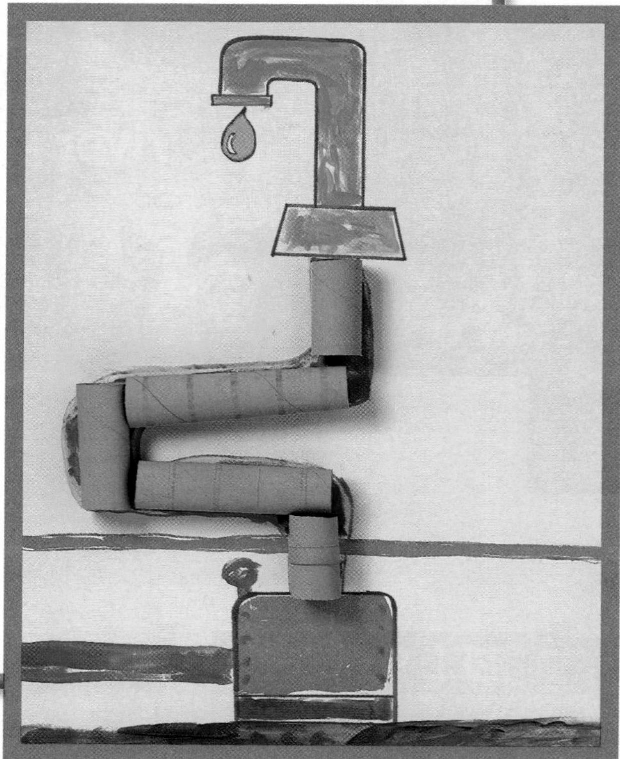

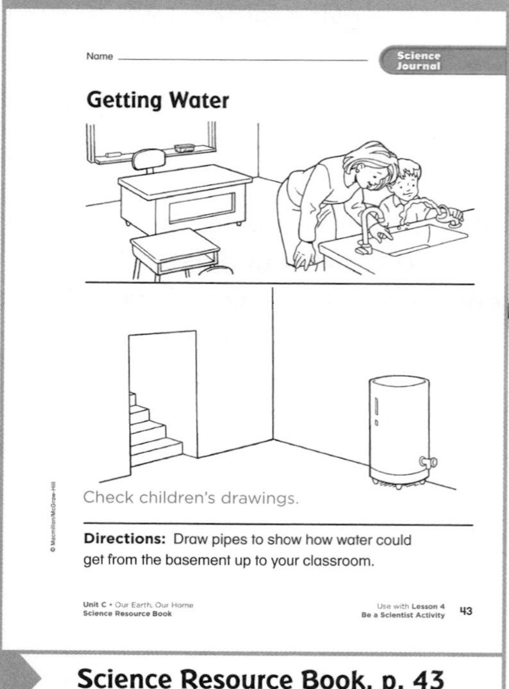

Name ______ Science Journal

Getting Water

Check children's drawings.

Directions: Draw pipes to show how water could get from the basement up to your classroom.

Unit C • Our Earth, Our Home Science Resource Book — Use with Lesson 4 Be a Scientist Activity 43

Science Resource Book, p. 43

Investigate More

Investigate Ask: **Where are the water pipes in your home? Are the pipes at school different from those at home?**

Music

Water All Around

Objective: Compare different bodies of water.

Inquiry Skills: observe, compare

- **Talk and Listen** Show children the song Flipbook page while listening to "Water All Around." Ask: **Which body of water can have big waves? Which body of water can be long and windy? Which body of water has land all around it?**
- **Sing and Chime In** Play the song again, inviting children to chime in as you follow the text with your finger. Sing the song a third time, inviting volunteers to come up and point to each body of water.
- **Locate It** Show children a physical map of the United States. Point to where they live and ask them if there is an ocean, lake, or river nearby.
- See TR28 for printed music for "Water all Around".

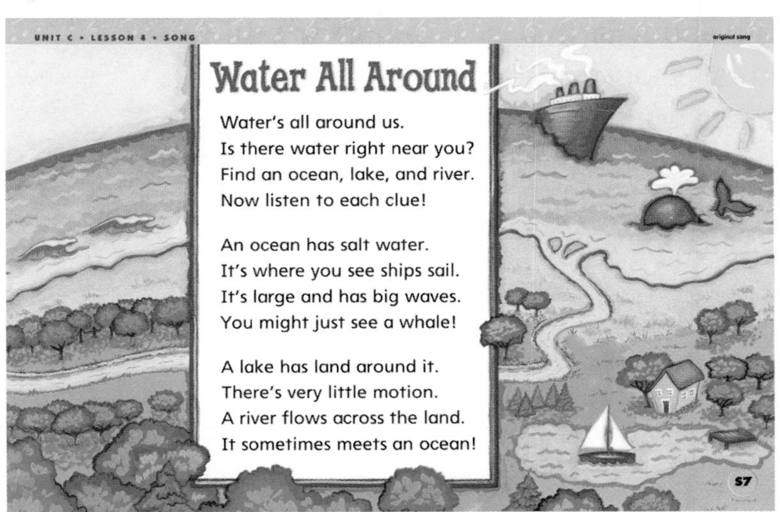

Flipbook p. S7
CD Tracks 13–14

Drawing and Writing

Hello, Ocean

Objective: Read about and discuss oceans.

Inquiry Skill: communicate

You need
- *Hello Ocean* by Pam Muñoz Ryan or another book about oceans
- paper
- crayons

- Show the book's cover and ask: **Where is this girl? What things are around her?**
- Read the book aloud and ask children to help complete the rhymes on each page and identify the five senses that the girl uses at the ocean. List them on chart paper.
- Have children draw and write about something they like to do at the beach.

Water Table

Water Flows

Objective: Observe that water flows down.

Inquiry Skills: predict, observe

You need
- wax paper
- eyedroppers
- water

- Give each child a piece of wax paper or coated paper and an eyedropper. Have them drip water at the top of the paper. Ask: **Where does it go?**
- Have children hold the paper vertically (like a wall) and ask: **Will the drops flow faster or slower? Why?**
- Then have children hold the paper like a small slide and ask: **Will the drops flow faster or slower? Why?**

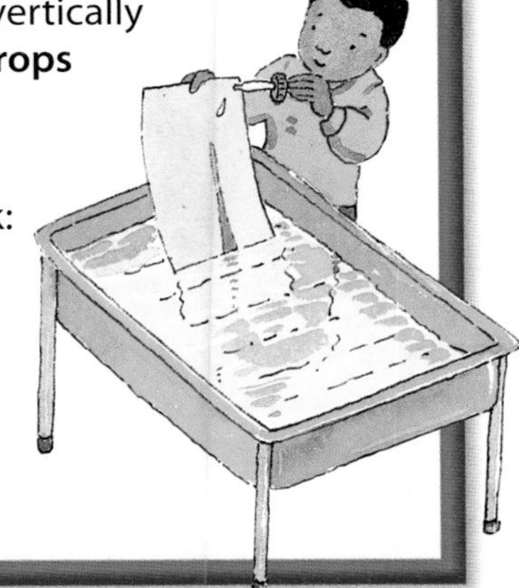

For more on water, use pp. 35–36 in the **Activity Book.**

▶ **Essential Question**

What resources from Earth do we use?

▶ **Objective**

Learn about Earth's natural resources that are used in everyday life and that resources can be conserved.

▶ **Vocabulary**

resource
firefighter

Resources

Flipbook, pp. 34–35

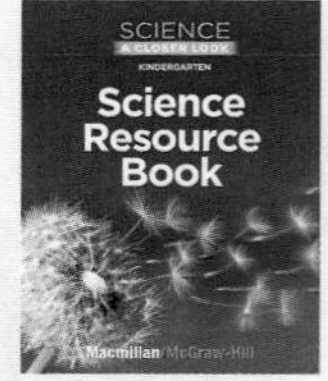

Science Resource Book, p. 44

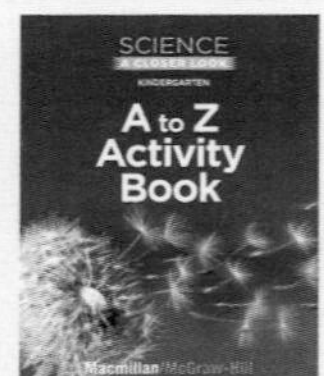

A to Z Activity Book, pp. 46–47

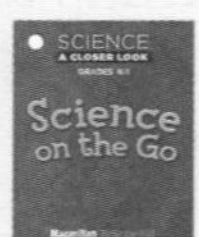

Science on the Go, pp. 65–68

Technology

Earth's Resources

Circle Time — WHOLE CLASS

Plants are Resources

You need

- *Apple Farmer Annie* by Monica Wellington
- chart paper
- marker

Inquiry Skills: observe, communicate, infer

- Read *Apple Farmer Annie* to children. Revisit the pictures and talk about what Annie does throughout the book.
- Invite children to recall where Annie gets her apples. Ask children if they have been apple picking and to share their experiences. Ask: **Where do apples come from?** trees
- Help children generate a list of products that are made from apples, including applesauce, apple pie, shampoo, and soap.

Infer Ask: **What other items do we get from trees?** Possible answers: furniture, paper, pencils, syrup

Time to Move!

Hand each child a ribbon or streamer to move like a windmill. Then have children follow directions such as *circle around fast* (for a very windy day) and *circle slowly* (for a gentle, breezy day).

Be a Writer Activity

You need
- paper
- pencils
- marker or crayons

These are Made from Trees

Objective: Create a class book about items that come from trees.

Inquiry Skill: communicate

- Hold up a piece of paper and a pencil. Ask: **Where does the wood for our pencils come from? Where does paper come from?** Discuss that wood and paper come from trees.
- Invite children to look around the room and identify other things that come from trees, such as boxes, tissues, or blocks. Ask each child to draw a picture of one of these objects. Have children write labels for their drawings by sounding out the name of the object.
- Compile their drawings in a class book titled *These Are Made from Trees*. Read the book aloud and provide children time to explore it with a classmate.

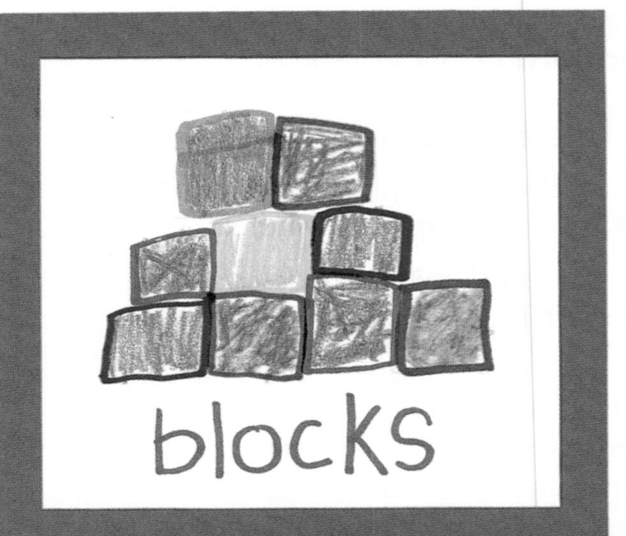

Be a Math Wiz Activity

You need
- small containers of varying dimensions
- measuring cups

Water Volume

Objective: Explore the volume of water containers.

Inquiry Skills: predict, measure, communicate

- Hold up two containers, one tall and thin, the other short and stout. Ask children to predict which one will hold more water.
- Help children fill each container with water and then pour into separate measuring cups. Discuss the results.
- Repeat this exercise with different-sized containers. Ask children to make predictions each time and invite them to explain their ideas. Ask: **How can we tell which container will hold more water? Do you think any of these containers might hold the same amount of water?**

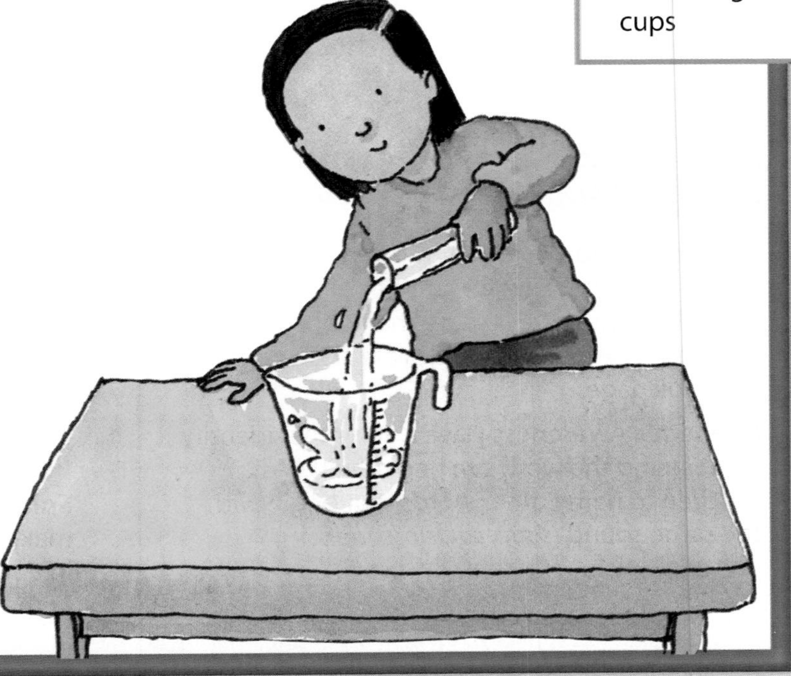

Read Together and Learn

How do we use these resources?

▶ Build on Prior Knowledge

Ask: **What do you think makes lights go on?** See what children know about electricity. Ask: **What is electricity?** Accept all responses.

▶ Use the Visuals

- Infer Point to the picture of the windmills. Ask: **What is making these windmills move?**
- Observe Ask children to describe what they see in each image.
- Communicate Prompt children to discuss how they think the wind, the Sun, and the water are being used. Explain that the windmills, the Sun's light, and the water from the dam are used to give us electricity and to make the energy that makes different machines work.

▶ Develop Vocabulary

resource Write the word *resource* on the board. Say: **Water is a resource. How do we use water?** to drink, to clean ourselves, to water plants, to put out fires, for play **Wind and the Sun are also resources. How do we use them?**

Flipbook

UNIT C • LESSON 5

How do we use these resources?

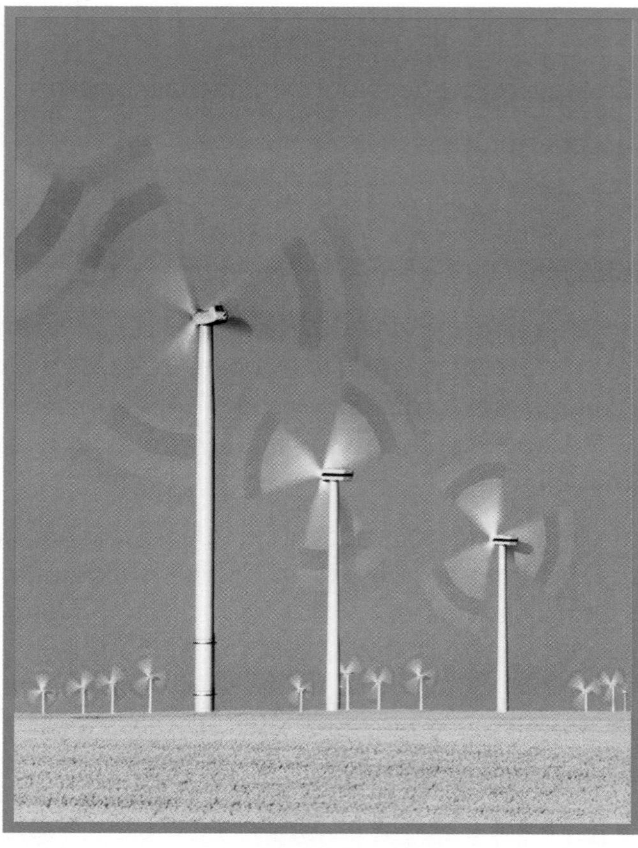

Reading Strategy

Comprehension Point out the letter *s* at the end of *resources*. Explain that *s* at the end of a word means there is more than one. Ask children how many resources are shown on the Flipbook page.

Phonemic Awareness Have a volunteer identify what sound the word *sun* begins with. Ask children to name other words that begin with the same sound *(soup, sat, sip, sand)*.

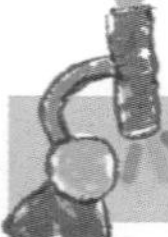

Science Facts

Windmills use wind power to make electricity, pump water, or grind grain. Windmills that make electricity are called *wind turbines*. They typically have two or three blades, or sails, which turn in the wind.

Solar Energy uses sunlight to make heat or electric power. Cells in a solar watch or calculator take in the Sun's energy and store it in batteries. Another kind of solar energy uses sunlight to heat water.

Dams are used to stop and control the flow of water. The movement of water through the dam can be used to generate electricity. The largest dam in the United States is the Grand Coulee Dam, located in central Washington.

Professional Development For more Science Facts and resources from NSDL visit http://nsdl.org/refreshers/science

34

▶ Ask Questions

- **What do you think happens to the windmill when it is not very windy?** It turns less; it makes less energy.
- **What kind of weather would be helpful in making solar energy?** sunny **What kind of weather would not be helpful?** cloudy, rainy
- **Why is water important to us? What do people have to do to make sure they do not use up all the water?** Accept all responses.

Write on It

Invite children to use a dry-erase marker to help write labels for the resources being used in each picture—*wind, sun,* and *water*.

Think and Talk

Review the pictures on the Flipbook page. Invite children to share ideas about ways to conserve, or save, these resources.

Differentiated Instruction

Extra Support Revisit the Flipbook page. Write labels beneath each picture: *wind, sunlight, water.* Read each label and have children repeat after you. Invite children to draw a picture that illustrates these resources and match them to the appropriate picture on the page.

More to Read

Why Should I Save Water?, by Jen Green (Barron's Educational Series, 2005)

Activity Read this book with children and discuss ways that people can help to conserve water. Use the note at the back of the book for further suggestions.

Read Together and Learn

How do we use water?

▶ Build on Prior Knowledge

Before looking at the Flipbook page, ask children to talk about how they use water in their lives.

▶ Use the Visuals

- **Observe** Ask children to describe what is happening in the pictures.
- **Communicate** Invite them to talk about when they have watered plants, been in a boat, or visited a firehouse.
- **Infer** Ask: **What would happen in each picture if there was no water?** Possible answers: The plants would not grow. The girl would be thirsty. The firefighters would not be able to put out a fire.

▶ Develop Vocabulary

firefighter Write the word *firefighter* and help children read it. Ask: **What two words are in *firefighter*?** Explain that it is a compound word, or a word made of two other words. Point out that we now usually call people firefighters, not firemen, because both women and men can do this job.

Flipbook

UNIT C • LESSON 5

How do we use water?

Reading Strategy

Phonological Awareness Have children clap once on each syllable of each word in the sentence *How do we use water?* Explain that syllables are the sound parts of a word. Then go through the sentence again and have them count the syllables.

Phonics Invite children to say their names, and have them clap on each syllable. Then have them count the syllables in each of their names.

Science Facts

Water Every day, the United States uses about 180 gallons of water per person. We use it to drink, cook, wash, and do laundry. We also use water to put out fires and clean streets. Water comes to us from wells, lakes, rivers, and reservoirs. Water travels away from us through pipelines, canals, or closed tunnels, to water treatment centers. Treatment centers filter the water and implement a process, such as adding chlorine, to take out bacteria and kill viruses. From there the water travels in pipes to our homes and businesses.

Professional Development For more Science Facts and resources from NSDL visit http://nsdl.org/refreshers/science

35

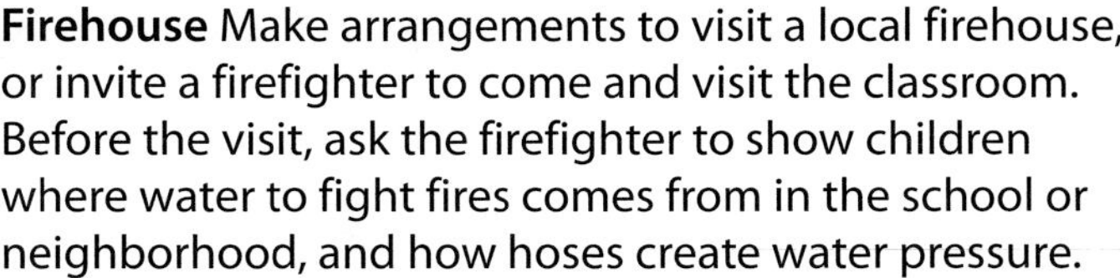

Take a Trip

Firehouse Make arrangements to visit a local firehouse, or invite a firefighter to come and visit the classroom. Before the visit, ask the firefighter to show children where water to fight fires comes from in the school or neighborhood, and how hoses create water pressure.

Back in the Classroom Have children draw pictures that show what they learned from the firefighters. Write what they tell you about their drawings and compile their drawings and stories into a class book.

▶ Ask Questions

- **Where does the firefighters' water come from?** fire hydrant **Where is the water before it is in the fire hydrant?** pipes under the ground
- **How are these people using this lake?** for canoeing
- **Do you think the water in every picture would be safe to drink?** No, some water is not clean enough to drink.
- **Look at the watering can and the firefighter's hose. What is different about how the water is coming out? What do you think makes it come out faster from the firefighter's hose?**

Write on It

Have children use a dry-erase marker to write their first initial on a picture that shows a way they have used water. Ask: **Which pictures have the most initials? Which have the fewest?**

Think and Talk

Look again at the pictures on the Flipbook page. Encourage children to talk about what they might do if they did not have running water at home.

Formative ASSESSMENT

Observe if children understand the concept of resources being used to help people. If they are having difficulty understanding this concept, review the Flipbook pages and invite volunteers to explain what is happening in the pictures.

Be a Scientist Inquiry Investigation

Conserve Water

Objective: Understand one way we can conserve water.

Inquiry Skills: communicate, predict, investigate, compare, draw a conclusion

You need **plastic toy dishes, sink, two large plastic containers, chart paper, marker, Science Journal p. 44**

Teacher Tip

If you do not have a sink in your classroom, plan a trip to the cafeteria or kitchen ahead of time. This activity can be adjusted to work well as a whole group activity, if necessary.

1. **Communicate** Engage children in a discussion about the uses of water. Ask: **How do you and your family use water every day?** Record children's responses.
2. **Predict** Work with small groups of children at the sink. Label the containers *1* and *2*. Place Container *1* at the bottom of the sink. Ask children: **What might happen if we leave the water running while we wash the dishes?**
3. **Investigate** Invite a volunteer to wash a few toy dishes while leaving the water running. After, remove the container. Now place Container *2* at the bottom of the sink. Repeat the process, but turn off the water during washing. Afterwards, remove the container.
4. **Compare** Ask children: **Why did the first container have more water? Why did the second container have less water?** Have them record their findings on their Science Journal page.

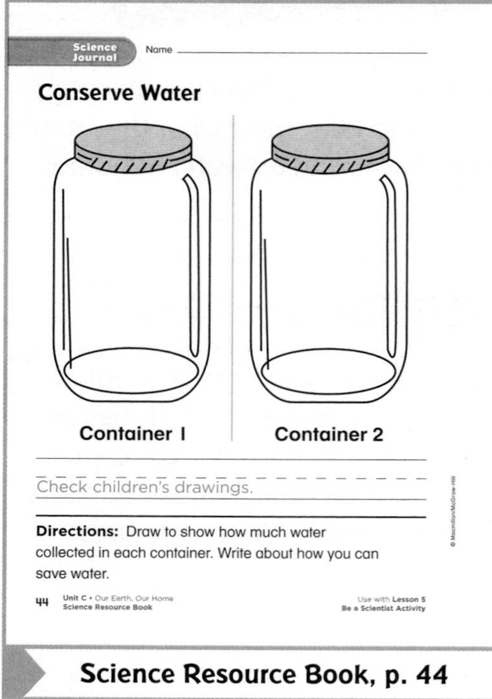

Science Journal Name

Conserve Water

Container 1 | Container 2

Check children's drawings.

Directions: Draw to show how much water collected in each container. Write about how you can save water.

44 Unit C • Our Earth, Our Home Science Resource Book — Use with Lesson 5 Be a Scientist Activity

Science Resource Book, p. 44

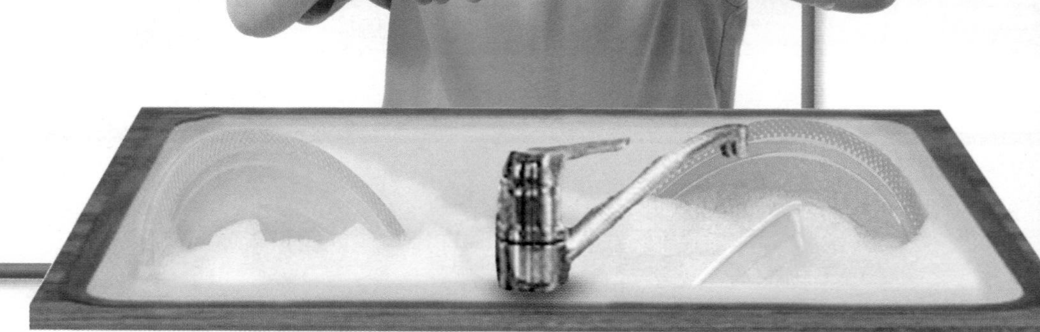

Investigate More

Draw a Conclusion Ask: **Why do you think it is important to turn off the water while doing the dishes? What else can you do to help conserve water?**

centers

Art

You need
- paint brushes
- watercolor paints
- pictures of landscapes

Water Colors

Objective: Use watercolors to paint.

Inquiry Skill: observe

- Place paper and watercolors for children to use in the Art Center. Display pictures of landscapes for children to look at for inspiration.
- Encourage children to tell you about what they are doing while they are painting. Allow children time to share their paintings with the class. Display their work on a bulletin board.
- Revisit the bulletin board on another day, and discuss with children how they used water to create their painting. Ask: **How did the water change?** Accept all responses.

Library

You need
- books about Earth and its resources

Read About Natural Resources

Objective: Explore books about Earth's resources.

Inquiry Skills: observe, communicate

- Gather books about water, soil, trees, and other natural resources.
- During Circle Time, tell children that you are placing the books in the Library Area.
- Encourage children to look at the books in pairs, and to talk about what they see.

Blocks

Water Works

Objective: Use blocks to show how water is used.

Inquiry Skill: make a model

- Explain to the group that they will create buildings that use water. Brainstorm different buildings and ways they use water. For example, a restaurant uses water to wash dishes; a laundromat uses water to clean clothes; a firehouse uses water to put out fires; a zoo uses water to wash and feed the animals.
- Help children decide what details, props, and shapes they will create to indicate the water in each building.
- Encourage children to play with their structures and create scenarios related to how the people in each kind of structure would use water.

For more on Earth's resources, use pp. 37–38 in the **Activity Book.**

▶ **Essential Question**
How can we take care of Earth?

▶ **Objective**
Learn different reasons for and ways of recycling.

▶ **Vocabulary**
recycle
reduce
reuse

Resources

Flipbook, p. 36

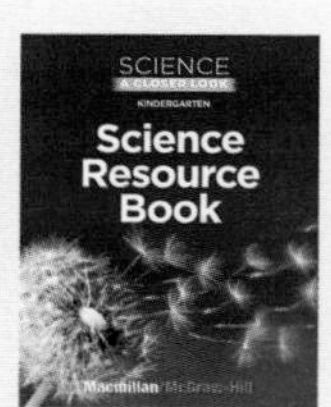

Science Resource Book, p. 45

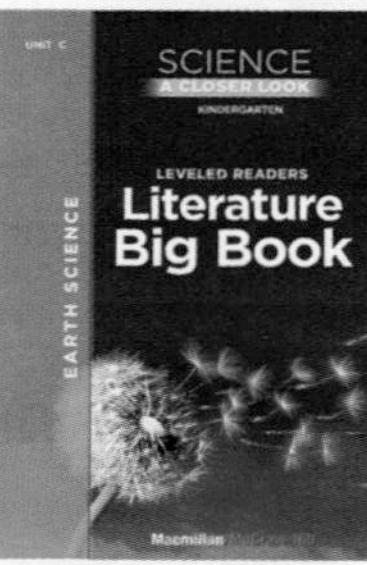

Unit C Literature Big Book: *Time for Kids*

Leveled Reader: *Recycle, Reduce, Reuse!*

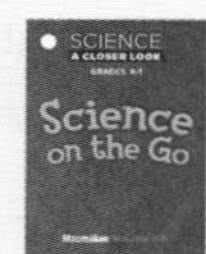

Science On the Go pp. 69–70

Technology

www.macmillanmh.com

NSDL

Recycle, Reuse

Circle Time — WHOLE CLASS

What's Recycled?

Inquiry Skills: communicate, infer

- Bring containers to the circle. Ask: **Where do empty containers belong?** If children say in recycling bins, ask why, and record their responses. If they say in the garbage, ask where things will go after that.
- Have children look for the recycling symbol on containers. Explain that *recycling* is using things again.

You need
- empty plastic and glass containers (with the recycling symbol)
- paper

Infer Ask: **Why would we want to recycle things?**
to keep Earth clean

Why Recycle?

So it doesn't get messy on Earth.

So we don't have to buy a new Earth.

Time to Move!

Give children a used piece of paper to be crumpled up tightly. Encourage children to toss the paper into a pail or bin, each time moving further away from it. Remember to recycle the paper when the game is finished!

Be a Reader

Read the Big Book

Objective: Understand that some materials can be reused and/or made into something else.

Inquiry Skills: predict, observe, communicate

You need
- Unit C Literature Big Book
- Leveled Reader: *Recycle, Reduce, Reuse!*
- chart paper
- marker

- Read the title and point out the picture on the cover. Ask children what they think this book might be about.
- Read the book. Then reread it. Invite children to tell what each child is doing. Ask: **What is the girl on page 5 doing? How do you reduce, or use only what you need? How do you reuse things?**
- **Guided Reading:** See the inside back cover of the Leveled Reader version of *Recycle, Reduce, Reuse!* for activities.

Be a Writer

School Recycling

Objective: Record information from an interview.

Inquiry Skills: investigate, communicate

You need
- chart paper
- marker
- crayons
- pencils
- paper

- Have children come up with a list of questions for the custodian about recycling at your school. Then invite the custodian in for an interview.
- Help children ask their questions. Write down the custodian's responses.
- Afterward, have children draw a picture of something they learned and have them write or dictate to you about their drawing.

Be a Math Wiz Activity

Penny Toss

Objective: Make a game using reused materials.

Inquiry Skills: compare, communicate

You need
- one egg carton per child
- markers
- pennies

- Give each child an egg carton cut in half and help them write the numbers *1–6* to label the bottom of carton sections.
- Have each child stand three feet away, toss a penny in, and record the number of the section that the penny lands in.
- After children have tossed the penny ten times, have them compare their results.

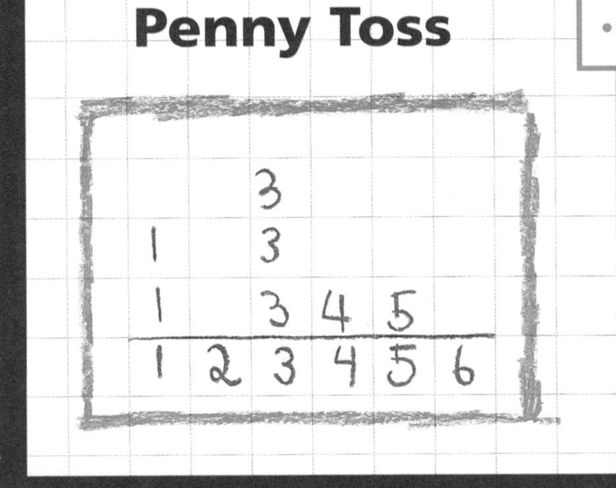

Read Together and Learn

How can we reuse things?

▶ Build on Prior Knowledge

Ask: **Have you ever been to a river, an ocean, or a waterfall and seen trash lying or floating around? Why do you think this happens?**

▶ Use the Visuals

- Communicate Have children talk about what they see in the pictures. Invite children to discuss which images look familiar.
- Ask if anyone has ever been on a tire swing or played at a playground that uses tires.
- Infer Ask: **Why would someone want to build a sculpture that uses tires? Why do you think there are old tires on the boat?**

▶ Develop Vocabulary

recycle, **reduce**, **reuse** Write *recycle, reduce,* and *reuse* on the board and underline *re–* in each word. Explain that when *re–* starts a word, it means "again." Tell children that to cut down on waste we can *recycle* some garbage. Things like tin cans, plastic containers, and paper can be recycled and made into other things. To *reduce* garbage, we bring our own bags to the grocery store. We *reuse* other things, such as old clothing pieces, to make a quilt.

Flipbook

UNIT C • LESSON 6

How can we reuse things?

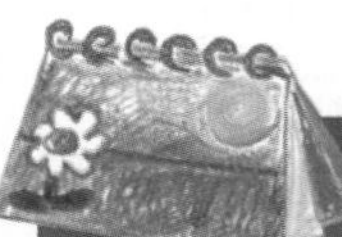

Reading Strategy

Phonics Have children circle every *e* in the question on the page. Then ask: **What sound does *e* make in *we*? What sound does it make in the first part of *reuse*? What sound does it make in the second part?**

Phonemic Awareness Write the word *can*. Have children list words that rhyme with *can (man, fan, tan, ran, plan, pan, ban, an)*. Have children use these words to dictate rhyming sentences.

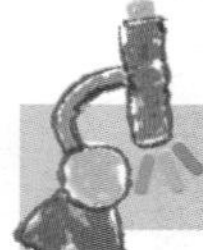

Science Facts

Recycling Countries used to get rid of old tires by burning them. However, that created a lot of air pollution. Because it is expensive for factories to break down worn-out tires and make new ones, people have come up with creative recycling ideas. For example, tires are reused as bumpers on vehicles and boat docks. Certain states have begun mixing rubber from old tires into asphalt to pave highways. The rubber minimizes skidding, and makes the highways quieter. Companies now use recycled tires to make soft surfaces for playgrounds. People also use old tires as swings, or turn them into sandals.

Professional Development For more Science Facts and resources from NSDL visit http://nsdl.org/refreshers/science

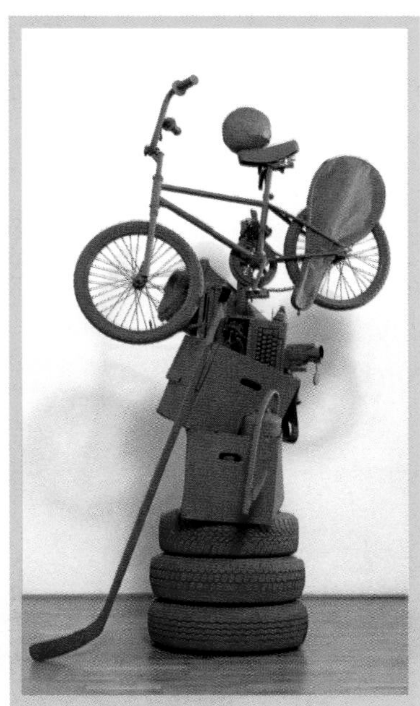

36

▶ Ask Questions

- **Do you think these are new tires or old ones? Why?** old, because they are being reused
- **Why do you think they cannot fix old tires and use them on cars?** It costs too much.
- **What could happen if nobody thought about helpful ways to reuse old tires?** We would have too much garbage on Earth; we would have nowhere to put the old ones.

Write on It

Invite children to use a dry-erase marker to help you write labels for the items on the page.

Think and Talk

Review the pictures on the Flipbook page. Ask children: **How are these tires being reused? Why are they being reused?** Invite children to share other ideas of ways to recycle and reuse things.

Formative ASSESSMENT

If children have difficulty understanding how we recycle and reuse materials, invite them to create a collage using found materials and discuss what they reused.

TIME FOR KIDS

Shared Reading Read *Save the Earth! Recycle a Bottle* with the class. Use this magazine, found in the **Unit C Literature Big Book**, as a photographic review of recycling.

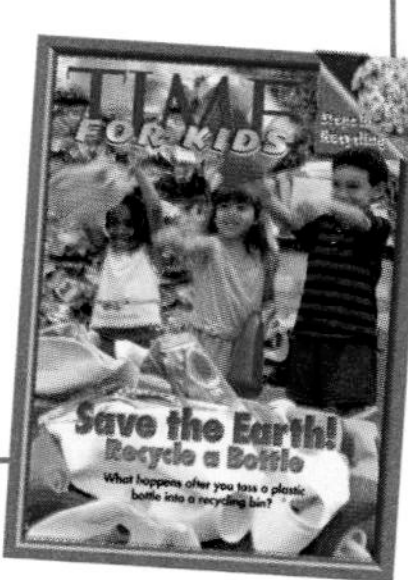

More to Read

Where Does the Garbage Go?, by Paul Showers (HarperCollins, 1994)

Activity Read this book with children to introduce what happens to things that we use and throw away. Afterward, go back through the book, discussing each step in the process of getting rid of garbage.

Be a Scientist **Inquiry Investigation**

Recycling Center

Objective: Make a classroom recycling center.

Inquiry Skills: communicate, classify, investigate, infer

You need **yogurt containers, coffee containers, egg cartons, boxes, newspaper, old magazines, paint, scissors, glue, markers, pencils, Science Journal p. 45**

1. **Communicate** Have children discuss materials in the classroom that could be recycled. List their ideas. Explain that they will help to create a classroom recycling center.
2. **Classify** As children build the center using empty boxes or plastic crates, help them decide what will go in each container. Help children make labels for the containers.
3. **Investigate** Brainstorm ways that different objects could be remade with new uses. For example, a yogurt container could become a piggy bank, a shoe box could become a storage box, a piece of paper could become a card for someone.
4. **Communicate** Invite another class to come in and learn about your recycling center. Have your class explain how and why they built the center, and how they use it. Have them record additional ideas on their Science Journal page.

Before beginning this project, encourage families to send in yogurt containers, coffee containers, egg cartons, boxes, large pieces of scrap paper, old magazines, used envelopes, and other items for recycling.

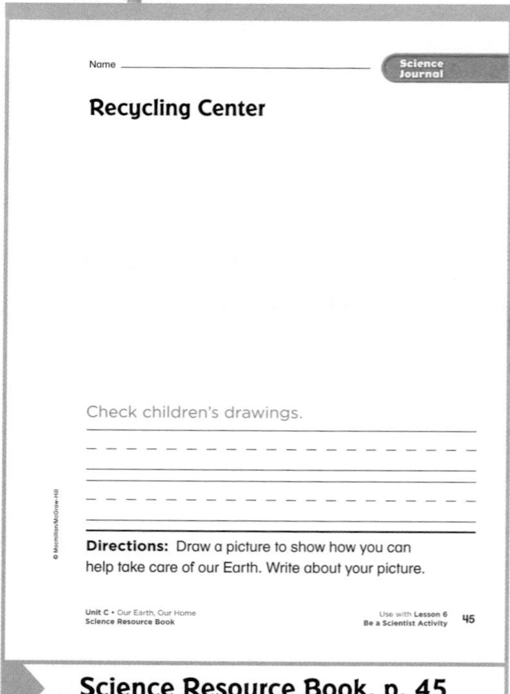

Name ____________ Science Journal

Recycling Center

Check children's drawings.

Directions: Draw a picture to show how you can help take care of our Earth. Write about your picture.

Unit C • Our Earth, Our Home
Science Resource Book

Use with Lesson 6
Be a Scientist Activity 45

Science Resource Book, p. 45

Investigate More

Infer Ask: **What happens to trash if it is buried in the ground over a period of weeks? How could we find out?** Possible suggestion: You could bury some plastic and paper materials in the schoolyard. After three months you could dig up the materials and see what has happened.

centers

Drawing and Writing

"Do It!" Posters

You need
- paper
- pencils
- markers

Objective: Create posters about how to recycle.

Inquiry Skill: communicate

- At Circle Time, invite children to make posters to put up outside the classroom that will encourage people to recycle.
- Have children draw one way that they have recycled, either in the classroom recycling center or at home.
- Help children decide what they want to write on their poster. Help them write it, or offer to take dictation.

Technology

More About Our Earth

- For games and activities about our Earth, help children log onto www.macmillanmh.com.
- Children can use these fun games to review and reinforce what they have learned about our Earth and recycling.

Art

Recycled Art

You need
- yogurt containers
- coffee containers
- egg cartons
- boxes
- newspaper
- magazines
- paint
- scissors
- glue
- markers

Objective: Use recycled items to create art.

Inquiry Skill: make a model

- Have children use objects from the classroom recycling center to make new things. For example, they can make math games from egg cartons, instruments from boxes or coffee containers, decorative storage boxes, or cards.
- Encourage children to discover and share new ways to use recycled items.

Cooking

Honey Yogurt

You need
- honey in bear container
- plain yogurt
- spoons
- cups
- marker
- chart paper

Objective: Eat food from containers and then find ways to recycle them.

Inquiry Skill: communicate

- Have children dish yogurt into cups for each person in the class, topping the yogurt with a squirt of honey.
- When they are finished making and eating their snack, have children brainstorm uses for the empty containers.
- List children's ideas on chart paper and post them in the Art Center.

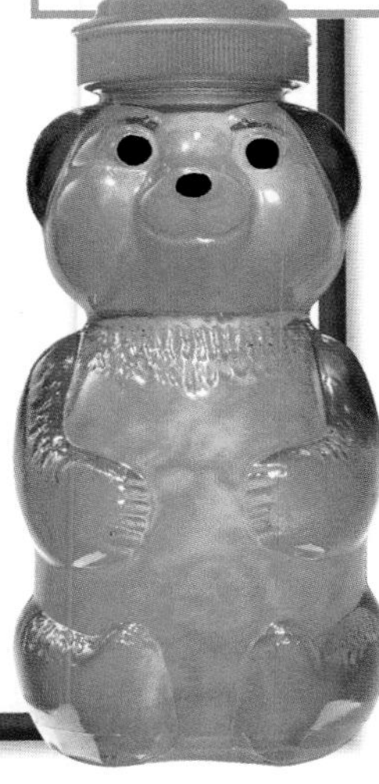

For more on recycling, use pp. 39–40 in the **Activity Book.**

Performance ASSESSMENT

DOK I

Geographic Features

- Use sticky notes to cover the labels on the Photo Sorting Cards for landforms and bodies of water, such as a mountain, a plain, a waterfall, and an ocean.
- Place the cards for landforms face up. Ask: **Can you point to the mountain? The plain?**
- Place the cards for bodies of water faceup. Ask: **Can you point to the ocean? The waterfall?**

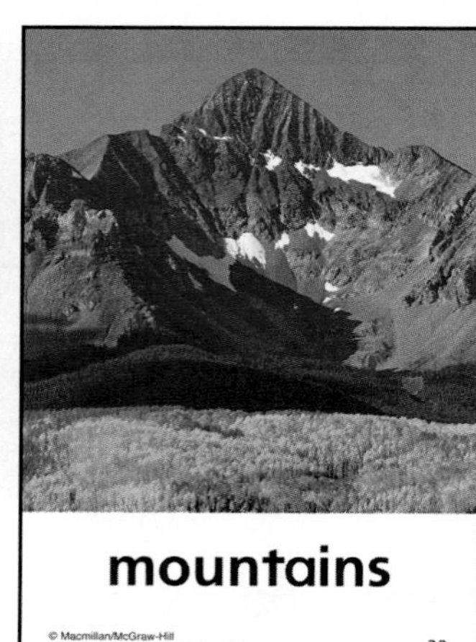

You need
- Photo Sorting Cards 21–30
- sticky notes

Scoring Rubric

Use the Scoring Rubric to evaluate each child's completion of the Performance Assessment task.

3 points = Child can identify each landform and body of water.

2 points = Child can identify at least one landform and one body of water.

1 point = Child can identify only one of the Photo Sorting Cards.

Summative ASSESSMENT

DOK I

Use the Unit C Assessment mini-book pages from the **Science Resource Book** pages 67–68 to assess what children have learned about rocks, landforms, bodies of water, and recycling.

- Help children fold their page in half to create a book and write their name. Guide children through each page, having them draw on each page to illustrate each sentence.
- Invite children to share their work with the class before taking their books home.

(fold here)

Name: ____________

Our Earth

child drawing of a mountain

Our Earth has **mountains**.

I

child drawing of rocks

Our Earth has **rocks**.

2

(fold here)

© Macmillan/McGraw-Hill

Unit C Our Earth, Our Home Assessment

child drawing of an ocean

Our Earth has **oceans**.

3

Science Resource Book 68

© Macmillan/McGraw-Hill

Science Resource Book

Unit C Our Earth, Our Home Assessment 67

child drawing of something people recycle

We **recycle** to help our Earth.

4

Science Resource Book, pp. 67–68

Portfolio ASSESSMENT

- Encourage children to select work from this unit to include in their portfolio. Discuss their choices and have children share which is their favorite and why.
- Use the Unit C Checklist from the **Science Resource Book** page 79 to record children's progress. Place checklists and notes in each child's portfolio.

Teacher's Notes

UNIT D

Weather and Sky

How do the weather and the sky change?

Essential Questions

Lesson 1 **What are some different kinds of weather?**

Lesson 2 **How can clouds be different?**

Lesson 3 **What happens when the seasons change?**

Lesson 4 **How does the sky change from night to day?**

Lesson 5 **How does the Sun make shadows?**

Visit www.macmillanmh.com for online resources.

Science Leveled Readers

All included in the Leveled Readers Literature Big Book.

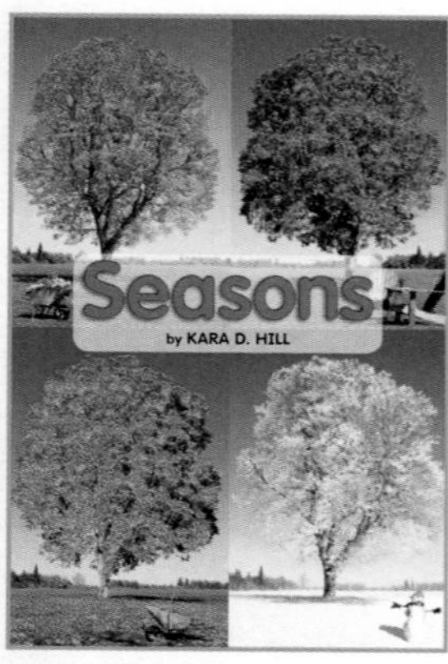

APPROACHING

Seasons
Seasons have different weather. There are lots of fun activities to do outside no matter what season it is!

ISBN: 978-0-02-284595-7

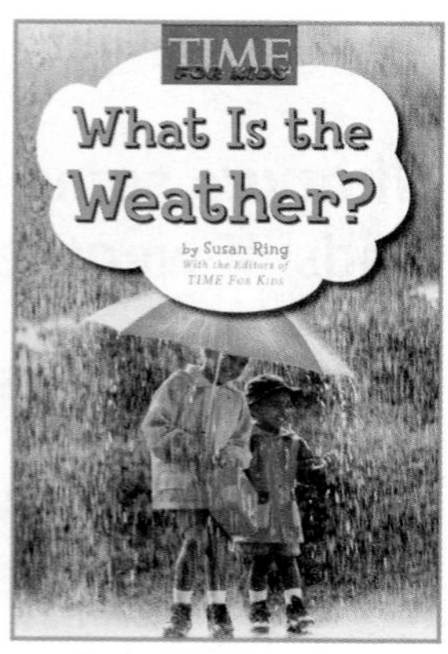

ON LEVEL

What Is the Weather?
Weather affects what we do every day. There are various types of weather, and weather changes every day.

ISBN: 978-0-02-284596-4

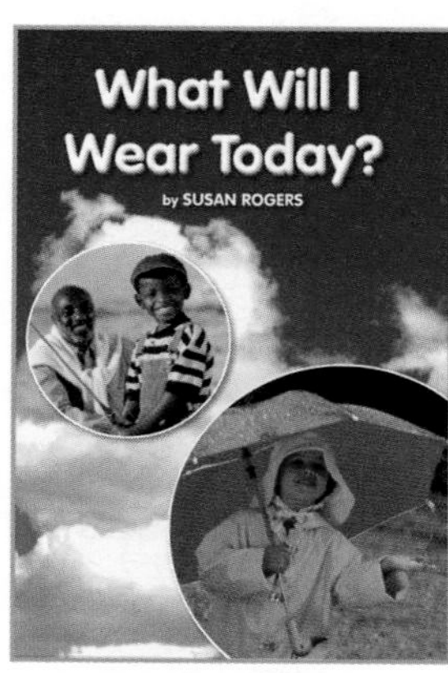

BEYOND

What Will I Wear Today?
The seasons and weather determine what people wear each day.

ISBN: 978-0-02-284597-1

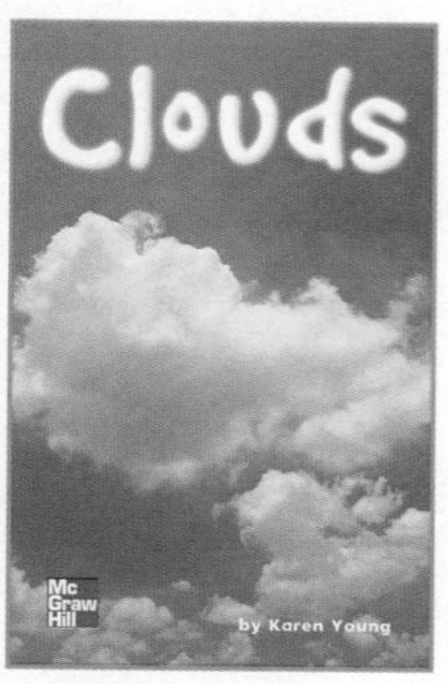

APPROACHING

Clouds
There are so many kinds of clouds! Clouds change their shape and color and different weather accompanies them.

ISBN: 978-0-02-281082-5

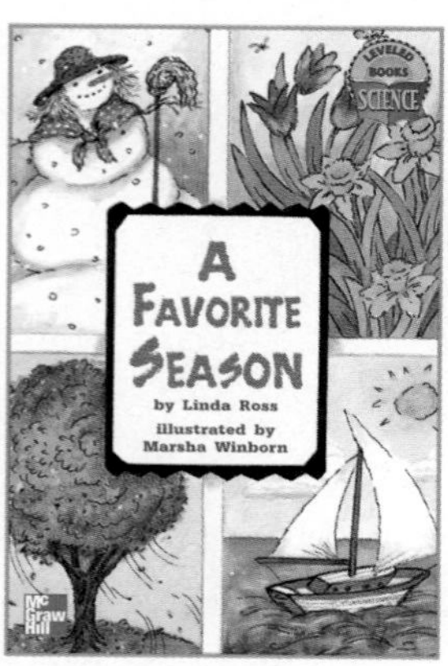

ON LEVEL

A Favorite Season
A rhyming story about members of the same family that each like a different season and why.

ISBN: 978-0-02-278973-2

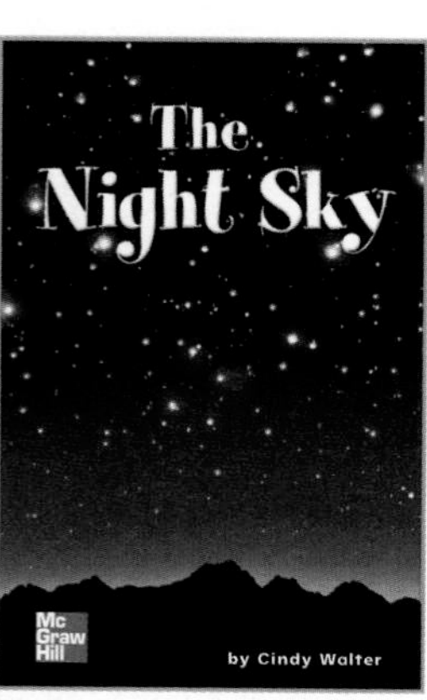

BEYOND

The Night Sky
The night sky has so much to see! This book talks about what can be seen up above after the Sun goes down.

ISBN: 978-0-02-281083-2

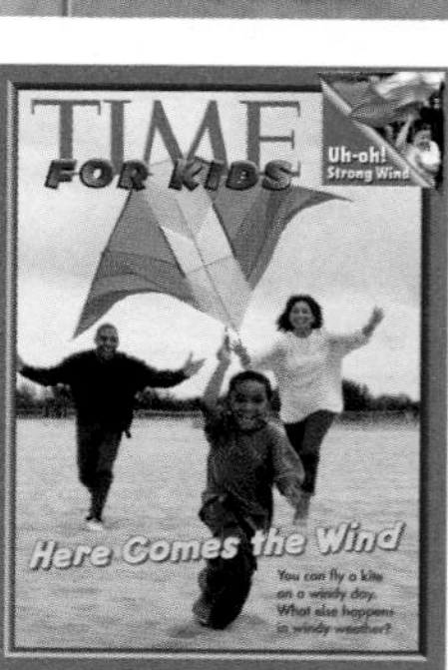

Includes *Time for Kids* magazine about windy weather.

Bibliography

Lesson 1 Exploring Weather

Like a Windy Day, by Frank Asch and Devin Asch (Harcourt Children's Books, 2002)

Rain Romp: Stomping Away a Grouchy Day, by Jane Kurtz (Greenwillow Books, 2002)

Snowflake Bentley, by Jacqueline Briggs Martin (Houghton Mifflin, 1998)

Snow Day!, by Patricia Lakin (Dial, 2002)

Super Storms, by Seymour Simon (SeaStar, 2002)

Lesson 2 Look at Clouds

It Looked Like Spilt Milk, by Charles G. Shaw (HarperTrophy, 1988)

This Is the Rain, by Lola M. Schaefer (Greenwillow Books, 2001)

Clouds, by Marion Dane Bauer (Aladdin, 2004)

Twister on Tuesday (Magic Tree House #23), by Mary Pope Osborne (Random House Children's Books, 2001)

Lesson 3 The Seasons

Here Comes the Year, by Eileen Spinelli (Henry Holt, 2002)

Moon Glowing, by Elizabeth Partridge (Dutton, 2002)

Pieces: A Year in Poems and Quilts, by Anna Grossnickle Hines (Greenwillow Books, 2001)

Spring Thaw, by Steven Schnur (Viking, 2000)

Lesson 4 Night and Day

Big Bear Ball, by Joanne Ryder (HarperCollins, 2002)

Some from the Moon, Some from the Sun: Poems and Songs for Everyone, by Margot Zemach (Farrar, Straus, Giroux, 2001)

Common Ground: The Water, Earth, and Air We Share, by Molly Bang (Blue Sky, 1997)

Lesson 5 Sun and Shadows

Light: Shadows, Mirrors, and Rainbows, by Natalie M. Rosinsky (Picture Window Books, 2002)

Shadows, by Carolyn B. Otto (Scholastic, 2001)

The Sun: Our Nearest Star, by Franklyn M. Branley (HarperCollins, 2002)

Units C and D Earth and Weather

Moon Glowing, by Elizabeth Partridge (Dutton, 2002)

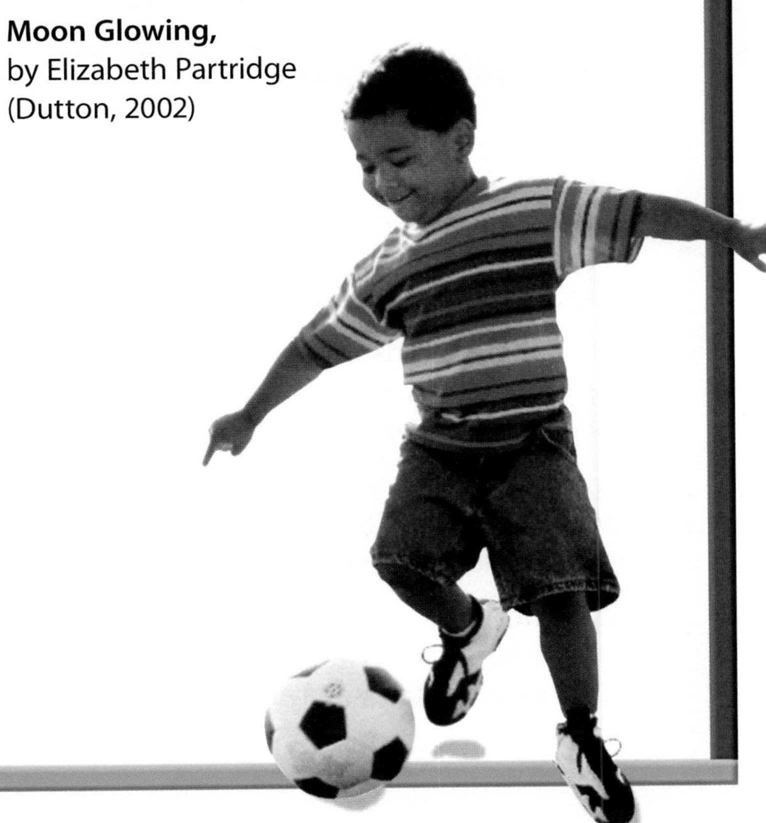

For more information on these or other books, visit **www.bankstreetbooks.com** or your local library.

Vocabulary Activities

Science Vocabulary

Vocabulary

- rainy
- windy
- snowy
- sunny
- cloud
- winter
- spring
- summer
- fall
- shadow
- heat
- Moon
- stars

The Seasons

Explore what children know about weather and seasons by creating a four-column language experience chart.

- Have children name the four seasons of the year. Write the seasons on chart paper.
- Invite children to share different types of weather, clothing, or seasonal changes related to each season and write their ideas in the appropriate column of the chart. Revisit and add to the chart throughout the unit.
- Use the routine on the **Vocabulary Cards** for further vocabulary development.

Seasons

Winter	Spring	Summer	Fall
cold	rain	hot	leaves
sweater	flowers	swim	cool
		sunny	

ELL Support

Weather Words

Write the vocabulary words on the board. Ask children to repeat them as you say them. Say them again and ask children to clap when they hear words beginning (or ending) with /s/.

- **Beginning** Use pantomime to reinforce the vocabulary words as you say each word aloud. Invite children to repeat your actions and echo the words.
- **Intermediate** Provide children with paper and crayons. Say a vocabulary word and invite them to draw pictures to illustrate it. Invite children to tell about their drawings.
- **Advanced** Ask children to say a sentence about weather based on the information in the four-column chart you created in the Science Vocabulary activity.

A to Z Activity Book

Look for additional cross-curricular activities about weather and sky in the **A to Z Activity Book.**

N is for Nature, pp. 28–29
Writing, Social Studies, Music

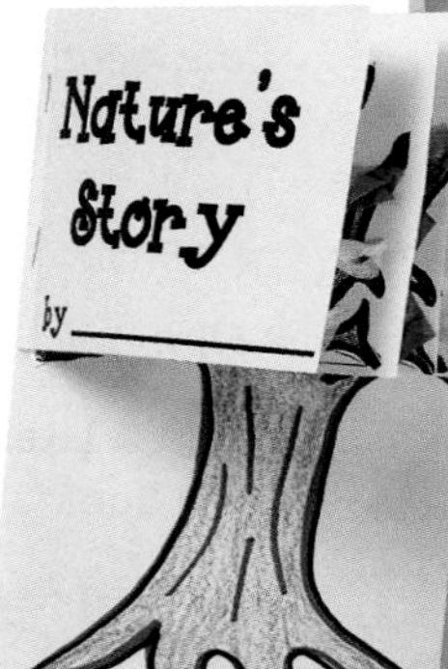

S is for Shadows, pp. 38–39
Reading, Art, Writing, Science

U is for Universe, pp. 42–43
Science, Reading, Writing

Z is for Z-z-z at Night, pp. 52–53
Science, Music, Reading, Math

Unit Project

Track the Temperature

Objective: Record the daily high and low temperatures over a period of time.

You need
- daily newspaper
- pencil
- paper
- thermometer

Step 1 Invite children to keep track of the outside temperature every day for a month. Discuss what temperature means and explain that it can be higher or lower at different times of the day. Showing your classroom thermometer, explain how to use it to read the temperature.

Step 2 Have children look in a local newspaper to find the previous day's high and low temperatures. Encourage children to take turns looking up and recording the temperatures. You may want to have children add the temperatures to the monthly weather graph.

Step 3 At the end of the month, help children analyze the temperature data.

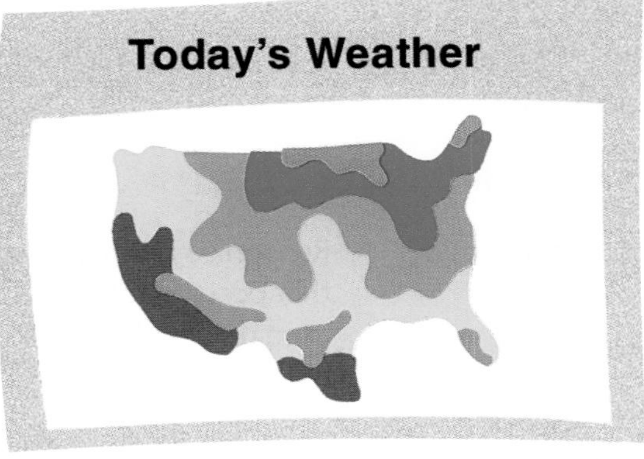

Technology

IWB INTERACTIVE WHITEBOARD READY

www.macmillanmh.com

Visit Macmillan/McGraw-Hill Science online for projects and activities for students, teachers, and parents.

www.bankstreet.edu

Visit Bank Street online for teacher resources and more activities.

www.timeforkids.com

Visit Time for Kids online to read about current science news and explore additional science activities.

UNIT D Planner

TeacherWorks™ Plus CD-ROM has interactive unit planner, Teacher's Edition, and worksheets.

Lesson	OBJECTIVES	VOCABULARY	RESOURCES and TECHNOLOGY	
1 Exploring Weather PAGES 170–177 PACING: 3–5 days	■ Recognize the characteristics of different kinds of weather, such as wind, sun, rain, and snow.	**rainy** **windy** **snowy** **sunny**	**Flipbook,** pp. 38–39 **Unit D Literature Big Book:** *Time for Kids* **Leveled Readers:** *What is the Weather?* and *What Will I Wear Today?*	■ **Science Resource Book,** p. 46 ■ **Photo Sorting Cards 31–40** ■ **A to Z Activity Book,** pp. 28–29
2 Look at Clouds PAGES 178–183 PACING: 3–5 days	■ Describe clouds and how they change.	**cloud**	**Flipbook,** p. 40 **Unit D Literature Big Book** **Leveled Reader:** *Clouds* ■ **Science on the Go,** pp. 75–76	■ **Science Resource Book,** p. 47 ■ **Floor Puzzle:** *Landforms*
3 The Seasons PAGES 184–189 PACING: 3–5 days	■ Identify what occurs in nature and what people do in different seasons.	**winter** **spring** **summer** **fall**	**Flipbook,** pp. 41, G1 **Unit D Literature Big Book** **Leveled Reader:** *A Favorite Season* and *Seasons* ■ **Science Resource Book,** p. 48	■ **A to Z Activity Book,** pp. 28–29 ■ **Science on the Go,** pp. 77–78
4 Night and Day PAGES 190–197 PACING: 3–5 days	■ Recognize changes that occur in the sky from day to night and night to day.	**day** **night** **Moon** **Sun** **stars** **patterns**	**Flipbook,** pp. 42–43, S8 **Unit D Literature Big Book** **Leveled Reader:** *The Night Sky* ■ **Science Resource Book,** p. 49	**Science Songs CD** ■ **Photo Sorting Cards 31–40** ■ **A to Z Activity Book,** pp. 42–43, 52–53 ■ **Science on the Go,** pp. 79–82
5 Sun and Shadows PAGES 198–203 PACING: 3–5 days	■ Recognize that the Sun creates shadows and appears to move through the sky.	**shadows** **shade** **heat**	**Flipbook,** pp. 44, S9 ■ **Science Resource Book,** p. 50	**Science Songs CD** ■ **A to Z Activity Book,** pp. 38–39, 42–43 ■ **Science on the Go,** pp. 83–86

PACING Assumes a day is a 20–25 minute session.

www.macmillanmh.com for more planning resources and http://nsdl.org/refreshers/science for science resources from NSDL

BE A SCIENTIST Activities

Wind Effects p. 176

PACING: 30 minutes 

Skills predict, observe, classify, draw a conclusion, infer

Materials fans, feathers, paper, plastic bags, aluminum foil, plants, blocks

Observe Clouds p. 182

PACING: 30 minutes 

Skills communicate, observe, compare, infer

Materials chart paper, markers

Nature Walk p. 188

PACING: 30 minutes 

Skills predict, observe, communicate, compare

Materials chart paper, markers, paper, paint, glue, clipboards, pencils

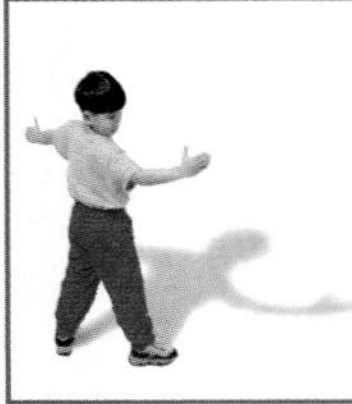

The Night Sky p. 196

PACING: 30 minutes 

Skills classify, compare, observe, communicate

Materials Photo Sorting Cards 31–40, daily newspaper, paper

Change Shadows p. 202

PACING: 30 minutes 

Skills observe, compare, measure, investigate, infer

Materials overhead projector, large white paper, markers, classroom objects, masking tape

UNIT ASSESSMENT

Formative Assessment, pp. 175, 181, 187, 195, 201
Performance Assessment, p. 204
Summative Assessment, p. 205
Portfolio Assessment, p. 205

SCIENCE KIT MATERIALS

- rocks
- wooden blocks
- small clear containers
- small fabric pieces
- modeling clay
- toy boats
- flashlights
- string

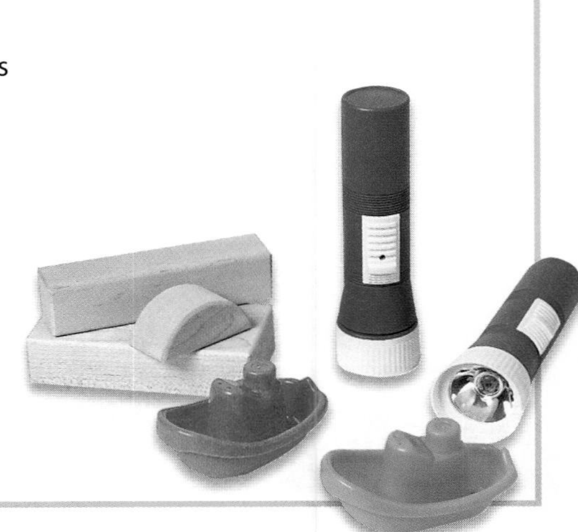

COLLECTIBLES

Begin collecting these items for Unit D:

- feathers
- plastic bags
- egg cartons
- craft sticks
- chenille stems
- aluminum foil

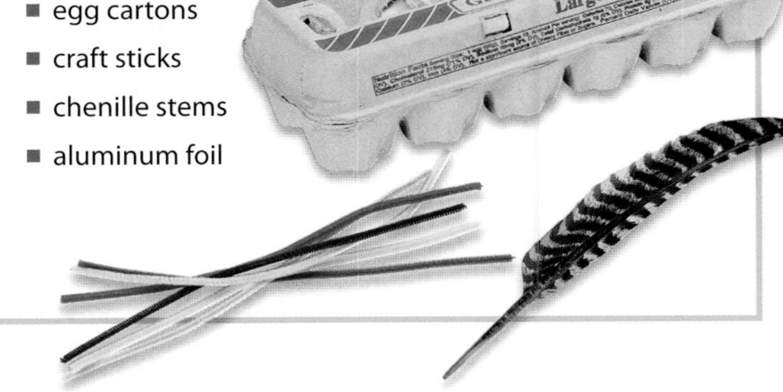

English Language Learner Support

Technology

For additional language support and oral vocabulary development, go to www.macmillanmh.com

Science in Motion
The Seasons

Vocabulary Games

Academic Language

English language learners need help in building their understanding of the academic language used in daily instruction and science activities. The following strategies will help to increase children's language proficiency and comprehension of content and instruction words.

Strategies to Reinforce Academic Language

- **Use Context** Academic Language should be explained in the context of the task. Use gestures, expressions, and visuals to support meaning.
- **Use Visuals** Use charts, transparencies, and graphic organizers to explain key labels that help children understand classroom language.
- **Model** Use academic language as you demonstrate the task to help children understand instruction.

Language Transfers

Phonics Transfer
Hmong and Vietnamese do not have the sound /w/.

Grammar Transfer
Spanish speakers use the verb *to have* when talking about being warm or cold.
In winter I have cold.

Academic Language Vocabulary Chart

The following chart shows unit vocabulary and inquiry skills as well as some Spanish cognates. **Vocabulary** words help children comprehend the main ideas. **Inquiry Skills** help children develop questions and perform investigations. **Cognates** are words that are similar in English and Spanish.

Vocabulary		Inquiry Skills	Cognates	
			English	Spanish
rainy, p. 170	day, p. 190	predict, p. 176	predict, p. 176	*predecir*
windy, p. 170	night, p. 190	observe, p. 176	observe, p 176	*observar*
snowy, p. 170	Moon, p. 190	classify, p. 176	classify, p. 176	*clasificar*
sunny, p. 170	patterns, p. 191	draw a conclusion, p. 176	conclusion, p. 176	*conclusión*
cloud, p. 178	stars, p. 191	infer, p. 176	infer, p. 176	*inferir*
winter, p. 184	shadow, p. 198	communicate, p. 182	communicate, p. 182	*comunicar*
spring, p. 184	shade, p. 200	compare, p. 182	compare, p. 182	*comparar*
summer, p. 184	heat, p. 200	measure, p. 202	investigate, p. 202	*investigar*
fall, p. 184		investigate, p. 202		

Vocabulary Routine

Use the routine below to discuss the meaning of each word on the vocabulary list. Use gestures and visuals to model all words.

Define *Winter* is the season that comes after fall.

Example Sometimes it snows in *winter*.

Ask What else happens in *winter*?

Children may respond to questions according to proficiency level with gestures, one-word answers, or phrases.

Vocabulary Activities

Help children identify the seasons of the year.

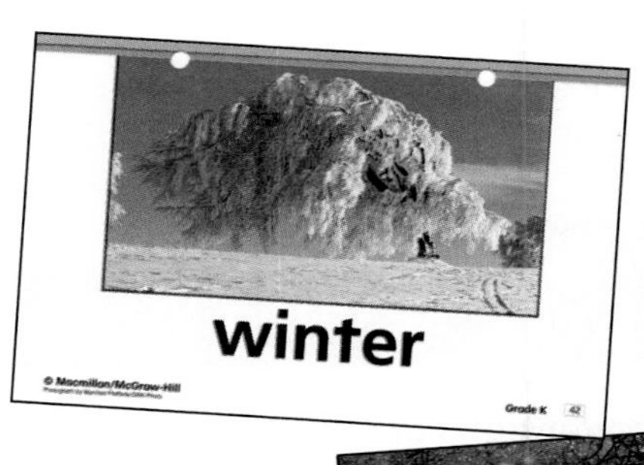

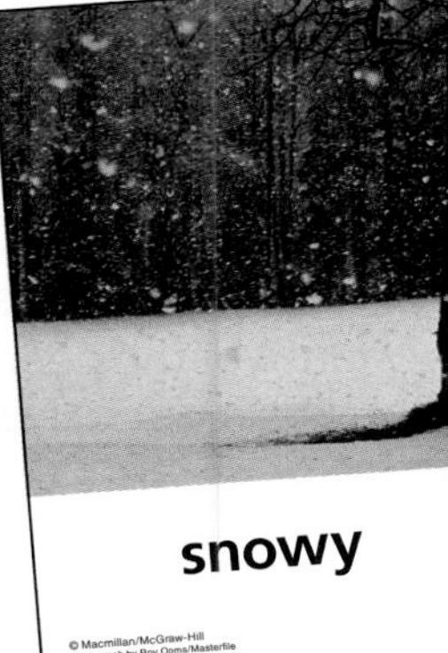

BEGINNING Show winter clothes. Say: *We wear these clothes in winter. It is very cold in winter. These clothes keep us warm*. Write, say, and have children repeat, *winter*. Then ask: *What season is it? Is it winter?*

INTERMEDIATE Turn to page 41 of the Flipbook. As you point to the pictures, say: *The year has four seasons. This is winter. This is spring. This is fall. This is summer*. Have children repeat the words. Ask: *What are some things you like to do in winter?*

ADVANCED Form two groups. Give one group Sorting Card 31 and the other group Sorting Card 33. Have children keep the cards facedown. Say: *Turn your cards over. Let's see what kind of weather they show*. Guide students to identify snowy weather with winter and sunny weather with summer.

UNIT D

Objective: Children will identify and describe various characteristics of weather and how people and places are affected by weather.

Before Reading

- Invite children to share what they know about the images that appear on the Flipbook page.
- Explain that in this unit, children will be scientists as they observe, describe, and compare various kinds of weather.

During Reading

- Read the poem, pointing to the words as you read.
- Read again, pausing to invite children to point to images on the page that correspond to the lines in the poem.
- Reread the poem two to three times, encouraging children to chime in as you read.

Flipbook

UNIT D

Weather and Sky

Thunder and lightning!
The rain pounds on the window.
Swirls of white and icy flakes
Dust the trees with snow.

On days with rain and Sun,
Rainbows may fill the sky.
On nights with no thick clouds,
The Moon glows bright and high.

by Laura Sedlock

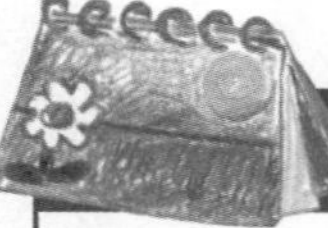

Reading Strategy

Conventions of Print Write *rain* and ask children to help you find how many times it appears in the poem. If no one mentions *rainbow*, cover up *bow* with a piece of paper and point out that *rain* is a smaller word inside the word *rainbow*.

Phonological Awareness Read the second stanza of the poem aloud and ask children to listen carefully for words that rhyme. List responses on a chart.

Science Facts

Blizzard A blizzard is a severe snowstorm that involves a combination of freezing temperatures, very strong winds, and blowing snow. By definition, a blizzard creates winds that are 35 miles per hour or greater and snowfall that significantly reduces visibility for at least three hours.

Rainbow A rainbow is formed when sunlight meets rain droplets in the air. The raindrops act like a prism, bending the sunlight so that its white light is separated into its component colors. The raindrops then act like a mirror and reflect this refracted light back towards the Sun, creating a rainbow.

37

After Reading

- Ask children if they have ever experienced or seen the kinds of weather shown in these pictures. **Where were you? What did it feel like?**
- Label the images *snow, lightning, rainbow, Moon*. Then reread the poem, and have children listen for the words used to describe the weather. **What words would you use to describe snow, lightning, rainbows, or the Moon?**

Write on It

Have children use a dry-erase marker to write their first initial on their favorite kind of weather on the page. Ask them to explain why they like that kind of weather.

▶ Revisit

Review the poem on another day, and have the group create a movement and/or sound to accompany each image on the page.

Science Background For more information on weather go to www.macmillanmh.com.

School to Home

At the beginning of Unit D, give children copies of the Home Letter on page 13 from the **Science Resource Book.** Read the letter aloud and help children write their names. Have children bring the letter home to share with their families.

At the end of Unit D, send children home with a copy of the Home Letter on page 15 to reinforce what they learned about weather and sky.

See pages 14 and 16 in the Science Resource Book for Spanish versions of these letters.

Lightning can occur when the particles in clouds separate into two sections, one with a negative charge and one with a positive charge. Lightning is a visible electrical discharge produced by a separation between the positive and negative portions of a cloud.

The Moon orbits Earth in a slight oval shape called an ellipse. As seen from Earth, the Moon appears to rise, move across the sky in a curved path, and then set. This apparent motion of the Moon across the sky is caused primarily by Earth's rotation.

Professional Development For more Science Facts and resources from NSDL visit http://nsdl.org/refreshers/science

▶ **Essential Question**
What are some different kinds of weather?

▶ **Objective**
Recognize the characteristics of different kinds of weather, such as wind, sun, rain, and snow.

▶ **Vocabulary**
rainy **snowy**
windy **sunny**

Resources

Flipbook, pp. 38–39

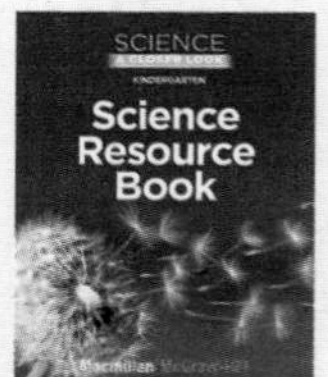
Science Resource Book, p. 46

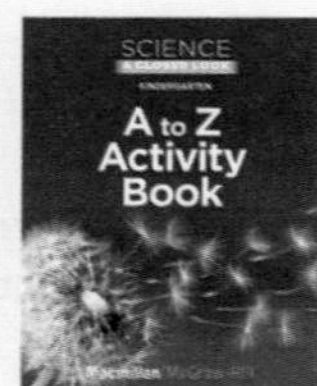
A to Z Activity Book, pp. 28–29

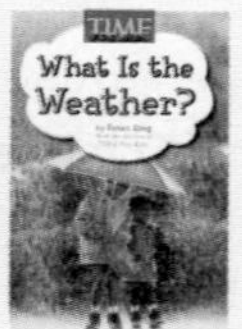
Leveled Reader: *What Is the Weather?*

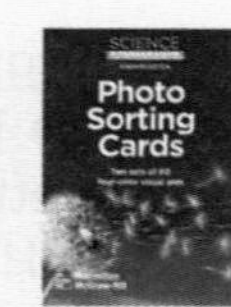
Photo Sorting Cards 31–40

Technology

www.macmillanmh.com

NSDL

Exploring Weather

Circle Time

WHOLE CLASS

What's the Weather?

Inquiry Skills: observe, infer, communicate

- Show children the weather cards and then put them in a bag. Have a volunteer choose one and act out how the weather might feel.
- Invite the class to guess what kind of weather the child is acting out. Repeat until all children have had a turn.

Observe Ask: **What types of weather can you see?** rain, snow, wind **What weather cannot be seen?** cold, hot

You need

- Photo Sorting Cards 31–40
- bag

Time to Move!

Assign movements to types of weather. *Sunny: jump up and down; Rainy: stomp your feet,* and so forth. Encourage children to make the proper movement when you call out a type of weather.

Be a Reader

You need
- Unit D Literature Big Book
- Leveled Reader: *What Is the Weather?*
- chart paper
- marker

Read the Big Book

Objective: Explore different weather.

Inquiry Skills: observe, communicate

- Read the title and point to the photo on the cover. Ask: **What type of weather do you see?**
- Read the book. Then reread it, inviting children to chime in. Ask children to find words on pages 4 and 5 that are the same. Today, is, a, day
- **Guided Reading:** Use the inside back cover of the Leveled Reader version of *What Is the Weather?* for activities.
- For more to read on weather, read *What Will I Wear Today?* available in the Unit D Literature Big Book and as a Leveled Reader.

Be a Writer

You need
- paper
- pencils or markers

Our Favorite Weather

Objective: Create a visual image of one type of weather.

Inquiry Skill: communicate

- Brainstorm a list of weather words with the class. Then explain to the group that they will make a class book about their favorite types of weather.
- Ask children to draw a picture of their favorite kind of weather and what they like to do in that type of weather. Have children write about their picture, or dictate one or two sentences to a teacher.
- After each child has made a page, bind the pages together to make a book. Ask the group for suggestions for the title. Read the book to the group at story time.

I like to swim when it is hot.

Be a Math Wiz

Activity

You need
- Photo Sorting Cards 31–40

Weather Match

Objective: Collect and count pairs of matching sorting cards.

Inquiry Skills: classify, compare, observe

- Arrange the Photo Sorting Cards facedown in rows.
- Have children take turns flipping over two cards. If they do not get a match, they should turn both cards over again. If they do get a match, have them take the pair of cards.
- When all of the cards have been paired up, ask children: **How many matches did you find?** Have children count their pairs of matching cards and compare their findings.

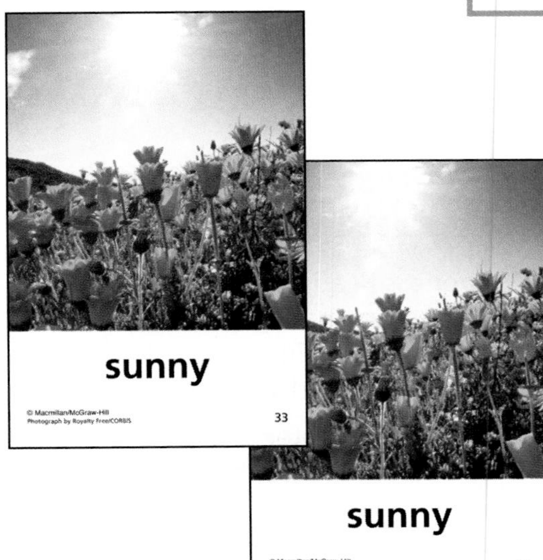

Read Together and Learn

What do you know about snow?

▶ Build on Prior Knowledge

Ask volunteers to share something they have done in the snow. If children have never experienced snow, ask them what they might do if it snowed.

▶ Use the Visuals

- Observe Ask a volunteer to describe what the family in the image is doing. Have children talk about other activities that people can do in the snow.
- Communicate Have the group generate a list of words to describe what snow looks and feels like. Record their ideas.
- Point to the photo of snow falling and ask: **What do you know about snowflakes?**

▶ Develop Vocabulary

snowy, **sunny** Write *snowy* and *sunny* on chart paper. Ask: **When you go outside on a snowy (sunny) day how might you feel? What do you think snow feels like?** cold **How do you feel when the Sun is shining on you?** warm, hot

Flipbook

UNIT D • LESSON 1

What do you know about snow?

Reading Strategy

Phonemic Awareness Read the question on the Flipbook page aloud, and have children listen for the rhyme (*know, snow*). Encourage children to think of other words that rhyme with *snow*.

Print Awareness Ask a volunteer to point to the words *know* and *snow* and describe what is the same about them. Underline *ow* in each word. Write other words that follow the same pattern, such as *flow, grow,* and *low*.

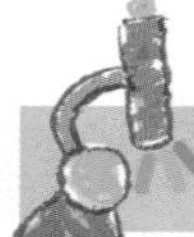

Science Facts

Snow Formation Snow is formed by ice crystals that join together as they fall from clouds toward Earth. If snowflakes do not pass through a layer of warm air that causes them to melt, they reach the ground as snow.

Snowflakes are made up of many ice crystals that are joined to form a six-sided (hexagonal) shape. Snowflakes come with an infinite number of variations, which are determined by the temperature when the crystal forms as well as the humidity in the air.

Professional Development For more Science Facts and resources from NSDL visit http://nsdl.org/refreshers/science

▶ Ask Questions

- **What is the weather usually like when it snows?** very cold, cloudy **Does it ever snow in a warm place?** Have children explain their answers. No, it must be cold to snow.
- **How are rain and snow the same or different?** Possible answers: they both fall from the sky; they are both wet; snow is frozen and rain is not.
- Ask children if they have ever looked at a snowflake with a hand lens or seen a close-up picture of a snowflake. **What do you know about the shapes of snowflakes?**

Write on It

Discuss how all snowflakes have six points, but come in many different shapes. Have children use a dry-erase marker to draw and label their own snowflake on the Flipbook page.

Think and Talk

Invite children to look at the image of the family on the Flipbook page. Ask: **What is this family doing?** Prompt children to discuss whether or not this family would be able to go sledding if there was no snow on the ground.

TIME FOR KIDS

Shared Reading Read *Here Comes the Wind* with the class. Use this magazine, found in the **Unit D Literature Big Book,** as a photographic review of wind and extreme weather.

More to Read

Snowflake Bentley, Jacqueline Briggs Martin (Houghton Mifflin, 1998)

Activity Read this book with children to explore snowflakes and how each one is unique. After reading the book, provide black paper and white crayons and invite each child to create their own snowflake shape.

Read Together and Learn

What can wind do?

▶ Build on Prior Knowledge

Ask children to describe what happens outside when it is very windy. Record their responses.

▶ Use the Visuals

- Observe Point to each picture and ask volunteers to describe what is happening and why. For example, if children say that the umbrella is inside out, ask them how they think that might have happened.
- Communicate Explain that there are terms used to describe different kinds of windy weather. As you point to each picture, say the word that describes that kind of weather.
- Ask children if they have ever experienced a big storm. **Did it look like one of these pictures? How was it the same or different?**

▶ Develop Vocabulary

rainy, **windy** Have children come up with sentences that use *rainy* or *windy*. Encourage them to create sentences that show their understanding of the word's meaning. For example, "On a *windy* day, my hair blows all around."

Flipbook

UNIT D • LESSON 1

What can wind do?

Reading Strategy

Phonics Ask the group to identify words on the page that begin with *w*. Ask children what sound the letter *w* makes, and have them name other words that begin with that letter.

Concepts of Print Ask a volunteer to count how many words are on the page, having them point to each word as they count. Have them explain how they know when one word ends and a new one begins.

Science Facts

Wind is caused by the movement of air from an area of high pressure to an area of low pressure. These pressure changes in the atmosphere are due to the warming and cooling of the air.

Hurricanes are tropical cyclones that circulate about their centers, with winds greater than 74 miles per hour and diameters of around 100 miles.

In order for thunderstorms to grow into hurricanes, the surface temperature of the ocean must be at least 82°F and have overhead rotating winds.

Rainstorm Rain forms when cloud particles become too heavy to remain in the cloud and fall toward Earth's surface. Heavy rainstorms are often accompanied by strong winds.

Professional Development For more Science Facts and resources from NSDL visit http://nsdl.org/refreshers/science

39

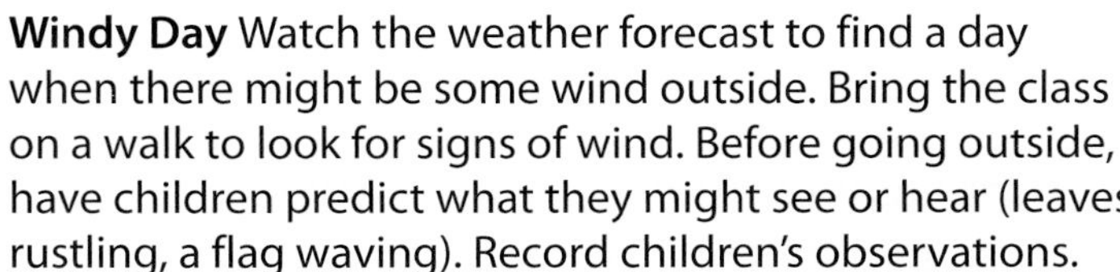

Take a Trip

Windy Day Watch the weather forecast to find a day when there might be some wind outside. Bring the class on a walk to look for signs of wind. Before going outside, have children predict what they might see or hear (leaves rustling, a flag waving). Record children's observations.

Back in the Classroom Read aloud the list of what the children observed outside on their walk. Invite volunteers to choose something from this list and act it out, and have the group try to guess what they are doing.

▶ Ask Questions

- **How is the wind changing things?** The trees are blowing; the kite is able to fly; the dog's fur is blowing around.
- Have children imagine that they are in these places. **What do you think the wind might feel like?** strong, cool, fast
- **How can wind be helpful?** It helps sailboats move, creates energy with turbines or windmills, helps a kite to fly, helps to carry seeds from one place to another.

Write on It

Have children use a dry-erase marker to label various elements of weather that appear in the pictures, such as clouds, wind, and rain. Help them identify the sounds in each word and their corresponding letters.

Think and Talk

Look again at the pictures on the Flipbook pages. Ask children to imagine that it snowed or rained where they live. **How would the temperature feel? What would the wind do? How could you protect yourself from it?**

Formative ASSESSMENT

Notice if children are able to describe the various ways that wind can make things move and change. If they are having difficulty explaining what wind can do, take a trip outside and engage them in a conversation about the weather.

Wind Effects

Objective: Understand the effect of wind on various materials.

Inquiry Skills: predict, observe, classify, draw a conclusion, infer

You need **battery-powered or hand-operated fans, materials with various properties (feathers, paper, plastic bags, aluminum foil, plants, wooden blocks), Science Journal p. 46**

1. **Predict** Show the objects and ask: **What do you think will happen to each of these objects when wind blows on them?** Record children's responses.
2. **Observe** In pairs, have one child use the fan while his or her partner holds various objects in front of the fan. Have children record what happens to each object on their Science Journal pages.
3. **Classify** After each pair has explored a number of materials, ask them to describe what happened: **Which materials moved? Which ones did not?**
4. **Draw a Conclusion** Ask: **Why did the wind move some objects, but not others?**

Teacher Tip

This activity is more fun and more challenging if you provide a variety of materials that have different properties—stiff or flexible, heavy or light. As children discuss how these materials respond to the wind in different ways, you will be supporting their higher-order thinking skills.

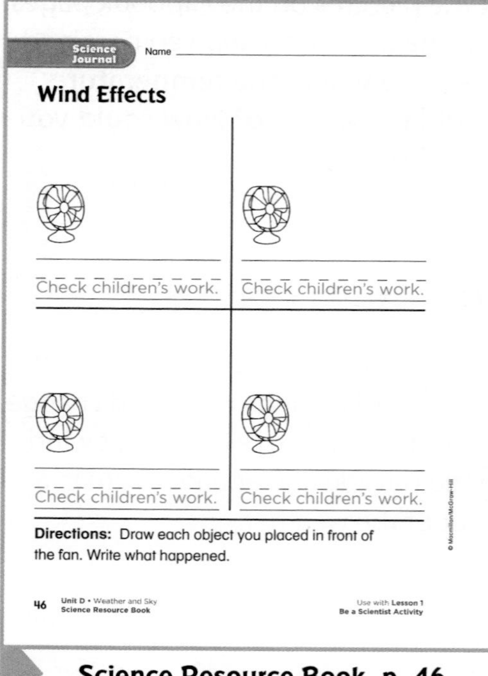
Science Journal Name

Wind Effects

Check children's work. Check children's work.

Check children's work. Check children's work.

Directions: Draw each object you placed in front of the fan. Write what happened.

46 Unit D • Weather and Sky Science Resource Book   Use with Lesson 1 Be a Scientist Activity

Science Resource Book, p. 46

Investigate More

Infer Say: **Think about what happens when there is wind outside. What moves in the wind? What does not move? What can happen if there is very strong wind outside?** Possible answers: leaves on trees blow; branches fall down; my hair blows around.

centers

Art

Moving with Wind

Objective: Create model sailboats.

Inquiry Skill: make a model

- Help children explore how they can use materials to create small sailboats that will float in the Water Table.
- You may want to show children some pictures of sailboats to help them think about the design of their boat.
- Encourage children to share their ideas with classmates as they work to create their boats. Then have them test their boats in the Water Table.

You need
- small plastic containers
- plastic egg cartons or other recyclable materials that float
- craft sticks
- paper
- fabric
- chenille stems
- tape
- glue
- scissors

Technology

More About Weather and Sky

- For games and activities about weather and sky, help children log onto  www.macmillanmh.com.
- Children can use these fun games throughout this unit to reinforce what they learn about weather and sky. Encourage children to revisit these games throughout the school year to review the concepts.

LOG ON www.macmillanmh.com for more science online.

Movement

Wild Weather

Objective: Act out different types of extreme weather.

Inquiry Skill: communicate

- Have children share some types of weather that they noticed on the Flipbook page or extreme weather that they know.
- Invite children to act out the different types of weather, such as tornado, lightning, strong wind, or snowstorm. Ask questions like these: **Is the wind very strong? Is it raining? How does the sky look?**

Water Table

Boats on the Water

You need
- toy boats or child-made boats
- hand fan

Objective: Understand how wind moves objects on the surface of water.

Inquiry Skills: observe, predict

- Have children blow air to move toy boats or boats they created across the surface of the water.
- Ask children to predict what direction the boat will move in if they wave a fan on it from different sides.
- Have children choose a place in the Water Table that they want the boat to reach, and wave the fan to push the boat over to that point.

For more on weather, use pp. 41-44 in the **Activity Book.**

▶ **Essential Question**
How can clouds be different?

▶ **Objective**
Describe clouds and how they change.

▶ **Vocabulary**
cloud

Resources

Flipbook, p. 40

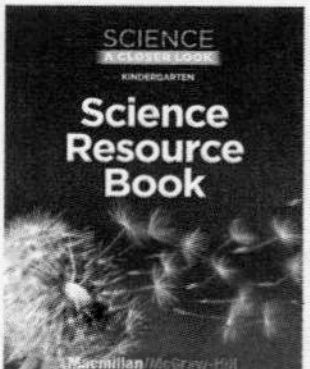
Science Resource Book, p. 47

Leveled Reader: *Clouds*

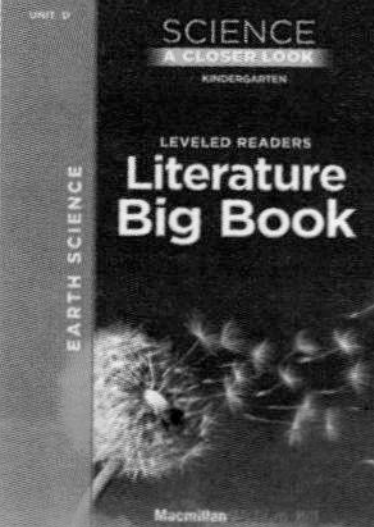
Unit D Literature Big Book

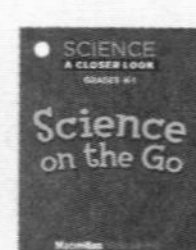
Science on the Go, pp. 75–76

Technology

www.macmillanmh.com

NSDL

Look at Clouds

Circle Time
WHOLE CLASS

Cloud Words

You need
- chart paper
- marker

Inquiry Skills: communicate, compare

- Copy the poem shown below onto chart paper. Read the poem aloud and ask children to help you decide on a good title for it.
- Reread the poem and have children listen for words that describe clouds. List those words, and then have children think of more words that describe clouds. Write their responses.

Compare Ask: **Do clouds always stay in the same place? What makes clouds move and change shape?** wind, rain

What's fluffy white and floats up high,
Like piles of ice cream in the sky?
And when the wind blows hard and strong,
What very gently flows along?

What seems to have lots of fun,
Peek-a-booing with the sun?
When you look up in the sky,
What do you see floating by?

Time to Move!

Have children form a circle and hold a parachute or large sheet. Pull tightly on the sheet and move slowly in a circle to be a cloud on a sunny day. Raise the sheet up and down quickly for a cloud on a stormy day.

Be a Reader Activity

You need
- Unit D Literature Big Book
- Leveled Reader: *Clouds*

Read the Big Book

Objective: Use words and illustrations to explore clouds.

Inquiry Skills: compare, communicate

- Read *Clouds* in the Unit D Literature Big Book with the class. As you read each page, stop to allow children to share their observations about the clouds.
- Reread the story, having children chime in with the predictable text. Invite a volunteer to find the word *clouds* on each page. As you reread the story, have a discussion about the clouds' similarities and differences.
- **Guided Reading:** See the inside back cover of the Leveled Reader version of *Clouds* for activities.

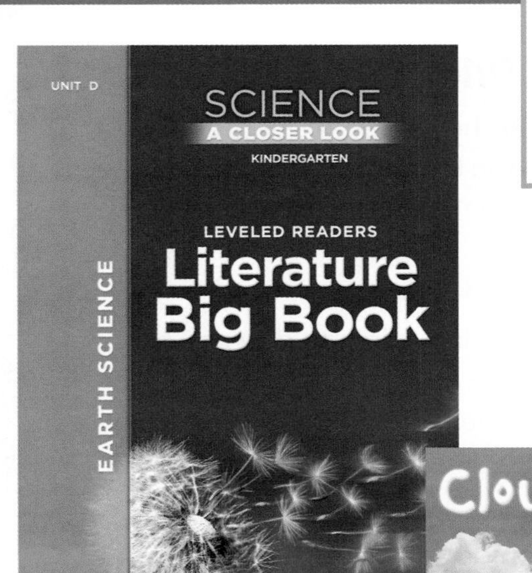

Be a Math Wiz Activity

You need
- pictures or photographs of various types of clouds
- string or rope

Sorting Clouds

Objective: Sort clouds by categories.

Inquiry Skills: communicate, compare, classify

- Have children look through magazines to find and cut out pictures of clouds.
- Encourage children to choose two categories to sort the cloud pictures. If they need guidance, suggest creating categories based on color (white, gray) or shape (puffy, flat).
- Once children have finished, encourage them to think of a new set of categories and sort the pictures again.

Read Together and Learn

What do you notice about these clouds?

▶ Build on Prior Knowledge

Ask children what they know about the shape and color of clouds. Ask: **Do clouds always look the same? Why or why not?**

▶ Use the Visuals

- Communicate Ask children to describe the clouds. List all the adjectives they use. Refer back to the list made in the Circle Time Activity and add any additional words to the list.
- Compare Have children describe what they think is happening in the top picture. Then encourage children to describe the similarities and differences between the three cloud pictures at the bottom.

▶ Develop Vocabulary

cloud Ask children to think of sentences that include the word *cloud*. Write the sentences on chart paper. Explain that scientists use *cirrus, cumulus,* and *cumulonimbus* to identify certain types of clouds. Ask children how they would describe each type of cloud.

Flipbook

UNIT D • LESSON 2

What do you notice about these clouds?

cirrus

cumulus

Reading Strategy

Phonemic Awareness Have children clap for each syllable in the words *cirrus, cumulus,* and *cumulonimbus*.

Print Awareness Have children find words that begin with *c* and words that end with *s*. Then have them count the number of times they see the letter *u* on the page.

Science Facts

Clouds are made of tiny water droplets and/or ice crystals. Rain and snow are cloud particles that fall to Earth because they are too heavy to stay in the air.

Cirrus From the Latin root meaning "curl of hair," cirrus clouds are at high altitudes where temperatures are cold. They are composed mainly of ice crystals.

Cumulus From the Latin root meaning "heap," cumulus clouds are mid-level clouds. Because of the warmer temperatures at this altitude, they are composed primarily of water droplets.

Cumulonimbus The Latin root *nimbus* means "rain." These clouds are thunderclouds and bring rain and storms.

Professional Development For more Science Facts and resources from NSDL visit http://nsdl.org/refreshers/science

cumulonimbus 40

▶ Ask Questions

- **What kind of clouds might you see on a sunny day?** white, puffy, or thin clouds **What happens to the Sun on a cloudy day?** it gets hidden behind clouds; it shines through clouds
- **Does it always rain when there are clouds in the sky?** no **What does the sky look like on rainy days?** gray, very cloudy, overcast
- **Which of these clouds might produce rain?** cumulonimbus **Why do you think they would?** Possible answer: Because they are dark and gray.

Write on It

Have children use a dry-erase marker to trace the shapes of the clouds on the page. Invite children to help you write descriptive terms below each label.

Think and Talk

Prompt a discussion by reviewing what children know about clouds. Children will learn from one another as they share their ideas.

ELL Support

Display a variety of clothing: sunglasses, raincoat, mittens. Name each item and discuss when you might wear it.

Beginning Have children select something to wear on a (rainy) day.

Intermediate Ask yes/no questions: **Do you wear (sunglasses) on a rainy day?**

More to Read

This Is the Rain,
by Lola M. Schaefer
(Greenwillow, 2001)

Activity After reading this book, go back through the text and have children identify the rhymes. Have them describe how the clouds in the illustrations are alike and different.

Formative ASSESSMENT

If children are having difficulty finding the language needed to describe specific features of clouds, take them outside and describe the clouds you see to help them learn descriptive terms.

Be a Scientist **Inquiry Investigation**

Observe Clouds

Objective: Observe and record information about clouds and the weather.

Inquiry Skills: communicate, observe, compare, infer

You need **chart paper, markers, Science Journal p. 47**

1. **Communicate** On chart paper, create a table with a row for each day of the week, and two vertical columns labeled *Clouds* and *Weather*. Show children the chart and explain that they will use it to record observations of the clouds and weather.
2. **Observe** Each day, ask a child to draw a picture of what the clouds look like.
3. **Compare** Discuss the cloud observation and then draw a symbol on the chart to represent the weather on each day. (For example, use a sun for sunny, a raindrop for rainy.) After one week, ask children to look for connections between the clouds and the weather. Ask: **Are there clouds in the sky on sunny days? What do they look like? What are the clouds like on rainy days?**

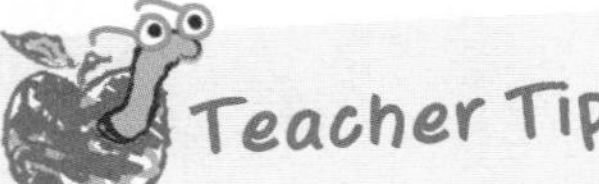

Developing theories based on observation is an important part of being a scientist. In this activity, children will develop ideas about the relationship between clouds and the weather based on observations made over time.

	Clouds	Weather
Monday		
Tuesday		
Wednesday		
Thursday		
Friday		

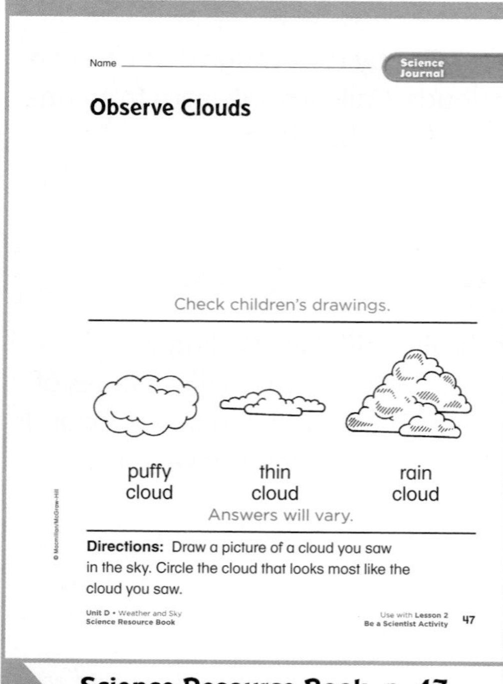
Name ______________ Science Journal

Observe Clouds

Check children's drawings.

puffy cloud | thin cloud | rain cloud

Answers will vary.

Directions: Draw a picture of a cloud you saw in the sky. Circle the cloud that looks most like the cloud you saw.

Unit D • Weather and Sky Science Resource Book | Use with Lesson 2 Be a Scientist Activity 47

Science Resource Book, p. 47

Investigate More

Infer Ask: **Does it ever rain on a sunny day?** yes **Why do you think rain and snow fall from some clouds, but not all clouds?** Rain clouds often look gray and thin; rain and snow often fall from these clouds, not white ones.

centers

Drawing and Writing

Cloud Books

Objective: Create a book that shows different kinds of cloud shapes.

Inquiry Skill: communicate

- Read *It Looked Like Spilt Milk* to the group and have children discuss the shapes of the clouds in the book. Explain that they will have a chance to create their own cloud books.
- Have each child draw three differently shaped clouds on separate pieces of paper. Encourage them to think about what each cloud shape reminds them of, such as a cat or a tree.
- Help them to write words to accompany their drawings, using the patterned language of text as used in the book.

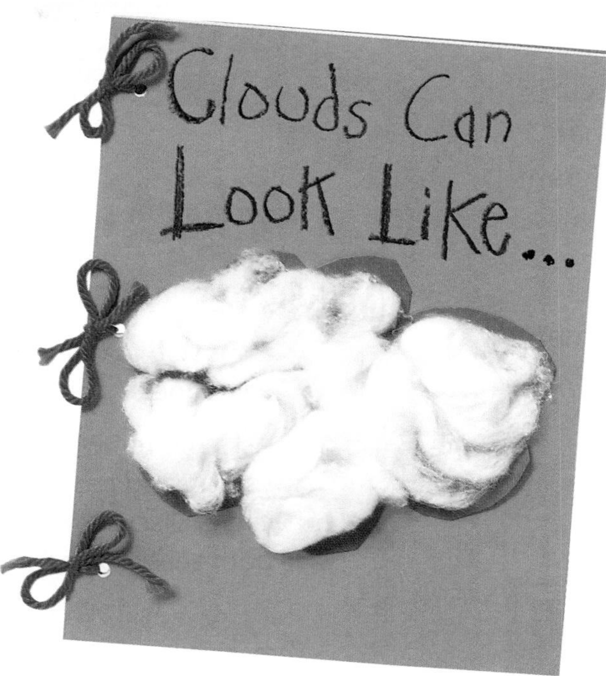

You need
- *It Looked Like Spilt Milk* by Charles G. Shaw
- paper
- crayons
- colored pencils
- cotton

Movement

Storm Clouds

Objective: Act out storm clouds.

Inquiry Skill: communicate

- Have children suppose they were clouds as they move to the music.
- Encourage children to join together by holding hands to make larger clouds.
- Ask: **How could you show rain coming from your cloud?** Invite children to show the rain with their movements.

You need
- music that suggests stormy weather

Art

Paint the Clouds

Objective: Make paintings that convey the shape and color of clouds.

Inquiry Skills: observe, communicate

- Invite children to paint clouds on blue background paper.
- Discuss what kinds of brush strokes they will use to make a particular kind of cloud. For example, they might use thick brush strokes to make a large, heavy-looking cloud.

You need
- blue or purple paper
- paint
- different-sized paint brushes

The Seasons

▶ **Essential Question**
What happens when the seasons change?

▶ **Objective**
Identify what occurs in nature and what people do in different seasons.

▶ **Vocabulary**
winter summer
spring fall

Resources

Flipbook, pp. 41, G1

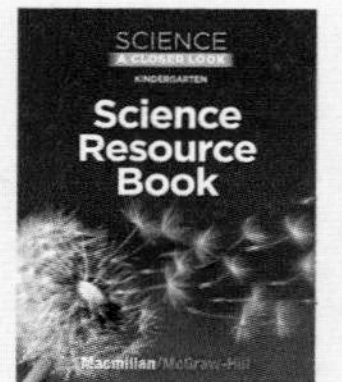

Science Resource Book, p. 48

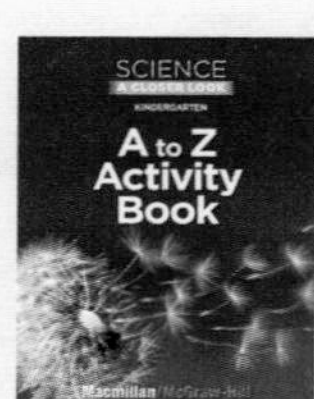

A to Z Activity Book, pp. 28–29

Leveled Reader: *Seasons*

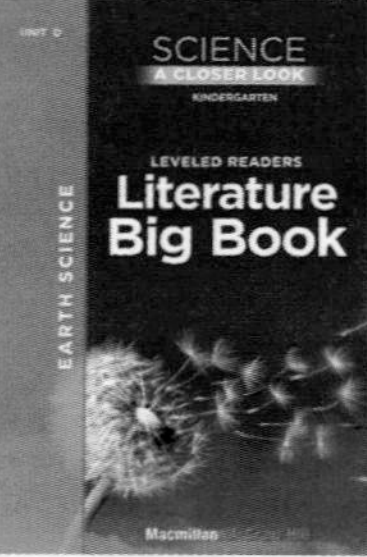

Unit D Literature Big Book

Technology

www.macmillanmh.com

NSDL

Circle Time

WHOLE CLASS

What to Wear?

You need
- Venn diagram, Flipbook page G1
- dry-erase marker

Inquiry Skills: communicate, compare

- Choose two different seasons and write the name of each one in a circle on the Venn diagram.
- Ask children what types of clothing they wear in each season and record their responses in the appropriate circle. Discuss whether some items might be worn in both seasons and record these items in the overlap.

Compare Ask: **What other seasons do you know? How are they similar to or different from the ones we discussed?**

Time to Move!

Have children act out things people do in different seasons, such as ice skating (winter), picking flowers (spring), tossing leaves in the air (fall), and fanning themselves (summer).

Be a Reader Activity

> **You need**
> - Unit D Literature Big Book
> - Leveled Reader: *Seasons*

Read the Big Book

Objective: Recognize that weather changes across the seasons.

Inquiry Skills: observe, compare, communicate

- Show children the cover of *Seasons*. Invite them to tell what is different about the tree and its surroundings in each picture.
- Read the book aloud, pointing to the text. Reread the story, inviting children to chime in. Read the story again, encouraging children to finish the sentences with their own ideas. For example, *Spring is windy.*
- **Guided Reading:** See the inside back cover of the Leveled Reader version of *Seasons* for activities.
- For more to read, see *A Favorite Season,* available in the Unit D Literature Big Book and as a Leveled Reader.

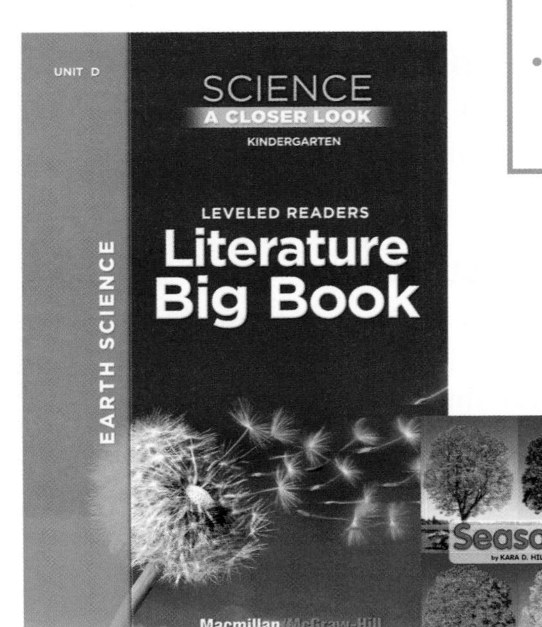

Be a Math Wiz Activity

> **You need**
> - paper
> - pencil
> - clipboard

Take a Survey

Objective: Count using one-to-one correspondence.

Inquiry Skills: classify, compare, communicate

- Explain that children are going to take a survey of their friends' favorite seasons. Help them write *winter, spring, summer,* and *fall* on a piece of paper. Children may also draw pictures to represent each season.
- Have each child ask at least eight friends what their favorite season is. Show them how to make a tally mark next to the appropriate season to record each response.
- Have children add up the tally marks and write the corresponding numerals. Ask: **Which was the most popular season? Which was the least popular?**

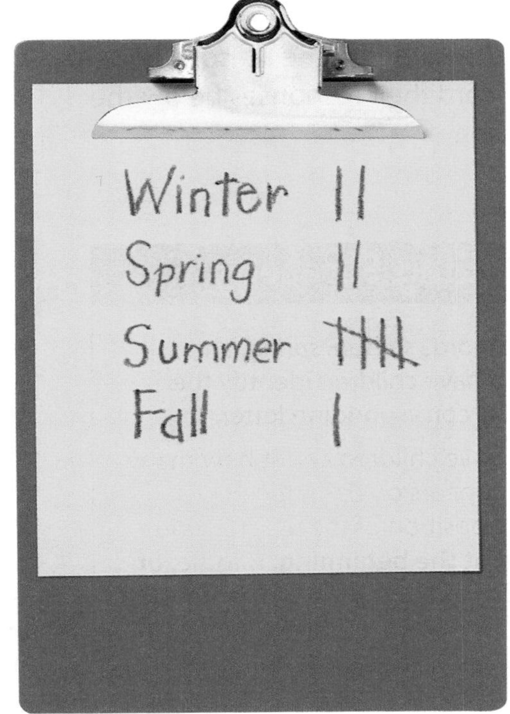

Read Together and Learn

What happens in each season?

▶ Build on Prior Knowledge

Ask children what their favorite season is. Have them explain what they like about that particular season.

▶ Use the Visuals

- **Communicate** Have volunteers describe what the child is doing in each picture. Ask: **What season do you think this child is in? Why?**
- **Observe** Ask children what they notice about the weather in each of these pictures. Invite them to share what each season is like where they live.
- **Infer** Point to the images of the trees. Ask children to tell why they think the leaves look different in different seasons.

▶ Develop Vocabulary

winter, **spring**, **summer**, **fall** Write these words on index cards, ask volunteers to pick a card and tell the group something that happens during that season. Write the names of the seasons on chart paper, and record their responses below the corresponding season.

Reading Strategy

Phonics Read the words *season*, *spring*, and *summer* aloud and have children identify the initial sound and its corresponding letter.

Print Awareness Have children count how many times the letter *s* appears on the page. Have them describe the position of the letter within the word. Ask: **Is it at the beginning, middle, or end of the word?**

Flipbook

UNIT D • LESSON 3

What happens in each season?

fall

Science Facts

Winter In winter, the Northern Hemisphere is tilted away from the Sun. The Sun appears in the sky for fewer hours a day and its energy is less concentrated, resulting in colder temperatures.

Spring During spring, Earth is tilted neither towards nor away from the Sun, resulting in more moderate weather and progressively longer days.

Summer In summer, the Northern Hemisphere is tilted towards the Sun. The Sun appears higher in the sky and results in more hours of daylight, intense sunlight, and warmer temperatures.

Fall In certain areas of the United States, fall is marked by the changing colors of the foliage. Temperature, light, and water supply all influence fall colors.

Professional Development For more Science Facts and resources from NSDL visit http://nsdl.org/refreshers/science

spring

mer

41

▶ Ask Questions

- Point to each picture and read the name of the season. **Is this what (name the season) is like where we live? What is the same or different?**
- **What do you think animals do in each season?** Some animals hibernate or change color in winter. Many birds move to other places.
- **Why are the leaves different colors in the picture labeled fall?** Because seasons affect plants. Some leaves change color during fall. **Why do you think the children are wearing raincoats in the picture labeled spring?** There is more rain during the spring season.

Write on It

Have volunteers use a dry-erase marker to draw and label a picture of something that happens where they live. Children can draw their pictures next to the corresponding seasonal picture on the page.

Think and Talk

Review the pictures on the Flipbook page. Have children look at the directions of the arrows. Ask: **Why is there an arrow from winter to spring? What changes as winter turns into spring?** Repeat the questions with other seasons.

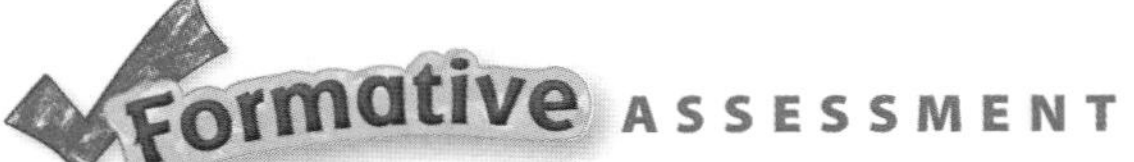

Formative ASSESSMENT

Notice if children can identify differences among the seasons in terms of what happens in nature, the activities people do, or the clothes we wear. If children are having difficulty with this concept, have them do the extra support activity.

Differentiated Instruction

Extra Support Collect pictures from magazines that show different kinds of weather, people doing various activities outside, or things that happen in nature in a specific season. Have children sort the pictures according to what season they happen in and make a collage for each season.

More to Read

Here Comes the Year, by Eileen Spinelli (Henry Holt , 2002)

Activity Before reading, ask children to notice what changes in nature occur in each month. After reading, ask: **Which months do you think were *winter, spring, summer,* or *fall*? How do you know?**

Be a Scientist Inquiry Investigation

Nature Walk

Objective: Identify what happens in nature during different seasons.

Inquiry Skills: predict, observe, communicate, compare

You need **chart paper, markers, paper, paint, glue, clipboards, pencils, Science Journal p. 48**

1. **Predict** Explain to children that they will take a nature walk outside to look for clues about the season. Ask them to make predictions about what they might see.
2. **Observe** As you walk, record children's observations. Encourage them to collect items that have already fallen to the ground, such as pine cones or autumn leaves.
3. **Communicate** When you return to the classroom, copy the list of clues that children found onto chart paper. Have children record their findings on their Science Journal page. Then have children make a mural of an outside scene that shows what season it is using the information they gathered.
4. **Compare** Ask children to imagine the same scene in a different season. Ask: **What would look different? What would stay the same?**

Teacher Tip

In this activity, children will look for clues outside that provide information about the current season. Help children to focus on their nature walk by asking them to think about where they will look before they go outside.

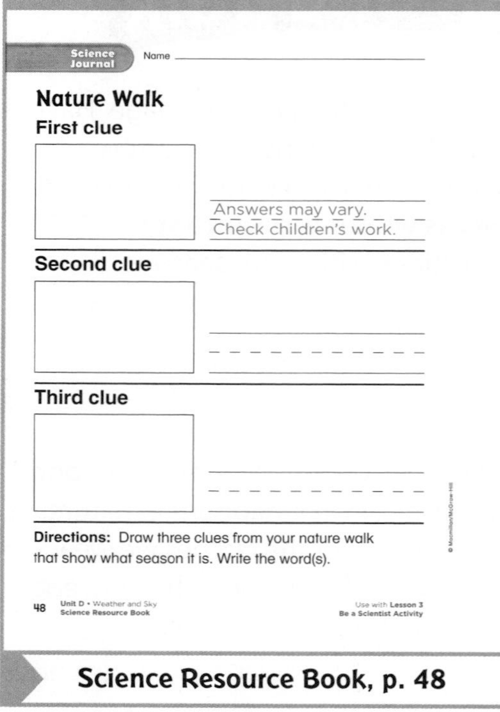

Science Journal Name ________

Nature Walk

First clue

Answers may vary. Check children's work.

Second clue

Third clue

Directions: Draw three clues from your nature walk that show what season it is. Write the word(s).

48 Unit D • Weather and Sky Science Resource Book   Use with Lesson 3 Be a Scientist Activity

Science Resource Book, p. 48

Investigate More

Compare Ask: **What things in nature remain the same in any season? What things change?** Possible answer: There are always trees, but their leaves change or disappear depending on the season.

Art

You need
- cardboard shoe boxes
- paper
- pencils
- crayons
- colored clay
- craft sticks
- index cards

Season Scenes

Objective: Create scenes to indicate a particular season.

Inquiry Skill: make a model

- Have children decide what season they will represent inside their shoe box dioramas.
- Ask children what materials they will need to show the seasons when they build their scenes.
- As children work, record their dictation on an index card and include this card with their scene when displayed in the classroom.

Blocks

You need
- paper
- crayons
- scissors
- tape

City Scene

Objective: Display various aspects of a particular season.

Inquiry Skills: communicate, make a model

- Explain to the group that they will create a small city in the Block Area. Have them decide which season it will be in their city.
- Help children decide what details, props, and shapes they will create to indicate the particular season they have chosen.
- Encourage children to play with their structures, creating scenarios related to that particular season.

Dramatic Play

You need
- scarves
- mittens
- raincoats
- sunglasses
- other seasonal items

Seasonal Activities

Objective: Act out what people do in a particular season.

Inquiry Skill: communicate

- Ask children to decide what season they would like it to be, and help them develop a scenario related to the kinds of activities people do in that season.
- Have children select the props that people use or wear in that season.
- As they play, ask questions to help children think about what happens in that season: **What events take place? What kinds of outdoor activities do people do?**

For more on seasons, use pp. 45-46 in the **Activity Book.**

Night and Day

▶ **Essential Question**
How does the sky change from night to day?

▶ **Objective**
Recognize changes that occur in the sky from day to night and night to day.

▶ **Vocabulary**

day **Sun**
night **stars**
Moon **patterns**

Resources

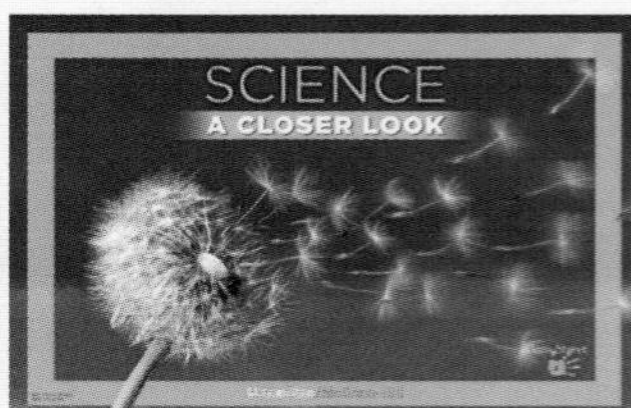
Flipbook, pp. 42–43, S8, G1

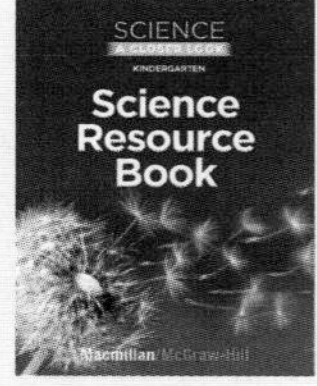
Science Resource Book, p. 49

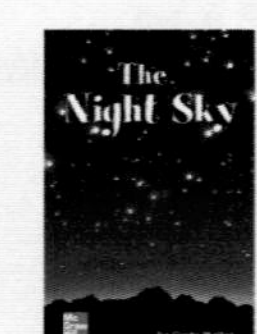
Leveled Reader: *The Night Sky*

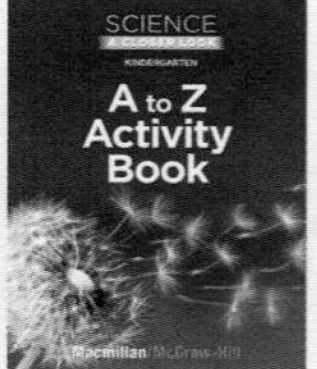
A to Z Activity Book, pp. 42–43, 52–53

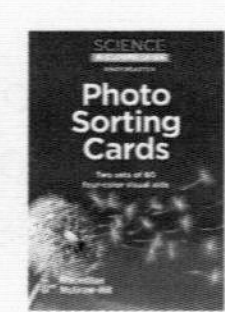
Photo Sorting Cards 31–40

Technology

Science Songs CD Tracks 17–18

www.macmillanmh.com

NSDL

Circle Time

WHOLE CLASS

Day Sky, Night Sky

You need
- Venn diagram Flipbook p. G1
- dry-erase marker

Inquiry Skills: observe, infer, communicate

- Ask: **What does the sky look like at night? During the day?** Record children's responses on a Venn diagram labeled *Night Sky* and *Day Sky*.
- Discuss what might appear during the day and at night. Record these items in the overlap.

Communicate Ask: **Does the Moon always look the same? Why do you think it looks different?**

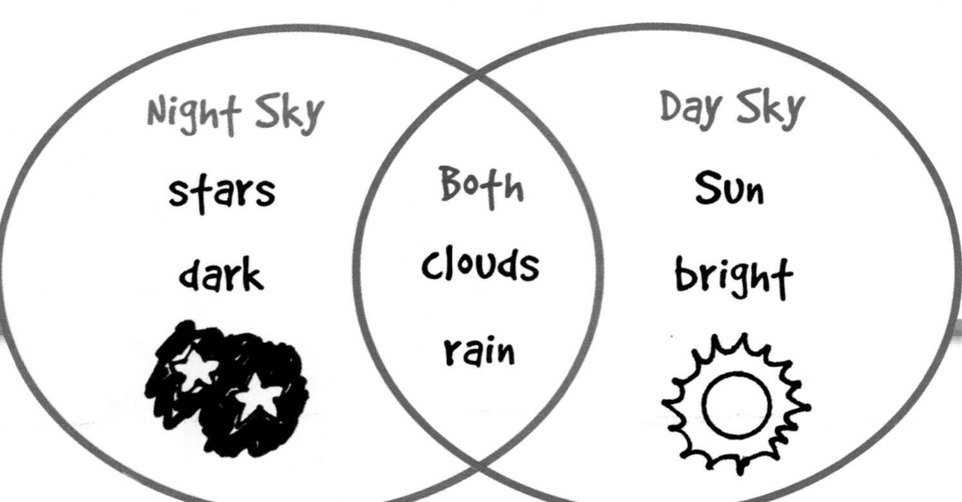

Time to Move!

Give some children white paper plates (Moons) and others yellow paper plates (Suns). Have children move their objects such as, the Sun goes up, the Sun goes down, the Moon is high, the Moon is low.

Be a Reader

Activity

You need
- Unit D Literature Big Book
- Leveled Reader: *The Night Sky*

Read the Big Book

Objective: Use illustrations to explore the night sky.

Inquiry Skills: communicate, compare

- Read *The Night Sky* in the Unit D Literature Big Book to the class. As you read, track the print with your finger.
- Reread the story, stopping at each spread to have children point to words that they know and answer the questions. On the last page, invite children to share star patterns that they see and explain that these are called *constellations.*
- **Guided Reading:** Use the Leveled Reader version of *The Night Sky* for small-group guided-reading instruction.

Be a Math Wiz

Activity

You need
- geoboards
- rubber bands
- constellation charts

Connect the Stars

Objective: Create shapes by making lines between points.

Inquiry Skills: communicate, compare

- Ask children to suppose that the pegs on the geoboards are stars in the sky. Ask: **What kinds of shapes can you make by connecting the stars?**
- After each child has made two or three shapes on his or her board with rubber bands, have them describe the shapes that they made. Ask: **How many stars (points) did you connect? How many sides does your shape have?**
- After doing this activity, show children some constellation charts. Encourage children to do the activity again, copying some of the constellations.

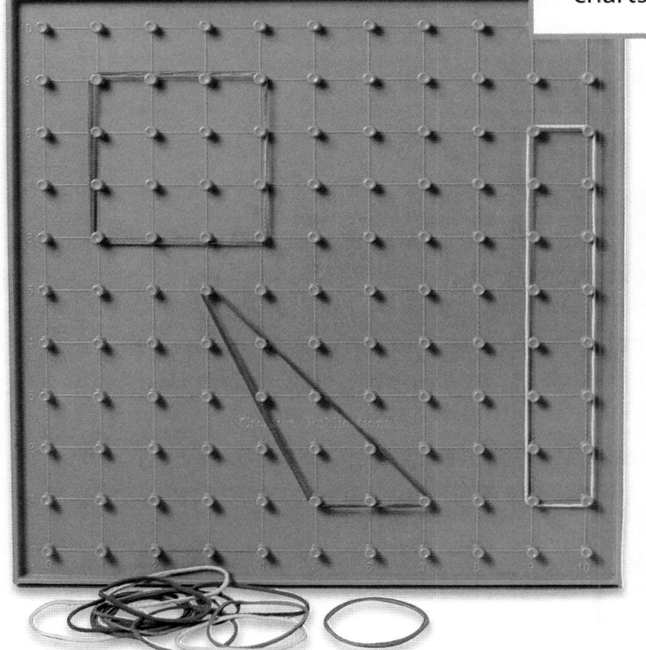

Read Together and Learn

How does the sky change?

▶ Build on Prior Knowledge

Ask children if they have ever looked at the sky at night. Have them recall what they saw. Ask: **How was the night sky different than the daytime sky?**

▶ Use the Visuals

- **Observe** Read the question on the Flipbook page. Encourage children to look at each picture closely. Prompt them to discuss what they notice. Record their ideas on a two-column chart labeled *Day Sky* and *Night Sky*.
- **Compare** Help children use the pictures and the chart to tell what is the same and different about the day sky and night sky.
- **Communicate** Read the question on the Flipbook page again. Help children determine if the pictures give them information to answer the question.

▶ Develop Vocabulary

Sun, **Moon**, **day**, **night** Make labels on index cards with these vocabulary words. Invite volunteers to tape each label to the appropriate place on the Flipbook page.

Flipbook

UNIT D • LESSON 4

How does the sky change?

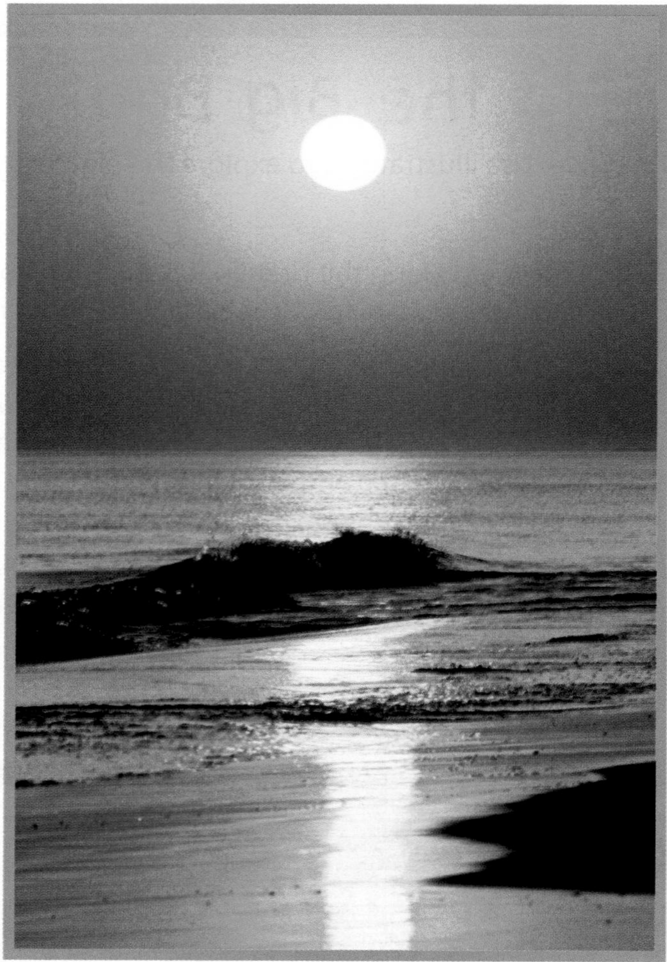

day

Reading Strategy

Phonological Awareness Invite a volunteer to find the word *day* on the Flipbook page. Encourage children to think of words that rhyme with *day*, such as *bay, hay, play, clay, say,* and *weigh*. Repeat with *night* and *sky*.

Comprehension Encourage children to tell about the setting in the pictures. Prompt them to use picture clues to identify the setting. Help them to understand that the setting of both pictures is the beach/ocean.

Science Facts

The Sun is a star, just like the ones you see in the sky at night. However, it is much closer to Earth than any other star, so it appears bigger and brighter. The Sun is 93 million miles away from Earth, and it is 4.5 billion years old. The Sun rises in the east, appears to move across the sky, and sets in the west.

The Moon orbits around Earth. It is so bright that it can be seen during the day and night. The Moon is approximately 250,000 miles away from Earth and is probably as old as the Sun.

Professional Development For more Science Facts and resources from NSDL visit http://nsdl.org/refreshers/science

night

42

Ask Questions

- **What do you think happens to the Sun at night? Where do you think it goes?** Accept all responses.
- **What do you think happens to the Moon and stars during the day?** Accept all responses.
- **Do you think the sky is always blue?** Accept all responses.
- **Do the Sun and Moon ever come down from the sky?** No, the Sun and Moon are always in the sky.

Invite children to circle the items in the sky in each picture with a dry-erase marker. Encourage them to name the items they circle.

Think and Talk

Review the idea that the sky looks different during the day than at night. Encourage children to share what they learned about the sky.

Take a Trip

The Sky Plan a trip to a planetarium to allow children to further explore the night sky. Have them observe the different ways the night sky changes. Provide each child with paper and clipboard (or paper stapled to cardboard) and a pencil. Encourage them to draw and label pictures to record their findings. If a trip to a planetarium is not possible, take the class outside and invite them to observe what they see in the day sky over a period of time. Encourage them to record their findings.

Back in the Classroom Encourage children to share their findings with each other. Compile their drawings into a class book and place in the Science Center.

Read Together and Learn

What do you see in the night sky?

▶ Build on Prior Knowledge

Ask children what the sky looks like at night. Record their responses.

▶ Use the Visuals

- Observe Have volunteers describe what they see in the large picture. Record their answers.
- Communicate Ask children if the sky always looks like this picture. **What is sometimes different about the night sky?**
- Compare Point out the inset pictures of the crescent, quarter, and full moons. Have children describe their shapes and how they are different from each other.

▶ Develop Vocabulary

patterns, **Moon**, **stars** Ask volunteers to give examples of patterns, or to explain what a pattern is. Explain that the Moon changes in a pattern—it appears to grow larger, then smaller, and then larger again. Also, people keep track of where stars are in the sky by identifying patterns of stars called *constellations*.

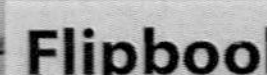

Flipbook

UNIT D • LESSON 4

What do you see in the night sky?

Reading Strategy

Phonemic Awareness Read the Flipbook question aloud. Repeat the word *night,* and ask children what words they know that rhyme with it (*light, bright, fight, sight, kite*).

Print Awareness Write the word *Moon* on chart paper. Have children count how many times that word appears on the Flipbook page.

Science Facts

The Moon is widely believed to have formed during a collision between Earth and a Mars-sized object about 4.5 billion years ago. It appears bright in the sky because it is reflecting the light from the Sun.

Phases of the Moon The orientation of the Sun and the Moon determines the phases of the Moon. Because of its spherical shape, one half of the Moon is always lit by the Sun.

Stars are giant balls of gas that shine through the darkness. Stars follow a seasonal pattern, so their locations for a given time of year are always the same. Constellations are patterns of stars that people have named.

Professional Development For more Science Facts and resources from NSDL visit http://nsdl.org/refreshers/science

crescent moon

quarter moon

full moon

43

▶ Ask Questions

- **Do you always see stars in the sky?** no **Why do you think stars are sometimes visible and sometimes not?** The Sun makes the sky too bright to see other stars.
- **Is the Moon always in the sky at night? Have you ever seen the Moon during the day? What did it look like?**
- Have children look at the inset pictures of the moon phases. **What do you notice about the color of the full moon? Why do you think some spots are darker than others?** Some craters may be bigger or deeper.

Write on It

Have children use a dry-erase marker to draw something they might see in the night sky on the Flipbook page.

Think and Talk

Review the pictures on the Flipbook pages. Ask: **What do you think the day/night sky would look like on a rainy night? Do you think you can see the Sun/stars on a cloudy day/night?**

ELL Support

Play a game of opposites by showing children pictures of *day/night, Moon/Sun, dark/light*, as well as *low/high*.

Beginning Hold up a picture and name it. Ask children to find its opposite and hold it up.

Intermediate Ask pairs to take turns choosing a picture and naming its opposite.

More to Read

Big Bear Ball,
by Joanne Ryder
(HarperCollins, 2002)

Activity After reading the book, ask children to suppose that they are going to meet their friends and play in the light of the full moon. Ask: **What would you like to do in the moonlight?**

If children are having difficulty understanding that what we see in the night sky changes, then encourage them to observe the sky each night for a week. Record their observations, and help them to notice if there are differences.

The Night Sky

Objective: Identify and name various elements of the night sky.

Inquiry Skills: classify, observe, compare, communicate

You need **Photo Sorting Cards 31–40, daily newspaper, paper, crayons, Science Journal p. 49**

1. **Classify** Give each child three cards and place the remaining cards in a pile facedown. Taking turns, have each child ask a classmate to match one of the cards in his or her hand. If there is not a matching card, the child should pick a card from the pile. Continue playing until all the cards are matched.
2. **Observe** Check the newspaper to see the times that the Moon will rise and set during that week. If the Moon is in the sky during the day, take the group outside to look for it.
3. **Compare** If the Moon is visible, have children draw it. Ask children to compare the shape of the Moon that they drew to those that appear on Flipbook page 43 or on the Photo Sorting Cards.
4. **Communicate** Have children take home the Science Journal page to draw what they see in the night sky on two different nights.

Teacher Tip

As children play this game, encourage them to use the proper terms for the different phases of the Moon when they ask each other for matches. If they cannot remember the exact name, they can describe what the Moon on their card looks like (a circle, skinny with pointy ends, and so on).

Name ____________ Science Journal

The Night Sky

Night 1

Check children's drawings.

clouds stars Moon

Night 2

Check children's drawings.

clouds stars Moon

Directions: Draw what you see in the sky on two different nights. Circle the word(s).

Unit D • Weather and Sky Science Resource Book — Use with Lesson 4 Be a Scientist Activity 49

Science Resource Book, p. 49

Investigate More

Observe Ask: **Does the sky look the same every night?** no **How does it change? Why does it change?** Possible answers: the Moon changes shape; the stars are sometimes brighter; sometimes it is too cloudy to see anything.

Music

What Do You See?

Objective: Sing about and describe the night sky.

Inquiry Skills: observe, communicate

- **Sing and Talk** Invite children to sing "Twinkle, Twinkle, Little Star." Then play the recording of "What Do You See?" Afterward, explain that a telescope is an instrument that is used to observe faraway objects. Have children look at the picture as you find and talk about the patterns that the stars make in the night sky. Then replay the song and encourage children to sing along.
- **Draw It** Distribute black construction paper and white chalk. Invite children to draw a picture of a night sky.
- See page TR29 for printed music for the song "What Do You See?"

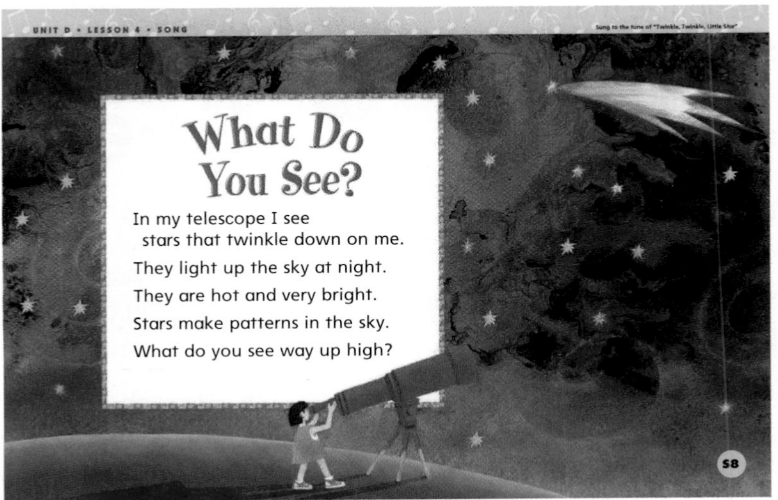

Flipbook p. S8
CD Tracks 17–18

Art

Sky Collages

You need
- construction paper
- small pieces of cardboard
- felt, or other fabric
- scissors
- glue

Objective: Make collages that show various elements of the night sky.

Inquiry Skill: communicate

- Have children make a collage that shows the night sky.
- Ask questions about the scene they will make: **What color background will you use? What will be in the sky? Will you show part of the ground? What shapes will you use?**
- Encourage children to share their creations with a classmate.

Dramatic Play

Traveling to the Moon

You need
- items that suggest space travel (such as an astronaut's helmet)

Objective: Use knowledge to act out taking a trip to the Moon.

Inquiry Skill: communicate

- Have children suppose that they are astronauts traveling to the Moon.
- Help them develop the scenario: **What will you bring with you? What will you do when you get there?**
- Have the group report back to the class about what happened on their trip.

Sun and Shadows

▶ **Essential Question**

How does the Sun make shadows?

▶ **Objective**

Recognize that the Sun creates shadows and appears to move through the sky.

▶ **Vocabulary**

shadows
shade
heat

Resources

Flipbook, pp. 44, S9

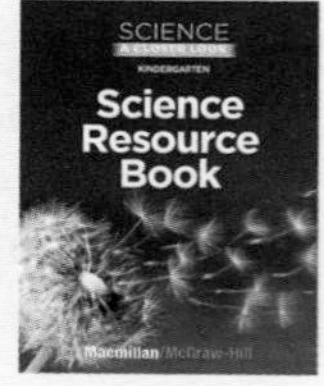

Science Resource Book, p. 50

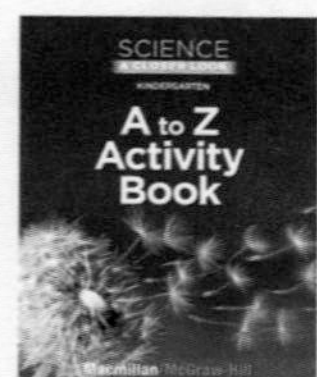

A to Z Activity Book, pp. 38–39, 42–43

Science on the Go, pp. 83–86

Technology

Science Songs CD Tracks 15–16

Science in Motion *Heat Energy*

www.macmillanmh.com

Circle Time

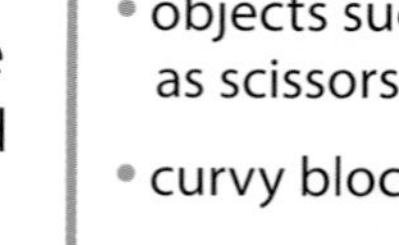

Guess the Object

You need

- paper
- objects such as scissors
- curvy block
- spoon

Inquiry Skills: observe, communicate, infer

- Hang paper in front of a window with light from outside shining through it. Hold an object behind the paper and invite children to guess what it is.
- After children have identified a number of objects, ask: **If the object is behind the paper, how can you tell what it is?** Explain that they are looking at a *shadow* of the object.

Infer Ask: **What do you think makes a shadow appear?** light

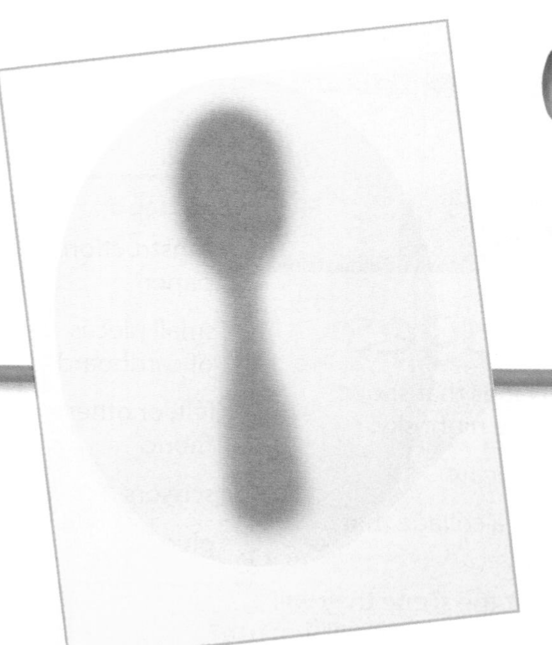

Time to Move!

Invite children to form pairs. Have each child take turns being his or her partner's shadow. The child who is the "shadow" needs to stand behind his or her classmate and imitate every movement the partner makes.

Be a Writer Activity

You need
- chart paper
- marker
- paper
- pencils
- crayons

Sun Activities

Objective: Draw and write about the experience of being in the sunlight.

Inquiry Skill: communicate

- Encourage children to share words that describe the Sun. Record their responses on chart paper.
- Have each child draw a picture of something he or she has done outside in the sunlight.
- Ask: **What did the sunlight feel like when you were (fill in the activity)?** Have children write or dictate a sentence or two about their experience.

Be a Math Wiz Activity

You need
- connecting cubes
- a stick (6 inches or shorter)
- recording sheet
- ruler

Measure Shadows

Objective: Measure the length of shadows using nonstandard units.

Inquiry Skills: observe, measure, compare

- Explain to children that they will measure a shadow at different times of the day.
- On a sunny day, take small groups outside to measure the shadow of a small stick placed upright in a spot with full sunlight.
- Have the group use connecting cubes or a ruler to measure the length of the shadow and record the time and length. Repeat later in the day. Encourage children to discuss why they think the length changed.

Read Together and Learn

What makes shadows? Why do shadows change?

▶ Build on Prior Knowledge

Ask children if they have ever seen their own shadow or made shadow puppets. Have volunteers describe what they noticed.

▶ Use the Visuals

- **Observe** Have volunteers describe what they see in each image. Ask them what differences they notice.
- **Communicate** Invite children to point to the shadows they see in each picture. Encourage them to describe the length of each shadow and its direction.
- **Compare** Have children compare each object to its shadow. Ask: **Which is longer, the object or its shadow? What is helping to make these shadows?** the Sun

▶ Develop Vocabulary

shadow, **shade**, **heat** Make labels with the vocabulary words. Ask volunteers to tape each label to the place on the page where they see the corresponding element.

Flipbook

UNIT D • LESSON 5

What makes shadows? Why do shadows change?

Reading Strategy

Concepts of Print Ask children how many questions there are on the page. Have them explain how they know.

Comprehension Point out that these questions begin with the words *what* and *why*. Ask children to think of other words that are used to begin questions. If children need help, go back and read previous questions. Make a list of words that appear frequently.

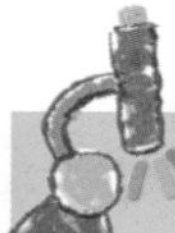

Science Facts

Shadows are formed when light is blocked by an opaque object. Even objects that appear transparent can create shadows if they reflect or absorb some of the light that hits them.

How Shadows Differ The shadow cast by an object farther from the ground will look fuzzier than that of an object closer to the ground. When light waves travel farther and hit an object, they bend. This causes shadows to appear fuzzy.

Shadows and the Sun As the Sun appears to move through the sky, outdoor shadows change in length and direction. The Sun seems to move across the sky due to the rotation of Earth on its axis.

Professional Development For more Science Facts and resources from NSDL visit http://nsdl.org/refreshers/science

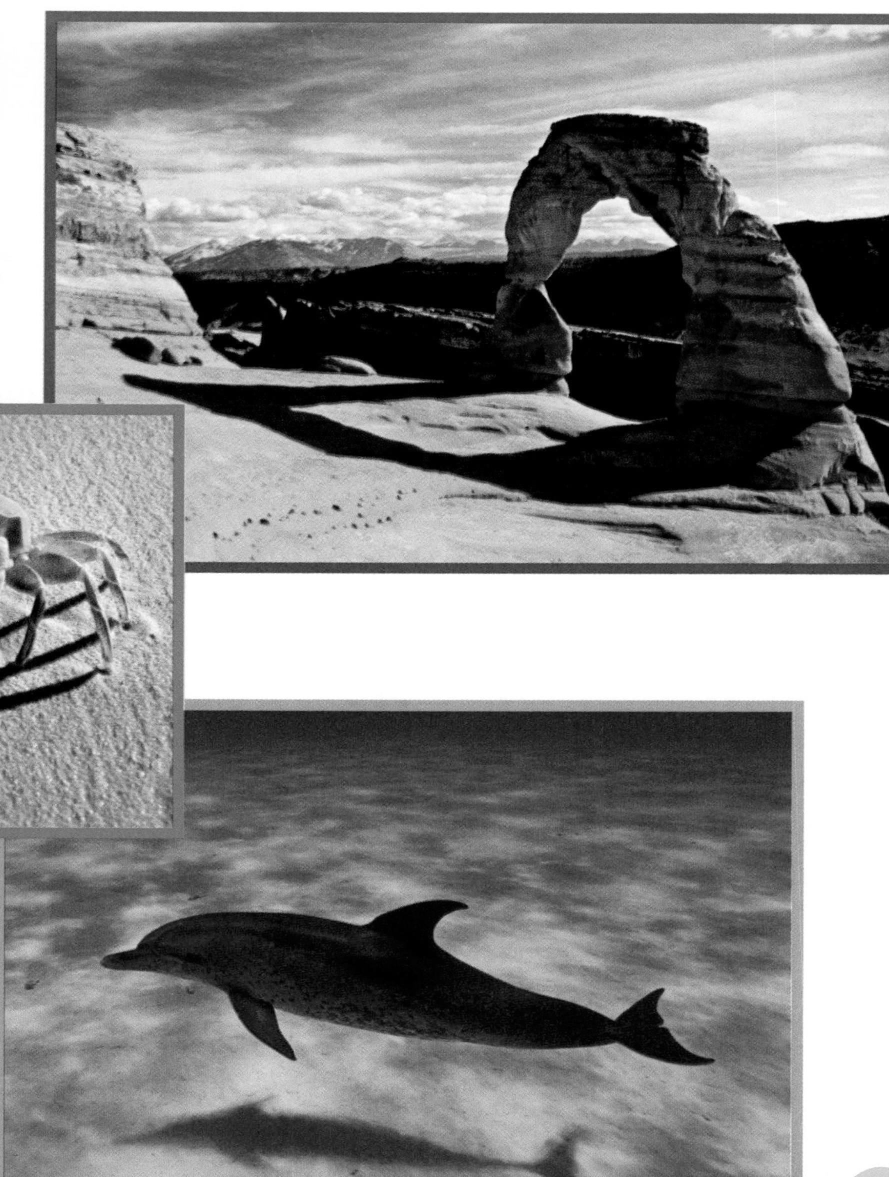

▶ Ask Questions

- **Why do you think shadows look different (or disappear) when the lights are off?** There is no light source. **What happens to shadows outside when the Sun goes behind a cloud? Why?** They change or disappear because light is needed to make shadows.
- **Where do you think is the best place to plant a flower outside?** In a spot that gets a good amount of sunlight, away from objects that might cast shadows on it.
- Have children locate the picture of the dolphin. Ask: **Why do you think the dolphin has a shadow?** Light is shining through the water and the dolphin's body is blocking the light. This makes a shadow.

Write on It

Have children use a dry-erase marker to trace an object and its shadow on each picture on the Flipbook page. Encourage them to notice how the object and its shadow are similar or different.

Think and Talk

Ask children to look at the objects and shadows on the Flipbook page and ask: **What side of the object do you think the light is coming from? Do you think it is high or low in the sky?**

Formative ASSESSMENT

Observe children as they discuss how shadows appear and change depending on the location of the light source. If they are having difficulty, have them create shadows with a flashlight and a standing object.

Differentiated Instruction

Enrichment Explain to children that they will plant something outside. Help them figure out where to plant it so that it will get the appropriate amount of sunlight. Discuss how the Sun and shadows will affect plant growth. Have children record the plant's growth over time.

More to Read

Shadows, by Carolyn B. Otto (Scholastic, 2001)

Activity Before reading, have children generate a list of what they know about shadows. After reading, have children describe what else they learned. Have the group choose experiments from the book that they would like to try.

Change Shadows

Objective: Recognize that shadows change depending on their proximity to a light source.

Inquiry Skills: observe, compare, measure, investigate, infer

You need **overhead projector, large white paper, markers, classroom objects, Science Journal p. 50**

Teacher Tip

Set up this activity so that there is space for children to move around between the overhead projector and the wall. It can be very exciting for children to see a projection of their own shadow, and to watch it change size as they move toward and away from the wall.

1. **Observe** Place a large piece of white paper on a wall. Set up an overhead projector facing this wall. Give each child a turn to stand between the projector and the wall and then walk closer to the wall. Ask: **What happens to your shadow as you move away from the projector?** Remind children to be safe. Tell them to be careful to not look directly into the light source. Also, remind them to be careful of their surroundings when moving around their area.

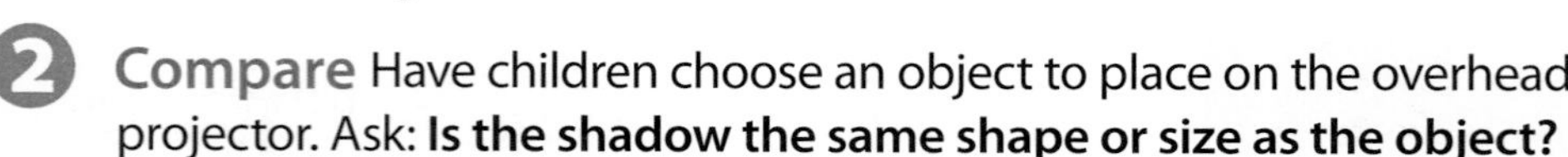

2. **Compare** Have children choose an object to place on the overhead projector. Ask: **Is the shadow the same shape or size as the object?**
3. **Measure** Have children trace the outline of the shadow on the paper and compare the sizes by placing the actual object next to its shadow outline. Have children use the Science Journal page to trace the object and describe their observations.
4. **Investigate** Ask: **How can you change what the shadow looks like?**

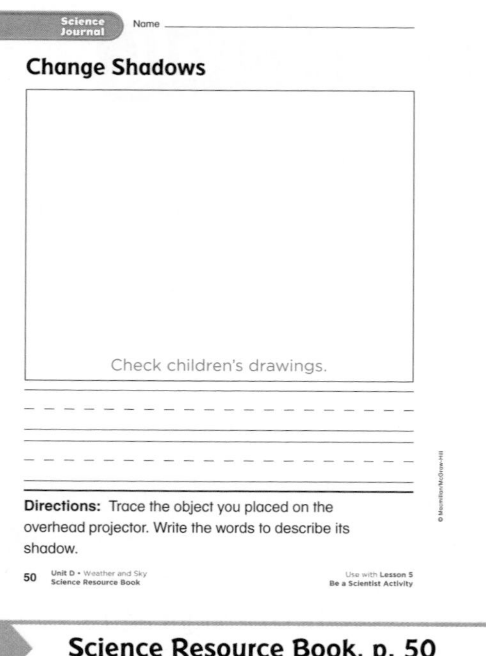
Science Journal Name ____________

Change Shadows

Check children's drawings.

Directions: Trace the object you placed on the overhead projector. Write the words to describe its shadow.

50 Unit D • Weather and Sky Science Resource Book

Use with Lesson 5 Be a Scientist Activity

© Macmillan/McGraw-Hill

Science Resource Book, p. 50

Investigate More

Infer Ask: **How do you think shadows are created outside?** by the Sun **Why do you think those shadows change size?** because the Sun is not always in the same place in the sky; because Earth moves

centers

Music

My Shadow

Objective: Sing about how shadow lengths change.

Inquiry Skills: measure, draw a conclusion

- **Listen and Talk** Display the Flipbook page for "My Shadow" and play the recording. Afterward, point to each picture and ask: **What does this child's shadow look like in the morning? What happens at noon? What happens at night when the Sun goes down?**
- **Act It Out** Have children make up actions to the song, such as pointing to themselves on the words *me* and *I*, bending down on the word *short*, and so on.
- **Move** Take children outside and have them jump, run, skip, and hop. Ask them to observe what their shadows do when they move. Point out that their shadows connect to their feet and follow them wherever they go.

My Shadow
Tell me, what is that I see?
It looks just like another me.
Sometimes it's short and then it grows.
It seems to meet me at my toes!
It's my shadow. This I know.
It follows me wherever I go.
It jumps and runs and moves with me.
It likes to keep me company!
In the morning, it's longer than me.
But by noon, it's small, you see.
Then it grows much longer through the day.
When the Sun goes down, it goes away!
S9

Flipbook p. S9
CD Tracks 15–16

Technology

More About Weather and Sky

- For games and activities about weather and sky, help children log onto www.macmillanmh.com.
- Children can use these fun games to review and reinforce what they have learned about weather and sky.

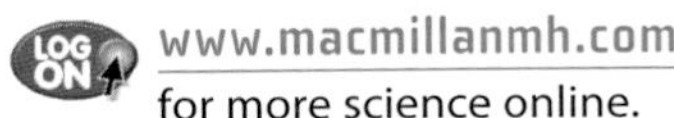

www.macmillanmh.com for more science online.

Art

Shadow Puppets

You need
- cardboard cut into shapes (circles, squares, rectangles, triangles)
- glue
- craft sticks

Objective: Create shadow puppets.

Inquiry Skills: observe, communicate

- With the cardboard shapes and craft sticks, have children make flat puppets. Explain that only the shadow of the puppet will be seen.
- Invite children to use their shadow puppets to put on plays for each other, either using the overhead projector or creating a shadow puppet theater with a large cardboard box and a thin white fabric screen covering one side.

Blocks

You need
- flashlights

Towering Shadows

Objective: Create shadows of tall block structures.

Inquiry Skills: make a model, communicate

- Have children create tall structures in the Block Area.
- Give children flashlights to use as the "Sun." Lower the overhead lights in the classroom and have children create shadows of their buildings.
- Encourage children to experiment. Ask: **How can you change the size and/or position of the shadow?**

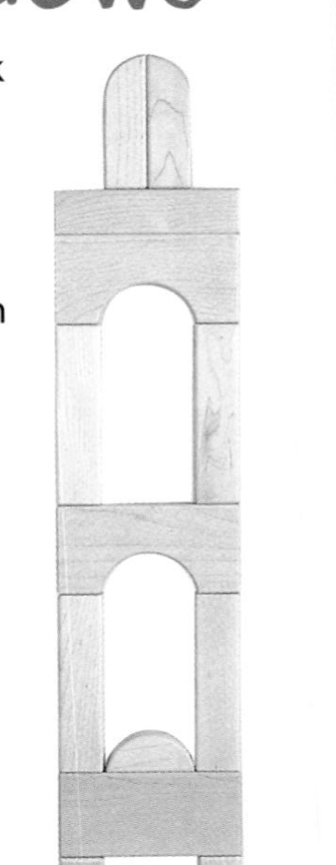

Performance ASSESSMENT DOK 2

Seasonal Weather and Activities

You need
- drawing paper
- crayons
- pencils

- Ask children to think of a season.
- Have children draw a picture about that season, showing what they know about what happens in that season both in nature and in the things people do or wear.
- Have children write or dictate what is happening in their picture.

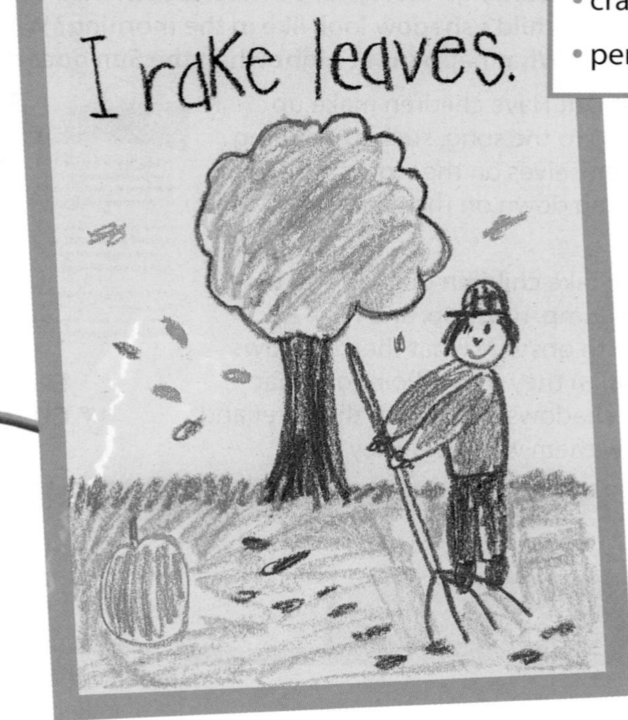

Scoring Rubric

Use the Scoring Rubric to evaluate each child's completion of the Performance Assessment task.

3 points = Child shows all of the following things: something that happens in nature during that season (snow, leaves falling, flowers growing), an activity that people do in that season (swimming, sledding), or something that people wear in that season (bathing suit, hat).

2 points = Child shows two of the three things mentioned above.

1 point = Child shows one of the three things mentioned above.

Summative ASSESSMENT

DOK I

You may choose to use minibook pages 69–70 from the **Science Resource Book** to assess what children have learned about weather and the sky in this unit.

- Help children fold their page in half to create a book and write their name.
- Guide children through each page, having them circle answers and write on each page to complete the book.

Science Resource Book, pp. 69–70

Portfolio ASSESSMENT

- Encourage children to select work from this unit to include in their portfolio. Discuss their choices and have children share which is their favorite and why.
- Use the Unit D Checklist from the **Science Resource Book** page 80 to record children's progress. Place checklists and notes in each child's portfolio.

Earth and Weather

Objective: Explore how weather varies depending upon where you live on Earth and which season it is.

What We Learned

- Ask children to share what they know about winter weather.
- Make a list of things that children see, feel, hear, and wear during winter.
- Review the four seasons with children and discuss how each season might be different depending upon where you live on Earth.

Flipbook

REVIEW TOGETHER • EARTH AND WEATHER

What is winter like where you live?

Floor Puzzle

Activity **Landforms** Revisit the floor puzzle and invite children to recreate the mountain image using Flipbook page 27. Introduce using the people and weather pieces and invite children to place them where they might be found. Ask: **Where will the people hike? Camp? What will the weather be like?**

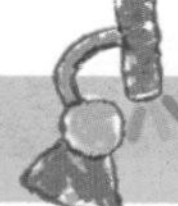

Science Facts

- **Northwest** States in the northwest tend to be cloudy in the winter. The bulk of the annual precipitation takes place during winter.
- **Southwest** During winter in states like Arizona, day temperatures may average 70°F, with night temperatures often falling to freezing or below in the lower desert valleys.
- **Northeast** Winters in the northeast tend to be characterized by snow, sleet, freezing rain, cold temperatures, and high winds.
- **Southeast** During winter, states such as Florida, have approximately double the amount of hours of sunlight than northeastern states, and far milder temperatures (50–60°F).

LOG ON **Professional Development** For more Science Facts and resources from  visit http://nsdl.org/refreshers/science

45

Take a Trip

Seasonal Study Take children outside during each season to observe the weather and how the environment around them looks. Give children paper stapled to cardboard so they can draw and write out in the field.

Back in the Classroom After each trip outside, compile children's findings on a cumulative chart that you can keep up all year. At the end of each season, review the chart and encourage children to compare each season.

Make Connections

What is winter like where you live?

▶ Use the Visuals

- Have volunteers describe what they see in each of the pictures.
- Ask children to notice the water and land formations in the pictures. Encourage them to think about how these features could affect the weather in each area.

Write on It

Invite children to use a dry-erase marker to circle one geographical feature, such as a mountain, valley, river, or lake.

▶ Ask Questions

- Explain that it is winter in each of these pictures. Ask: **What signs of winter do you see in each picture? Does it look similar to or different from winter where you live?**
- **Why do you think temperatures may be colder in a mountainous place?** It is colder up high.
- **How do plants and animals stay safe and warm during the winter?** They adapt in different ways. Some hibernate, others grow more fur, and many migrate, move, or stay underground.

Teacher's Notes

UNIT E

UNIT F

Physical Science

National Standards

The following K-4 National Science Education Standards are covered in Units E and F:

- Objects have many observable properties, including size, weight, shape, color, temperature, and the ability to react with other substances. Those properties can be measured using tools, such as rulers, balances, and thermometers.
- Objects are made of one or more materials, such as paper, wood, and metal. Objects can be described by the properties of the materials from which they are made, and those properties can be used to separate or sort a group of objects or materials.
- Materials can exist in different states—solid, liquid, and gas. Some common materials, such as water, can be changed from one state to another by heating or cooling.
- An object's motion can be described by tracing and measuring its position over time.
- The position and motion of objects can be changed by pushing or pulling. The size of the change is related to the strength of the push or pull.
- Sound is produced by vibrating objects. The pitch of the sound can be varied by changing the rate of vibration.
- Light travels in a straight line until it strikes an object. Light can be reflected by a mirror, refracted by a lens, or absorbed by the object.
- Magnets attract and repel each other and certain kinds of other materials.

Floor Puzzle

Workbench Puzzle Activity

- Introduce your study of Physical Science by showing Flipbook page 46. Discuss the image and have children share which tools they recognize, can identify, or have used.
- Place the floor puzzle pieces out for children and invite them to construct the puzzle, using the Flipbook image as a guide. Ask questions such as: **What do people use these tools for? Which tool can cut wood? Which tool might you use with nails? How can tools help people build things?**
- **Revisit** Encourage children to build, rebuild, and explore the floor puzzle over the course of your study of matter and motion in Units E and F.

Exploring Matter

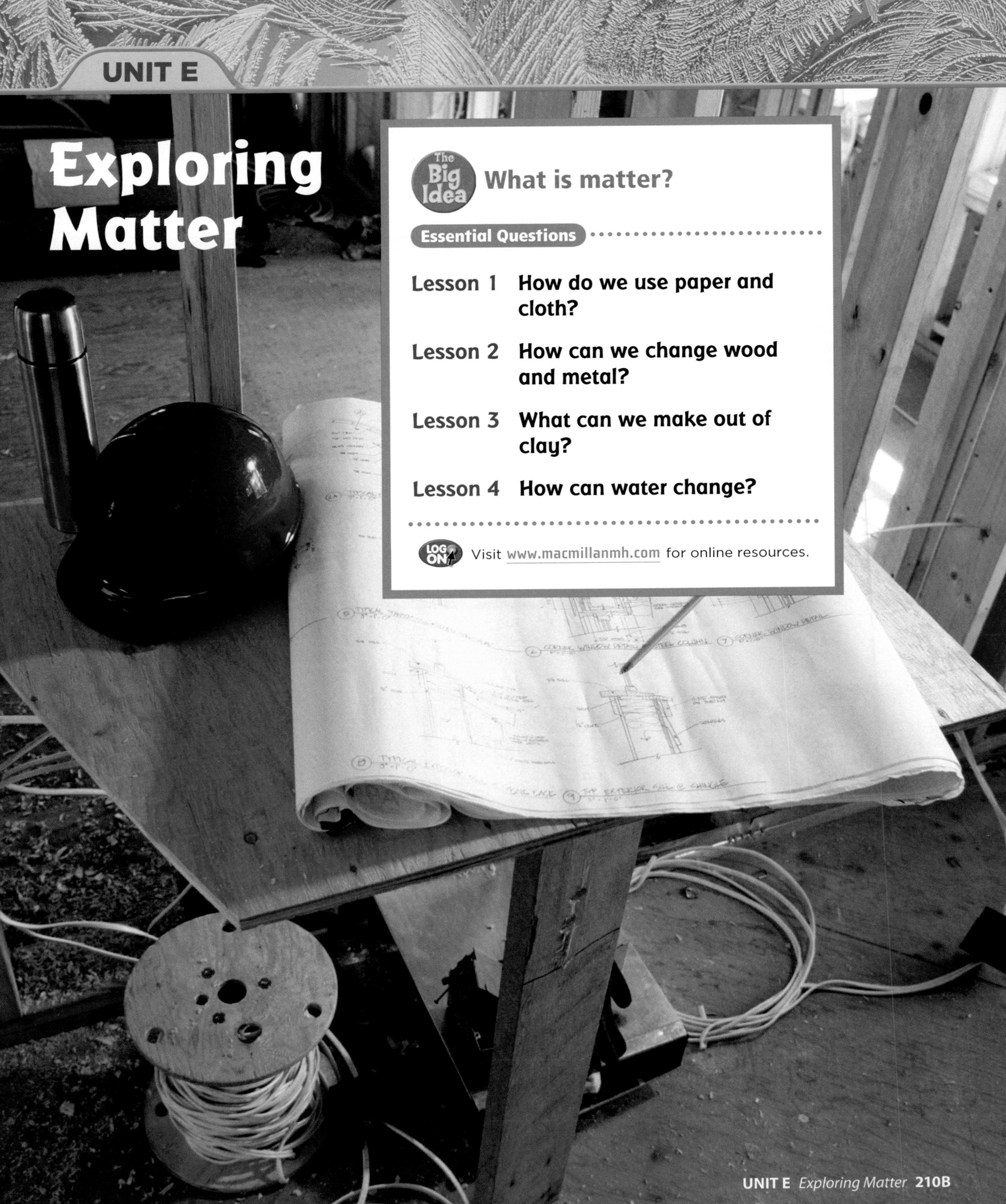

The Big Idea **What is matter?**

Essential Questions

Lesson 1 **How do we use paper and cloth?**

Lesson 2 **How can we change wood and metal?**

Lesson 3 **What can we make out of clay?**

Lesson 4 **How can water change?**

LOG ON Visit www.macmillanmh.com for online resources.

Science Leveled Readers

All included in the Leveled Readers Literature Big Book.

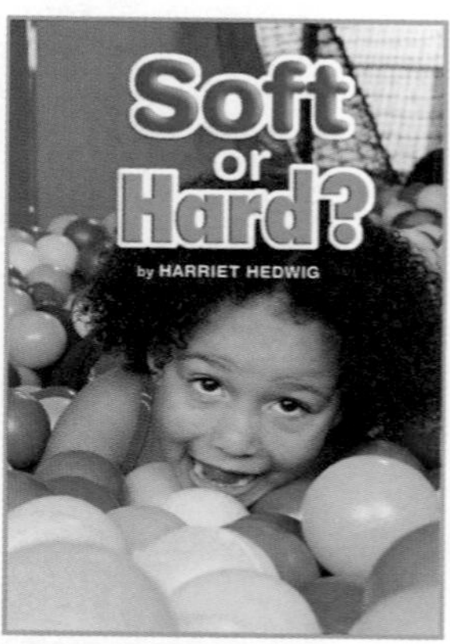

APPROACHING

Soft or Hard?
Different types of solid matter can feel soft or hard.

ISBN: 978-0-02-284598-8

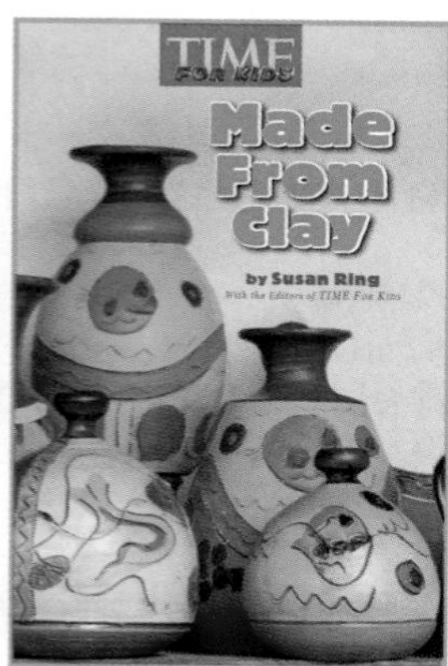

ON LEVEL

Made from Clay
So many items can be made from this natural resource that comes from the earth.

ISBN: 978-0-02-284599-5

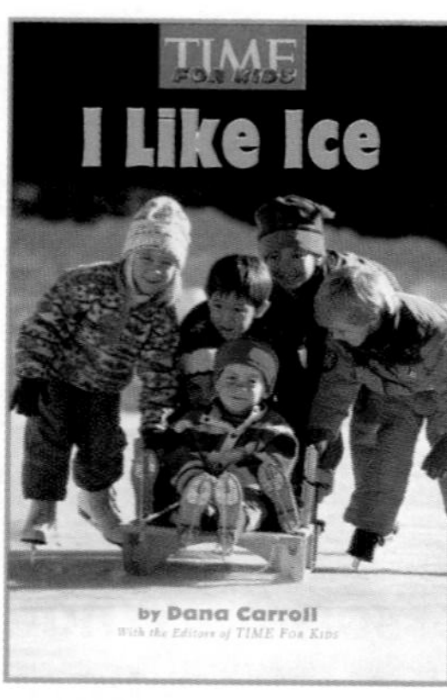

BEYOND

I Like Ice
Ice is nice! Ice is solid water and has many uses.

ISBN: 978-0-02-284603-9

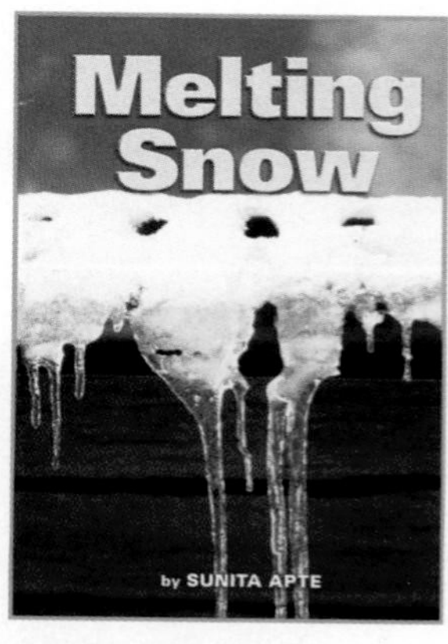

APPROACHING

Melting Snow
When the Sun warms the snow, it melts.

ISBN: 978-0-02-284602-2

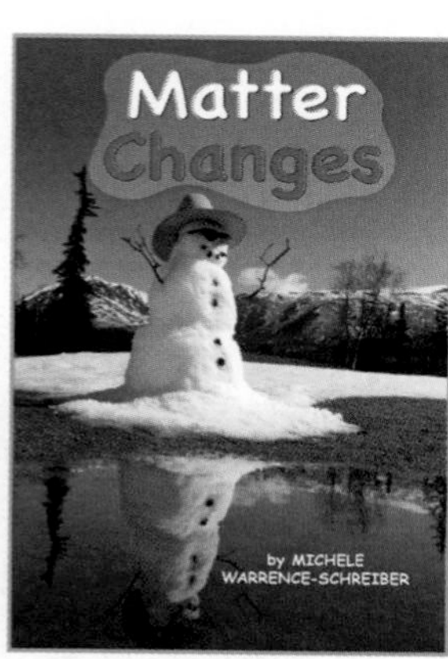

ON LEVEL

Matter Changes
There are three states of matter that can change from one form to another!

ISBN: 978-0-02-284604-6

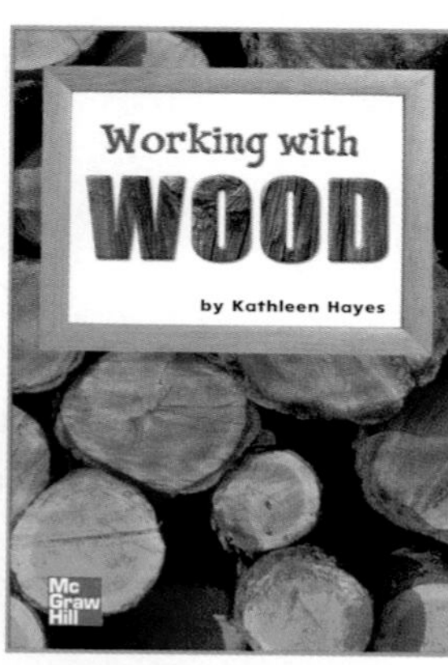

BEYOND

Working with Wood
Wood can be made into many objects.

ISBN: 978-0-02-281084-9

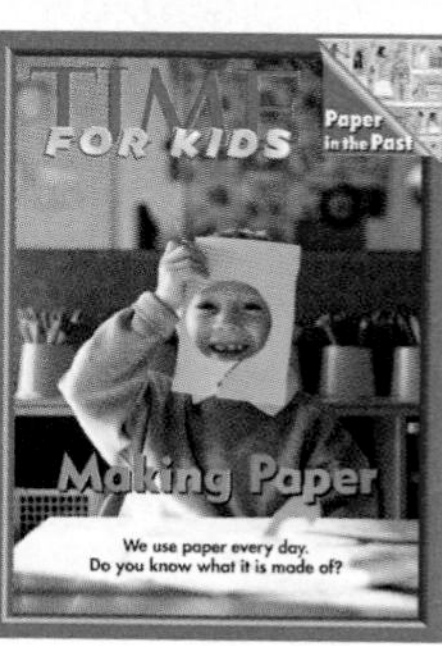

Includes *Time for Kids* magazine about the history of paper and how it is made.

Bibliography

Lesson 1 Paper and Cloth

Joseph Had a Little Overcoat, by Simms Taback (Viking, 1999)

The Jumbo Book of Paper Crafts, by Amanda Lewis (Kids Can Press, 2002)

Kid's Clothes: From Start to Finish, by Samuel G. Woods (Blackbirch Press, 2001)

Paper, by Sara Louise Kras (Capstone Press, 2003)

The Story of Paper, by Ying Chang Compestine (Holiday House, 2003)

What Is the World Made Of? All About Solids, Liquids, and Gases, by Kathleen Weidner Zoehfeld (HarperCollins, 1998)

Lesson 2 Wood and Metal

Metal, by Claire Llewellyn (Sea to Sea, 2006)

Screws, by Sally M. Walker and Roseann Feldmann (Lerner Publishing Group, 2001)

Tools, by Ann Morris (HarperCollins, 1998)

Turtle and Snake Fix It, by Kate Spohn (Scholastic, 2001)

Wood, by Claire Llewellyn (Sea to Sea, 2006)

Workshop, by Andrew Clements (Clarion Books, 1999)

Lesson 3 Working with Clay

Clay, by Mary Firestone (Capstone Press, 2005)

Clay (Let's Create! Series), by Dorothy L. Gibbs (Gareth Stevens Publishing, 2004)

Mud Is Cake, by Pam Muñoz Ryan (Hyperion, 2002)

The Pot That Juan Built, by Nancy Andrews-Goebel (Lee & Low Books, 2002)

Lesson 4 Investigate Water

Down Comes the Rain, by Franklyn M. Branley (HarperCollins, 1997)

Let's Try It Out in the Water, by Seymour Simon and Nicole Fauteux (Simon & Schuster, 2001)

Making Things Float & Sink, by Gary Gibson (Copper Beech Books, 1995)

Pop! A Book About Bubbles, by Kimberly Bradley (HarperCollins, 2001)

Rising Up, Falling Down, by Craig Hammersmith (Compass Point Books, 2002)

Snowflake Bentley, by Jacqueline Briggs Martin (Houghton Mifflin, 1998)

For more information on these or other books, visit **www.bankstreetbooks.com** or your local library.

Science Vocabulary

Vocabulary

bend
fold
tear
cut
paper
cloth
wood
metal
clay
solid
liquid
gas

Change Matter

Explore what children know about matter by creating a four-column language experience chart.

- Encourage children to think about different ways to change paper, clay, wood, and metal.
- Create a four-column language experience chart and write children's responses under the appropriate column of the graph.

Change It

Paper	Clay	Wood	Metal
fold	mold	saw	melt
tear	shape	sand	bend
cut			

ELL Support

You need

- paper
- clay
- wood
- metal objects
- cloth
- cup of water
- ice cubes

Exploring Matter

Have children repeat the vocabulary words as you read them. Show samples of wood, metal, clay, water, ice, cloth, and paper.

- **Beginning** Ask children to put sticky notes labeled *wood, metal*, *clay,* and so forth, on the appropriate sample.
- **Intermediate** Ask a volunteer to hold a sample behind his or her back. Encourage the others to ask: "Is it __?"
- **Advanced** Bend, fold, tear, or cut a variety of the samples. Ask children to say what you are doing.

A to Z Activity Book

Look for additional cross-curricular activities about matter in the **A to Z Activity Book.**

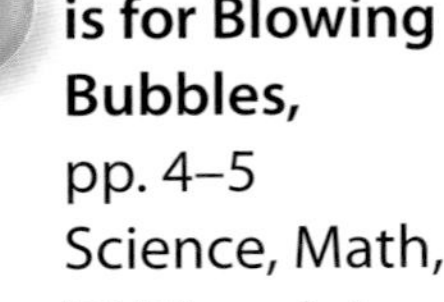

B is for Blowing Bubbles,
pp. 4–5
Science, Math, Writing, Art

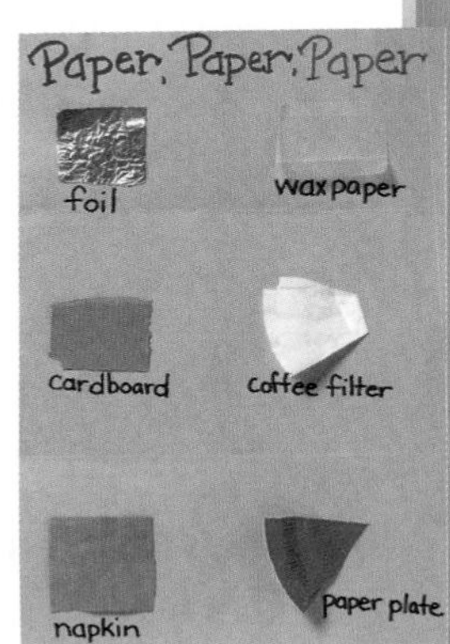

P is for Paper Fun,
pp. 32–33
Math, Science

W is for Water,
pp. 46–47
Reading, Science, Math

X is for X-ray,
pp. 48–49
Math, Music, Science

Unit Project

Make a Meal

You need
- cooking utensils
- paper
- cloth
- wood
- metal
- clay
- water
- chart paper

Objective: Understand that materials have different properties that can change and that matter's properties determine its usefulness.

Step 1 Invite children to brainstorm activities (design place mats, sew tablecloths, make tissue paper flowers, bake brownies) to prepare for a meal. Make these materials with children as you move through the lessons or do them as a culminating project.

Step 2 During cooking, encourage children to discuss the utensils they used and how materials change. Ask: **What happens when you boil water to cook vegetables? What happens to brownie mix after baking? Are the utensils made of wood or metal?**

Step 3 Encourage children to help prepare the room for the meal. After the meal, record how the class used paper, cloth, wood, metal, clay, and water.

Technology

IWB INTERACTIVE WHITEBOARD READY

LOG ON www.macmillanmh.com

Visit Macmillan/McGraw-Hill Science online for projects and activities for students, teachers, and parents.

Bank Street www.bankstreet.edu

Visit Bank Street online for teacher resources and more activities.

TIME FOR KIDS www.timeforkids.com

Visit Time for Kids online to read about current science news and explore additional science activities.

UNIT E Planner

TeacherWorks™ Plus CD-ROM has interactive unit planner, Teacher's Edition, and worksheets.

Lesson	OBJECTIVES	VOCABULARY	RESOURCES and TECHNOLOGY	
1 Paper and Cloth PAGES 212–217 **PACING**: 3–5 days	■ Identify and explore the ways we can use and change paper and cloth.	**bend** **fold** **tear** **cut**	**Flipbook,** p. 48 **Unit E Literature Big Book:** *Time for Kids* **Leveled Reader:** *Soft or Hard?* ■ **Science on the Go,** pp. 87–90	■ **Science Resource Book,** p. 51 ■ **Photo Sorting Cards 41–50** ■ **A to Z Activity Book,** pp. 32–33, 48–49
2 Wood and Metal PAGES 218–223 **PACING**: 3–5 days	■ Identify and explore the ways we can use and change natural resources, such as wood and metal.	**wood** **metal**	**Flipbook,** p. 49 **Unit E Literature Big Book** **Leveled Reader:** *Working with Wood* ■ **Science on the Go,** pp. 91–94	■ **Science Resource Book,** p. 52 ■ **Photo Sorting Cards 41–50** ■ **Floor Puzzle:** *Workbench*
3 Working with Clay PAGES 224–229 **PACING**: 3–5 days	■ Identify clay as a natural resource that can be manipulated to make things.	**clay** **kiln** **fire**	**Flipbook,** p. 50 **Unit E Literature Big Book** **Leveled Reader:** *Made from Clay*	■ **Science Resource Book,** p. 53 ■ **Science on the Go,** pp. 95–96
4 Investigate Water PAGES 230–237 **PACING**: 3–5 days	■ Identify and explore the properties and changing states of water and investigate objects that sink and float in water.	**solid** **liquid** **gas**	**Flipbook,** pp. 51–52, S10 **Unit E Literature Big Book** **Leveled Readers:** *Melting Snow, I Like Ice, Matter Changes* ■ **Science on the Go,** pp. 97–100	■ **Science Resource Book,** p. 54 **Science Songs CD** ■ **A to Z Activity Book,** pp. 4–5, 46–47

PACING Assumes a day is a 20–25 minute session.

www.macmillanmh.com for more planning resources and http://nsdl.org/refreshers/science for science resources from NSDL

BE A SCIENTIST Activities

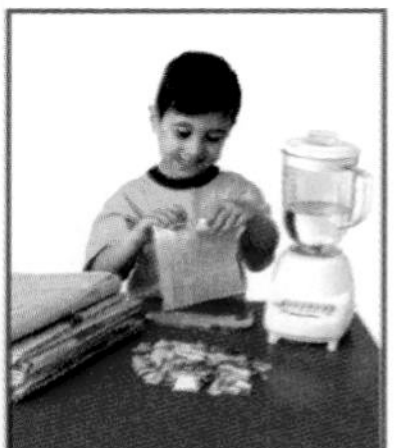

Making Paper *p. 216*

PACING: 30 minutes 

Skills predict, compare, observe, infer

Materials newspaper, blender, water, mesh screen, shallow pan

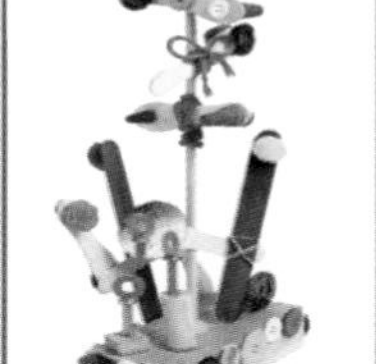

Sculpture Fun *p. 222*

PACING: 30 minutes  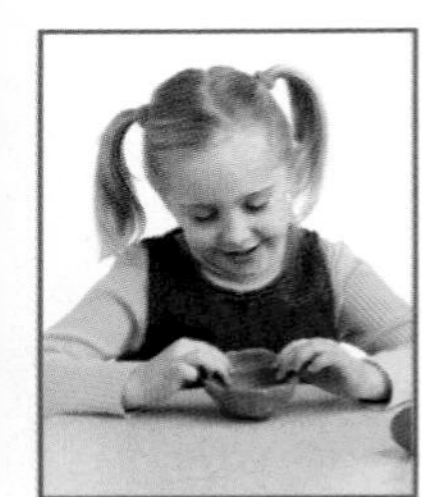

Skills classify, predict, communicate, infer

Materials wood scraps, metal scraps, glue

Pinching Pots *p. 228*

PACING: 30 minutes 

Skills observe, make a model, draw a conclusion, predict

Materials potter's clay

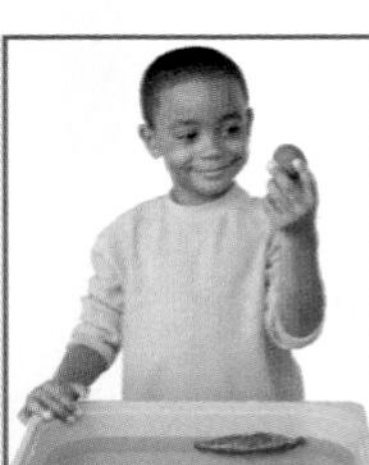

Make It Float *p. 236*

PACING: 30 minutes 

Skills predict, observe, make a model, infer, communicate

Materials clay, containers of water

UNIT ASSESSMENT

Formative Assessment, pp. 215, 221, 227, 235
Performance Assessment, p. 238
Summative Assessment, p. 239
Portfolio Assessment, p. 239

SCIENCE KIT MATERIALS

- large jars
- modeling clay
- balance scale
- clear containers
- corks
- sponge squares
- measuring cups
- chenille stems

COLLECTIBLES

Begin collecting these items for Unit E:

- newspaper
- paper bags
- cloth scraps
- craft sticks
- wooden objects
- metal objects
- ceramic objects

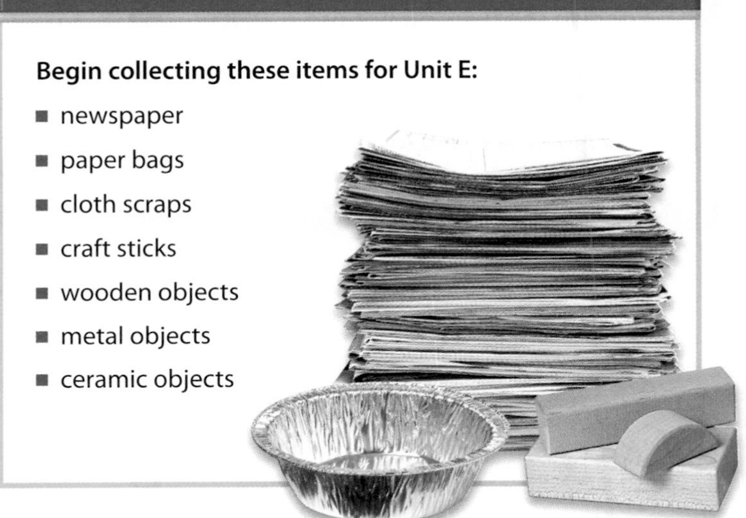

Technology

For additional language support and oral vocabulary development, go to www.macmillanmh.com

Science in Motion
Water Changes
Gas Changes Shape

Vocabulary Games

Language Transfers

Phonics Transfer
Cantonese, Hmong, and Vietnamese do not have the short vowel sound /e/.

Grammar Transfer
Cantonese, Korean, and Spanish speakers do not use helping verbs in negative statements.

I not have metal toy truck.

Academic Language

English language learners need help in building their understanding of the academic language used in daily instruction and science activities. The following strategies will help to increase children's language proficiency and comprehension of content and instruction words.

Strategies to Reinforce Academic Language

- **Use Context** Academic Language should be explained in the context of the task. Use gestures, expressions, and visuals to support meaning.
- **Use Visuals** Use charts, transparencies, and graphic organizers to explain key labels that help children understand classroom language.
- **Model** Use academic language as you demonstrate the task to help children understand instruction.

Academic Language Vocabulary Chart

The following chart shows unit vocabulary and inquiry skills as well as some Spanish cognates. **Vocabulary** words help children comprehend the main ideas. **Inquiry Skills** help children develop questions and perform investigations. **Cognates** are words that are similar in English and Spanish.

Vocabulary	Inquiry Skills	Cognates	
		English	**Spanish**
fold, p. 212	predict, p. 216	predict, p. 216	predecir
tear, p. 212	compare, p. 216	compare, p. 216	*comparar*
bend, p. 214	observe, p. 216	observe, p. 216	*observar*
cut, p. 214	infer, p. 216	infer, p. 216	*inferir*
wood, p. 218	classify, p. 222	metal, p. 218	*metal*
metal, p. 218	communicate, p. 222	classify, p. 222	*clasificar*
clay, p. 224	make a model, p. 228	communicate, p. 222	*comunicar*
solid, p. 230	draw a conclusion, p. 228	model, p. 228	*modelo*
liquid, p. 230	investigate, p. 236	conclusion, p. 228	*conclusión*
gas, p. 230		solid, p. 230	*sólido*
		liquid, p. 230	*líquido*
		gas, p. 230	*gas*
		investigate, p. 236	*investigar*

Vocabulary Routine

Use the routine below to discuss the meaning of each word on the vocabulary list. Use gestures and visuals to model all words.

Define *Metal* is a shiny material that can be melted and shaped.

Example This paper clip is made of *metal*.

Ask How does *metal* look and feel?

Children may respond to questions according to proficiency level with gestures, one-word answers, or phrases.

Vocabulary Activities

Help children identify objects made of metal.

BEGINNING Show a large paper clip. Say: *This paper clip is made of metal.* Pass it around for children to feel. Write, say, and have children repeat, *metal*. Continue: *What other objects in this room are made of metal?* Have children respond by pointing or saying the word.

INTERMEDIATE Form three groups. Take out Sorting Cards 44, 45, and 48. Give one to each group, facedown. Tell the groups: *Your cards have pictures of objects, but only one object is made of metal. Turn your cards over and tell me what objects you see.* Encourage children to describe the objects. Ask: *Which object is made of metal?*

ADVANCED Turn to page 49 of the Flipbook. As you point to the sets of objects, say: *On this page we see objects made of wood and objects made of metal. Show me who is using metal to make metal objects.* Encourage children to name or describe the objects that are made of metal.

UNIT E

Objective: Children will discover characteristics of different types of matter.

Before Reading

- Invite children to observe the differences between the ducks on the Flipbook page.
- Explain that in this unit, children will be learning about materials that we use to make objects. Ask them what they notice about what the ducks are made out of.

During Reading

- Read the poem, pointing to the words as you read.
- Read again, pausing to invite children to point to images on the page that correspond to the lines in the poem.
- Reread the poem two or three times, encouraging children to chime in as you read.

Flipbook

UNIT E

Exploring Matter

I see...
Ducks made out of rubber.
Ducks made out of wood.
A duck that makes a whistle.
Ducks that fly, if they could!

A duck made out of shiny silver.
Ducks marching in a row.
A duck that's soft and furry.
Ducks ready for a show.

by Susan Greenebaum

Reading Strategy

Phonemic Awareness Ask children if the poem rhymes. Discuss the words that make the poem rhyme. Write *wood* and *could* on chart paper. Ask children what they notice about the two words. Repeat with *row* and *show*.

Print Awareness Have children count the capital letters in the text. Ask: **Do you see a pattern where these letters are found?** Have children explain their answers.

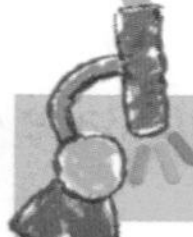

Science Facts

Raw material comes from nature in an unprocessed form. Raw materials include clay, wood, rubber, wool fleece, and some metals, and can be classified as renewable (regrown or reused) and nonrenewable (used and not regrown). For example, wood and rubber are renewable resources that come from plants that can be reused or replanted.

Processed materials have gone through some level of manufacturing to modify their natural form, but not their properties. Processed materials include bricks (processed clay), lumber (processed wood), paper (processed wood), yarn (processed wool fleece), and steel (processed iron).

After Reading

- Begin a class chart with the title *Ducks can be made of. . .* . Make a list of the materials children can name on the Flipbook page. If there are materials that the children do not name, help them out.
- Invite volunteers to point to the ducks that are the same sizes. Repeat for textures, shapes, and colors.

Write on It

Have volunteers use a dry-erase marker to draw an *x* on a particular duck when you name the material it is made out of.

▶ Revisit

Revisit the poem by writing it out on paper and copying one for each child. Invite children to draw their own illustrations to go along with the text.

Science Background For more information on our Earth go to www.macmillanmh.com.

School to Home

At the beginning of Unit E, give children copies of the Home Letter on page 17 from the **Science Resource Book.** Read the letter aloud and help children write their names. Have children bring the letter home to share with their families.

At the end of the unit, send children home with a copy of the Home Letter on page 19.

See pages 18 and 20 in the Science Resource Book for Spanish versions of these letters.

Properties Materials have properties, such as color, brittleness, transparency, reflectivity, permeability, density, flexibility, and rigidity. The material that makes up an object is selected to meet the specific requirement of the object and its function.

Changing matter Matter and materials can be changed in many ways. Materials can be changed in form and state. We can change matter with our hands by ripping, tearing, stretching, squashing, folding, and cutting. Matter can also change states through heating and cooling. Some materials are easily changed, and others less so.

Professional Development For more Science Facts and resources from NSDL visit http://nsdl.org/refreshers/science

▶ **Essential Question**
How do we use paper and cloth?

▶ **Objective**
Identify and explore the ways we can use and change paper and cloth.

▶ **Vocabulary**
bend
fold
tear
cut

Resources

Flipbook, p. 48

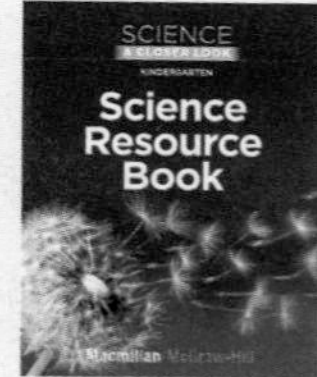

Science Resource Book, p. 51

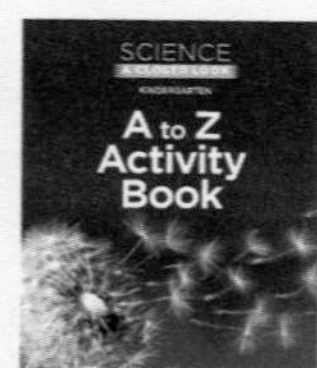

A to Z Activity Book, pp. 32–33, 48–49

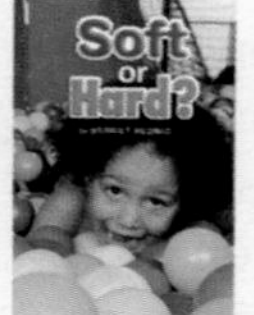

Leveled Reader: *Soft or Hard?*

Photo Sorting Cards 41–50

Technology

 www.macmillanmh.com

Paper and Cloth

Circle Time

WHOLE CLASS

Fold It

You need
- scrap paper
- cloth swatches
- chart paper
- marker

Inquiry Skills: observe, communicate, compare

- Write *Paper* and *Cloth* on chart paper. Give one piece of paper and one cloth swatch to each child. Have children try to crumple, fold, tear, and pull apart their paper and cloth. Record what happens on the chart.
- Invite children to name things that are made out of paper and cloth.

Compare Ask: **How are paper and cloth alike and different? Can light pass through them?**

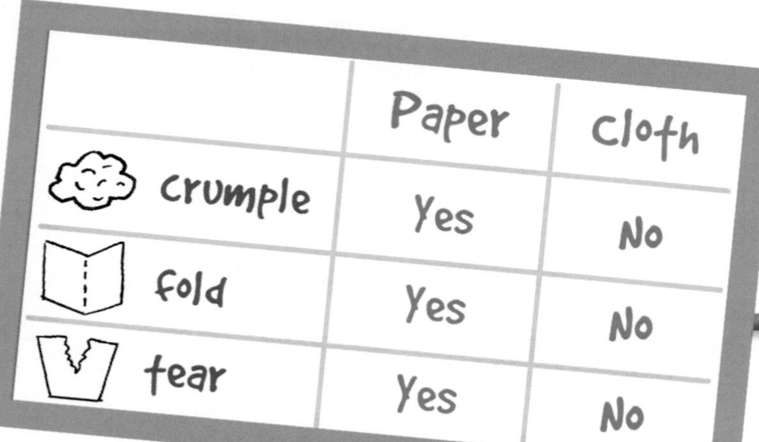

	Paper	Cloth
crumple	Yes	No
fold	Yes	No
tear	Yes	No

Time to Move!

Show children items made of paper and cloth. Tell them to hop up and down on one leg when you hold up an item made of paper, and jump up and down on two legs when you hold up an item made of cloth.

Be a Reader

You need
- Unit E Big Book
- Leveled Reader: *Soft or Hard?*
- chart paper
- marker

Read the Big Book

Objective: Compare properties of different objects.

Inquiry Skills: predict, observe, communicate

- Show children the cover of the book *Soft or Hard?* Read the title. Have children predict what the book might be about. Ask children to name soft and hard objects. Record their ideas on a two-column chart labeled *Soft* and *Hard*. Ask children how the objects are different.
- Read the book. Then reread it, inviting children to identify which objects are soft or hard.
- **Guided Reading:** See the inside back cover of the Leveled Reader version of *Soft or Hard?* for activities.

Be a Writer

You need
- different types of paper
- stapler

Make a Book

Objective: Understand how paper is used to make a book.

Inquiry Skills: make a model, communicate

- Show children a book and ask them how paper is used in books. Explain that a book has pages and a cover made of paper.
- Provide many colors and types of paper and allow children to choose three different papers for the inside pages and one for the cover. Show children how to fold the inside pages in half, fold the cover page over them, and then staple the spine.
- Invite children to use their books as a journal to draw and write in each day.

Stacy's Journal

Be a Math Wiz

Activity

You need
- 1″ x 6″ paper strips
- tape or stapler
- ruler

Paper Chains

Objective: Practice counting to 20 and measuring.

Inquiry Skills: predict, compare, measure

- Have children make paper chains to represent numbers up to 20. Encourage pairs to choose three numbers. Ask: **Which chain will be the longest? Shortest?**
- Have children count the strips of paper they need to make their chain and write a number on each one consecutively. Then have pairs attach the links in order to make a chain.
- Have children stop periodically, hold their chains side by side, and compare the lengths. Invite children to measure classroom items using their chains and then using a ruler. Compare the results.

Read Together and Learn

How can we change paper and cloth?

▶ Build on Prior Knowledge

Have children think of objects that are made out of paper or cloth. Record their ideas on a two-column chart labeled *Paper* and *Cloth*.

▶ Use the Visuals

- **Observe** Invite volunteers to point to images where they see paper or objects made with paper. Then invite volunteers to point to images where they see cloth or objects made out of cloth.
- **Infer** Point to the socks and the hat. Ask children to think about how cloth was changed to make these items.
- Point to the picture of the child cutting paper. Encourage children to share ideas about other ways we can change paper.

▶ Develop Vocabulary

bend, **fold**, **tear**, **cut** Demonstrate these action words with a piece of paper, naming each action. Then give each child a piece of paper to explore the different actions.

Reading Strategy

Phonemic Awareness Write *bend, fold, tear,* and *cut* on the Flipbook page. Have children locate the words that begin or end with the letter *t*. Then ask children to locate the words that end with the letter *d*.

Comprehension Point to the image of the sewing machine. Invite a volunteer to describe what the person in the picture is doing. Ask: **What might this person be making?**

Science Facts

Paper Americans use about 580 pounds of paper per person per year. Paper products make up 40% of our trash. Making recycled paper instead of new paper uses 64% less energy and 58% less water. Every year, less paper is sent to landfills because Americans continue to recycle more and more paper.

Cotton has been spun, woven, and dyed since prehistoric times. Native Americans skillfully spun and wove cotton into fine garments and dyed tapestries. The manufacturing of cotton cloth involves many processes including carding, combing, and spinning, which changes raw fiber into yarn or thread strong enough for weaving.

Professional Development For more Science Facts and resources from NSDL visit http://nsdl.org/refreshers/science

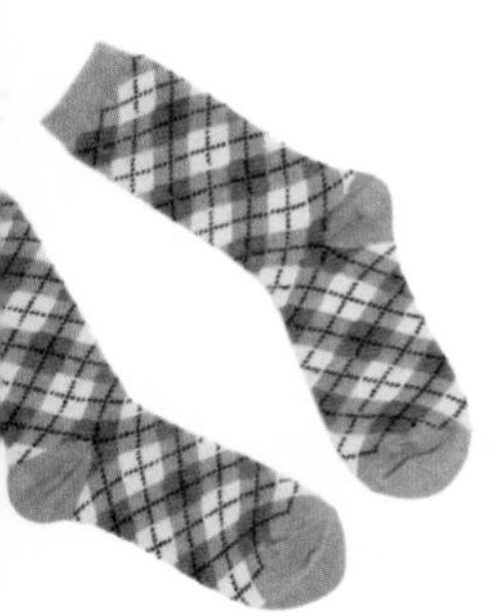

48

▶ Ask Questions

- **Why do you think there are different types of paper?** to make different types of things
- **Which is heavier — a tissue or a book?** a book **A sock or a quilt?** a quilt **Why?**
- **How many shapes do you see on the quilt? How do you think they were put together?**
- **Why do you think we make clothes out of cloth instead of paper?** Because cloth will last longer; If we made clothes from paper, they would rip easily and fall apart when washed.

Write on It

Use a dry-erase marker to write *paper* and *cloth* on the Flipbook page and invite children to draw a line from each image to the labels. Help children count and number how many cloth and paper items there are on the page.

Think and Talk

Review the ways that paper and cloth are used to make the objects shown on the Flipbook page. Have children share other ways that they have used or changed paper and cloth.

Formative ASSESSMENT

Observe children as they explore the concept of how and what things are made of using the Flipbook page and the Photo Sorting Cards.

TIME FOR KIDS

Shared Reading Read *Making Paper* with the class. Use this magazine, found in the **Unit E Literature Big Book,** as a photographic review of how paper is made from natural resources.

More to Read

The Story of Paper,
by Ying Chang Compestine
(Holiday House, 2003)

Activity After reading this story about the invention of paper, help children create their own class story about what school would be like without paper.

Making Paper

Objective: Understand how recycled paper is made.

Inquiry Skills: predict, compare, observe, infer

You need **newspaper, blender, water, 9" x 13" piece of mesh screen, 9" x 13" shallow pan, Science Journal p. 51**

1. **Predict** Have children tear newspaper into small pieces. Fill a blender halfway with water, add the shredded paper, and let it soak for a few minutes. Ask: **What do you think might happen to the newspaper?** It will get wet and mushy. Then blend to make an oatmeal-like mush.
2. Place the screen into the pan and have children help spread out some mush into a thin layer on the screen. Then place the screen between several layers of newspaper to dry.
3. **Compare** Press down hard on the newspaper, peel off the top layer, and remove the screen. The wet, flat mush will be left on the bottom layer of newspaper. Place it on a shelf to dry. Let the children feel it and ask: **How is this like the paper we use in school? How is it different?**
4. **Observe** Later in the day, ask: **What is happening to the mush?** Have children record this process on their Science Journal pages.

Teacher Tip

When making paper, do not throw any leftover pulp into the sink. Instead, reuse it. Cover the drain in the sink with the screen, and pour the remaining pulp on it. Remove the pulp from the screen. Dry the leftover pulp and place it back in your recycling container.

Name ____________________ Science Journal

Making Paper

First, we . . .

Check children's work.

Then, we . . .

Check children's work.

Finally, we . . .

Check children's work.

Directions: Draw pictures and write words to describe the steps in making paper.

Unit E • Exploring Matter Science Resource Book — Use with Lesson 1 Be a Scientist Activity 51

Science Resource Book, p. 51

Investigate More

Infer Ask: **Do you think light will be able to pass through our homemade paper? Why or why not?** Possible answer: No, because it is thick and I cannot see my hand through it.

centers

Drawing and Writing

Paper Story

Objective: Write about and summarize the paper-making process.

Inquiry Skill: communicate

- Help children make a book using blank paper and a stapler. Have children use their books to write the story of how they made recycled paper.
- As a class, review the steps of the process and write them down on chart paper.
- Have children illustrate the story and write words to describe each illustration. Invite children to use the Photo Sorting Cards about paper to help with their writing.

You need
- paper
- stapler
- colored pencils
- Photo Sorting Cards 41–50

Art

Sewing Station

Objective: Experience the process of sewing cloth.

Inquiry Skills: infer, make a model

You need
- burlap or felt with premade holes
- large plastic sewing needles
- yarn

- Explain that clothes and other things made from cloth are really small pieces of cloth sewn together with a tool called a needle.
- Thread needles with the yarn and tie a knot. Demonstrate how to do a stitch, but do not put too much emphasis on form.
- Encourage children to attach pieces of cloth together by sewing them with the yarn.

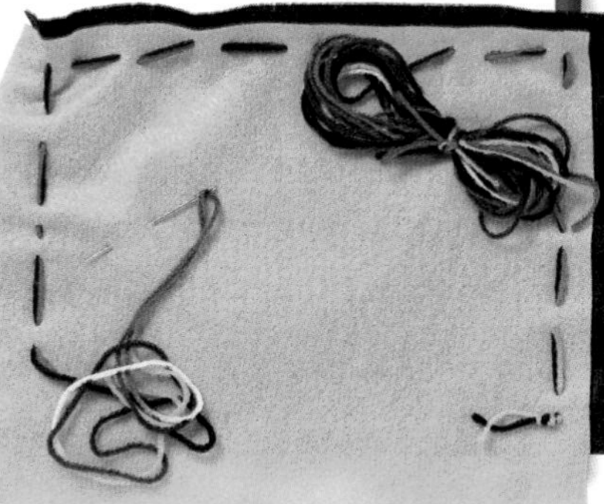

Dramatic Play

Paper Bag Puppets

Objective: Develop small motor skills.

Inquiry Skill: communicate

You need
- paper bags
- markers
- cloth scraps
- yarn
- glue

- Point out that bags are often made of paper. Have children examine paper bags of different sizes.
- Have children decide which bag they will use to make a puppet. Provide a variety of materials for children to use to make their puppets.
- Encourage children to use their puppets to put on a puppet show.

For more on paper and cloth, use pp. 47–48 in the **Activity Book**.

▶ **Essential Question**

How can we change wood and metal?

▶ **Objective**

Identify and explore the ways we can use and change natural resources, such as wood and metal.

▶ **Vocabulary**

wood

metal

Resources

Flipbook, p. 49

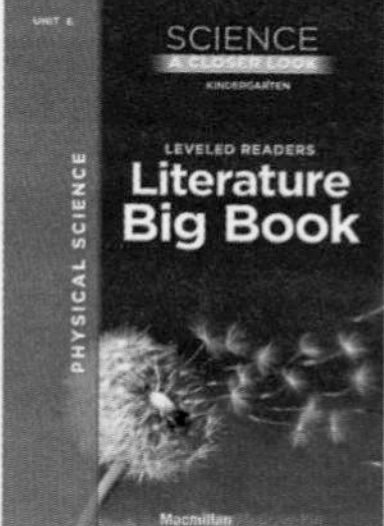
Unit E Literature Big Book

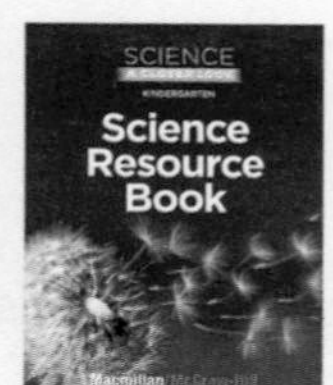
Science Resource Book, p. 52

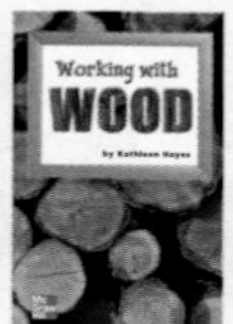
Leveled Reader: *Working with Wood*

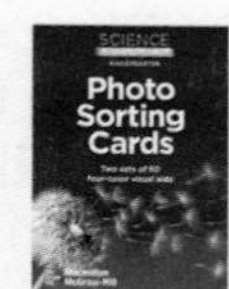
Photo Sorting Cards 41–50

Technology

www.macmillanmh.com

Wood and Metal

Circle Time

WHOLE CLASS

Feely Box

You need

- 8–10 objects made of wood or metal
- cardboard box with hole in one side

Inquiry Skills: observe, investigate

- Show the "Feely Box" and explain to children that they will try to determine if the objects inside are metal or wood using only their sense of touch.
- Have a child reach into the box, feel one object, and describe its texture. Have the child decide if it is made of metal or wood, then remove the object from the box to verify his or her description. Repeat until everyone has had a turn.

Investigate Ask: **Can light pass through wood or metal?** no **How could we find out for sure?** We can use a flashlight to test it.

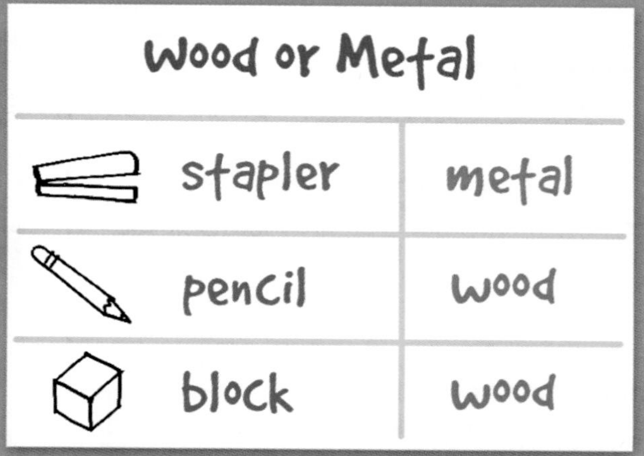
Wood or Metal

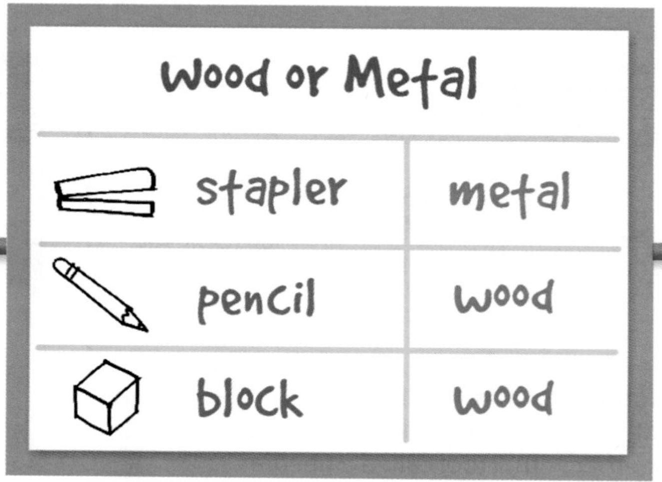

stapler	metal
pencil	wood
block	wood

Time to Move!

Have children march around the room and make sounds with wood and metal instruments. If you do not have instruments in your classroom, you can use wooden blocks, baskets, pots, pans, and spoons.

Be a Reader Activity

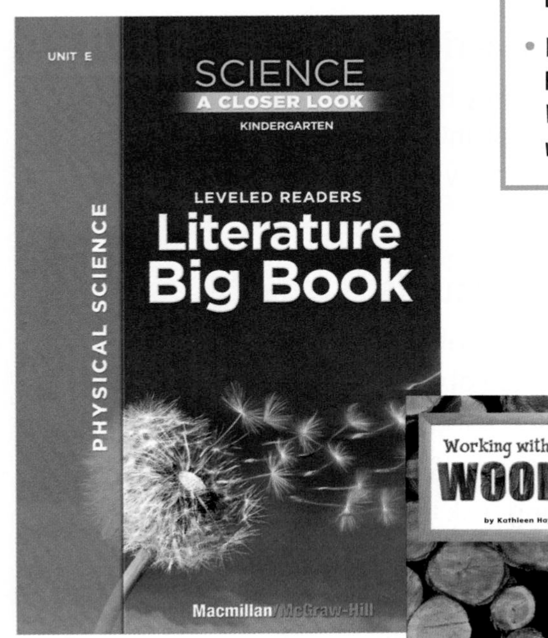

You need
- Unit E Literature Big Book
- Leveled Reader: *Working with Wood*

Read the Big Book

Objective: Use illustrations to infer what an unfamiliar word might be.

Inquiry Skills: communicate, infer

- Show children the cover of *Working with Wood* and have them describe the image.
- Read each page aloud, stopping to allow children to discuss what is happening and to point to words they can identify using the picture clues. Encourage children to share their experiences with wooden materials.
- **Guided Reading:** See the inside back cover of the Leveled Reader version of *Working with Wood* for activities.

Be a Math Wiz Activity

You need
- clear plastic containers
- a variety of screws
- small metal objects

Sorting Screws

Objective: Sort and count objects by tens.

Inquiry Skills: classify, compare

- Have children look closely at the screws and metal objects in the container. Have them sort the materials by type (such as screws, nuts, and paper clips).
- Ask: **How many different types of metal objects are there?** Have children count the number of groups they made.
- Then ask: **How many of each type do you have?** Have pairs count the objects in each group. Encourage them to create groups of 10 objects and then count by tens.

Read Together and Learn

How can we use wood and metal?

Build on Prior Knowledge

Ask children if they have ever built anything out of wood or metal. Invite them to look around the room and see if they can find anything made of wood or metal.

Use the Visuals

- Observe Invite volunteers to point to images where they see objects made out of wood and then metal.
- Infer Refer children to the images of the two men working. Ask them to notice and talk about the differences in the images.
- Point to the wooden and metal baskets. Ask: **What are these objects used for?** Possible answer: holding and carrying things

Develop Vocabulary

wood, **metal** Write *wood* on ten sticky notes and *metal* on ten sticky notes. Invite volunteers to choose a sticky note and put it on something in the room or on the Flipbook page that is made of either wood or metal.

Flipbook

UNIT E • LESSON 2

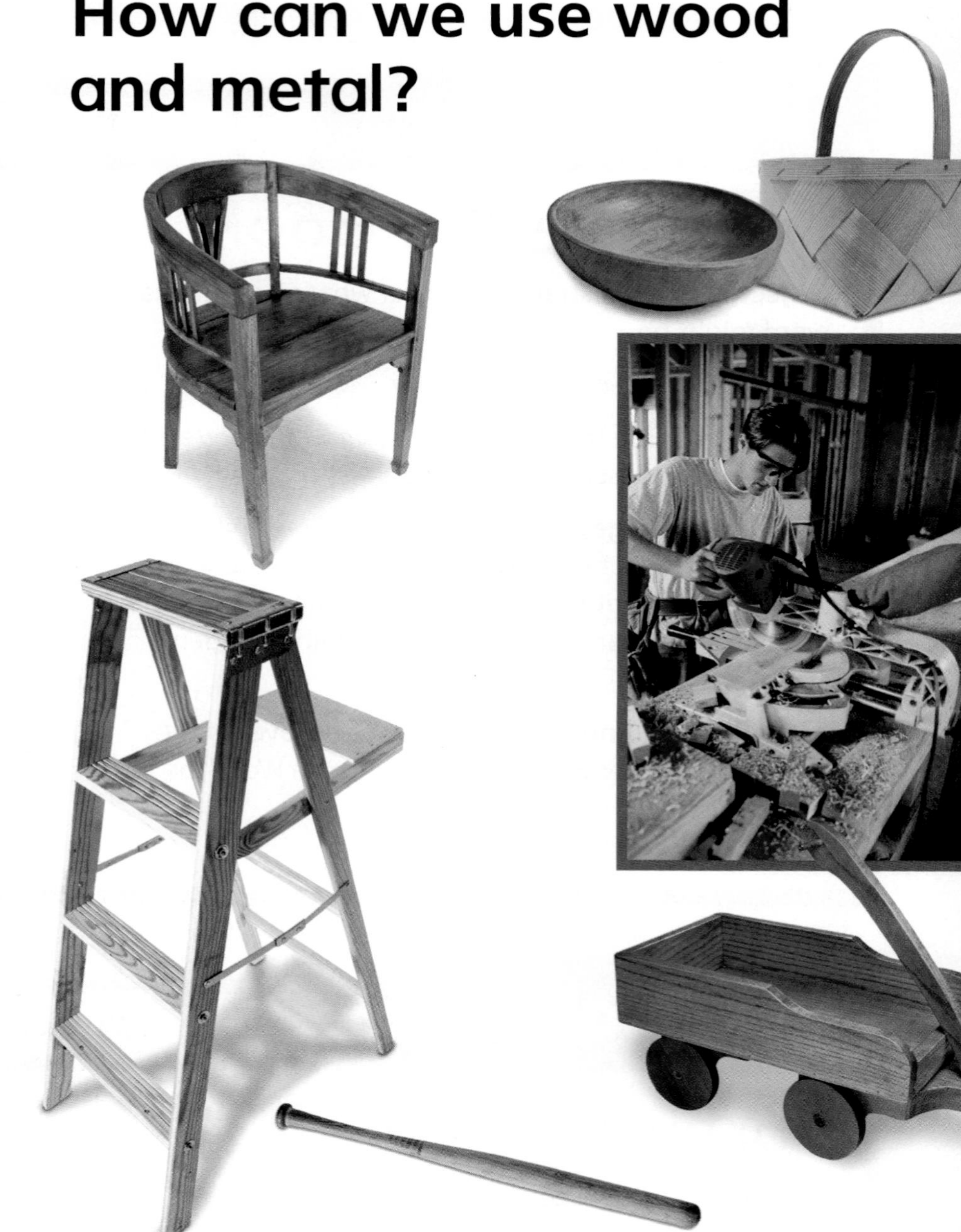

Reading Strategy

Phonemic Awareness Encourage children to help you create a word family using *-an* (*fan, man, ran, can, pan*).

Phonological Awareness Direct children's attention to the question on the Flipbook page. Ask children to count how many words have one syllable and how many have two syllables.

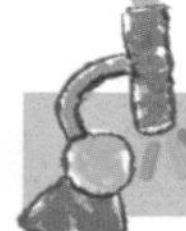

Science Facts

Welding is the most common way of permanently joining metal parts. Heat is applied to metal pieces, melting and fusing them to form a permanent bond. Because of its strength, welding is used in shipbuilding, automobile manufacturing and repair, and thousands of other manufacturing activities.

Woodworking Pieces of wood can be joined in a variety of ways, most commonly by using metal fasteners. Nails and screws are used for light-frame construction. Heavy-gauge hardware, such as bolts and side plates, is used to connect wood in heavy construction.

Professional Development For more Science Facts and resources from NSDL visit http://nsdl.org/refreshers/science

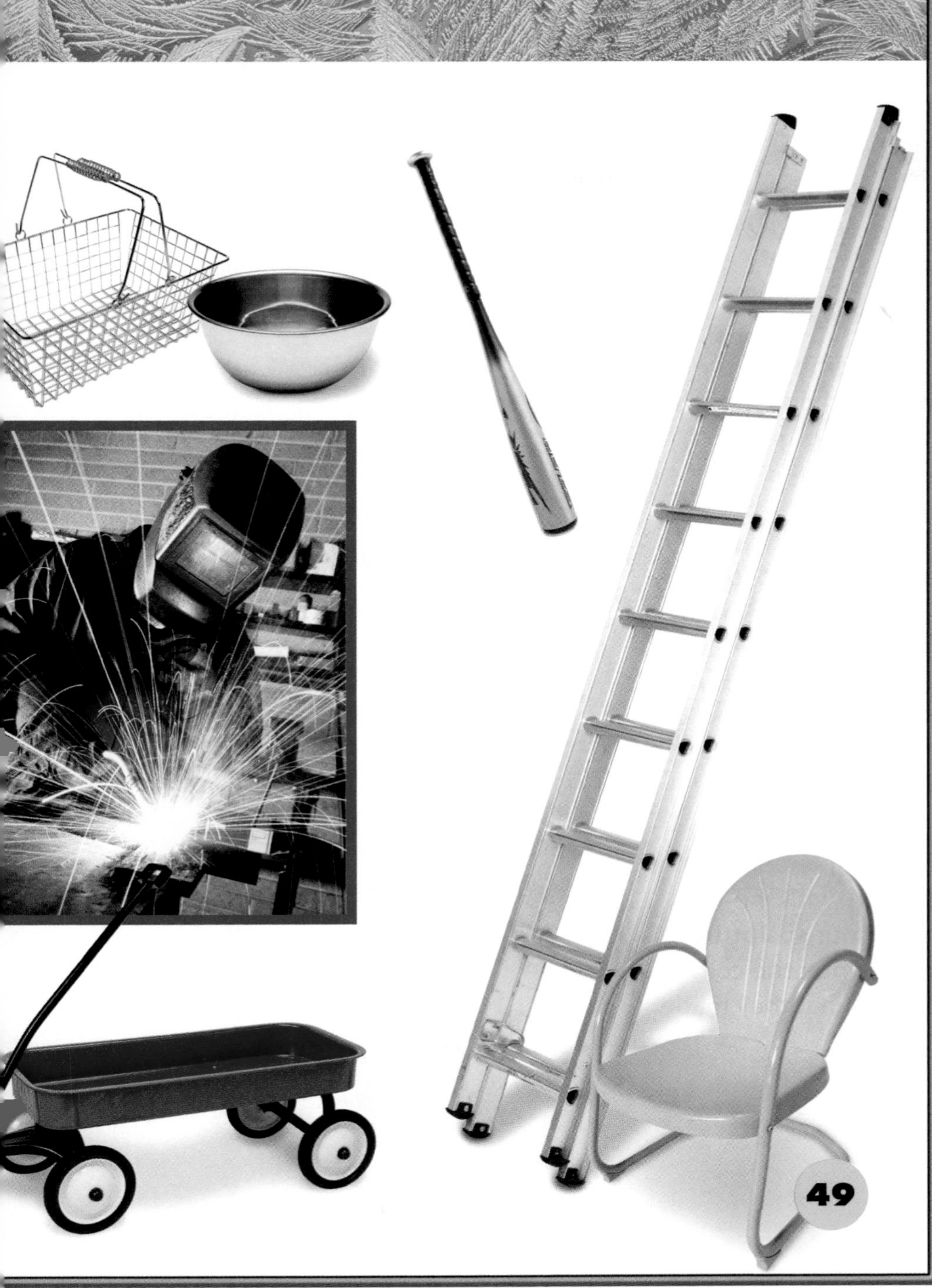

▶ Ask Questions

- **Which wood items could also be made of metal?** Answers may vary. **Why or why not?**
- **Which metal items could also be made of wood?** Answers may vary. **Why or why not?**
- **What do you think is heavier—the wooden basket or the wooden bat? The metal ladder or the metal wagon? Why?**

Invite children to use a dry-erase marker to label the objects made of wood with a *W* and the objects made of metal with an *M*.

Think and Talk

Review the ways that wood and metal are used to make the objects shown on the Flipbook page. Have children share other ways that they can use metal and wood. Ask: **Why are certain objects made of wood and others out of metal?**

Formative ASSESSMENT

As children discuss the Flipbook page, you will be able to assess what they understand and what they are confused about. If children need support, encourage them to play a concentration game using the Photo Sorting Cards.

ELL Support

Display six common items made of wood or metal. Invite children to remove them one at a time as you name each. Repeat.

Beginning Ask children to hold up an item as you name it.

Intermediate Hold up an item and ask children to name it.

More to Read

Metal, by Claire Llewellyn (Sea to Sea, 2006)

Activity Read the book with children, pointing out the ways that metal is transformed and used for everyday objects. After reading, have children make drawings for their own book of metal objects and help them with labeling.

Sculpture Fun

Objective: Understand how to manipulate wood and wire to create a sculpture.

Inquiry Skills: classify, predict, communicate, infer

You need **wood scraps of different sizes, different types of metal scraps (screws, nuts and bolts, washers, wire, aluminum foil, paper clips), glue, tools, Science Journal p. 52**

1. **Classify** Ask children if they have ever made a sculpture. Then show them the materials available to use for their sculptures. Help the group determine if items are wood or metal. Remind children about safety before building.
2. **Predict** Create an example of a sculpture with an unstable base. Ask: **What will happen if you make the base of your sculpture this way? Why?** It will fall; It may break because it cannot stand up.
3. **Communicate** When children have completed their sculptures, invite them to share their work with the group. Ask: **What did you use to make your sculpture? What could you do if you wanted to change the color of your sculpture?** Have children record what they made on their Science Journal pages.

Teacher Tip

The key to this activity is to make sure children create a good base for their sculptures. It also takes patience on their part to wait for glue to dry. You can suggest that they work on other parts of their sculpture while they are waiting for glue to dry. It may be helpful to have a premade sample on hand to spark their imaginations.

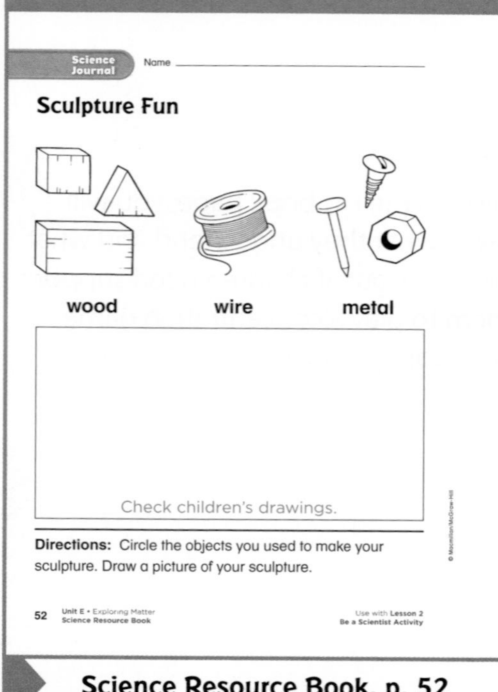
Science Journal Name

Sculpture Fun

wood wire metal

Check children's drawings.

Directions: Circle the objects you used to make your sculpture. Draw a picture of your sculpture.

52 Unit E • Exploring Matter Science Resource Book

Use with Lesson 2 Be a Scientist Activity

Science Resource Book, p. 52

Investigate More

Infer Ask: **How could you use tools to help build a sculpture?** Possible answers: I could use a saw to cut wood, a hammer to drive nails, a ruler to measure size, scissors to cut wire.

centers

Drawing and Writing

My Tool Book

Objective: Recognize and examine tools.

Inquiry Skills: observe, communicate

- Place an assortment of tools in the Writing Center and have children draw and write about them.
- If children need help with their drawings, ask questions such as: **What shape is the top? What part will you draw first? Where will you start on the page?**
- Have them identify what materials the tool is made of and label that part of their illustration. Encourage children to draw and write about several of the tools. Then bind their pages into a book with the title *My Tool Book*.

You need

- screwdriver
- wrench
- hammer
- nails
- saw
- paper
- crayons

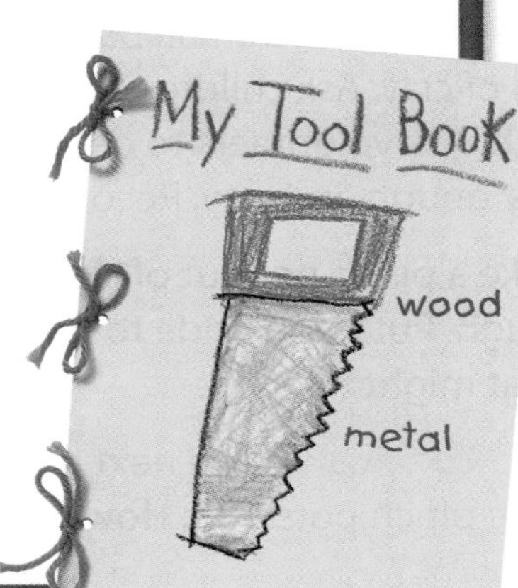

Technology

All Kinds of Tools

- For information about tools, see the lesson "We Use Tools" on pages 2–9 in **Technology A Closer Look**. This lesson will provide information about what a tool is, tools from the past and present, and special kinds of tools used for special jobs.

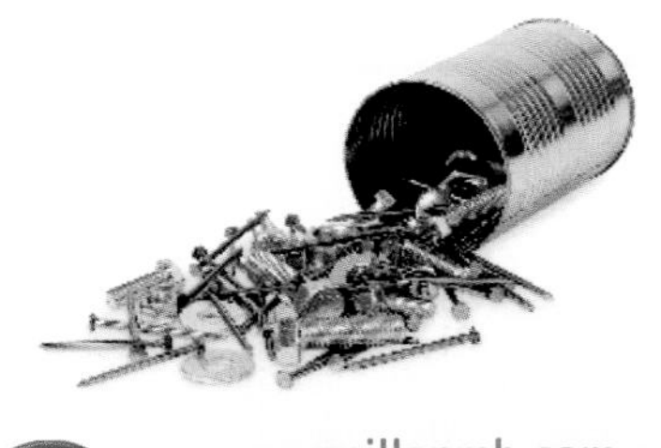

www.macmillanmh.com
for more science online.

Water Table

Floating Boats

Objective: Discover that wood floats.

Inquiry Skills: observe, make a model

- Invite children to make boats out of wood, wire, and paper. Ask: **Do you think the wood will float?** Encourage children to use wire and paper to make something that looks like a sail.
- Have children test their boats in the Water Table. Invite children to share how they made their boats and whether they sink or float.

You need

- balsa wood
- unit blocks
- wire
- paper

Art

Wire Work

Objective: Manipulate different forms of metal.

Inquiry Skills: communicate, make a model

- Help children make a base for their sculpture with corrugated cardboard and thick wire. Model how children can create their wire sculptures by bending wire pieces and attaching them to the base with staples, tape, or by sticking the ends through the cardboard.
- When children finish their sculptures, have them describe what they used and how they attached the pieces to the base.

You need

- wires of different thicknesses
- chenille stems
- corrugated cardboard
- small found objects
- stapler
- tape

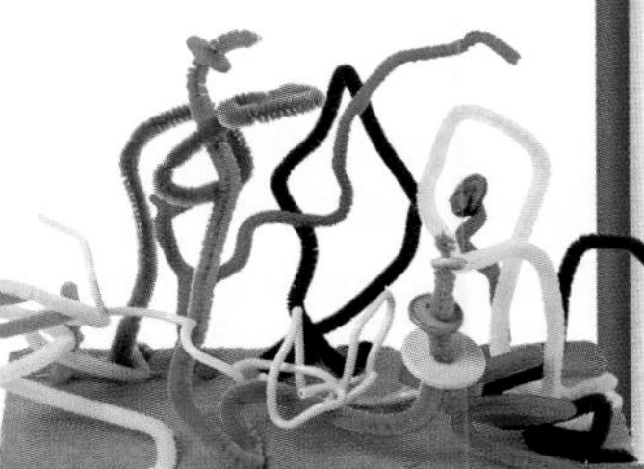

For more on wood and metal, use pp. 49–50 in the **Activity Book.**

▶ **Essential Question**
What can we make out of clay?

▶ **Objective**
Identify clay as a natural resource that can be manipulated to make things.

▶ **Vocabulary**
clay
kiln
fire

Resources

Flipbook, p. 50

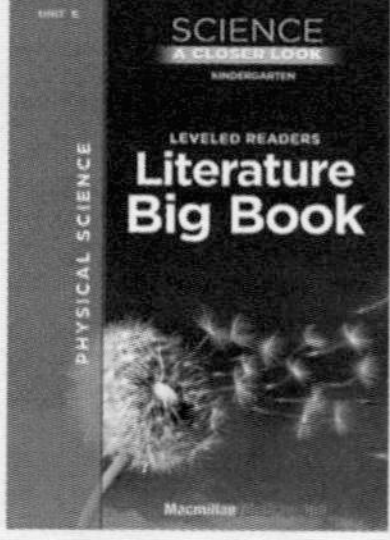

Unit E Literature Big Book

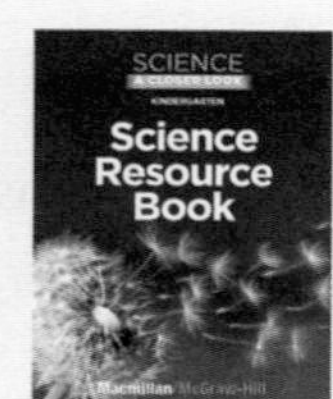

Science Resource Book, p. 53

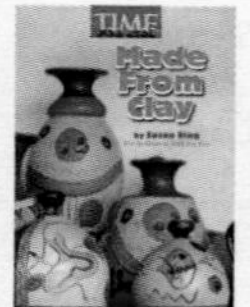

Leveled Reader: *Made from Clay*

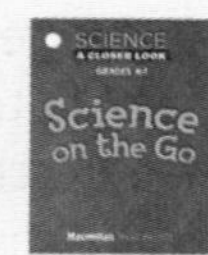

Science on the Go, pp. 95–96

Technology

www.macmillanmh.com

Working with Clay

Circle Time — WHOLE CLASS

Feel Clay

You need
- play dough
- potter's clay

Inquiry Skills: observe, communicate, infer

- Give each child a small ball of play dough and a small ball of clay. Ask children to feel, smell, and play with both. Ask volunteers to describe the difference between play dough and clay. Record their responses.
- Make a pinch pot out of clay and another out of play dough. Put them aside to dry and ask children to predict what might happen.

Draw a Conclusion The next day, invite children to hold and feel the pinch pots. Ask: **How do they feel? What happened?**

Time to Move!

Play "Pass the Clay." Have children sit in a circle and pass a ball of clay around to music. When the music stops, whoever has the clay ball needs to jump up, squash the clay, and make a new clay ball.

Be a Reader Activity

Read the Big Book

Objective: Describe the physical properties of clay.

Inquiry Skill: communicate

- Ask children to remember a time when they played with clay. Ask: **What did it feel like? What did you make?**
- Read the book *Made from Clay*. Reread it, pointing out the bold words *clay, pot,* and *mask*. Ask children to identify which of these words name things that can be made from clay. pot, mask **What other things can be made from clay?** houses **Where does clay come from?** the ground
- **Guided Reading:** Use the inside back cover of the Leveled Reader version of *Made from Clay* for activities.

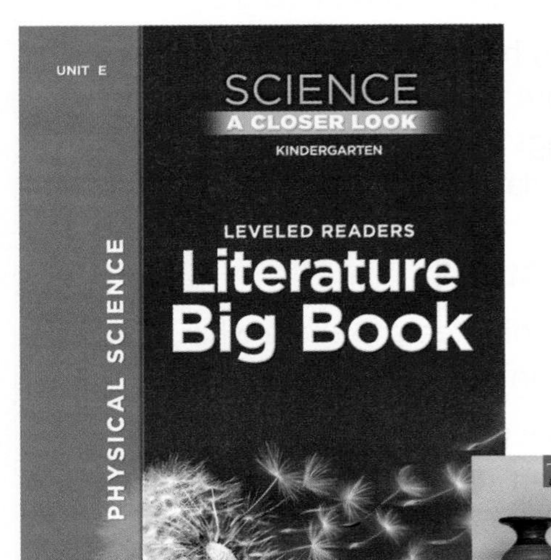

You need

- Unit E Literature Big Book
- Leveled Reader: *Made from Clay*

Be a Math Wiz Activity

Clay Patterns

Objective: Create and extend patterns.

Inquiry Skills: observe, communicate, put in order

- Explain to children that they will be making patterns using shapes and colors.
- Invite children to choose three colors and to mold each piece of clay into any shape they like. Have children first create an A-B-C pattern, repeating their pattern twice. Then have their partner extend it. Have partners switch roles and repeat the activity, creating different patterns.
- Invite children to draw their patterns on a piece of paper.

You need

- different-colored modeling clay or play dough
- paper
- crayons

Read Together and Learn

What can we do with clay?

▶ Build on Prior Knowledge

Ask children if they have ever worked with clay. Invite them to share their experiences about how it felt and the things they made. Ask children if they know where clay comes from.

▶ Use the Visuals

- **Observe** Invite volunteers to share what images they recognize on the Flipbook page.
- **Compare** Explain to children that the two images on the left side of the page show different types of clay. Help children notice what is the same and different about the two types of clay.
- **Communicate** Have children describe what they think is happening in the top row. Then have them describe what they think is happening in the bottom row.

▶ Develop Vocabulary

clay, **kiln**, **fire** Help children practice using these words as they apply to the pottery process. Write the words above the appropriate pictures and ask children to use the words as they describe the process.

Flipbook

UNIT E • LESSON 3

What can we do with clay?

Reading Strategy

Letter and Sound Recognition Have a volunteer identify what sound the word *clay* begins with. Ask children to name other words that begin with the *cl* sound (*class, clown, club, clap*).

Comprehension Have children use the Flipbook page to help them describe the process of making a clay pot using the terms *first, next,* and *last*.

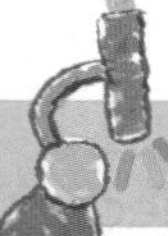

Science Facts

Modeling Clay does not dry out, so you can sculpt and form it without worrying about a time limit. Firing, the process that drys and hardens the clay, requires only low temperatures, low enough to use a home oven.

Clay Some clay is ready to use as dug, other clay may need refining or mixing with other components to be useful. Most potters work the clay on wheels, using their hands to manipulate it. When a piece is thoroughly dry, it is fired in a kiln. The temperature in a kiln can be as high as 2,000°F.

Professional Development For more Science Facts and resources from NSDL visit http://nsdl.org/refreshers/science

50

Differentiated Instruction

Extra Support While volunteers are describing the pottery process on the Flipbook page, have children act out the process of digging out the clay, spinning the pot or molding the clay, placing it in the kiln, taking it out, glazing it, and replacing it in the kiln.

More to Read

The Pot That Juan Built, by Nancy Andrews-Goebel (Lee & Low Books, 2002)

Activity Read the book, reviewing the process that a potter uses to make a pot. After reading, invite children to make a drawing or collage of a clay pot inspired by the colorful illustrations in this book.

Ask Questions

- **What do you think might happen if the clay was not fired?** It might not keep its shape.
- **What kinds of objects would be difficult to make out of clay? Why?**
- **What do you think the clay's texture is like before the potter works with it? How might it feel?** smooth, gooey, wet **During the process?** warm, soft **After it has been fired?** hard, cool, rough

Have volunteers use a dry-erase marker to number the images in order from *1–4* in the top row. Repeat for the bottom row.

Think and Talk

Review the difference between potter's clay and modeling clay. Review the process of sculpting, firing, and baking clay using Flipbook page 50.

Observe children as you explore the Flipbook page to see if they understand the characteristics of clay and the steps in the process of making a clay pot. If children are having difficulty, invite them to explore and work with clay.

Pinching Pots

Objective: Experience working with clay and manipulating it into different forms.

Inquiry Skills: observe, make a model, draw a conclusion, predict

You need **potter's clay, Science Journal p. 53**

Teacher Tip

In order to give children time to explore and manipulate clay, this activity should be done over several days, culminating in the making of a pinch pot. Prepare the clay ahead of time so that it is just right for little hands. It should not be too wet or too dry.

1. **Observe** Give children a ball of clay and encourage them to use their hands and fingers to change its shape. As they work, have children describe how the clay feels. Ask: **Is it getting drier? Is it easier or harder to use?** Encourage children to make a small sculpture.
2. **Make a Model** Show children how to make a pinch pot. Roll clay into a ball at least 2–3 inches in diameter, poke your thumb into the middle, and use your fingers to hollow out the middle. Invite children to make their own. Leave pots in a safe place to air-dry. Have children record their work on their Science Journal pages.
3. **Draw a Conclusion** Ask: **What do you think is the difference between making a thin-walled pot and a thick-walled pot?** Thin walls may break; thick walls are stronger.

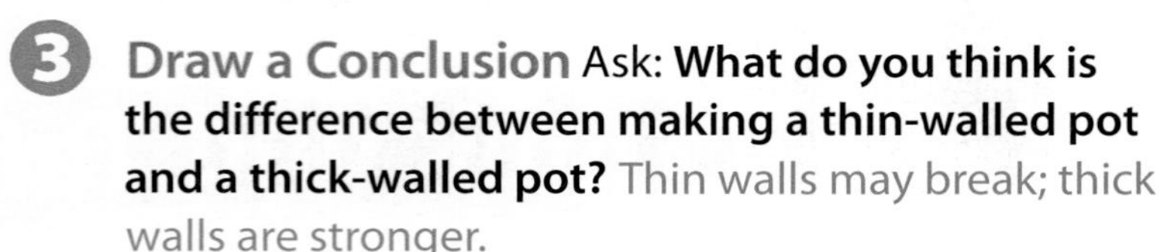

Name ______________ Science Journal

Pinching Pots

clay	pot
Check children's work.	Check children's work.
I started with a ball of clay.	I made a pot.

Directions: Draw pictures to show the clay you used and the pot you made. Complete each sentence.

Unit E • Exploring Matter Science Resource Book — Use with Lesson 3 Be a Scientist Activity 53

Science Resource Book, p. 53

Investigate More

Predict Ask: **What might happen when we leave the pots out to dry?** The clay will get hard.

centers

Drawing and Writing

Drawing Clay

Objective: Develop fine motor skills.

Inquiry Skills: observe, communicate

- Display some ceramic objects in the Drawing and Writing Center and invite children to examine them with their eyes and hands.
- Encourage children to draw the objects from observation rather than memory.
- Help children to write words to describe their drawings. Display children's work on a bulletin board in the classroom with the title *These Are Made of Clay.*

You need
- drawing paper
- pencils
- mug
- vase, bowl, or plate with decorative design

Cooking

Baker's Clay

Objective: Observe how matter can change form with heat.

Inquiry Skills: measure, make a model

You need
- bowl
- tempera paint
- cookie sheets
- see recipe

- Help children measure and combine the ingredients in a bowl to make baker's clay.
- Have children make shapes with the dough and choose one that they would like to bake and paint. Discuss the difference between baker's clay and potter's clay.
- Invite children to predict what will happen to the clay objects when they are in the oven and why.

Baker's Clay
4 cups flour
1 cup salt
1 1/2 cups water

Preheat oven to 350 degrees. Mix the flour, salt, and water. Knead dough until smooth. Shape dough as desired. Bake at 350 degrees for 1 hour.

Art

Painting Clay

Objective: Recognize that clay objects can be painted.

Inquiry Skills: infer, make a model

You need
- dried pinch pots
- tempera paints
- white glue
- paintbrushes
- newspaper

- Have children paint the dried pinch pot they made with clay.
- After the paint dries, have children cover their pot with a watered-down mixture of white glue.
- When the glue dries, send the clay pieces home, reminding children that they are very fragile.

For more on clay, use pp. 51–52 in the **Activity Book.**

Investigate Water

▶ **Essential Question**
How can water change?

▶ **Objective**
Identify and explore the properties and changing states of water and investigate objects that sink and float in water.

▶ **Vocabulary**
solid liquid gas

Resources

Flipbook, pp. 51–52, S10

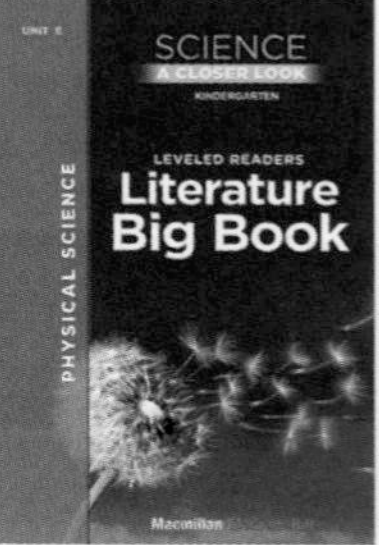
Unit E Literature Big Book

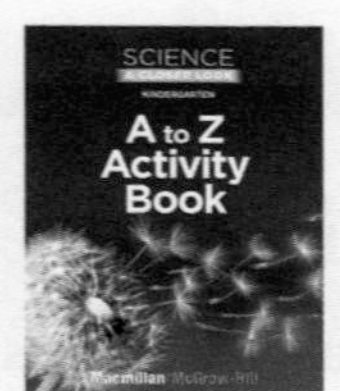
A to Z Activity Book, pp. 4–5, 46–47

Leveled Reader: *I Like Ice*

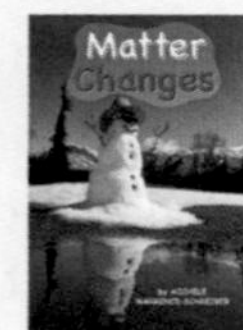
Leveled Reader: *Matter Changes*

Technology

Science Songs CD Tracks 19–20

 Science in Motion *Water Changes*

NSDL

Circle Time

WHOLE CLASS

Altered States

You need
- ice cubes
- transparent cup

Inquiry Skills: observe, predict, infer

- Show a cup with ice cubes and explain that ice is water in a solid form. Ask: **What will happen to the ice if we leave the cup on a shelf?** Discuss how a solid becomes a liquid.
- Check the cup of water each day. When the water is gone, explain that the water has evaporated into the air. Record the states of water on a chart.

Infer Ask: **What happens to puddles after it stops raining?** The water evaporates. **Where do you think the water goes?** into the air; up to the clouds

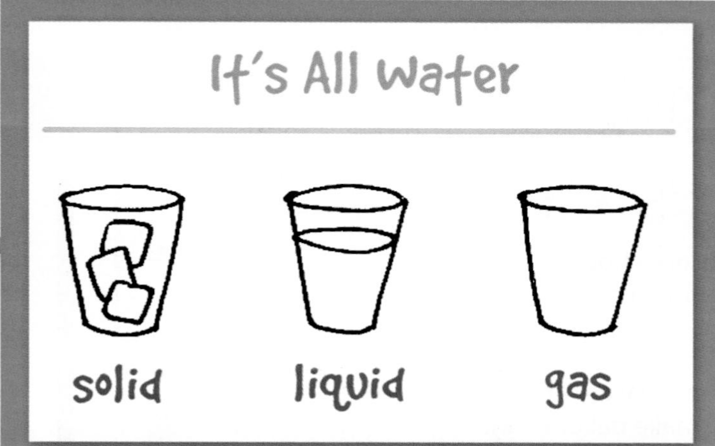

Time to Move!

Hand each child a plastic spoon and an ice cube. Encourage children to move around the room carefully balancing the ice cube on the spoon.

Be a Reader Activity

Read the Big Book

Objective: Compare two nonfiction text selections.

Inquiry Skills: communicate, infer

- Show children the covers of *Matter Changes* and *I Like Ice.* Read each book.
- Label the Venn diagram with the two book titles. Invite children to recall information from each story and talk about how they are alike and different. Record children's ideas.
- **Guided Reading:** See the inside back cover of the Leveled Reader versions of *Matter Changes* and *I Like Ice* for activities.
- For more to read on how water changes, read *Melting Snow* available in the Unit E Literature Big Book and as a Leveled Reader.

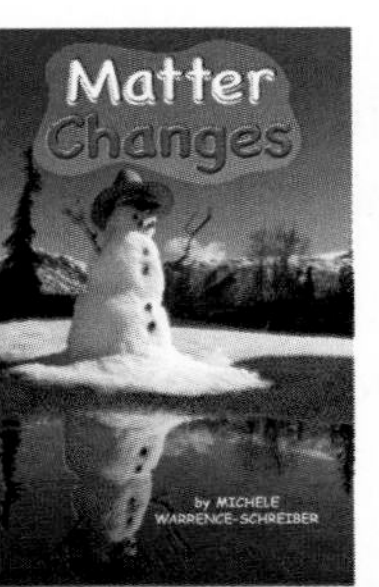

You need

- Unit E Literature Big Book
- Leveled Readers: *Matter Changes* and *I Like Ice*
- Flipbook p. G1
- dry-erase marker

Be a Math Wiz Activity

Weighing Water

Objective: Investigate the weight of water.

Inquiry Skills: observe, measure, compare, communicate

- Show children a jar of water and hold up a pencil. Ask: **Do you think the pencil weighs less or more than this jar of water?** Hold up a book and repeat the question.
- Have children work in pairs and use a balance scale to find something in the room that weighs about the same as the jar of water. Have children record their findings on paper.
- Discuss their findings as a class. Ask: **What have we learned about water?**

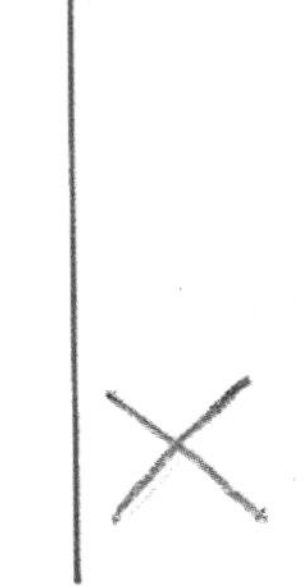

You need

- jar of water
- collection of classroom objects
- balance scale

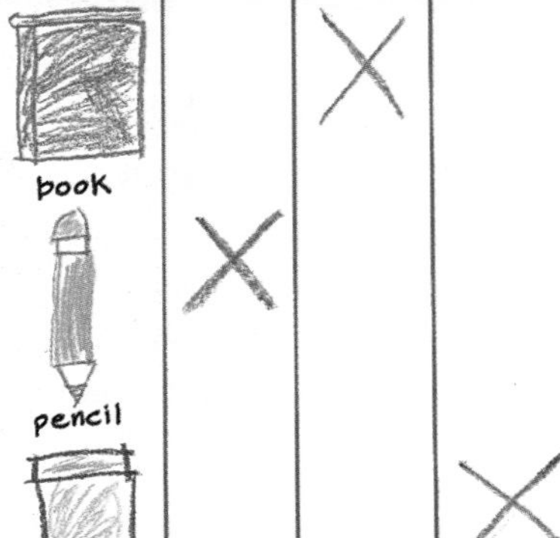

Read Together and Learn

Where is the water?

▶ Build on Prior Knowledge

Ask children if they have ever had a drink with ice cubes in it. Invite volunteers to describe what happens to the ice cubes. Allow children to share their experiences.

▶ Use the Visuals

- Communicate Invite volunteers to describe the picture on the Flipbook page.
- Invite a volunteer to point to all the places in the picture where there is water. Help children identify any of the sources of water that they missed (clouds, snow, ice, air, water).
- Infer Ask children to share ideas about why water can look so many different ways.

▶ Develop Vocabulary

solid, **liquid**, **gas** Invite children to place sticky notes labeled *solid* and *liquid* on the corresponding images on the Flipbook page. Explain what water looks like in both states. Remind children about the evaporating water in the Circle Time Activity and that when water evaporates it becomes a gas, which exists in the air, but cannot be seen.

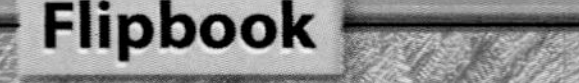

UNIT E • LESSON 4

Where is the water?

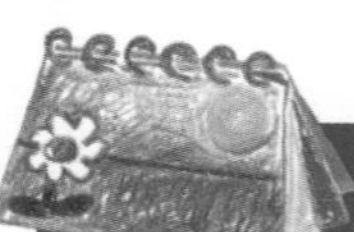

Reading Strategy

Comprehension Ask children to explain how water can be in so many states in one place. Notice their logic about how ice, water, snow, and clouds are related.

Print Awareness Refer children to the sticky notes labeled *solid* and *liquid*. Ask children to identify which word is *solid*. Ask them to explain how they figured it out.

Science Facts

Solid When water freezes it turns from a liquid to a solid. Water freezes at 32°F.

Liquid A liquid is defined as a substance that can be poured and that takes the shape of the container it is poured into.

Gas When water evaporates it turns into a gas. Gases are colorless, odorless, and invisible to the human eye.

Phase Changes When a substance, such as water, changes states, it is called a *phase change*. The physical properties change, but not the chemical properties. Water changing from a solid to a liquid is said to be *melting*. When it changes from liquid to gas, it is *evaporating*. Water changing from gas to liquid is called *condensation*.

Professional Development For more Science Facts and resources from NSDL visit http://nsdl.org/refreshers/science

51

▶ Ask Questions

- **Why do you think there can be ice and water in the same place?** Some areas are colder than others; some areas are exposed to the Sun which makes those areas warmer.
- **What do you think is the difference between snow and ice?** Possible answer: Snow is not as cold or frozen as ice.
- **Why does water change forms? Do you think any liquid can freeze? Why or why not?**

Help children to use a dry-erase marker to label the water, ice, snow, and clouds on the Flipbook page.

Use the Flipbook page to review the states of water and ask children to identify and describe what water looks like in each state.

ELL Support

Review what happened to the ice in the Circle Time Activity. Draw pictures as children describe the changes.

Beginning Ask yes/no questions such as: **Is this juice a solid?**

Intermediate Show objects and ask children if they are liquid, solid, or gas.

More to Read

Down Comes the Rain, by Franklyn M. Branley (HarperCollins, 1997)

Activity Read the book on a rainy day if possible. When the rain stops, bring groups of children outside to observe a puddle. Measure how deep it is with a yard stick and monitor the evaporation process.

Read Together and Learn

Does it sink or float?

▶ Build on Prior Knowledge

Ask children if they recognize any of the objects on the Flipbook page. Invite children to recall a time when they might have played with objects in a tub of water. Encourage them to share their experiences.

▶ Use the Visuals

- **Predict** Invite volunteers to predict if each object will sink or float. Encourage children to explain their predictions for each object.
- **Infer** Explore concepts of *heavy* and *light* by asking children to say which objects they think are heavy and which objects they think are light. Ask: **Do you think that weight has something to do with whether an object can float?**

▶ Additional Vocabulary

sink, **float** Ask children to demonstrate with their bodies what it looks like to *float in water* and what it looks like to *sink in water.*

Flipbook

UNIT E • LESSON 4

Does it sink or float?

Reading Strategy

Print Awareness Have children look at the word *float*. Ask: **What letters make up the word? How many letters are there in the word?** Repeat with the word *sink*.

Comprehension Explain to children that the word *sink* has more than one meaning. Point to the word *sink* on the Flipbook page and read it aloud. Ask children: **What does this word mean?** to fall to the bottom **What else can it mean?** the sink in the kitchen or bathroom

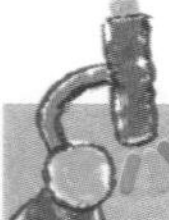

Science Facts

Density and Weight Water tries to support solid objects. If the objects are heavy for their size, the objects will sink. If the objects are light for their size, the objects will float. An object that is heavy for its size is said to have a high density. For example, if a box is empty, it has a low density. As it begins to fill with objects, its density increases, but its size stays the same. Both the density and the shape of an object affect whether it will float or sink in water. The shape of an object controls the amount of water that it pushes out of the way or displaces. If the amount of water that is displaced weighs more than the object, it will float. If the displaced water weighs less than the object, it will sink.

Professional Development For more Science Facts and resources from NSDL visit http://nsdl.org/refreshers/science

52

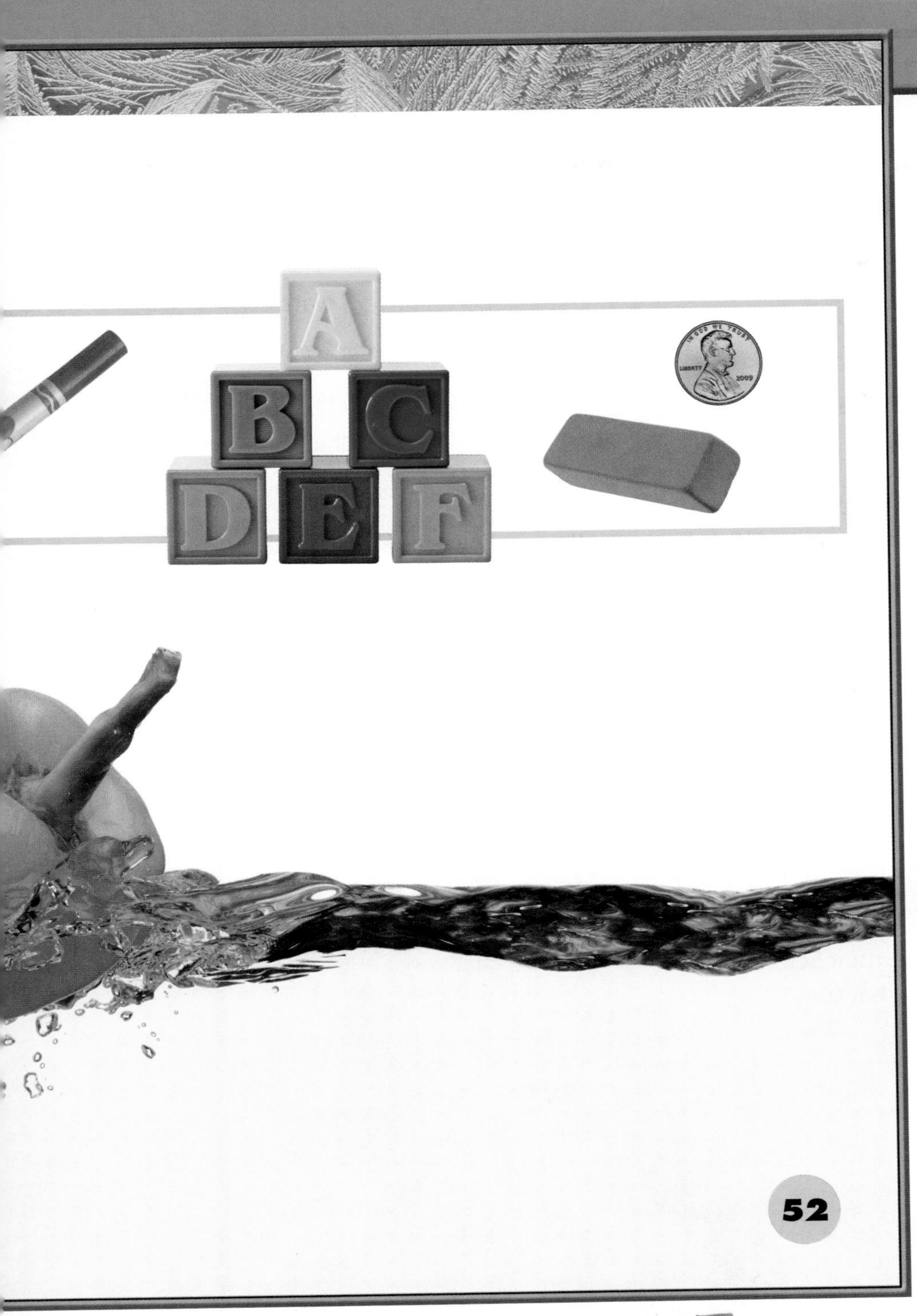

▶ Ask Questions

- **Do you think the pepper will sink or float? How do you know? Why is the water around the pepper splashing upward?**
- **What makes the objects that will float different from the objects that will sink?** Possible answer: the objects that float are lighter
- **Why do you think there are bubbles in the water?** Accept all responses.

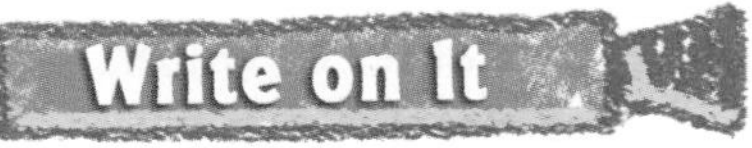

Invite children to use a dry-erase marker to write *F* on objects that float and *S* on objects that sink.

Think and Talk

Invite children to talk about what they learned about water. Prompt them to summarize their ideas about water and its different states, and why things sink or float in water.

Formative ASSESSMENT

As children discuss the Flipbook pages, note what they understand about the states of water and objects that sink or float and what they are still confused about. You may want to repeat some of the activities in this lesson to give children more opportunities to discover things about water.

Take a Trip

Water Watching Provide children with clipboards and paper and pencils. Bring the class to a nearby pond, stream, lake, or brook and invite children to record their observations about the water. Revisit this spot at different times during the year. Encourage children to notice if the water has changed. **Does there seem to be more or less water since the last visit? Is the water still a liquid? Has it become a solid?** Have children record their findings.

Back in the Classroom Each time the class returns from a visit, engage children in a discussion about what they observed.

Make It Float

Objective: Identify that some objects that sink can be made to float.

Inquiry Skills: predict, observe, make a model, infer, communicate

You need small balls of clay (2-inch diameter), containers of water, Science Journal p. 54

1. **Predict** Show children the ball of modeling clay and the container of water. Ask: **What might happen if I drop the ball of clay into the water?**
2. **Observe** Drop the ball in and have children observe that it sinks. Ask children if they have any ideas about how they could make the clay float. Gather ideas and record them on chart paper.
3. **Make a model** Have children work in pairs to see if they can make the clay ball into a "boat" that floats.
4. **Infer** Have children tell about why they think the clay boat floats and the clay ball sinks.
5. **Communicate** Have children use their Science Journal page to draw a picture of what they did to the clay to make it float.

Teacher Tip

One ounce of water takes up more space than one ounce of clay. Because clay is denser than water, it will sink. When the clay is made into a boat, the clay fills with air. Air and clay together are less dense than water, so this time the boat will float.

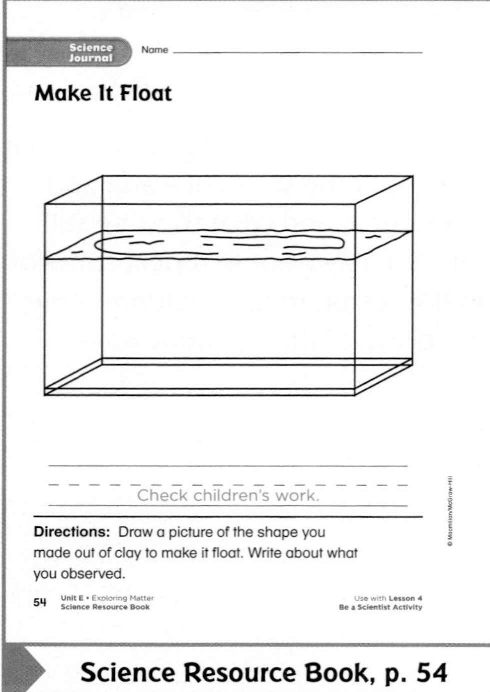

Science Journal Name

Make It Float

Check children's work.

Directions: Draw a picture of the shape you made out of clay to make it float. Write about what you observed.

54 Unit E • Exploring Matter Science Resource Book   Use with Lesson 4 Be a Scientist Activity

Science Resource Book, p. 54

Investigate More

Infer Ask children: **What other materials could you use to make a boat?** Possible answers: wood, cork **Would a sponge be a good choice? Why?** Possible answer: No, because it might fill up with water and sink.

Music

Water, Steam, and Ice

Objective: Identify what will happen when water is heated or frozen.

Inquiry Skills: observe, communicate

- **Sing and Act Out** Invite children to sing the original version of "I'm a Little Teapot," encouraging them to act out the song. Then play "Water, Steam, and Ice," inviting children to add actions to the song—point to their eyes on the word *Watch,* hold up one finger on the word *one,* point to themselves on the word *me,* and so forth.
- **Talk About It** Point to the water, the teapot, and the ice on the Flipbook page. Ask children what they think would happen if you boiled water in a teapot or froze water in an ice tray.
- See page TR31 for printed music to "Water, Steam, and Ice."

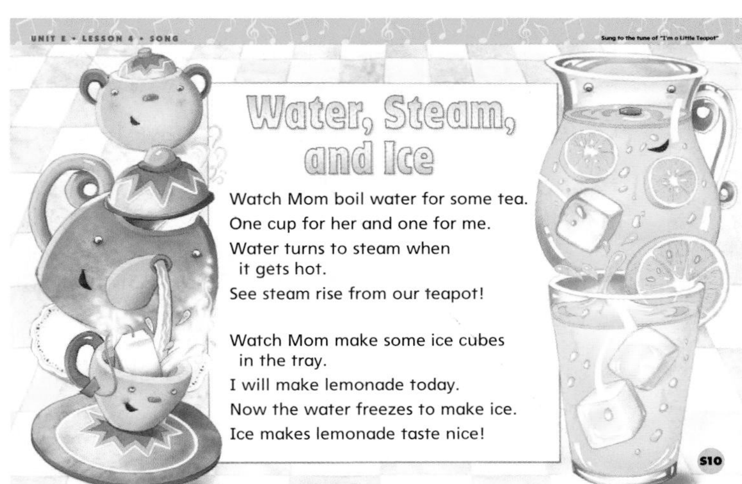

Flipbook p. S10
CD Tracks 19–20

Water Table

Salt and Float?

You need
- objects that float used in previous activities
- container with salt water

Objective: Discover that objects float more easily in salt water.

Inquiry Skills: investigate, infer

- Place a bin with salt water next to the water table. Explain that the water table has fresh water and the bin has salt water.
- Have children choose objects to test in the fresh water and then the salt water.
- Ask children what they noticed and for possible explanations. Ask: **Has anyone tried floating in the ocean? Is it easier or harder than in a pool or lake?**

Cooking

Ice Pops

You need
- powdered fruit juice mix
- small paper cups
- craft sticks
- freezer

Objective: Investigate if water mixed with drink mix will change properties.

Inquiry Skills: observe, predict

- Mix the powdered fruit juice with water. Have children predict if the juice will freeze. Ask: **How long do you think it might take?**
- Write each child's name on a cup and fill each with the juice mixture. Have each child place a craft stick in the cup. The craft stick will rest against the side of the cup.
- Record the time you put the cups in the freezer and invite children to periodically check on the progress.

For more on the states of water, use pp. 53–58 in the **Activity Book.**

UNIT E Assessment

Performance ASSESSMENT

DOK 2

Properties of Matter

You need
- cup
- water
- paper
- ceramic bowl

Objective: Assess what children have learned about matter and its properties.

- Show the child a cup of water. Ask: **How could you change this water from a liquid to a solid?**
- Show the child a piece of paper. Ask the child to name two things we can make out of paper.
- Show the child a ceramic bowl. Ask: **What material is this bowl made out of?**

Scoring Rubric

Use the Scoring Rubric to evaluate each child's completion of the Performance Assessment task.

points = Child can explain that you freeze water to make ice. Child can name two things made out of paper. Child can identify that the bowl is made of clay.

points = Child performs two out of three tasks mentioned above.

point = Child performs one out of three tasks mentioned above.

Summative ASSESSMENT

DOK 1

You may choose to use the following minibook pages from the **Science Resource Book** pp. 71–72 to assess what children have learned about matter. Help children fold the page in half to create a book and write their name.

- On page 1, read the sentence, have children circle the correct word that completes the sentence, and write the word on the line provided.
- Repeat these same directions for pages 2–4 of the minibook.

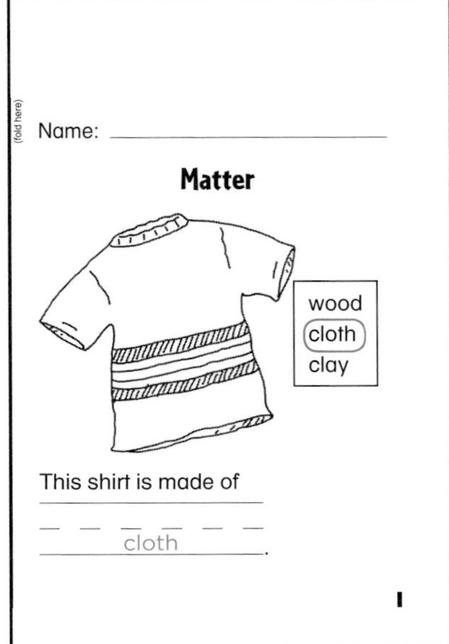

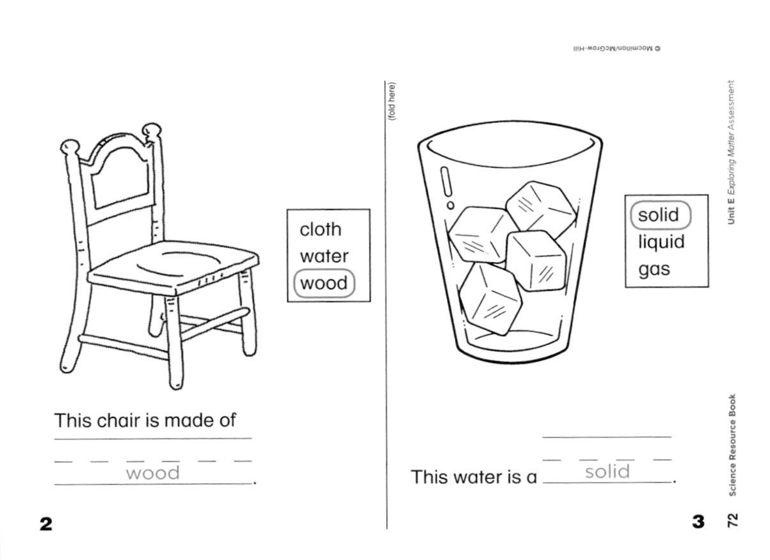

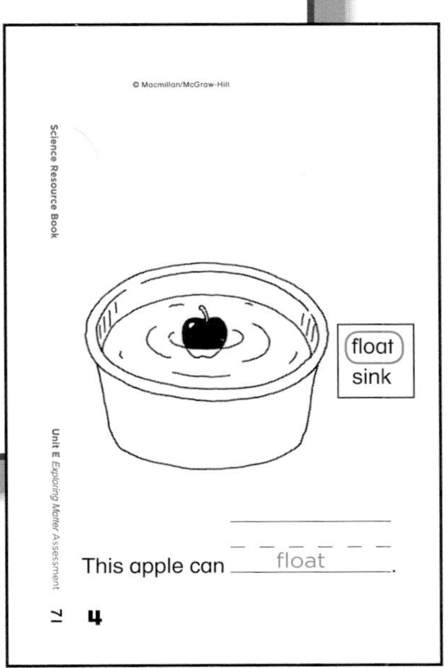

Science Resource Book, pp. 71-72

Portfolio ASSESSMENT

- Encourage children to select work from this unit to include in their portfolio. Discuss their choices and have children share which is their favorite and why.
- Use the Unit E Checklist from page 81 of the **Science Resource Book** to record children's progress. Place checklists and notes in each child's portfolio.

Teacher's Notes

Moving Right Along

What causes things to move?

Essential Questions

Lesson 1 How do wheels help us?

Lesson 2 What makes things move?

Lesson 3 How do we stay on the ground?

Lesson 4 How are sounds made?

Lesson 5 What objects will magnets move?

LOG ON Visit www.macmillanmh.com for online resources.

Science Leveled Readers

All included in the Leveled Readers Literature Big Book.

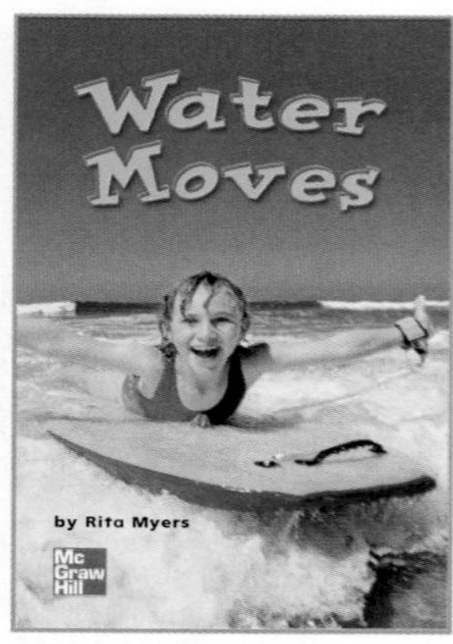

APPROACHING

Water Moves
This rhyming book shows the many ways water can be used.

ISBN: 978-0-02-281086-3

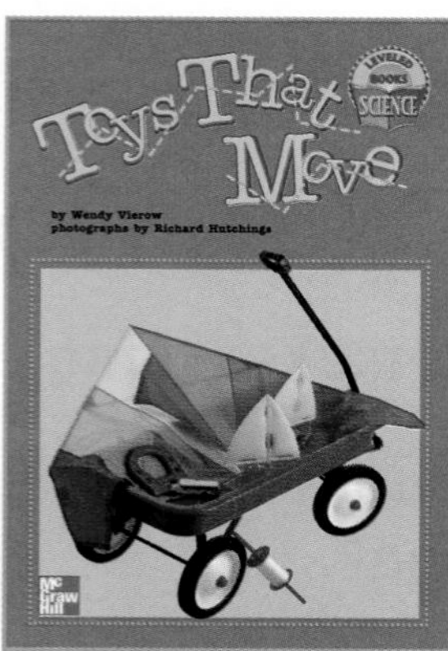

ON LEVEL

Toys That Move
This book talks about various ways that toys can move around.

ISBN: 978-0-02-278971-8

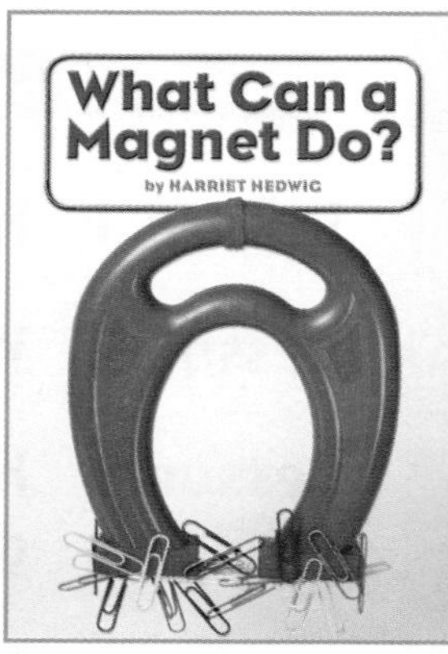

BEYOND

What Can a Magnet Do?
Magnets are magnificent! They may look different, but they all move metal.

ISBN: 978-0-02-284601-5

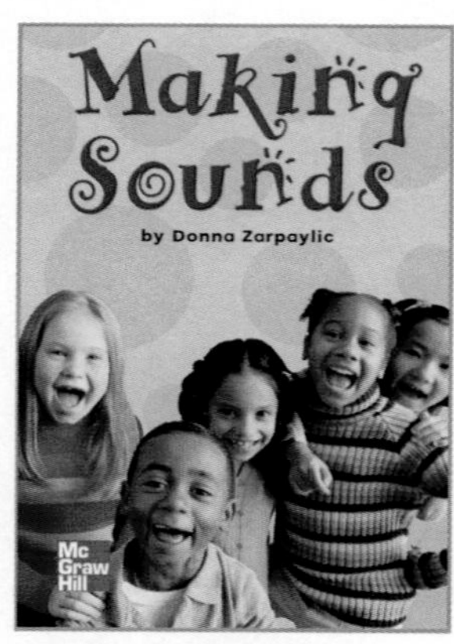

APPROACHING

Making Sounds
There are so many creative ways to make sounds using all sorts of objects!

ISBN: 978-0-02-282027-5

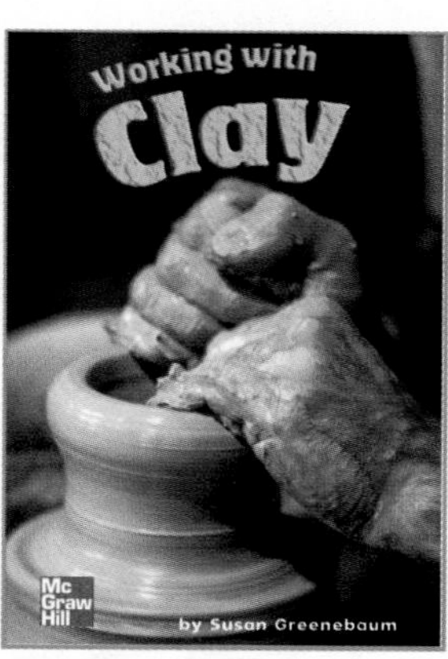

BEYOND

Working with Clay
This book shows the unique process used to make a clay pot.

ISBN: 978-0-02-281085-6

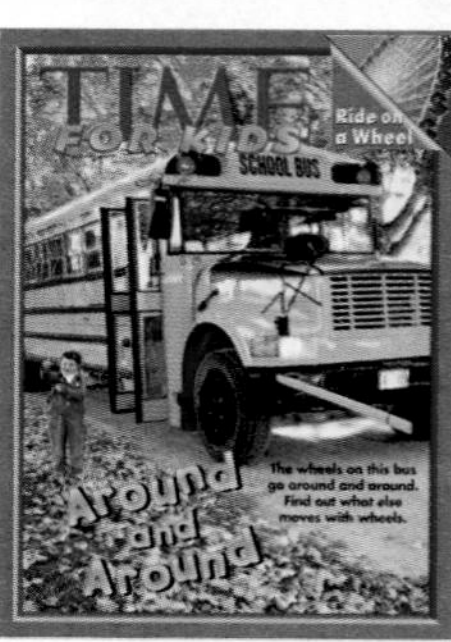

Includes *Time for Kids* magazine about objects that have wheels to help them move.

Bibliography

Lesson 1 Wheels

Mama Zooms, by Jane Cowen-Fletcher (Scholastic, 1996)

What Does a Wheel Do?, by Jim Pipe (Copper Beech Books, 2002)

Wheels and Axles, by Sally M. Walker and Roseann Feldmann (Lerner Publications, 2002)

The Wheels on the School Bus, by Mary-Alice Moore (HarperCollins Children's Books, 2006)

Lesson 2 How Things Move

Cinnamon's Day Out, by Susan L. Roth (Dial Books, 1998)

Construction Trucks, by Jennifer Dussling (Grosset & Dunlap, 1998)

Dig Dig Digging, by Margaret Mayo (Henry Holt, 2002)

In the Spin of Things, by Rebecca Kai Dotlich (Boyds Mill Press, 2003)

Inclined Planes and Wedges, by Sally M. Walker and Roseann Feldmann (Lerner Publications, 2002)

Kite Flying, by Grace Lin (Dragonfly Books, 2004)

Levers, by Sally M. Walker and Roseann Feldmann (Lerner Publications, 2002)

Screws, by Sally M. Walker and Roseann Feldmann (Lerner Publications, 2002)

The Way Things Move, by Heidi Gold-Dworkin (McGraw-Hill, 2001)

Lesson 3 Ups and Downs

I Fall Down, by Vicki Cobb (HarperCollins, 2004)

The Science of Gravity, by John Stringer (Raintree, 2000)

Lesson 4 Sounds All Around

The Listening Walk, by Paul Showers (HarperCollins, 1993)

Polar Bear, Polar Bear, What Do You Hear?, by Bill Martin Jr. (Henry Holt and Company, 1997)

Snow Music, by Lynne Rae Perkins (HarperCollins Children's Books, 2003)

Lesson 5 Magnets

Magnets (Science All Around Me) by Karen Bryant-Mole (Heinemann, 1998)

Mickey's Magnet, by Franklyn M. Branley and Eleanor K. Vaughan (Scholastic, 1999)

What Makes a Magnet?, by Franklyn M. Branley (HarperCollins, 1996)

For more information on these or other books, visit **www.bankstreetbooks.com** or your local library.

Science Vocabulary

Vocabulary

pulley
slide
roll
push
pull
vibration
loud
soft
magnet
force

Toys in Motion

Explore what children know about motion by creating a word web.

- Have children identify ways in which an object can move. Display some classroom toys and invite children to share how they can move each toy.
- Create a word web by drawing a line from a center circle and writing the children's responses.

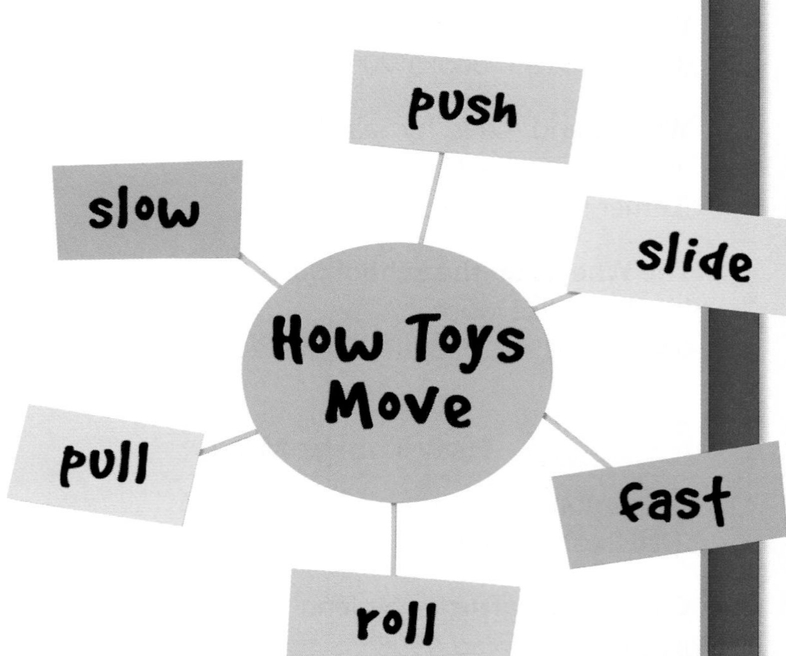

ELL Support

Things Move

Write the vocabulary words on the board. Ask children to repeat them as you say them. Discuss meanings and which words are *naming* words and which words are *doing* words.

- **Beginning** Ask children to pantomime a *doing* word.
- **Intermediate** Ask children to take turns miming and guessing words.
- **Advanced** Encourage children to take turns miming and guessing the word by asking: **Is it ____?**

Activity Book

Look for additional cross-curricular activities about motion in the **A to Z Activity Book.**

M is for Magnets and Much More,
pp. 26–27
Science, Math, Social Studies, Writing

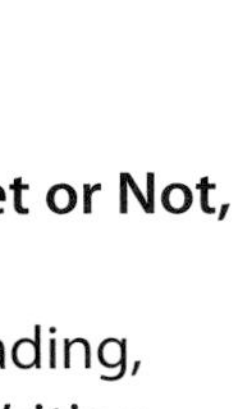

Q is for Quiet or Not,
pp. 34–35
Music, Reading, Science, Writing

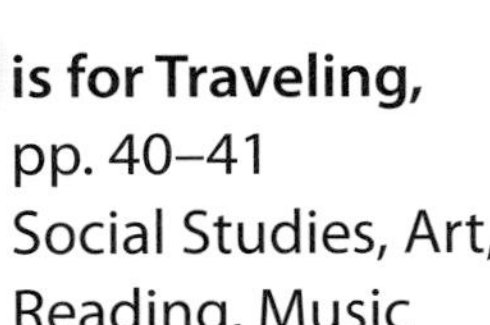

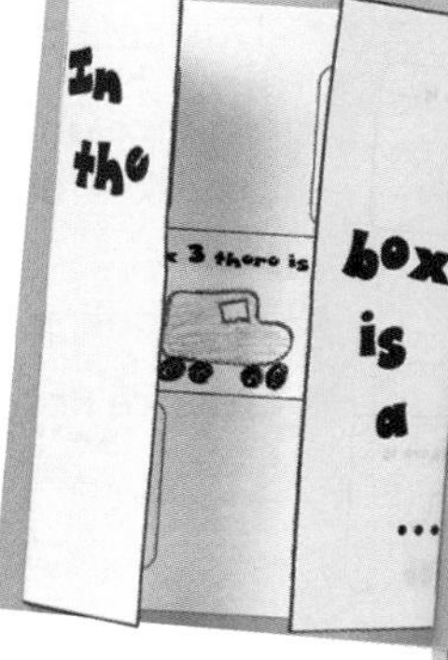

T is for Traveling,
pp. 40–41
Social Studies, Art, Reading, Music

Unit Project

Toys That Move Museum

Objective: Sort and classify toys that move.

You need
- toys from home and school
- construction paper
- crayons
- index cards
- tape
- blocks

Step 1 Invite children to work together to make a museum of toys that move. Have children describe different ways toys can move. Record their ideas on chart paper.

Step 2 Send letters home telling families about the museum and asking children to loan a toy to the museum. Label toys with the children's names. If a child cannot lend a toy, he or she can display a classroom toy. Have children describe how their toys move, and help them write their descriptions on index cards.

Step 3 Help children organize the toys by how they move. Display the groups of toys around the room. Invite other classes and parents to visit your museum.

Technology

IWB INTERACTIVE WHITEBOARD READY

LOG ON www.macmillanmh.com

Visit Macmillan/McGraw-Hill Science online for projects and activities for students, teachers, and parents.

Bank Street www.bankstreet.edu

Visit Bank Street online for teacher resources and more activities.

TIME FOR KIDS www.timeforkids.com

Visit Time for Kids online to read about current science news and explore additional science activities.

UNIT F Planner

TeacherWorks™ Plus CD-ROM has interactive unit planner, Teacher's Edition, and worksheets.

Lesson	OBJECTIVES	VOCABULARY	RESOURCES and TECHNOLOGY	
1 Wheels PAGES 242–247 PACING: 3–5 days	■ Recognize that wheels affect speed and motion and make moving easier.	**wheel** **pulley**	**Flipbook,** pp. 54, S11 **Unit F Literature Big Book :** *Time for Kids* **Leveled Reader:** *Working with Clay* **Science Songs CD** ■ **Science on the Go,** pp. 101–104	■ **Science Resource Book,** p. 55 ■ **A to Z Activity Book,** pp. 40–41 ■ **Photo Sorting Cards** 51-60
2 How Things Move PAGES 248–255 PACING: 3–5 days	■ Explore ways objects move and forces that cause movement.	**slide** **roll** **push** **pull** **force**	**Flipbook,** pp. 53, 55–56 **Unit F Literature Big Book** **Leveled Readers:** *Toys That Move* and *Water Moves* ■ **Science on the Go,** pp. 105–106	■ **Science Resource Book,** p. 56 ■ **Floor Puzzle:** *Workbench*
3 Ups and Downs PAGES 256–261 PACING: 3–5 days	■ Understand that certain objects, like the Sun and the Moon, stay in the sky, while others, like an airplane, are in the sky but return to Earth.	**gravity**	**Flipbook,** p. 57 ■ **Science Resource Book,** p. 57	■ **Science on the Go,** pp. 107–108
4 Sounds All Around PAGES 262–267 PACING: 3–5 days	■ Describe sounds and understand how they are made.	**vibration** **loud** **soft**	**Flipbook,** pp. 58, S12 **Unit F Literature Big Book** **Leveled Reader:** *Making Sounds* **Science Songs CD**	■ **Science Resource Book,** p. 58 ■ **A to Z Activity Book,** pp. 34–35 ■ **Science on the Go,** pp. 109–112
5 Magnets PAGES 268–273 PACING: 3–5 days	■ Recognize that magnets can be used to make some objects move without being touched.	**magnet**	**Flipbook,** p. 59 **Unit F Literature Big Book** **Leveled Reader:** *What Can a Magnet Do?* ■ **Science Resource Book,** p. 59	■ **A to Z Activity Book,** pp. 26–27 ■ **Science on the Go,** pp. 113–114

PACING Assumes a day is a 20–25 minute session.

LOG ON www.macmillanmh.com for more planning resources and http://nsdl.org/refreshers/science for science resources from NSDL

BE A SCIENTIST Activities

Pull with a Pulley *p. 246*

PACING: 30 minutes

Skills observe, infer, make a model, communicate, draw a conclusion

Materials pulleys, string, paper cups, pictures of pulleys, classroom objects, chart paper

Sliding and Rolling *p. 254*

PACING: 30 minutes

Skills observe, predict, investigate, compare, infer

Materials block, ball, crayon, book, pencils, clipboards

All Fall Down *p. 260*

PACING: 30 minutes

Skills observe, investigate, draw a conclusion, infer

Materials classroom items (book, eraser, ruler), area with a table

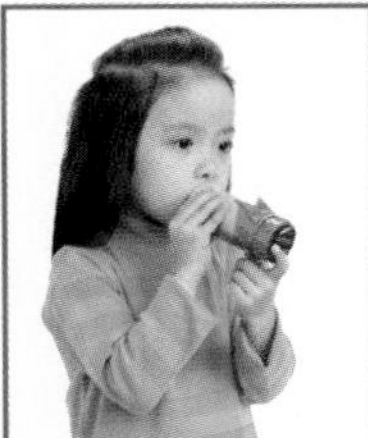

Sound Cylinders *p. 266*

PACING: 30 minutes

Skills observe, make a model, investigate, communicate, infer

Materials cardboard rolls, rubber bands, plastic bags

Moving Clips *p. 272*

PACING: 30 minutes

Skills predict, observe, investigate, measure

Materials magnets, paper clips, pencils

UNIT ASSESSMENT

Formative Assessment, pp. 245, 253, 259, 265, 271
Performance Assessment, p. 274
Summative Assessment, p. 275
Portfolio Assessment, p. 275

SCIENCE KIT MATERIALS

- modeling clay
- toy cars
- balance cubes
- fabric pieces
- magnets
- blocks
- string
- balls
- goggles
- pulleys

COLLECTIBLES

Begin collecting these items for Unit F:

- cardboard
- toy catalogs and advertisements
- magazines with pictures of toys
- cardboard rolls
- empty tissue boxes
- refrigerator magnets

English Language Learner Support

Technology

For additional language support and oral vocabulary development, go to www.macmillanmh.com

 Vocabulary Games

Language Transfers

Phonics Transfer
Cantonese does not have the short vowel sound /a/.

Grammar Transfer
Cantonese, Haitian Creole, Hmong, Korean, and Vietnamese do not use a plural marker.

This magnet picks up many paper clip.

Academic Language

English language learners need help in building their understanding of the academic language used in daily instruction and science activities. The following strategies will help to increase children's language proficiency and comprehension of content and instruction words.

Strategies to Reinforce Academic Language

- **Use Context** Academic Language should be explained in the context of the task. Use gestures, expressions, and visuals to support meaning.
- **Use Visuals** Use charts, transparencies, and graphic organizers to explain key labels that help children understand classroom language.
- **Model** Use academic language as you demonstrate the task to help children understand instruction.

Academic Language Vocabulary Chart

The following chart shows unit vocabulary and inquiry skills as well as some Spanish cognates. **Vocabulary** words help children comprehend the main ideas. **Inquiry Skills** help children develop questions and perform investigations. **Cognates** are words that are similar in English and Spanish.

Vocabulary	Inquiry Skills	Cognates	
		English	Spanish
wheel, p. 242	observe, p. 246	observe, p. 246	*observar*
pulley, p. 246	infer, p. 246	infer, p. 246	*inferir*
roll, p. 248	make a model, p. 246	model, p. 246	*modelo*
push, p. 248	communicate, p. 246	communicate, p. 246	*comunicar*
pull, p. 248	draw a conclusion, p. 246	conclusion, p. 246	*conclusión*
slide, p. 252	predict, p. 254	pulley, p. 246	*polea*
force, p. 252	investigate, p. 254	force, p. 252	*fuerza*
gravity, p. 256	compare, p. 254	predict, p. 254	*predecir*
loud, p. 262	measure, p. 272	investigate, p. 254	*investigar*
soft, p. 262		compare, p. 254	*comparar*
vibration, p. 264		gravity, p. 256	*gravedad*
magnet, p. 268		vibration, p. 266	*vibración*

Vocabulary Routine

Use the routine below to discuss the meaning of each word on the vocabulary list. Use gestures and visuals to model all words.

Define A *magnet* is an object that can make metal objects stick to it.

Example Metal paper clips stick to a *magnet*.

Ask What else can stick to a *magnet*?

Children may respond to questions according to proficiency level with gestures, one-word answers, or phrases.

Vocabulary Activities

Help children understand what a magnet does.

BEGINNING Have a pile of metal paper clips and a pile of torn pieces of paper. Show a magnet. Say: *This is a magnet. Metal objects will stick to it*: Write, say, and have children repeat, *magnet*. Have children demonstrate how the paper clips stick to the magnet, but not the pieces of paper.

INTERMEDIATE Show children Sorting Card 60. Ask children questions about the magnet: *What is sticking to the magnet? What is it made of?* Have children name one thing that sticks to magnets and one thing that doesn't.

ADVANCED Show Flipbook page 59. Pointing to the pictures, say: *These pictures show different objects that stick to a magnet: paper clips, nails, pins, and safety pins. How are all these objects alike? Why don't the other objects stick to the magnet?*

UNIT F

Objective: Children will discover that a push or a pull causes motion by observing toys and comparing how they move.

Before Reading

- Have children share what they know about the toys that appear on the Flipbook page.
- Explain that in this unit, children will be scientists as they observe, compare, and classify how things move.

During Reading

- Read the poem, pointing to the words as you read.
- Read again, pausing to invite children to point to images on the page that correspond to the lines in the poem.
- Reread the poem two to three times, encouraging children to chime in and mimic the action words as you read.

Flipbook

UNIT F

Moving Right Along

I pull my train both fast and slow.
I push my truck to make it go.
I move my toys as you can see,
But my best toys move all of me!

I roll on skates, I pump on swings.
I jump up high on trampolines.
I skate on ice, I slide on snow.
I slide on sleds, how fast I go!

by Sharon White

Reading Strategy

Print Awareness Have children help you count how many times the word *I* appears in the poem. Reread the poem, pointing to the word *I*, asking children to read it each time.

Phonemic Awareness Ask children to point to the word *train* in the first sentence and *truck* in the second sentence. Write both words on the board and invite a child to circle the parts of the words that are alike.

Science Facts

Roller Skates An unknown European invented roller skates in the early 1700s by nailing wooden spools to strips of wood and attaching them to shoes. There have been many variations of roller skates. Rolling friction is less than sliding friction.

Sled The first sleds were probably flat sleds, such as toboggans. Sleds with runners decrease the surface area in contact with snow so sleds glide more easily.

Dump Truck This toy is among many that are modeled after construction vehicles. In the early nineteenth century, dump trucks were created to haul heavy loads. The bed of a dump truck tilts, making it easy to unload its contents onto the ground at a construction site.

After Reading

- Write *Toys That Roll* and *Toys That Slide* at the top of a piece of chart paper. Ask children to help you decide in which column each toy pictured on the Flipbook page belongs.
- Invite children to name sliding and rolling toys not yet on the list and tell where they belong.

Write on It

Have volunteers use a dry-erase marker to draw an *x* on all the toys that help people move. Have them circle all the toys people move.

▶ Revisit

Invite children to replace names of toys in the poem with others. You might want to refer children to the *Toys That Roll* and *Toys That Slide* chart for ideas. Write names of replacement toys on sticky notes and cover appropriate toy words.

Science Background For more information on motion go to www.macmillanmh.com.

53

Train These toy train cars are linked together with magnetic couplings. For the magnets to attract, opposite poles must be face to face.

The Windup Toy is one of many mechanical toys that perform simple motions. A small knob on the outside of the toy is used to wind it up so that it can perform various actions, such as walking, rolling, and spinning.

School to Home

At the beginning of Unit F, give children copies of the Home Activity Letter on page 21 from the **Science Resource Book**. Read the letter aloud and help children write their names. Have children bring the letter home to share with their families.

At the end of the unit, send children home with a copy of the Home Letter on page 23.

See pages 22 and 24 in the Science Resource Book for Spanish versions of these letters.

Professional Development For more Science Facts and resources from NSDL visit http://nsdl.org/refreshers/science

Wheels

▶ Essential Question
How do wheels help us?

▶ Objective
Recognize that wheels affect speed and motion and make moving easier.

▶ Vocabulary
wheel pulley

Resources

Flipbook, pp. 54, S11

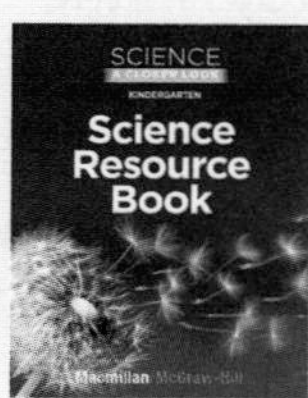
Science Resource Book, p. 55

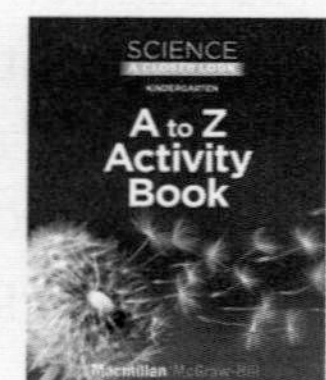
A to Z Activity Book, pp. 40–41

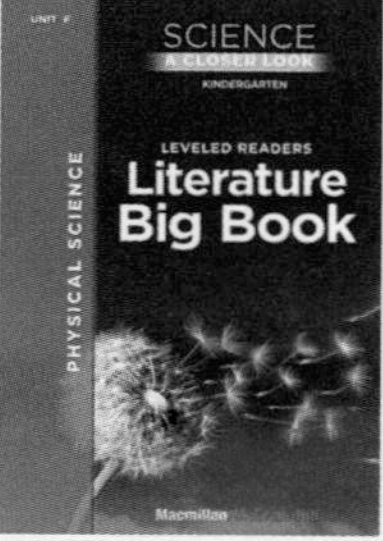
Unit F Literature Big Book: *Time for Kids*

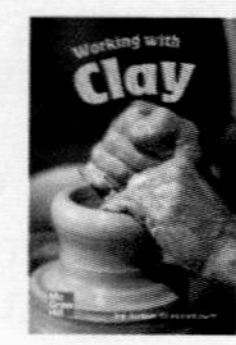
Leveled Reader: *Working with Clay*

Technology

Science Songs CD Tracks 21–22

www.macmillanmh.com

On the Move

You need
- Photo Sorting Cards 51–60
- yarn
- chart paper
- marker

Inquiry Skills: communicate, classify, infer

- Have children share how they get to school. Record their responses on a two-column chart labeled *With Wheels* and *Without Wheels*. Help children brainstorm other ways to travel with and without using wheels.
- Use yarn to make two sorting circles on the floor, one for *wheels* and one for *no wheels*. Show a card and ask a volunteer to place it in the correct circle. Continue until all cards have been sorted.

Infer Ask: **Why do you think we use wheels to move things?** things are heavy; they help move things faster

Time to Move!

Place pictures of objects with wheels (such as cars, airplanes, and buses) in a container. Have volunteers choose a picture to act out, then move around and make sounds so the class can guess what they are.

Be a Reader Activity

You need

- Unit F Literature Big Book
- Leveled Reader: *Working with Clay*

Read the Big Book

Objective: Understand that wheels are used for different things.

Inquiry Skills: observe, communicate, draw a conclusion

- Show children the cover of *Working with Clay*. Have them share their ideas about what they see. Ask: **What do you think this person is doing?**
- Read the book aloud, pointing out that the potter is using a wheel. Point out the word *wheel* on pages 3 and 4. Invite children to count and to name the letters in the word.
- Ask: **Does the potter use the wheel to travel? How is the wheel she uses help her to do a job? How is this wheel different from other wheels?**
- **Guided Reading:** See the inside back cover of *Working with Clay* for activities.

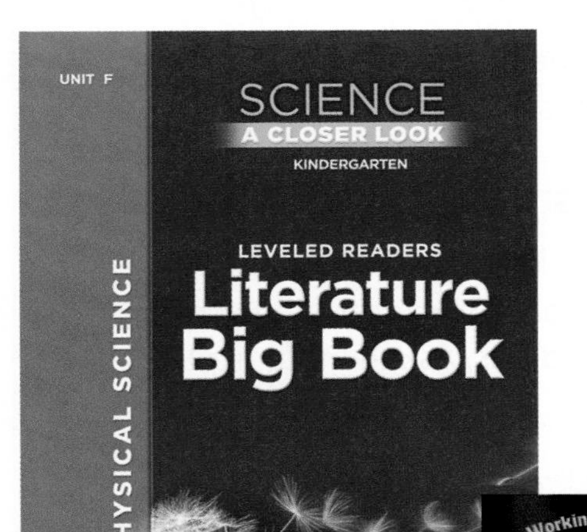

Be a Math Wiz Activity

You need

- toy cars
- hardcover books
- blocks
- connecting cubes
- paper
- pencils
- ruler

How Far?

Objective: Measure distances toy cars travel using ramps.

Inquiry Skills: predict, measure, compare

- Ask: **What would happen if we put a toy car at the top of our playground slide?**
- Model the ramp activity by using a book and one block. Put the car at the top of the ramp and gently let it go. Measure the distance from the bottom of the ramp to the back of the car with connecting cubes.
- Have children use a one-block and a two-block ramp in the experiment. Have children record their findings. Then show them how to measure the distance using a ruler.

How Far

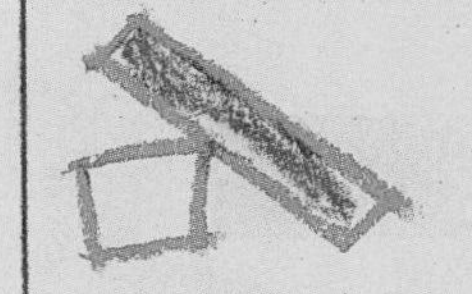

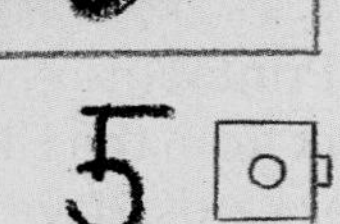

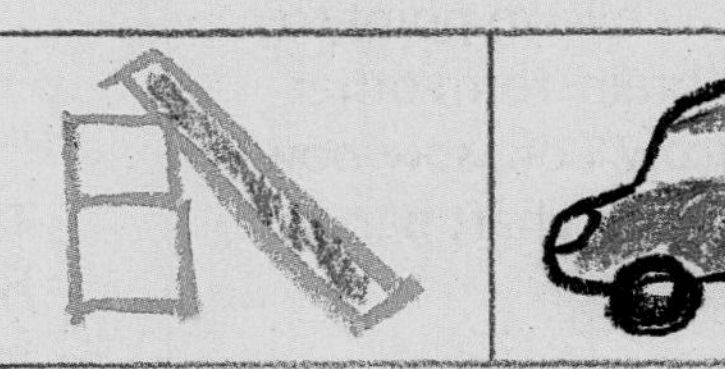

Read Together and Learn

How do we use wheels?

▶ Build on Prior Knowledge

Ask children to name things that have wheels. Record their responses.

▶ Use the Visuals

- **Communicate** Ask the group to guess why wheels are used on the wagon, wheelchair, wheelbarrow, and bus. Ask them to explain their thinking.
- **Compare** Point to the Ferris wheel. Ask a volunteer to describe it. **How is it like the other objects on the page? How is it different?**
- Ask a child to find the wheelchair. **How is the wheelchair helping the boy in the activity? How is it the same as/different from other wheelchairs you have seen?**

▶ Develop Vocabulary

wheel, **pulley** Ask children to identify things that have *wheels*. Have children describe how the wheels help make something move. Display a photograph of a pulley. Explain that a *pulley* is a simple machine that has a wheel and a rope or chain. Point out that pulleys are used to lift heavy objects. Both wheels and pulleys make moving things easier.

Flipbook

UNIT F • LESSON 1

How do we use wheels?

Reading Strategy

Conventions of Print Write each word and punctuation mark from the Flipbook question on separate index cards. Have children sequence the cards to make the question.

Phonemic Awareness Invite a child to point to the word *how*. Have children brainstorm other words that rhyme with *how (cow, bow, sow, now, pow, vow)* and write their ideas on chart paper.

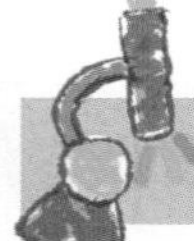

Science Facts

Wheel A simple machine, the wheel and axle, increases force to make work easier. Wheeled transportation may have come from the use of logs for rollers, but the oldest known wheels were wooden disks.

The **Ferris wheel** was conceived by G. W. Ferris and introduced in Chicago at the 1893 World's Columbian Exposition.

The first Ferris wheel had a diameter of 250 feet and a circumference of 825 feet. Its 45-foot-long axle was the largest piece of forged steel in the world at the time.

The **wheelbarrow** is a one-wheeled cart designed to be pushed by one person. A wheelbarrow makes lifting and moving heavy loads easier.

Professional Development For more Science Facts and resources from NSDL visit http://nsdl.org/refreshers/science

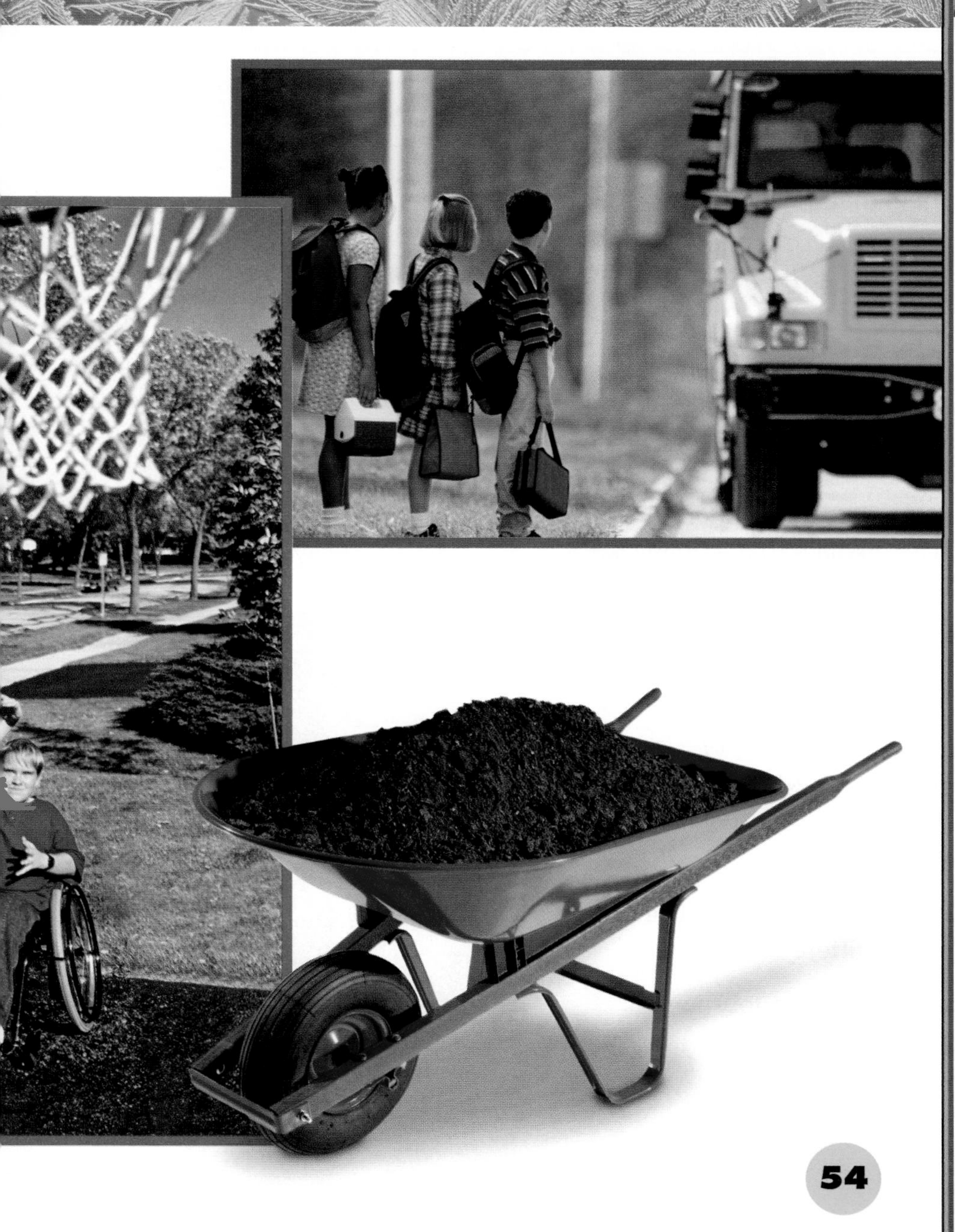

▶ Ask Questions

- **Do you think the wagon, wheelchair, wheelbarrow, and bus could move without wheels?** no; not well
- **Do all vehicles that move people have wheels? Why do some not have wheels?** No, boats do not have wheels because they float in water.
- **What other things do people push to make their work easier?** Possible answers: shopping cart, stroller, lawn mower

Write on It

Invite children to use a dry-erase marker to circle all the wheels on the page. Have them count and record the number of wheels on each item.

Think and Talk

Ask a volunteer to name something in the classroom that has wheels and to explain how the wheels are used. Continue until all children have had an opportunity to respond. Record responses on a *Where Are the Wheels?* chart.

TIME FOR KIDS

Shared Reading Read *Around and Around* with the class. Use this magazine, found in the **Unit F Literature Big Book**, as a photographic review of wheels.

More to Read

Mama Zooms,
by Jane Cowen-Fletcher
(Scholastic, 1996)

Activity After reading, ask children to illustrate where they would like to go if they were riding on Mama's lap. Write the following sentence starter for children: *Mama zooms me to the ________ in her wheelchair.*

Formative ASSESSMENT

Notice if children understand that wheels make moving easier. If they are having difficulty, use the Photo Sorting Cards and center activities to review how wheels help move people and things.

Pull with a Pulley

Objective: Recognize that a pulley is a wheel that makes moving easier.

Inquiry Skills: observe, infer, make a model, communicate, draw a conclusion

You need **chart paper, pulleys, string, paper cups, pictures of pulleys (cranes, flagpoles), classroom objects, Science Journal p. 55**

1. **Observe** Show children a pulley and write the word on the board. Point out the word *pull* in *pulley*. Explain that a pulley is a kind of wheel used to help move things. Have a volunteer hold the pulley, tie one end of the string to a paper cup (through holes made near the rim), loop the string over the pulley's wheel, and have another volunteer pull on the string.
2. **Infer** Ask: **Have you seen pulleys at work? Where?** Write children's responses on chart paper. If children have not, show them pictures of pulleys at work.
3. **Make a Model** Give each group a pulley, a cup, and string. Have them choose a material to load into their cup, and then use the pulley. Invite groups to experiment with different loads.
4. **Communicate** Have children use their Science Journal page to draw and write about the pulley.

Teacher Tip

You can arrange for children to see a working pulley. For example, you might be able to participate in raising or lowering the school flag. If your classroom blinds use pulleys, let children help you raise and lower them.

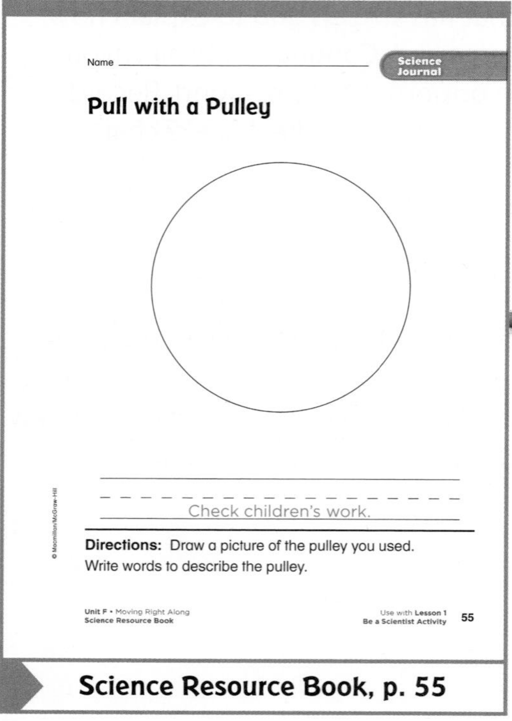

Name ______________________ Science Journal

Pull with a Pulley

Check children's work.

Directions: Draw a picture of the pulley you used. Write words to describe the pulley.

Unit F • Moving Right Along
Science Resource Book

Use with Lesson 1
Be a Scientist Activity 55

Science Resource Book, p. 55

Investigate More

Draw a Conclusion Ask children: **Is it easier to move a heavy load by pulling down on a pulley's rope or lifting up the load? Why?** It is easier to move a load by using a pulley because the pulley makes the load seem lighter.

Music

Roll Along!

Objective: Compare and contrast different kinds of transportation.

Inquiry Skills: observe, compare

- **Listen, Name, and Compare** Play the song "Roll Along!" Then ask children to name the different kinds of transportation mentioned. Ask: **How are the kinds of transportation the same? How are they different?**
- **Act It Out** Invite children to act out the lyrics, moving like they are skating, driving a bus or car, taking off in a jet, and riding on a train.
- **Think and Draw** Encourage children to think of other types of transportation. Have children draw two objects with wheels that they use.
- See page TR32 for printed music to "Roll Along!"

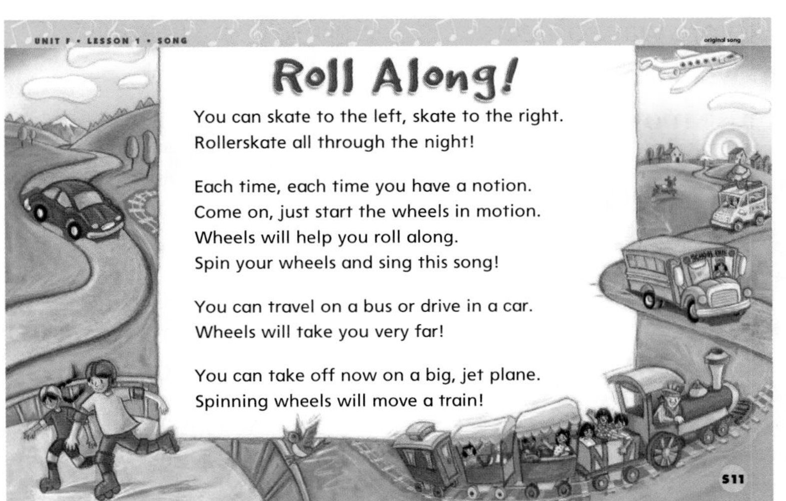

Flipbook p. S11
CD Tracks 21–22

Drawing and Writing

You need
- paper
- crayons
- colored pencils

Toys with Wheels

Objective: Draw and write about toys that move.

Inquiry Skills: classify, communicate

- Ask children to draw a picture of themselves going somewhere using toys that have wheels such as bikes, in-line skates, or scooters.
- Have children write sentences or offer to take dictation about their pictures.

Blocks

You need
- toys with wheels
- sturdy cardboard
- tunnels

Roads and Ramps

Objective: Create structures for toys to roll on.

Inquiry Skills: make a model, communicate

- Explain to children that this week the Block Area will include several toys with wheels.
- Help children brainstorm what they will build for the toys. Ask: **What are some of the things we can build for our toys to travel on?**
- Adding different-sized vehicles and materials for ramps and roads will encourage wheel investigations.

How Things Move

▶ **Essential Question**
What makes things move?

▶ **Objective**
Explore ways objects move and forces that cause movement.

▶ **Vocabulary**
slide
roll
push
pull
force

Resources

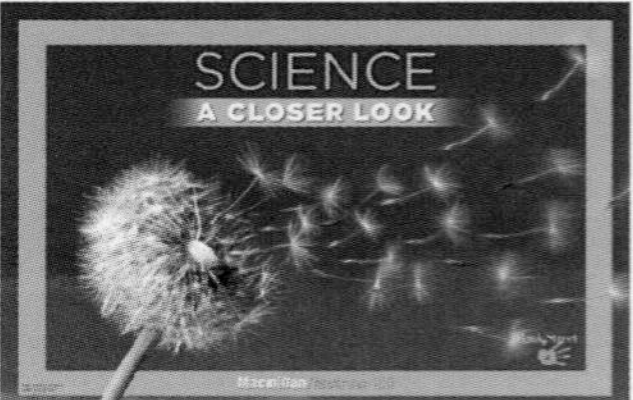
Flipbook, pp. 53, 55–56, G1

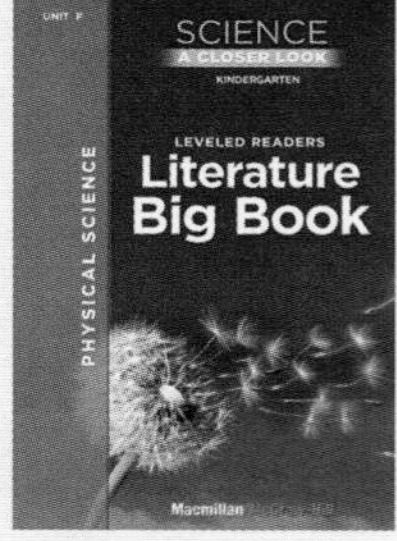
Unit F Literature Big Book

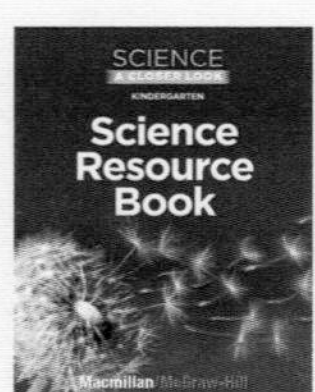
Science Resource Book, p. 56

Leveled Reader: *Toys That Move*

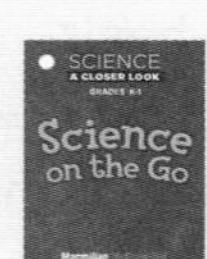
Science on the Go pp. 105–106

Technology

Circle Time
WHOLE CLASS

Make It Move

You need
- toy car
- tape
- Flipbook p. 53

Inquiry Skills: observe, classify, communicate

- Show children a toy car. Put a piece of tape a few inches from the car on the floor. Ask: **How can the car move to the tape? Will it move by itself?**
- Have children act out each suggestion. Possible answers: push it; pull on it; roll it; kick it; blow on it. Revisit Flipbook page 53, and have children explain how each toy moves.

Classify Ask: **What are some toys you can move in more than one way?** toy cars, toy boats, ball

Time to Move!

Play some music and invite children to dance. While they are dancing give directions such as *Put your hands up in the air; Shake your body down; Dance around the table;* and *Roll your arms in front of your body.*

Be a Reader Activity

You need
- Unit F Literature Big Book
- Leveled Reader: *Toys That Move*
- sticky notes

Read the Big Book

Objective: Use illustrations to read words.

Inquiry Skill: infer

- Before reading the story, cover the words *push* and *pull* with sticky notes. Read the title with children and have them name the toys they see.
- As you read, stop before each covered word and ask children what it might be. Before rereading the text and inviting children to join in, discuss ideas of how the wind, magnets, etc. can push or pull things.
- **Guided Reading:** Use the Leveled Reader version of *Toys That Move* for guided-reading instruction.
- For more to read on motion, see *Water Moves* available in the Unit F Literature Big Book and as a Leveled Reader.

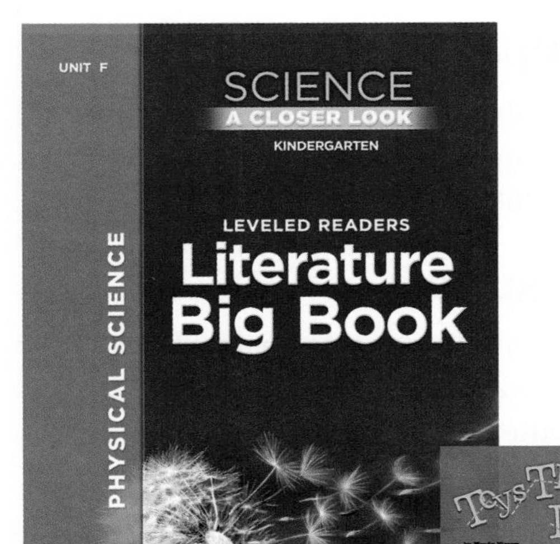

Be a Math Wiz Activity

You need
- geometric solids
- hardcover books
- Flipbook p. G1

Rollers and Sliders

Objective: Sort geometric solids that roll and slide.

Inquiry Skills: classify, compare

- Show children a set of geometric solids. Ask: **Can you find a block that will roll?** Set up a book as a ramp and have a volunteer test the block. Repeat, asking for a block that will slide and a block that will both slide and roll.
- Give partners a collection of solid figures. Have them test their blocks on ramps and sort them into groups of sliders, rollers, and blocks that both slide and roll.
- After children have tested the solid figures with their partner, discuss their findings together as a class. Use the Venn diagram from the Flipbook to record the results.

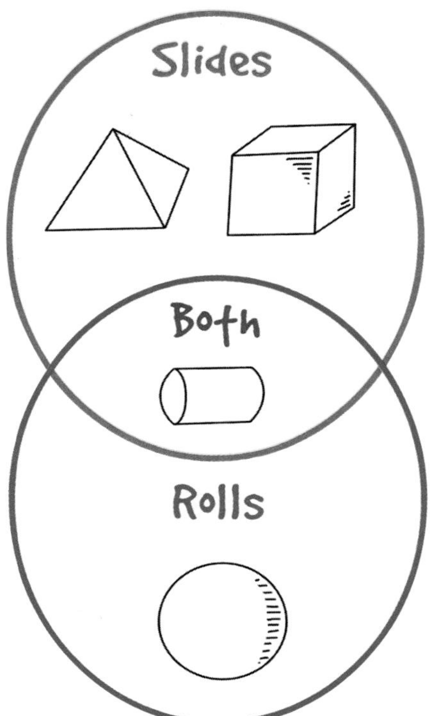

Read Together and Learn

How many ways can the gerbils move?

▶ Build on Prior Knowledge

Ask volunteers to talk about gerbil habitats they have seen. Encourage them to describe the equipment in the habitat. Record children's words.

▶ Use the Visuals

- **Observe** Ask a volunteer to trace with a finger one tunnel pathway a gerbil could take and describe the route. Repeat with other routes.
- **Predict** Ask the group to think about which tunnel routes might take less effort for each gerbil to travel and which might require more. Ask them to explain their thinking.
- **Compare** Point to the exercise wheel. Ask: **How is this part of the habitat alike or different than the other parts?**

▶ Additional Vocabulary

up, **down**, **around**, **across** Write *up, down, around,* and *across* on the board. Using their fingers, have volunteers show how a gerbil could travel up, down, around, and across the habitat.

Flipbook

UNIT F • LESSON 2

How many ways can the gerbils move?

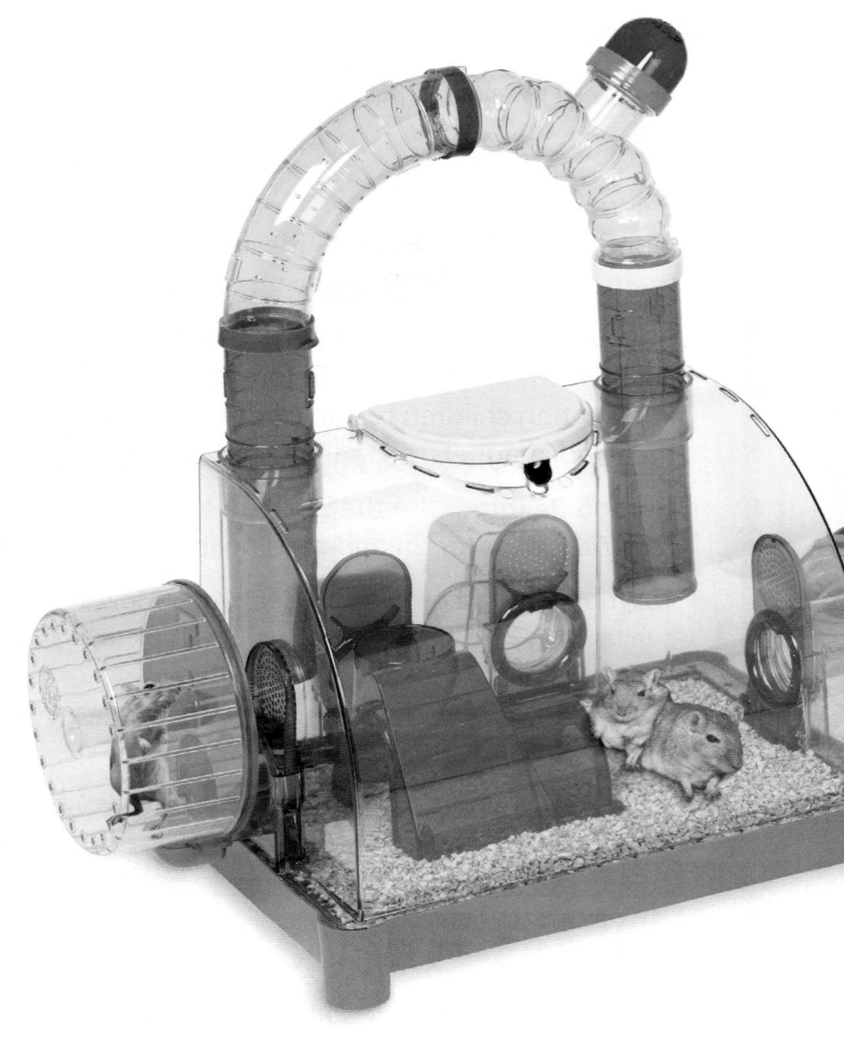

Reading Strategy

Print Awareness Ask a volunteer if the sentence on the Flipbook page is an asking sentence or a telling sentence. Ask how he or she knows.

Phonics Ask a volunteer to point to the word *What*. Ask him or her to find something in the habitat that begins with the same *wh* sound. Ask children for other *wh* words and have them circle the two letters.

Science Facts

Gerbil In the wild, gerbils live in colonies inside underground tunnels up to three feet long. These tunnels have several rooms and entryways. The best way to keep a pet gerbil happy is to replicate this environment in its habitat by including a layer of bedding that the gerbil can dig, as well as plenty of materials for climbing and tunneling.

Ramp A ramp is an inclined plane. The word *inclined* means "sloped" and a plane is a flat surface. A gradual slope makes it easier to push or pull loads.

Force is required to get an object, such as an exercise wheel, in motion. The motion of the gerbil's feet provides the force necessary to rotate the wheel.

LOG ON **Professional Development** For more Science Facts and resources from  visit http://nsdl.org/refreshers/science

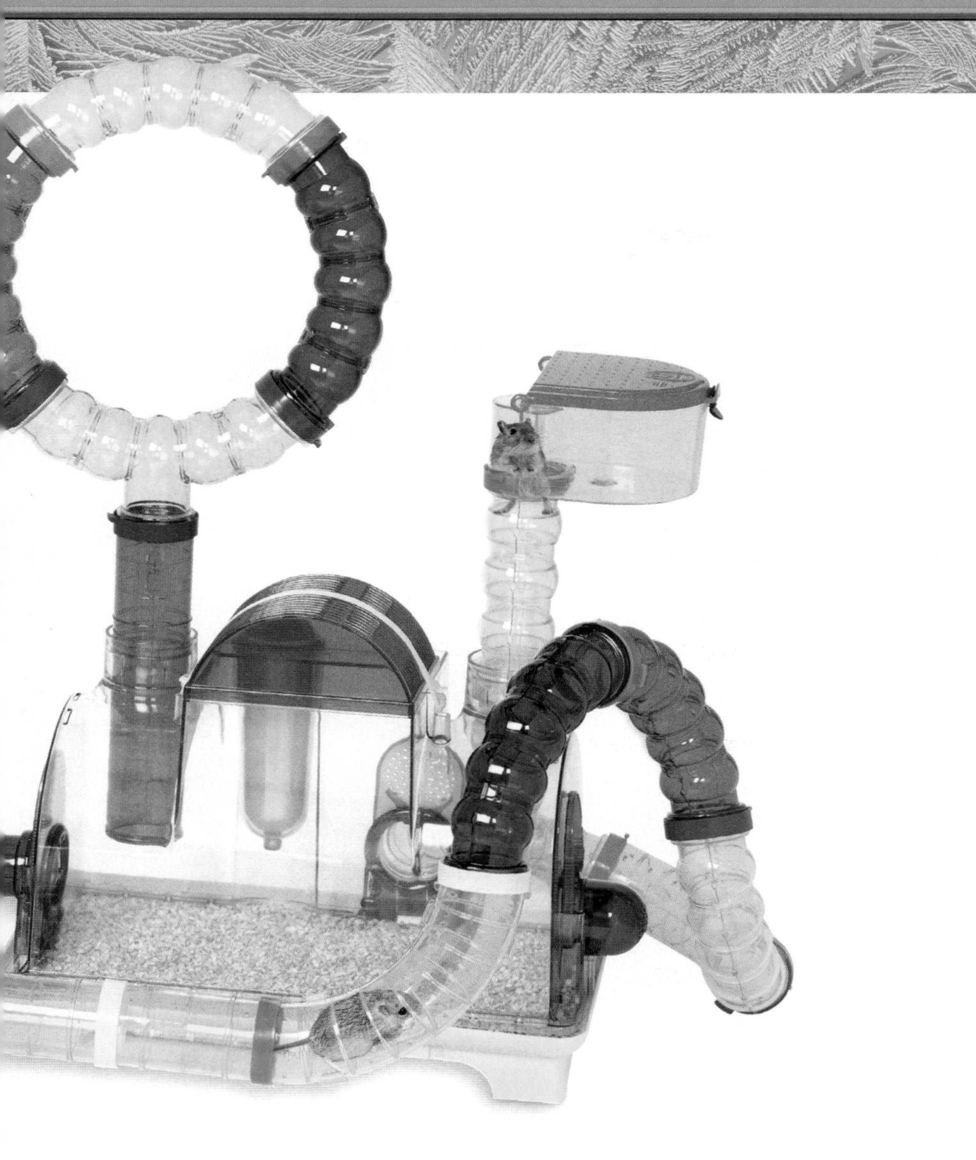

▶ Ask Questions

- **Can you find a ramp? How could you make more ramps for the gerbils to climb?**
- **Look at the exercise wheel. What makes the wheel go round and round?** the gerbil's feet **What are some wheels that spin inside your house?** clothes dryer, blender, salad spinner **Outside?** merry-go-round, cars, bicycles
- **Can you find a tunnel a gerbil might slide down? Have you ever been on a tunnel slide? Why do you slide faster on some tunnel slides than on others?** Some are steeper than others and make me go faster.

Write on It

Invite children to use a dry-erase marker to trace several gerbil routes of approximately the same length. Have them write an *E* on the route they think would be easiest for the gerbil to travel and an *H* on the route they think would be hardest.

Think and Talk

Revisit Flipbook page 55. Encourage children to use words to describe how the gerbils might move around in the habitat. Prompt children to use the words *straight, zigzag, left, right, sideways, up, down, back,* and *forth*.

ELL Support

Write *up, down, forward,* and *backward* on cards. Show and say the words as you act out movements.

Beginning Show and read a word and ask children to demonstrate the action.

Intermediate Demonstrate a direction and ask children to name it.

More to Read

Cinnamon's Day Out,
by Susan L. Roth
(Dial Books, 1998)

Activity After reading this book, have children compare Cinnamon's movements outside and inside the cage. Then have children make a collage using the book's illustrations as inspiration.

Read Together and Learn

What makes these toys move?

▶ Build on Prior Knowledge

Invite children to describe their own moveable toys. Help them describe exactly how they move.

▶ Use the Visuals

- **Observe** Have volunteers point to toys they have seen before and describe how they move. Write motion words (*spin, slide, roll*) on chart paper.
- **Communicate** Ask the group to think about what sets each toy in motion. Record the action words (*push, pull, drop*).
- After discussing all the toys on the page, have children determine which toys require the same motion to move. push toys: truck, clown, top **Which toys move in other ways?**

▶ Develop Vocabulary

push, **pull**, **roll**, **slide**, **force** Have volunteers demonstrate a *push* and *pull*, using a classroom object, such as a door. Then have children identify other classroom objects that they can *roll* or *slide*. Explain that each movement requires a *force* or power to start the movement.

Flipbook

UNIT F • LESSON 2

What makes these toys move?

Reading Strategy

Phonemic Awareness Ask a volunteer to point to the word *toys*. Ask which toy on the Flipbook page has the same beginning sound. top Have volunteers name other words that begin the same way.

Conventions of Print Read the Flipbook question with the class. Ask children if this is an asking sentence or a telling sentence. **How do you know?** Encourage volunteers to make an asking sentence and a telling sentence about a toy on the page.

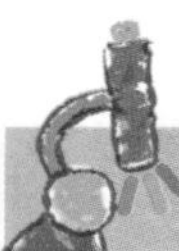

Science Facts

Force and Motion Over 300 years ago, Sir Isaac Newton discovered principles of force and motion and stated three laws of motion. An object speeds up or slows down only when a force acts on it. The amount by which an object changes speed depends on its mass and how big the force is that acts on it. Whenever a force acts upon an object, there is an equal and opposite reaction.

Force of Gravity When something is dropped, it falls due to the pull of Earth's gravity. The more mass an object has, the stronger gravity pulls on it.

Center of Gravity The clown toy returns to an upright position because its weight is concentrated in its base.

LOG ON **Professional Development** For more Science Facts and resources from  visit http://nsdl.org/refreshers/science

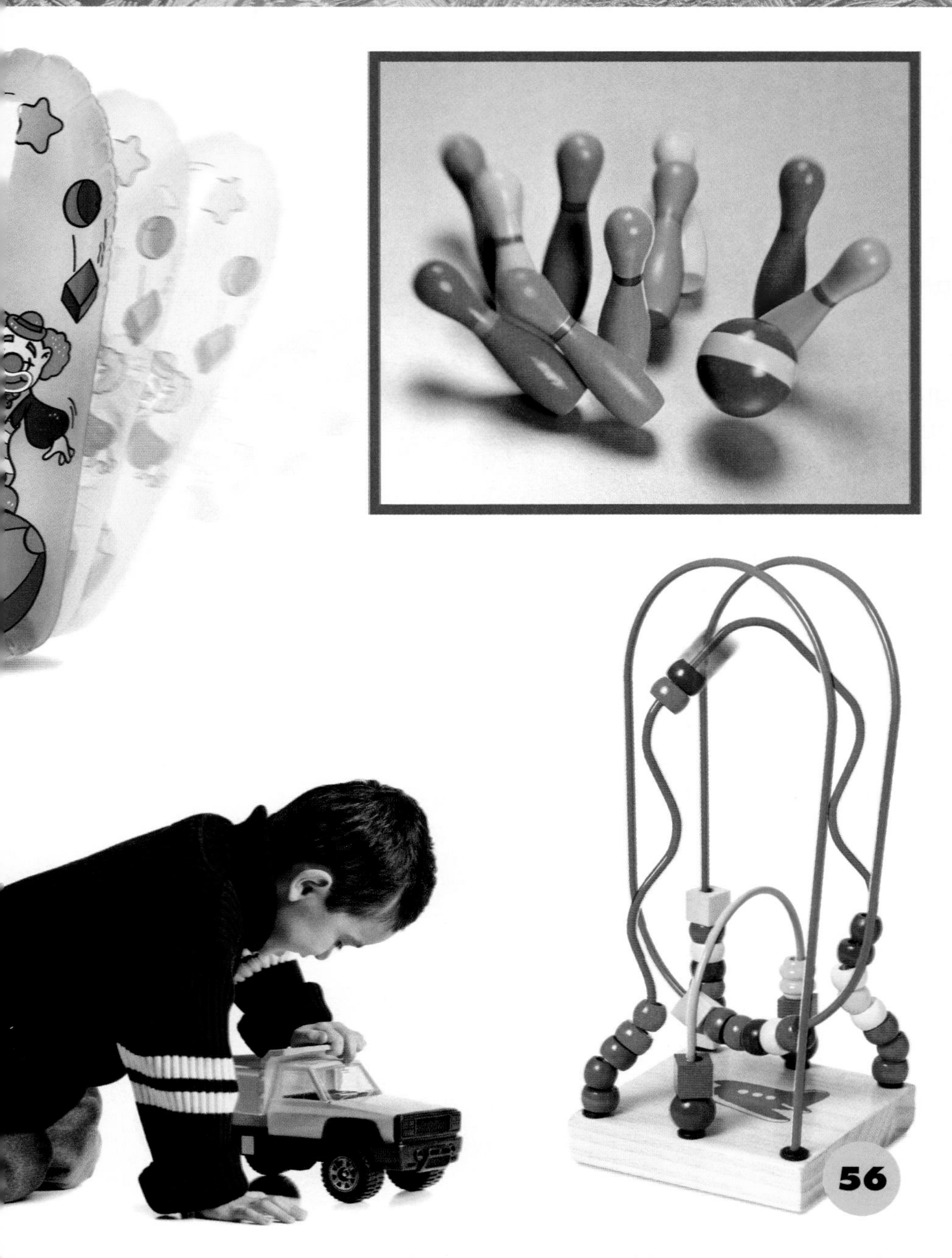

▶ Ask Questions

- **Do you think that any toys on this page can move by themselves? Why or why not?** No, I would have to make them move.
- **Are certain toys easier to move than others? Why or why not? Can you name some toys that are motionless?**
- **How can a toy move faster? Slower?** I can push or pull it harder; I can run while pushing or pulling the toy; I can use more or less force.

Write on It

Invite children to use a dry-erase marker to write a vocabulary word next to each toy. Ask them to explain their choices.

Think and Talk

Have children review Flipbook pages 55 and 56. Ask a volunteer to pick a toy from the Flipbook images and describe what makes it move. Repeat the activity using each image. Invite children to name a favorite toy that moves and tell what causes its motion.

Take a Trip

Playground Bring the class to the school playground. Have them bring clipboards, paper, and pencils to record the equipment and toys that move. If your playground has no play equipment, look around the neighborhood to find examples of things that move.

Back in the Classroom Have children share their findings. Ask: **What makes the toys or other objects move?** Make a series of word webs using the words *push, pull, slide,* and *roll* and use the information gathered on the trip to complete the webs.

Formative ASSESSMENT

Observe if children understand that an object's motion is caused by a force (push or pull) and that motion can be affected by inclined planes and wheels. If they are having trouble with the concept, use the Block Center to provide hands-on practice.

Be a Scientist | Inquiry Investigation

Sliding and Rolling

Objective: Explore things that slide and things that roll.

Inquiry Skills: observe, predict, investigate, compare, infer

You need **block, ball, crayon, book, pencils, clipboards, Science Journal p. 56**

1. **Observe** Show children the rectangular block, ball, crayon, and book. Have them watch as you push the block and then push the ball. Ask: **Why do you think the ball went farther?**
2. **Predict** Ask volunteers to predict and test which of the remaining items would move like a ball or like a block. Have children look for things in the classroom that slide (blocks, books) and that roll (balls, crayons).
3. **Investigate** Give each child a Science Journal page, a clipboard, and a pencil. Have children explore the room in pairs to find, test, and record items for each category. Ask: **Where in the diagram will you put the things that both slide and roll?** in the middle
4. **Compare** Have children share and discuss their findings and compare the results with other classmates.

Teacher Tip

It may be helpful to reinforce position words when talking about motion. For example, you might want to ask: "Can you slide the block next to the crayon? Roll the pencil in front of the block?"

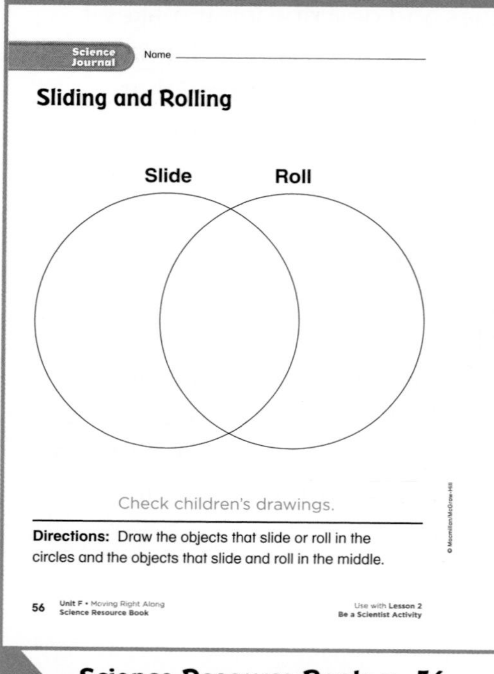

Science Journal Name ______

Sliding and Rolling

Slide Roll

Check children's drawings.

Directions: Draw the objects that slide or roll in the circles and the objects that slide and roll in the middle.

56 Unit F • Moving Right Along Science Resource Book

Use with Lesson 2 Be a Scientist Activity

© Macmillan/McGraw-Hill

Science Resource Book, p. 56

Investigate More

Infer Ask: **What do you notice about the shape of the objects that slide?** flat sides **The shape of the objects that roll?** round **The shape of the objects that slide and roll?** round and flat sides

centers

Drawing and Writing

Push and Pull

Objective: Create a class book about toys that move.

Inquiry Skills: observe, classify, communicate

- Explain that children will make their own pages to put in a class book called *Toys That Move*.
- Show children the prepared pages with the writing *This ________ moves with the ________ of ________*. Review the pattern of text. Ask children to decide on a push or pull toy and help them to write in the spaces.
- Have children illustrate their stories.

You need

- Leveled Reader: *Toys That Move*
- crayons
- colored pencils
- pencils

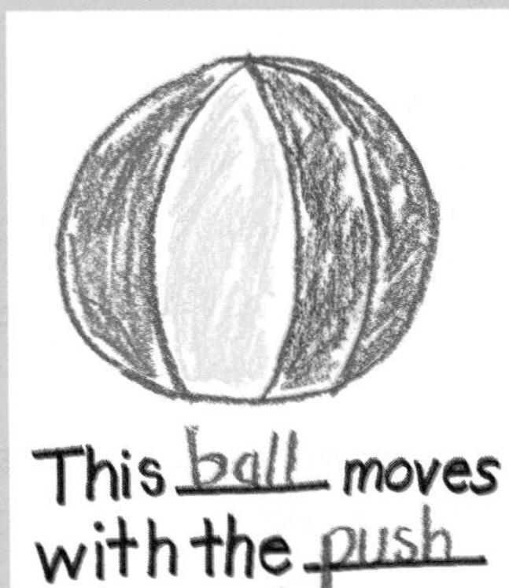

Technology

More About Motion

- For games and activities about how things move, help children log onto www.macmillanmh.com.
- Children can use these fun games throughout this unit to reinforce what they learn about motion. Encourage children to revisit these games throughout the school year to review the concepts.

LOG ON www.macmillanmh.com for more science online.

Blocks

Does It Roll?

Objective: Investigate whether an item rolls or slides.

Inquiry Skills: predict, investigate, observe

- Collect a basket of items that roll and a basket of items that slide. Put the baskets in the Block Area.
- Have children explore surfaces and ramps to investigate the way objects slide and roll.
- As children explore, encourage them to talk about differences between things that slide and things that roll.

You need

- classroom items

Art

Push/Pull Mural

Objective: Compare how toys move.

Inquiry Skills: observe, communicate, compare

- Have children cut out pictures of toys that are pushed and toys that are pulled.
- Ask them to sort the toy pictures and glue them onto a class mural with one side labeled *Push* and the other *Pull*.
- Share the completed collage and let children explain why they included specific toys in each group.

You need

- toy catalogs and advertisements
- magazines with pictures of toys
- glue
- scissors
- poster board

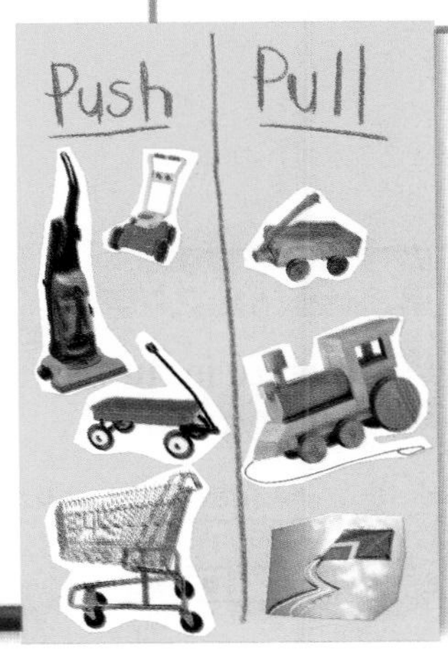

▶ **Essential Question**

How do we stay on the ground?

▶ **Objective**

Understand that certain objects, like the Sun and the Moon, stay in the sky, while others, like an airplane, are in the sky but return to Earth.

▶ **Vocabulary**

gravity

Resources

Flipbook, p. 57

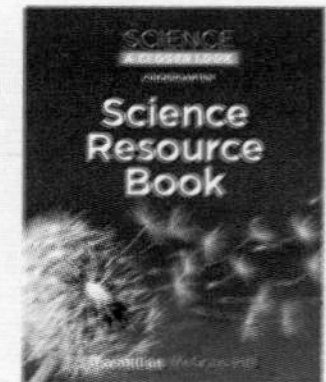

Science Resource Book, p. 57

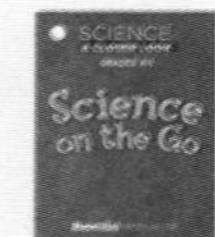

Science on the Go, pp. 107–108

Technology

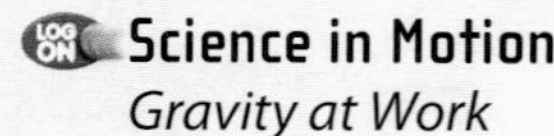

Science in Motion
Gravity at Work

www.macmillanmh.com

Ups and Downs

Toss Up

You need

- bean bag
- chart paper
- marker

Inquiry Skills: predict, communicate, infer

- Show children the bean bag. Ask them what might happen when you throw it up into the air. Prompt children to make predictions by asking: **Will it stay up in the air? Where will it go?** Record their ideas.
- Invite volunteers to throw the bean bag up in the air to see if the same thing happens each time.

Infer Ask: **Why do you think the bean bag does not stay up in the air? What makes it come down? How come it does not fall to the left or right?**

Time to Move!

Invite children to move around the room as if there were no gravity. Encourage them to move like they were floating in the air.

Be a Writer Activity

You need
- chart paper
- marker
- drawing paper
- crayons
- pencils

Up in the Sky

Objective: Recognize that some objects stay in the sky and others do not.

Inquiry Skills: observe, communicate

- Invite children to share ideas about what they see in the sky. Record their ideas on chart paper. Ask children which of the objects they mentioned stay in the sky and which come down.
- Have children draw a picture to show what they see in the sky. Help them label or write sentences about their drawings. Remind children to use the list of ideas to help them write words.
- Encourage children to share their drawings with a partner. Prompt them to discuss which objects in their drawings stay in the sky and which come down. Bind children's drawings together to make a class book titled *Up in the Sky.*

Be a Math Wiz Activity

You need
- light and heavy objects
- paper
- pencils

Going Down

Objective: Compare how quickly objects fall to the ground.

Inquiry Skills: predict, investigate, compare

- Give pairs of children one light items and one heavy item, for example, a feather and a small ball, a piece of paper and an eraser, and a paper clip and a pencil. Invite children to predict which item of each pair might fall to the ground first when dropped from where they are standing.
- Invite children to test each pair of items. Have them record which item fell to the ground first by drawing pictures on a two-column chart. Encourage partners to take turns dropping the items and recording the results.
- Have groups compare and discuss their results.

Read Together and Learn

What will come down?

▶ Build on Prior Knowledge

Have children recall what they see in the sky at night and during the day. Ask: **What else do you see up in the sky?**

▶ Use the Visuals

- **Observe** Invite children to look at the images on the Flipbook page. Have them describe what they see.
- **Compare** Encourage children to tell how the airplane and birds are the same and different. Both fly in the sky. The airplane is nonliving and the birds are living things. Have them tell how the Sun and the Moon are different.
- **Infer** Ask: **Which images show objects that stay up in the sky? Which show objects that come down from the sky?**

▶ Develop Vocabulary

gravity Ask children if they have ever heard the word *gravity*. Have them tell what they think it means. Accept all responses. Explain to children that *gravity* is what causes people and objects to have weight on Earth, and it is what makes objects fall toward Earth.

Flipbook

UNIT F • LESSON 3

What will come down?

Reading Strategy

Print Awareness Invite volunteers to find, point to, and name the capital letters on the Flipbook page. Then ask volunteers to find, point to, and name the lowercase letters.

Phonemic Awareness Point to the word *sun*. Invite children to brainstorm words that rhyme with *sun (bun, fun, ton, one, won, run)*. Write the words on chart paper. Ask: **Do all the words end with -*un*, like *sun*?** Repeat with the word *moon*.

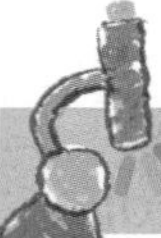

Science Facts

Gravity is a force of attraction that pulls things toward Earth. Gravity is produced by every object in the universe and can be found just about everywhere. That is why people, as well as objects of all sizes, stay on the ground. If we or the objects around us did not exert gravitational force, we would float up in the sky. Gravity keeps Earth and the other planets in orbit around the Sun, and the Moon in its orbit around Earth. Gravity is the reason why the Sun and the Moon do not come down from the sky.

Professional Development For more Science Facts and resources from NSDL visit http://nsdl.org/refreshers/science

Moon

57

▶ Ask Questions

- **Why do you think the Sun and the Moon stay up in the sky?** Allow children to share and discuss their ideas. The concept of gravity is a difficult one that children may not realize or understand at this level. However, foster their ideas and help them to understand that the Sun and the Moon stay up in the sky.
- **Why must airplanes and birds come down from the sky and return to Earth?** airplanes need to get more gas, let off passengers; birds need to build nests and get food

Write on It

Invite volunteers to use a dry-erase marker to trace the Sun and the Moon and then compare their shapes. Repeat for the airplane and birds.

Think and Talk

Review the ideas that the Sun and the Moon stay in the sky while the airplane and birds are in the sky, but come down. Prompt children to share and discuss their ideas about where the Sun goes at night and where the Moon goes during the day.

Differentiated Instruction

Extra Support Have children who have a stronger understanding of this topic help children who are having difficulty. Invite partners to work together to draw and label pictures and talk about what is happening in their pictures.

More to Read

The Science of Gravity, by John Stringer (Raintree, 2000)

Activity Read this book with children to offer a more scientific explanation of how gravity works. After reading, help children write a class story about what life would be like without gravity.

Formative ASSESSMENT

If children are having difficulty understanding the concept that the Sun and the Moon stay in the sky, discuss the idea further. Ask: **What do you think would happen if the Sun or the Moon fell out of the sky?**

All Fall Down

Objective: Investigate what happens when items are dropped.

Inquiry Skills: observe, investigate, draw a conclusion, infer

You need **classroom items (such as a book, an eraser, a ruler), area with a table, Science Journal p. 57**

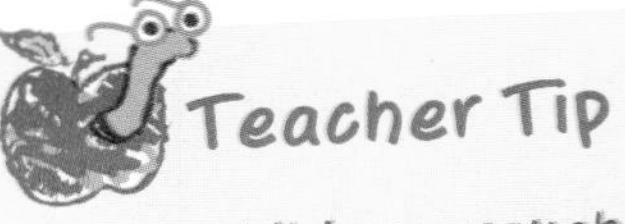

Teacher Tip

Allow children enough time to investigate with the items. After they have repeated this activity a few times using the table, invite them to see what happens when they use a chair or a desk.

1. **Observe** Display classroom items on a table. Allow small groups of children to take turns working in the area. Invite children to examine the items and make observations about their shape, weight, and size.

2. **Investigate** Invite children to investigate what happens when they hold an item up, away from the table, and then drop it. Then have them hold the same item up, over the table, and drop it. Ask: **What happened when you dropped the item away from the table?** It fell to the ground/floor. **What happened when you dropped the item over the table? Did the item fall to the floor? Why or why not?** Have children repeat this investigation with other items. Encourage them to observe what happens each time. Have children record their observations on their Science Journal page and discuss the results of their investigations.

3. **Draw a Conclusion** Ask: **Why did the object not fall to the floor when you held it above the table and dropped it?** Because the table stopped it. **When you stepped away from the table, why did the object hit the floor?** Because there was nothing in the way to stop it.

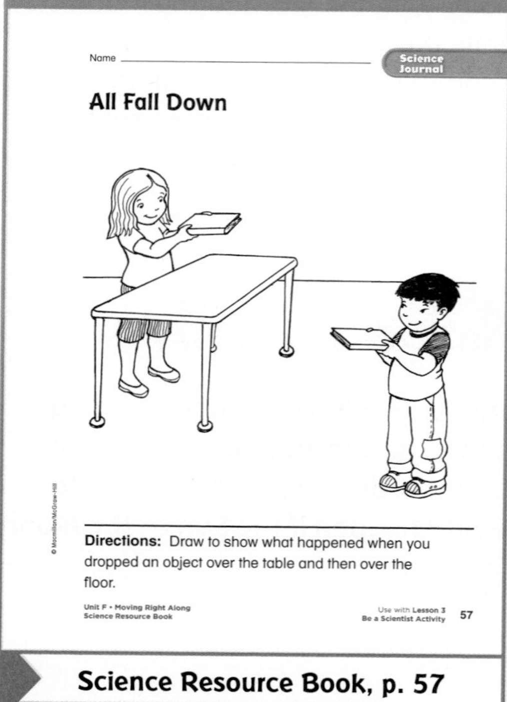

Name ______________________ Science Journal

All Fall Down

Directions: Draw to show what happened when you dropped an object over the table and then over the floor.

Unit F • Moving Right Along Science Resource Book | Use with Lesson 3 Be a Scientist Activity | 57

Science Resource Book, p. 57

Investigate More

Infer Ask: **Do you think an item falls or drops faster if it is heavy or light? Why?**

centers

Art

Sky Scenes

Objective: Create scenes to indicate objects that remain in the sky and others that return to Earth.

Inquiry Skill: make a model

You need
- cardboard shoe boxes
- paper
- pencils
- crayons
- colored clay
- craft sticks
- index cards
- glue
- scissors

- Have children create shoe box dioramas and decide if they will represent a day or night sky. Then have them decide which objects they will place in the sky and on land.
- Ask children what materials they will need to show the sky and land when they build their scene.
- As they work, invite them to describe what they are making. Record their dictation on an index card and include this card with their scene when displayed in the classroom.

Movement

What's Up?

You need
- bean bags

Objective: Develop gross motor skills and balance to keep a bean bag up.

Inquiry Skills: communicate, infer

- Have children take turns placing a bean bag on their head.
- Encourage children to move about the room trying their best to keep the bean bag from falling down without using their hands.
- Ask: **What keeps the bean bag up? What makes it fall down? What can you do so that it does not fall down to the floor?**

Blocks

Breaking Down

You need
- paper
- crayons
- scissors
- tape

Objective: Create tall block structures that will stay up.

Inquiry Skills: make a model, communicate

- Invite children to create tall structures in the Block Area. Have them add details, props, and shapes.
- Encourage children to discuss how they can keep their structures from falling down and experiment with different ways to keep it up.
- When a structure does break, ask: **What caused your building to break and fall down? How can you stop the blocks from falling to the floor?**

Sounds All Around

▶ **Essential Question**
How are sounds made?

▶ **Objective**
Describe sounds and understand how they are made.

▶ **Vocabulary**
vibration
soft
loud

Resources

Flipbook, pp. 58, S12

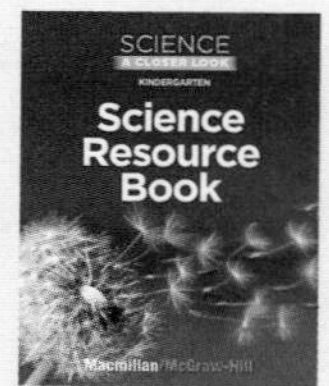

Science Resource Book, p. 58

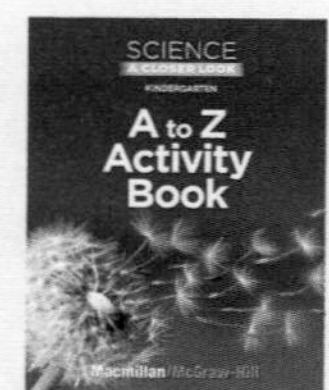

A to Z Activity Book, pp. 34–35

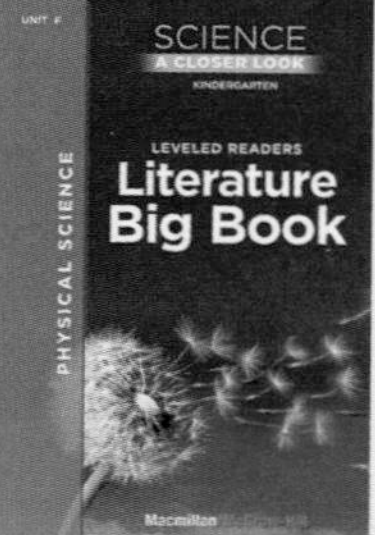

Unit F Literature Big Book

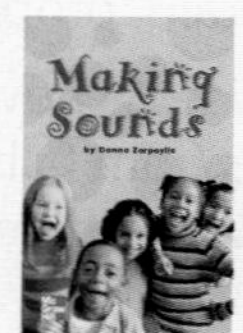

Leveled Reader: *Making Sounds*

Technology

Science Songs CD Tracks 23–24

www.macmillanmh.com

Circle Time

WHOLE CLASS

Sound Detective

You need

- chart paper
- marker

Inquiry Skills: observe, communicate, compare

- Have children close their eyes and concentrate on sounds for a minute. As children name a sound they heard, have the group listen for that sound. Record responses on chart paper.
- Review the sounds and ask children to discuss the characteristics of each sound they heard.

Compare Ask: **Which sounds are loud? Soft? High? Low?**

Time to Move!

Have an assortment of items handy that make sounds. When children hear a loud sound have them move and shake their bodies quickly. When children hear a soft sound have them sway their bodies gently side to side.

Be a Reader Activity

You need
- Unit F Literature Big Book
- Leveled Reader: *Making Sounds*

Read the Big Book

Objective: Use illustrations and words to explore different types of sounds.

Inquiry Skills: communicate, compare

- Show children the cover of *Making Sounds* and have them predict what they might find in the story. Read each page aloud, stopping for children to discuss the sounds they might hear.
- Reread the story, encouraging children to chime in as you read. Invite volunteers to point to beginning letters and words that they know on each page.
- **Guided Reading:** See the inside back cover of the Leveled Reader version of *Making Sounds* for activities.

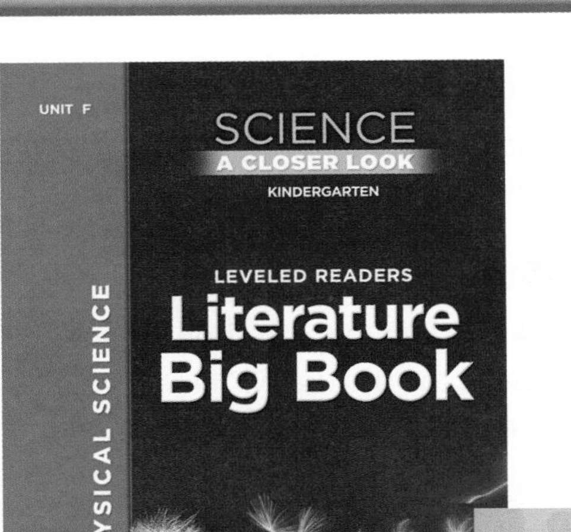

Be a Math Wiz Activity

You need
- crayons
- paper
- musical instruments

Sound Patterns

Objective: Recognize, predict, and create sound patterns.

Inquiry Skills: observe, predict, communicate

- Invite children to join you in an A-B-B clapping pattern, such as loud-soft-soft. Model additional patterns for the group to copy.
- Then have children work in pairs, each taking turns creating and copying sound patterns. Vary the activity by using musical instruments such as drums, triangles, rhythm sticks, and shakers.
- Some children may be able to record their patterns. For example, a black line could represent a loud clap and a blue line a soft clap.

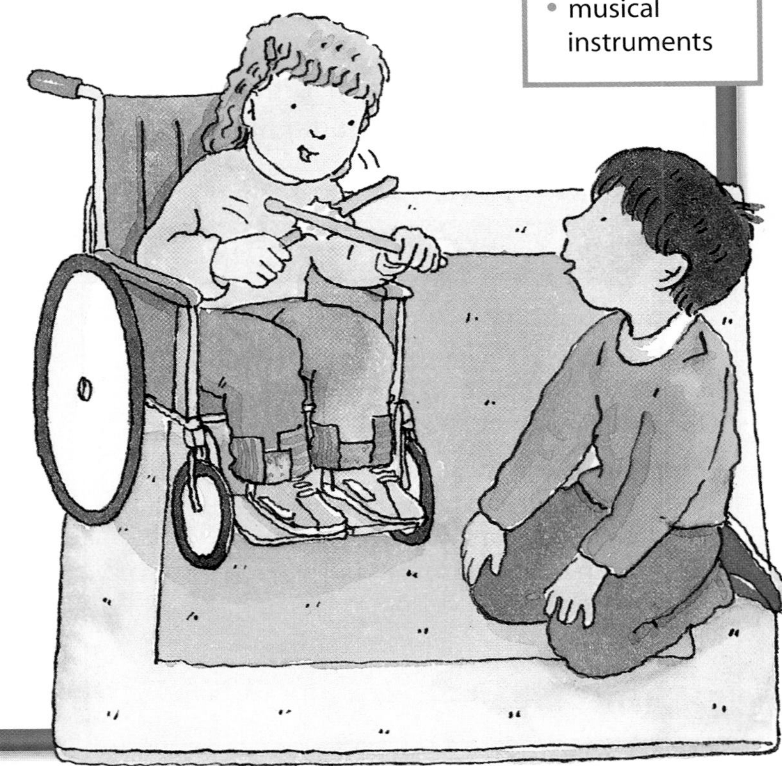

Read Together and Learn

What sounds might you hear?

▶ Build on Prior Knowledge

Ask children to name quiet sounds. Record their responses. Then repeat with loud sounds.

▶ Use the Visuals

- Communicate Ask children to think about which sounds represented in the pictures would be loud. Record their ideas and repeat for soft sounds.
- Ask the group to consider ways to voice each represented sound. Invite volunteers to imitate the sounds.
- Compare Invite children to discuss which sounds are pleasant and which sounds are not. Ask: **How are these sounds alike? How are they different?**

▶ Develop Vocabulary

loud, **soft**, **vibration** Write two sentence starters. *The moose makes a_________ sound. Sweeping makes a _________ sound.* Read aloud the sentences. Write *loud* and *soft*, and read them aloud. Invite volunteers to complete the sentences. Explain that a *vibration* is when an object, such as a guitar string, moves back and forth quickly. Ask: **What kind of sound does a guitar make?**

Flipbook

UNIT F • LESSON 4

What sounds might you hear?

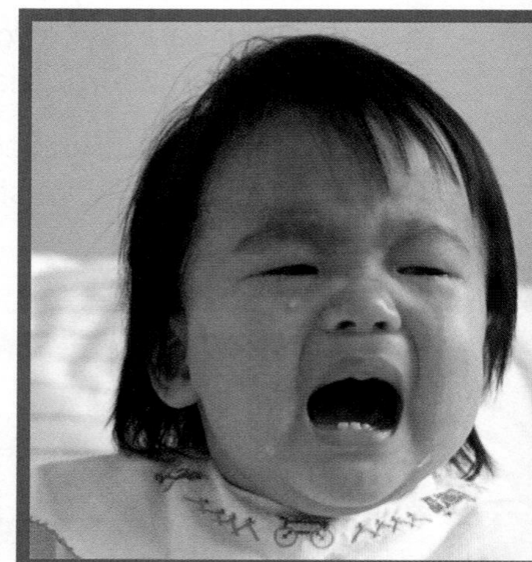

Reading Strategy

Print Awareness Cover the final *s* in *sounds* with a sticky note. Then invite children to read the sentence. Ask how covering the letter changes the meaning of the question.

Phonemic Awareness Ask a volunteer to find the word *might* and write it on the board. Invite children to make a rhyming word by covering the *m* with other consonants written on sticky notes.

Science Facts

Sound is created when air is moved. When a baby cries or a lion roars, vocal cords vibrate and make the air around them vibrate. These movements travel through the air in waves to our ears.

Decibels measure sound intensity. The sound of a jackhammer measures about 100 decibels. The sound of a girl brushing her teeth may only be 20 decibels.

Frequency Sound has a high frequency, or pitch, when waves vibrate quickly. Slow vibrations cause a low frequency.

Hertz Frequency is measured in hertz (Hz), the number of waves passing a point each second. Humans hear sounds between 20 and 20,000 Hz. Bats can hear over 120,000 Hz, and pigeons can hear as low as 1 Hz.

 **Professional Development** For more Science Facts and resources from NSDL visit http://nsdl.org/refreshers/science

58

▶ Ask Questions

- **Look at the fire truck. What sound might you hear? What does the loud sound tell you?** Sirens are used to warn people to move out of the way.
- Explain that animals use sounds to send messages. **What message might the baby be sending? The moose?** The baby may be hungry. The moose may be protecting his home.
- **Which pictures show someone who needs ear protection? What other workers might need ear protection?** firefighters, construction workers

Write on It

Invite children to use a dry-erase marker to write a *1* next to the softest sound. Have them continue to order the Flipbook sounds numerically using an *8* for the loudest sound.

Think and Talk

Ask a volunteer to pick a sound from the Flipbook page and describe it. Repeat until all the sounds have been described.

Observe if children understand that sounds are made in many ways. If children need more practice with the concept, review the Circle Time Activity.

Differentiated Instruction

Enrichment Explain that sound is measured in *decibels* and that a jackhammer measures about 100 decibels and brushing teeth may only be 20. Have children work in pairs to estimate the number of decibels for each Flipbook image, using a 100 chart to help them.

More to Read

The Listening Walk, by Paul Showers (HarperCollins, 1993)

Activity After reading, take a "listening walk." Provide children with clipboards, pencils, and paper and have them walk around the school and record things they hear.

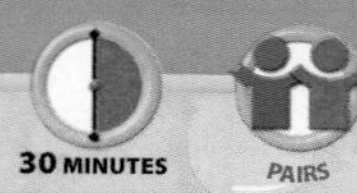

PAIRS

Sound Cylinders

Objective: Understand that sound is made from vibrations.

Inquiry Skills: observe, make a model, investigate, communicate, infer

You need **cardboard rolls, rubber bands, plastic bags cut in squares, Science Journal p. 58**

1. **Observe** Have children gently place their hands on their throats and say their name in a loud voice. Ask: **What do you feel?** Have children repeat using a soft voice and ask: **Can you feel a difference?** Explain that what they feel is air moving back and forth very fast, called vibrations, and that sound is made when something vibrates.
2. **Make a model** Model using a rubber band to fasten plastic over one end of a cardboard roll. Next, have a child talk into the open end of the roll while gently touching the plastic.
3. **Investigate** Have children work in pairs to make sound cylinders, talk into them, and feel the sound vibrations. Encourage partners to experiment with different sounds.
4. **Communicate** Have children draw pictures about sound on their Science Journal page.

You might want to help children visualize how vibrations travel. One way to do this is to jerk on one end of a loosely stretched coil toy. The children will be able to watch the wave travel the length of the coil.

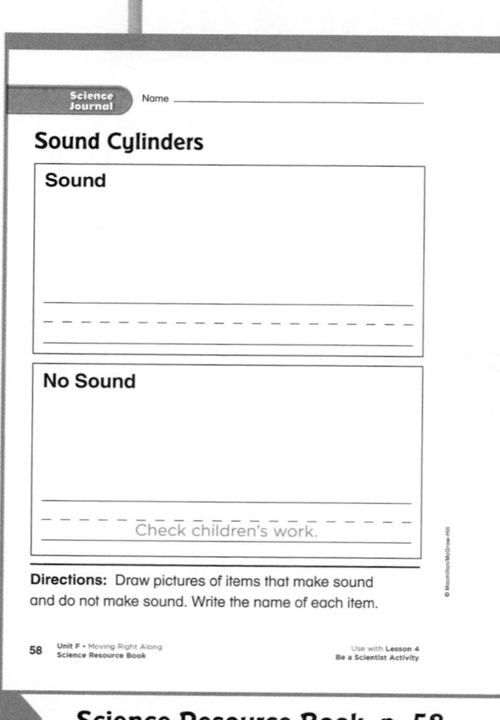
Science Journal Name ______

Sound Cylinders

Sound

No Sound

Check children's work.

Directions: Draw pictures of items that make sound and do not make sound. Write the name of each item.

58 Unit F • Moving Right Along Science Resource Book

Use with Lesson 4 Be a Scientist Activity

© Macmillan/McGraw-Hill

Science Resource Book, p. 58

Investigate More

Infer Refer to the list of classroom sounds from the Circle Time Activity. Ask: **Would the vibrations made by these sounds feel the same? Why or why not?**

Music

Here Comes the Band

Objective: Listen to how a variety of large and small instruments make different sounds.

Inquiry Skills: observe, compare

- **Listen and Talk About It** Play the recording of "Here Comes the Band." Point out the different types of sounds the bass and snare drums make. Ask children to name the instrument that made the highest sound (the piccolo) and the lowest sound (the tuba).
- **Sing** Play the song again and invite children to sing the sound each instrument makes.
- **March and Act It Out** Have children line up and pretend to play an instrument as they march around the room to the music.
- See pages TR33–TR34 for printed music to "Here Comes the Band."

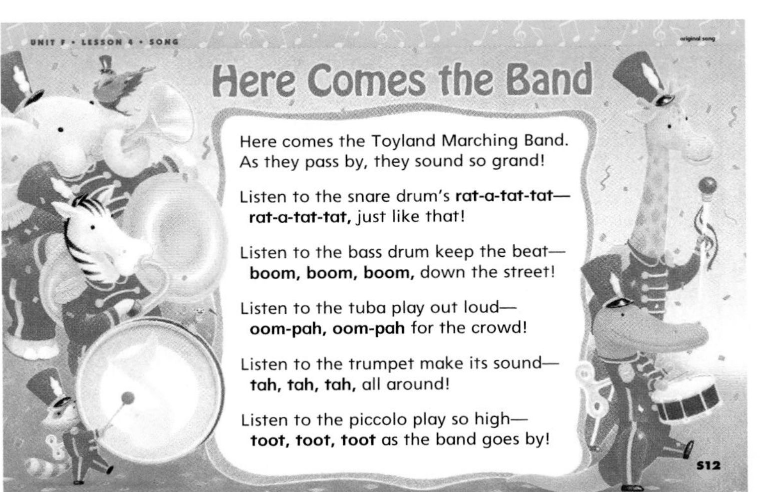

Flipbook p. S12
CD Tracks 23–24

Drawing and Writing

You need
- construction paper
- crayons
- markers

Loud and Soft

Objective: Identify and compare sounds in the environment.

Inquiry Skills: compare, communicate

- Have children fold a piece of construction paper in half like a card and draw a picture of something that makes a loud sound on one half. On the other half, have them draw a picture of something that makes a soft sound.
- Help children write *loud* and *soft* to label each half of their paper.
- Bind the pictures together to make a *Loud and Soft* class book.

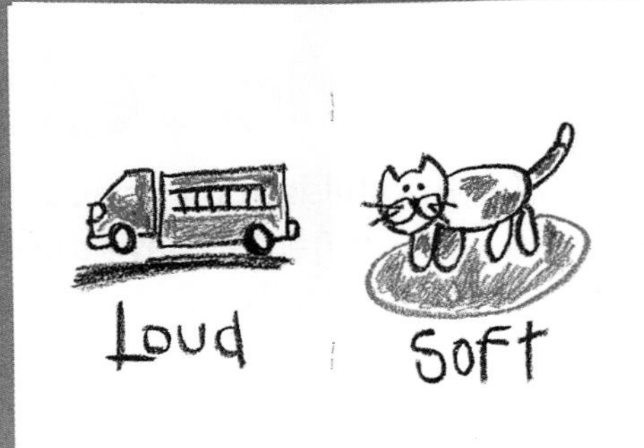

Art

You need
- empty tissue boxes
- rubber bands of varying widths
- goggles
- paint
- paintbrushes

Music Boxes

Objective: Investigate the sounds made by vibrations.

Inquiry Skills: make a model, observe, compare

- Have children paint or decorate a tissue box. Then slip three or four different widths and lengths of rubber bands around a box.
- Have children pluck each rubber band and listen to the sound it makes. Encourage children to discuss how and why the rubber bands make different sounds. Ask: **Can you see something vibrate?**

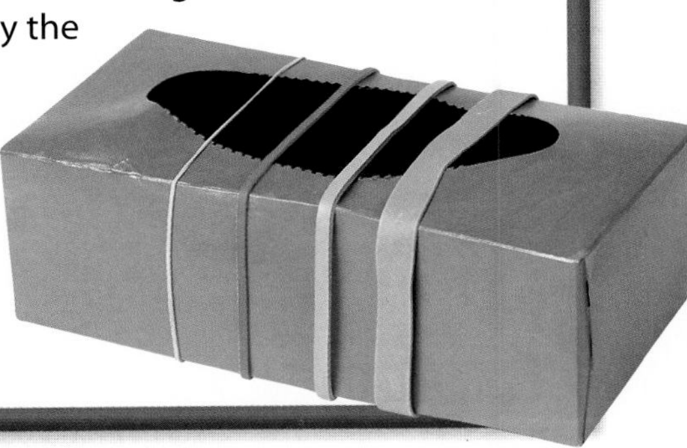

Magnets

▶ **Essential Question**
What objects will magnets move?

▶ **Objective**
Recognize that magnets can be used to make some objects move without being touched.

▶ **Vocabulary**
magnet

Resources

Flipbook, p. 59

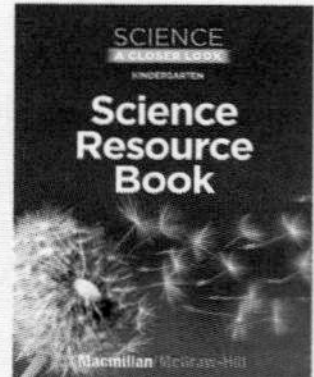
Science Resource Book, p. 59

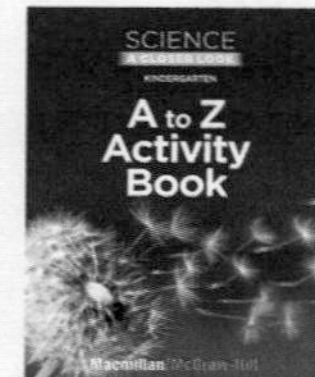
A to Z Activity Book, pp. 26–27

Leveled Reader: *What Can a Magnet Do?*

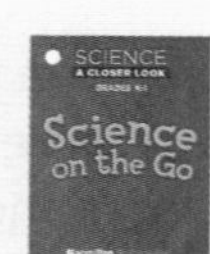
Science on the Go, pp. 113–114

Technology

www.macmillanmh.com

NSDL

Circle Time
WHOLE CLASS

Try Magnets

Inquiry Skills: observe, predict, classify

You need
- refrigerator magnets
- horseshoe magnet
- classroom objects

- Show children the refrigerator magnets and the horseshoe magnet. Explain that they may vary in size, but they do the same thing.
- Use string to make two circles on the floor, one for items that the magnet holds and the other for items it does not hold. Have children predict and test whether each item is magnetic and sort the items accordingly.

Observe Ask: **What do you notice about the things the magnet picked up?** They have metal on them.

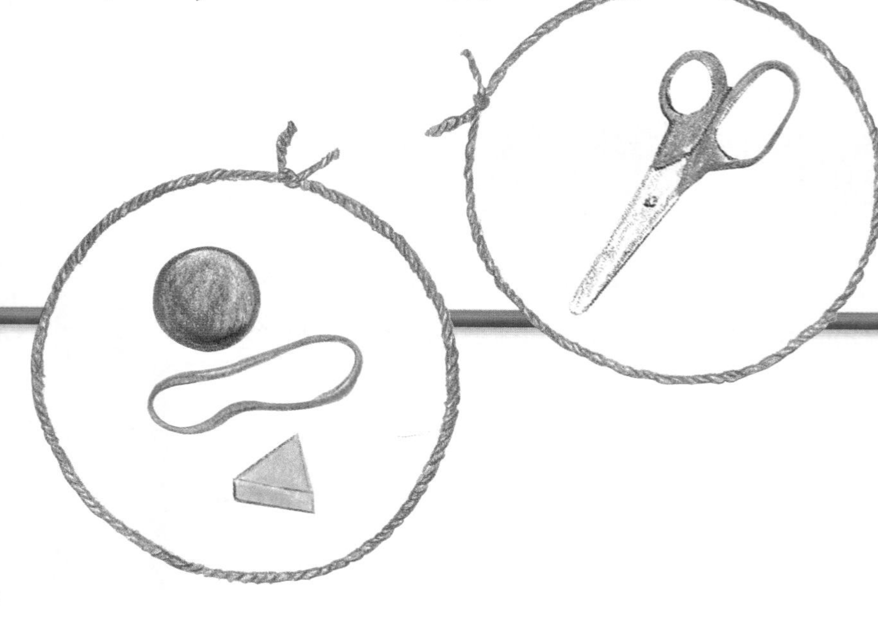

Time to Move!

Have children get in pairs. Tell the children they need to act like magnets and stick their two palms together. Have the pairs move around the room doing their best not to break apart.

Be a Reader

You need
- Unit F Literature Big Book
- Leveled Reader: *What Can a Magnet Do?*

Read the Big Book

Objective: Learn about magnets.

Inquiry Skills: observe, communicate

- Encourage children to discuss what they think magnets can do. List their ideas on chart paper.
- Read the title as you track the print. Then read the book. On page 6, encourage children to predict whether the magnet will pull the paper clip and whether it will pull the paper. On page 7, invite children to point out objects on the page that the magnets might attract.
- **Guided Reading:** See the inside back cover of the Leveled Reader version of *What Can a Magnet Do?* for activities.

Be a Writer

You need
- masking tape
- magnets
- paper
- pencils
- clipboards

Magnetic Walk

Objective: Discover and record magnetic and nonmagnetic items.

Inquiry Skills: investigate, compare, communicate

- Decide which classroom objects you want children to test and mark them with wide masking tape. Write the name of the object on the tape for children to use as a reference.
- Model how to fold a piece of paper in half and write *Yes* on one side and *No* on the other. Have children write the names of the items that stick to the magnet under *Yes* and things that do not stick under *No*.
- Have children work alone or in pairs and meet as a class to discuss their results.

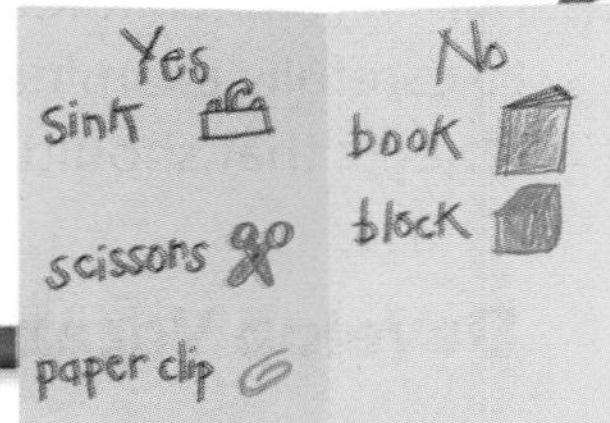

Be a Math Wiz Activity

You need
- paper clips
- horseshoe magnet
- bar magnet
- paper
- pencils

Paper Clip Chains

Objective: Use one-to-one correspondence to compare numbers.

Inquiry Skills: observe, compare, communicate, infer

- Explain that when an iron or steel item sticks to a magnet, it becomes a magnet too. Model how the first paper clip sticks to a magnet and holds a second paper clip.
- Have pairs build a paper clip chain on each magnet. Then have children draw a clip for each one that stays on the chain, and record the total number of clips on a piece of paper. Have children circle the larger number.
- Discuss their results, asking: **Why were your results the same or different?**

Read Together and Learn

What do you notice about the magnets?

▶ Build on Prior Knowledge

Show children a horseshoe magnet. Have them share what they know about magnets.

▶ Use the Visuals

- **Observe** Ask volunteers what they notice about the objects that the magnets hold. **How are they alike? Different?**
- **Communicate** Invite children to discuss the objects the magnets do not hold. Ask what they notice about these objects.
- Write *Magnetic* and *Not Magnetic* on chart paper. Help children read the words. Ask them to carefully examine the Flipbook page to find objects that should be recorded in each list.

▶ Develop Vocabulary

magnet Explain that a *magnet* is an object that attracts certain metals (iron, nickel, and cobalt). Write *magnetic* on the board, read the word, and then have children read it with you. Ask what smaller word is part of *magnetic*. Invite a volunteer to circle *magnet* in the word. Explain that if something is pulled toward a magnet, it is said to be magnetic.

Flipbook

UNIT F • LESSON 5

What do you notice about the magnets?

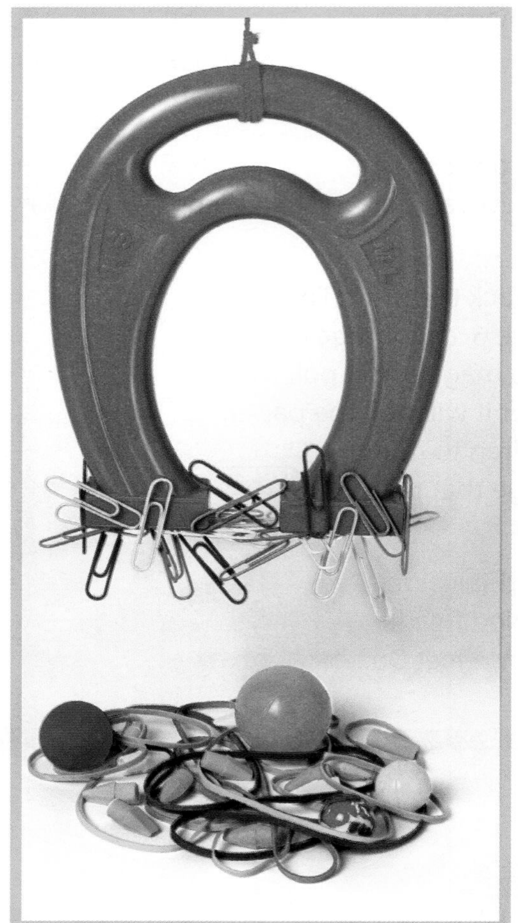

Reading Strategy

Phonological Awareness Read the text with the children and then have them clap the syllables in each word. Ask: **Which words have one clap? Two?**

Phonemic Awareness Write *what* on the board and have children add and remove letters to create new words *(hat, sat, where, when)*.

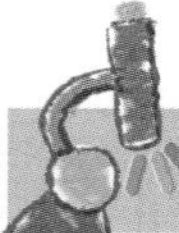

Science Facts

Magnet Any substance that can attract or repel iron or steel is a magnet. Now most magnets are artificially made, but the first magnets were naturally occurring lodestones. A magnet's force is strongest at its ends. The ends of magnets are called poles. One end is the north-seeking pole and the other end is the south- seeking pole. Opposite poles attract, and like poles repel each other. The word *magnet* comes from Magnesia, a place in Turkey where lodestone is found.

Professional Development For more Science Facts and resources from NSDL visit http://nsdl.org/refreshers/science

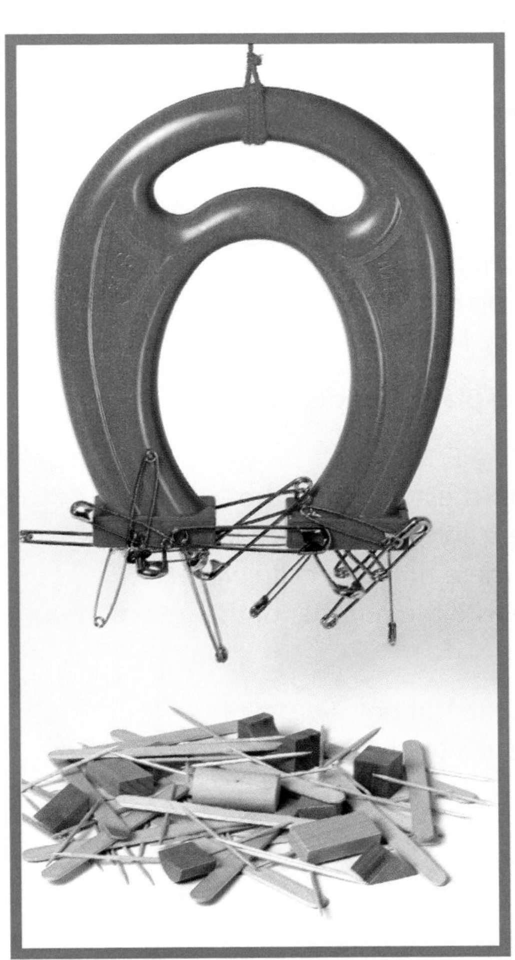

59

ELL Support

Collect and label with sticky notes common objects made of wood, plastic, metal, glass, cloth, and rubber.

Beginning Ask children to give you something metal, wood, and so on.

Intermediate Ask pairs to take turns asking for objects (metal, wood, and so on).

More to Read

Mickey's Magnet,
by Franklyn M. Branley and Eleanor K. Vaughn
(Scholastic, 1999)

Activity As you read, record the items Mickey picks up with his magnet. After reading, ask children what items from their own homes Mickey's magnet could pick up and add these to the list.

▶ Ask Questions

- **What are paper clips made of?** metal **Do you think a magnet would pick up an eraser?** No, erasers are not metal.
- **Find the magnet holding nails. What are nails made of?** metal **Are all metals attracted to magnets?** No, copper and aluminum are not.
- **Look at the pins. What do you think they are made of?** metal **Why are there more items on this magnet than on the others?** There are more metal objects in this picture.

Write on It

Invite children to use a dry-erase marker to draw other nonmagnetic items in each group. Ask them to explain their choices.

Think and Talk

Ask volunteers to identify one magnet fact that the page shows. Record responses on a *Magnet Facts* chart. While looking at the Flipbook page with the class, review these facts and how a magnet can be a useful tool.

Formative ASSESSMENT

If children have difficulty understanding what is magnetic and what is not, invite them to do the activities shown in each image at the Science Center to explore magnetism concretely.

Moving Clips

Objective: Understand that magnets can pull objects and stronger magnets can pull objects greater distances.

Inquiry Skills: predict, observe, investigate, measure

You need **three different magnets labeled 1–3, paper clips, pencils, Science Journal p. 59**

1. Show children the Science Journal page and the three magnets. Demonstrate the experiment by positioning the first magnet on the wide line at the top of the page so that the bottom of the magnet touches the line.
2. **Predict** Ask: **What will happen if we place a paper clip on a line near the magnet?**
3. **Observe** Place a paper clip on the farthest line from the magnet. Continue to move the paper clip closer on each line. When children have found the line where the paper clip will jump to the magnet, model how to write the number of the magnet on that line.
4. **Investigate** Ask children to do the experiment with each magnet and record the results on the Science Journal page.

Teacher Tip

You may want to vary this activity by having children use a variety of magnets, such as disk magnets and refrigerator magnets.
You can also try to find someone who works in construction who may have a very powerful magnet that you could borrow and use.

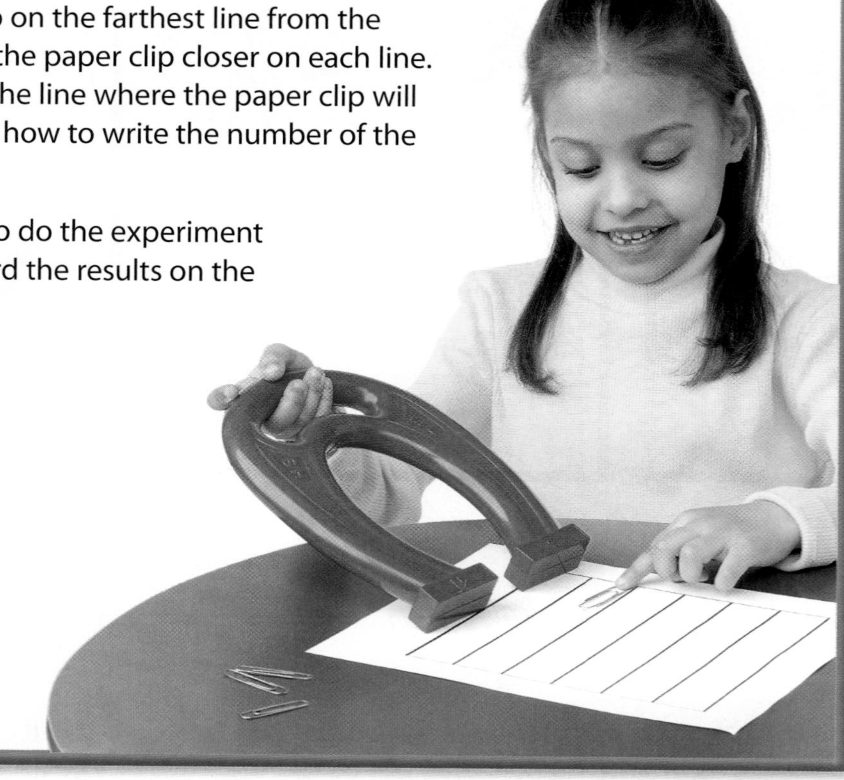

Name ______________________ Science Journal

Moving Clips

Magnet
Check children's work.

Directions: Write the magnet's number on the line from which the clip jumped. Repeat with other magnets.

Unit F • Exploring Matter
Science Resource Book

Use with Lesson 5
Be a Scientist Activity 59

Science Resource Book, p. 59

Investigate More

Measure Ask: **What did you notice about the different strengths of magnets and how far the clips traveled?** Some magnets are stronger and can pull from farther distances. Show children how to measure the distances using a ruler.

centers

Drawing and Writing

Pulling Through

Objective: Recognize, draw, and write about the materials magnets can pull through.

Inquiry Skills: observe, classify, communicate

- Have children put a magnet on the table, cover it with colored paper, and put another magnet on top of the paper. Then have children try to lift the first magnet with the second magnet.
- Have children try materials other than paper to see what magnets can work through.
- Have children draw and write about what the magnets can and cannot pull through.

You need
- magnets
- colored paper
- block
- piece of fabric
- book
- cardboard
- pencils
- crayons

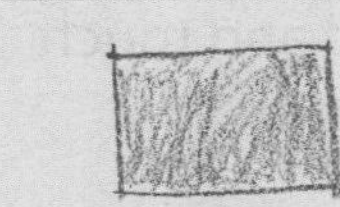

The magnets worked through paper.

The magnets did not work through my hand.

Technology

More About Motion

- For games and activities about how things move, help children log onto www.macmillanmh.com.
- Children can use these fun games to review and reinforce what they have learned about motion.

www.macmillanmh.com for more science online.

Art

Make Your Own

Objective: Develop fine motor skills.

Inquiry Skills: observe, make a model, investigate

You need
- air-drying clay
- paint
- paintbrushes
- magnet strips

- Show a variety of refrigerator magnets and explain to children that they will be making their own magnets to use on their refrigerators at home.
- Have children make small clay figures and press a magnetic strip into the back.
- Let the figures air-dry. When dry, you may need to reattach the magnet with glue if it has fallen off. Then have children paint their magnet.

Blocks

Magnetic Blocks

You need
- magnet strips
- paper

Objective: Create structures and toys that employ magnetism.

Inquiry Skills: infer, investigate, make a model

- Attach pieces of magnet strip to faces of selected blocks. Show children that you have attached magnets to some of the blocks.
- Help children brainstorm how magnetic blocks could be used to build things and to help things move.
- Keep magnetic blocks in the center so children can continue experimenting with magnetic force.

For more on magnets, use pp. 59–60 in the **Activity Book.**

Performance ASSESSMENT

DOK 2

Things Move and Make Sounds

Objective: Children will identify a toy that rolls and what makes it easy to move, demonstrate force of a push, classify toys in terms of magnetism, and recognize that sound is caused by vibrations.

- Place the toys on a table. Ask the child to push the toy that rolls and name what makes the toy easy to move.
- Ask the child to point to the toy that makes sound and touch the part that vibrates. Ask which toy a magnet might pick up.

You need

- toy guitar or music box
- magnetic toy car

Scoring Rubric

Use the Scoring Rubric to evaluate each child's completion of the Performance Assessment task.

3 points = Child can identify the wheels on the toy car and push it to make it roll. The child can point to guitar strings that vibrate and tell which toy is magnetic.

2 points = Child is able to push the toy that rolls and tell that wheels make it easier to move. Child can either identify what vibrates to make sound, but cannot distinguish between magnetic and nonmagnetic items, or cannot identify what vibrates to make sound, but can distinguish between magnetic and nonmagnetic items.

1 point = Child is able to push the toy that rolls, but cannot tell what makes it easier to move. Child can identify the toy that makes sound, but not its vibrating part. Child cannot distinguish between magnetic and nonmagnetic items.

Summative ASSESSMENT

DOK I

Use the Unit F Assessment booklet on pages 73–74 of the **Science Resource Book** to assess what children have learned about movement and sound.

- Help children fold their page in half to create a book and write their name.
- On pages 1–3, help children read the sentences and have them circle the correct picture and write the word to complete the sentence. On page 4, invite children to draw a picture of something that they can pull and write the word on the line.

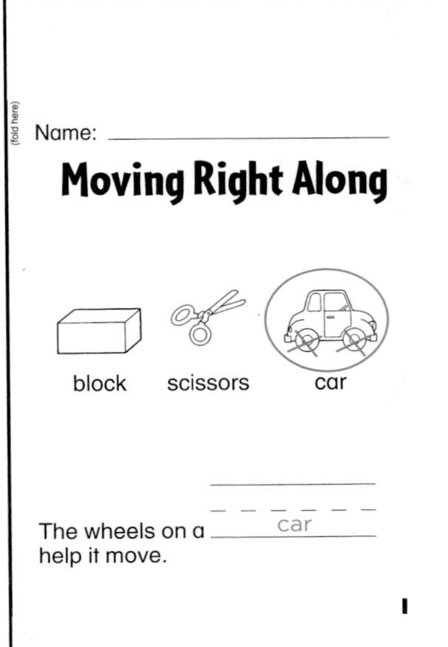

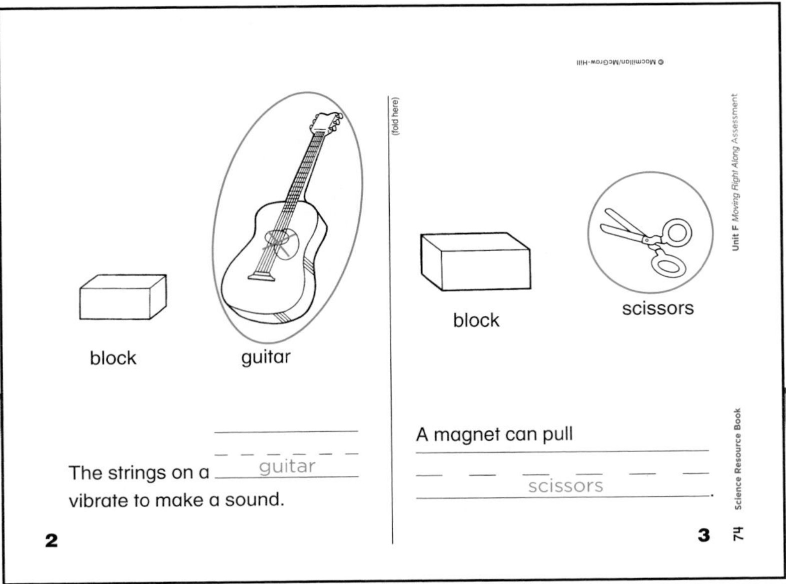

Science Resource Book, pp. 73-74

Portfolio ASSESSMENT

- Ask children to choose one piece of work from their Science Journal that they want to place in their portfolio. Ask them to describe what they did on the page they chose and why they chose this piece of work to be included in their portfolio.
- Use the Unit F Checklist from the **Science Resource Book** page 82 to record children's progress. Place checklists and notes in each child's portfolio.

Matter and Motion

Objective: Explore ways matter can be described, changed, and set into motion.

What We Learned

- Before showing children the Flipbook page, ask them to describe any homes they have seen under construction.
- Have children think of materials that might be needed to build a home. Record their ideas.
- Ask children to name the tools a construction worker might use to build a house. Make a list.

Flipbook

REVIEW TOGETHER • MATTER AND MOTION

Floor Puzzle

Activity **Workbench** Revisit the floor puzzle and invite children to recreate the image using Flipbook page 46 as a reference. Introduce the action pieces and invite children to match tools, place them where they might belong, and act out scenarios using the tool pieces.

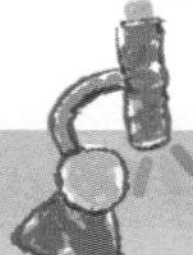

Science Facts

Saw Sharp teeth on one edge of the saw cut and tear through wood and other materials.

Hammer and Nails People use these tools to put materials together. A hammer is a hand tool that has a long handle and a heavy, metal head. A nail is a thin, pointy fastener made of metal. A hammer is used to hit the head of a nail through materials to hold them together.

Ladders are structures with two long upright pieces connected by several parallel bars, or steps. Used for climbing, ladders come in many lengths.

A **Hard Hat** is worn to prevent injury to a construction worker's head. These can be made of metal or fiberglass.

LOG ON **Professional Development** For more Science Facts and resources from  NSDL visit http://nsdl.org/refreshers/science

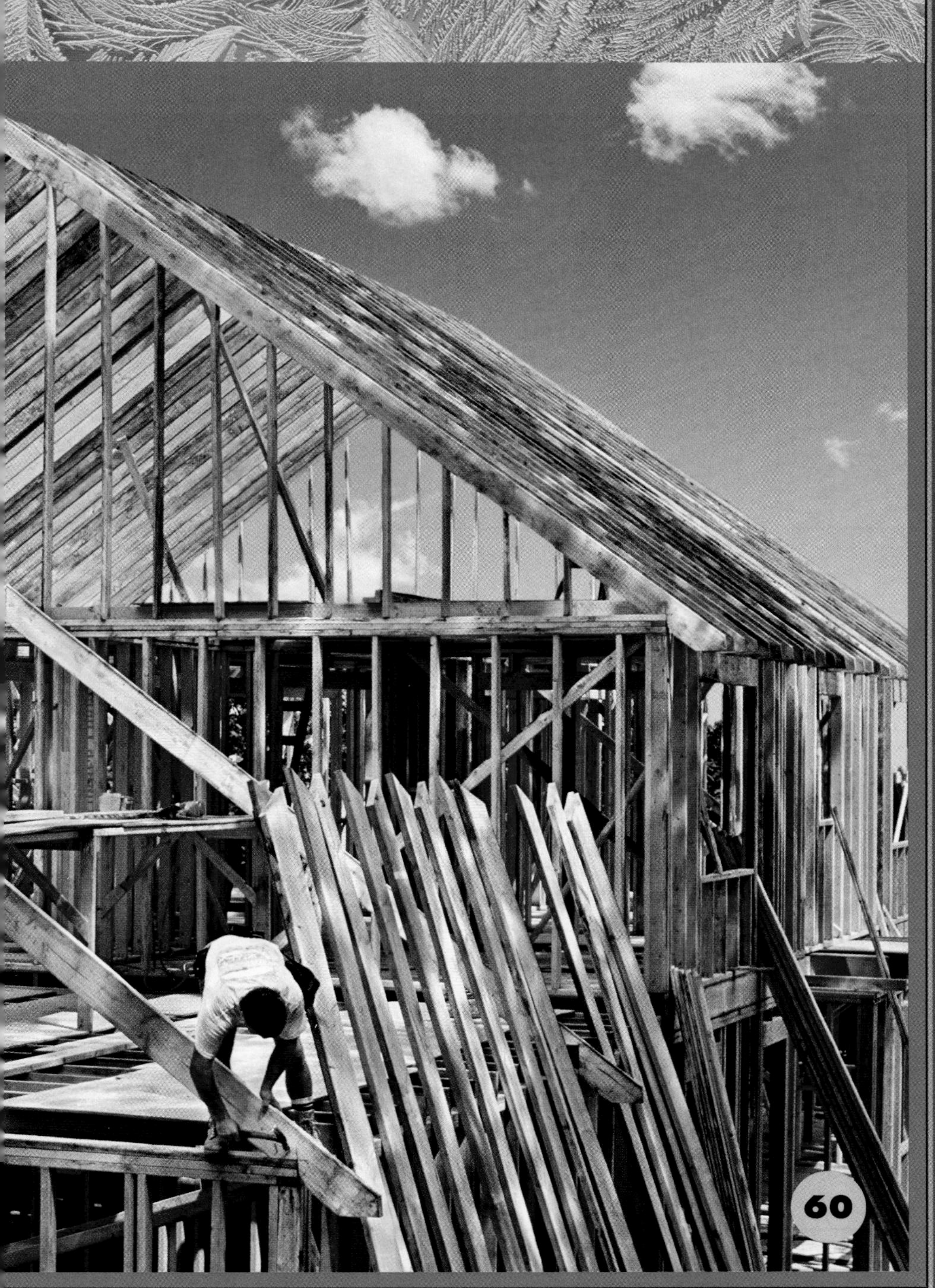

Make Connections

What do you need to make a house?

▶ Use the Visuals

- **Observe** Ask children to locate the materials used to build the house. Compare and contrast these with the materials they previously discussed.
- **Communicate** Have volunteers circle materials on your list that change during the construction process. Invite them to describe the changes.
- Then make a list of the machines and tools children find. Ask how the machines help people use and change the materials.

Write on It

Have children use a dry-erase marker to circle the objects that move. Ask them to write *push* and/or *pull* next to each circled item to show the force that moves it.

▶ Ask Questions

- **What do you think happened at the site before the Flipbook scene?** A hole was dug for the foundation.
- **How does the saw make work easier?** Saws make cutting and shaping wood easier.
- **Why do you think the construction worker is wearing a hard hat?** to protect his head

Take a Trip

Construction Site Visit a nearby construction site so children can investigate the materials, tools, and process of building. Give children pencils, paper, and clipboards to record what they notice. If possible, make repeat visits to note ongoing changes in the work. You can also invite a construction worker to visit the class.

Back in the Classroom Have children examine real tools (such as hammer and pliers) and then create a class book to document what they learned about the site and tools.

Teacher's Notes

Teacher Resources

Additional Teacher Resources

The **Teacher's Desk Reference** offers information tailored to today's science teacher.

It includes practical, hands-on support in several areas of *Professional Development,* including: Managing Inquiry Based Science and Scaffolding Inquiry Instruction, the 5-E Instructional Model, Differentiated Instruction, Gender Equity, Why Kindergarten Science Is Important, and Assessment in the Science Classroom.

This book also traces how science is taught across the grades. Beginning with First Grade, the *Science Across the Grades* section compares the key concepts of each chapter and lesson with the concepts in the corresponding chapters and lessons of the previous and subsequent grades. Teachers can, therefore, place the material they will be teaching in the greater context of the students' overall learning path from first to sixth grade.

Introduce Plants

Taking care of plants in the classroom is a great way for children to get a sense of responsibility while also teaching them about the needs of plants. Children will explore how to grow plants inside and establish routines for maintaining them.

Choosing Seeds

You need
- seeds
- soil

Inquiry Skills: observe, communicate

- Ask children what kind of plant they would like to grow. Alfalfa, radish, and grass seeds are great for a classroom since they grow fast. Lima and mung beans are also good choices.
- Sometimes mold grows on seeds. If this happens, discard the molding seeds and start over again with fresh seeds. In a cup, wash the fresh seeds with a solution of 1 part bleach in 8 parts water. Using a sieve, pour off the water-bleach solution and plant the seeds.
- Encourage children to observe the growth of the plant.

Planting

You need
- planter (or milk container)
- saucer
- soil
- seeds
- water

Inquiry Skill: observe

- When choosing a pot for your classroom plants, be sure to have a planter with drainer holes. Place a saucer under the planter to catch excess water.
- You can buy potting soil from a nursery, or you can make your own potting soil by mixing topsoil and construction sand. Sand from the beach cannot be used since the salt in it is harmful for plants.
- Invite children to place seeds just under the surface of the soil. Encourage children to take turns watering the plant. If the plant's soil is cool to the touch, then it has adequate moisture.

Sprouting Seeds

You need
- alfalfa and lima bean seeds
- paper towels
- jar
- tray
- mesh cover
- resealable bags

Inquiry Skills: observe, investigate

- Soak seeds in water and set them on a damp paper towel. Then remove any seeds that are hard and have not expanded, as they will not sprout.
- Help children place alfalfa seeds in a jar, put a mesh cover on the jar, and roll it so the seeds cling to the sides. Invite children to spread the lima bean seeds out on a tray, cover the tray, and place it in a dark area.
- When the alfalfa seeds have sprouts about one inch long, move the jar to a place with indirect sunlight for 3 hours. Sprouted lima bean seeds can be stored in a refrigerator in a resealable bag on a damp paper towel for about one week before planting.

Taking Care

You need
- pot
- soil
- water
- plant

Inquiry Skills: observe, measure

- Encourage children to take care of plants in the classroom by adding plant-care tasks to your job chart. Show them how to water plants without saturating them. Generally, yellow leaves are an indicator of too much water.
- Set the plant in a place with adequate sunlight. Encourage children to notice how the plant grows toward the light. Have children feed the plant about once every six months with plant food.
- As the plant grows, help children to transplant it into a larger pot. Place a saucer under the pot to catch excess water if there are drainage holes. If there are no drainage holes, place pebbles or gravel at the bottom of the pot.

Introduce Animals

Classroom pets are an exciting and educational part of a classroom. Children will set up animal habitats and establish routines for maintaining them in the classroom, such as feeding the animals or cleaning their tanks.

Earthworms

You need
- clear container to use as a tank
- soil
- dry leaves
- grass
- earthworms

Inquiry Skills: observe, compare, make a model

- Invite children to collect earthworms. When children dig for earthworms in soil, they will understand the needs of earthworms as well as what kind of habitat they live in.
- Encourage children to collect some soil from where they found the earthworms and place it in the tank to recreate their habitat. Have children collect dry leaves and grass clippings to put in the tank as well.
- When the tank is set up, place it where children can easily look into it. Children learn by touching the earthworms and observing them closely.

Hermit Crabs

You need
- 5-gallon clear tank
- sand
- dried coconut shells
- hermit crabs

Inquiry Skill: communicate

- Before introducing hermit crabs to your classroom, read *A House for Hermit Crab* by Eric Carle with children. Then help children fill the bottom of the tank with sand and place dried coconut shells inside the tank. Remind children that hermit crabs adopt a protective shell.
- Since hermit crabs like a moist environment, encourage children to take turns misting the tank once a day. Have children feed them small amounts of fruit and crackers or hermit crab food from the pet store. Encourage children to change the food daily.
- Place the tank where children can see into it from the top and the sides. Encourage children to observe the hermit crabs and record what they notice.

Guinea Pigs

You need
- cage or large clear tank
- objects for the tank
- guinea pig(s)

Inquiry Skills: observe, communicate

- You can create a home for your guinea pig in a clear tank. Generally, guinea pigs need about 2 square feet each, so if you have two guinea pigs you will need 4 square feet. Since guinea pigs are friendly animals that need care and attention, they are great pets to send home with children on weekends.
- When setting up the tank, encourage children to create ramps for the guinea pig. Invite children to add toys to the tank, such as balls and blocks. Avoid small objects that the guinea pig could swallow.
- Guinea pigs need vitamin C in their diet. Encourage children to feed their guinea pig hay and a variety of fresh vegetables daily. Guinea pig pellets can also be purchased at the pet store. Show children how to carefully hold and play with the guinea pigs.

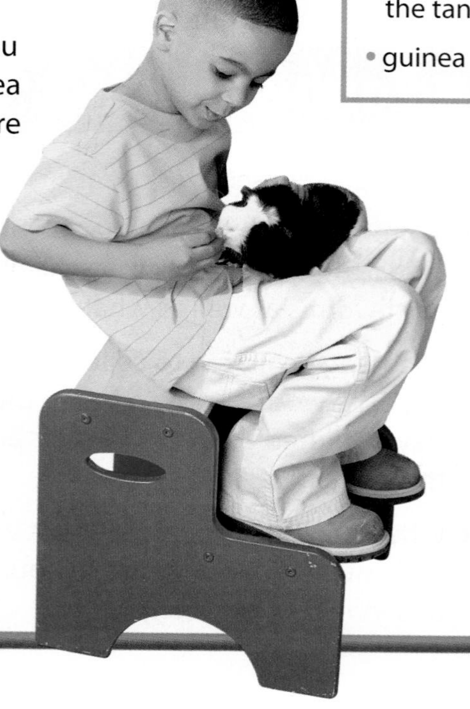

Fish

You need
- fish tank
- air filter
- pebbles
- plants
- water
- fish
- fish food

Inquiry Skills: observe, make a model

- Have children create a freshwater aquarium. Help children place small pebbles and plants at the bottom of the tank. Fill the tank with tap water. Attach an air filter to insure a quality environment for the fish.
- Fish food can be bought at local pet stores. Show children how to feed the fish sparingly, because overfeeding can harm the fish.
- Help children clean the tank when the water starts to become cloudy. Have children wash the pebbles in the tank and replace the water. Allow the water to become room temperature before putting the fish into it.

Measurement

Measuring is an important skill that is used by scientists to collect data. Children will explore how to use a variety of nonstandard units to find length, temperature, weight, and capacity. In addition, children will recognize that tools used for measuring, such as rulers, thermometers, and balance scales provide more consistent results.

Measure Length

You need
- connecting cubes
- paper clips
- crayons
- linking loops
- ruler

Inquiry Skills: predict, investigate, measure, compare

- Use nonstandard units such as connecting cubes to model how to measure the length of a classroom object (e.g., a book). Then invite children to help you count how many cubes you used.
- Have children work with a partner to measure the length of various objects in the classroom. Encourage groups to use different nonstandard units and then a ruler to measure the same objects. Help children record their findings.
- Invite children to compare results. For example, the length of the stapler was 9 connecting cubes, 2 crayons, and 7 inches long.

Investigate Weight

You need
- balance scale
- collection of objects

Inquiry Skills: classify, infer, measure, compare

- Show a collection of objects and have children work in small groups to sort them into two piles—light and heavy—holding objects in their hands to compare the weights.
- Ask a child to choose one object from the heavy group and one from the light group. Then have another child guess which is heavier.
- Invite children to place the objects on a balance scale to check their guess. Allow all children to have a turn.

Explore Capacity

Inquiry Skills: predict, measure, compare, communicate

- Show children the containers and invite volunteers to fill each with connecting cubes. Then have children help you count how many cubes it took to fill up each container.
- Place the containers in the Sand Table and show children the measuring cup. Model how to fill and level the measuring cup with sand and pour it into a container. Invite a volunteer to make a tally mark each time you add ½ cup of sand to the container.
- Then have partners take turns filling the containers and recording tally marks. Encourage children to compare the capacity of the different containers.

You need
- Sand Table
- 5 containers of various sizes
- connecting cubes
- ½-cup measuring cup
- clipboard
- pencil
- paper

Temperature

Inquiry Skills: infer, investigate, observe, compare

- Ask: **How do you know if something is hot or cold?** Invite children to feel the water in each container and describe the temperatures.
- Encourage children to discuss how they know if it is hot or cold outside and inside. Pass around a thermometer with a piece of tape showing children where to look, and tell them the temperature inside the room.
- Ask children if they think the temperature of the air outside will be warmer or colder. Then take children and the thermometer outside and note any changes.

You need
- container of ice water
- container of warm water
- indoor and/or outdoor thermometer
- tape

Safety Tips

Tell the teacher about accidents and spills right away.

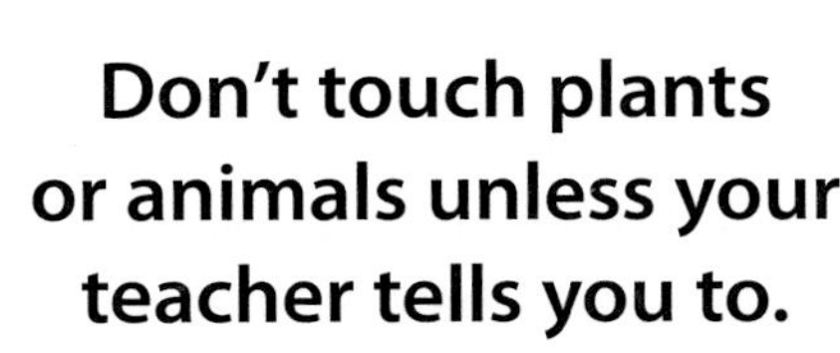

Don't touch plants or animals unless your teacher tells you to.

Be careful with glass and sharp objects.

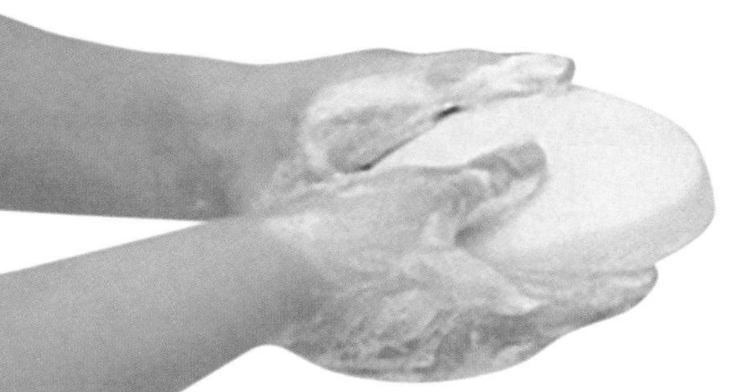

Wash your hands after each activity.

Keep your workplace neat. Clean up after you are done.

Never throw your trash on the ground.

Your Body

It is important to help children understand the parts that make up their bodies, both inside and out. Children will explore the basic body parts, how we use them, and how we take care of them.

Body Tracing

You need **butcher paper, crayons, scissors, glue**

Inquiry Skills: observe, communicate, make a model

- Invite children to do the following: place their hand over their heart to feel it beat; place their hands on their ribs to feel how they move; feel their skull; feel their arm muscles as they make a fist; listen to stomach noises with a stethoscope.
- Have children help each other trace their bodies on paper and cut them out. Invite children to draw the parts of their body.
- Help children label body parts including the eyes, nose, ears, tongue, and even skin and talk about how they work with our senses. Help them discover additional parts that they may not know.

Look Your Best

You need **construction paper, white paper, markers, stapler**

Inquiry Skill: communicate

- Invite children to think about three things that they did in the morning before they came to school. Help children to discuss that washing your face and hands, brushing your teeth, brushing your hair, dressing, and eating all help you to look your best.
- Give children three pieces of white paper to illustrate each activity. Help children to record their words for each page. Use construction paper to make front and back covers and use a stapler to bind each child's book.
- Place children's books in the Library Center.

Be Active, Rest, and Stay Healthy

Children require more rest and activity than adults, and are exposed to many germs in the school environment. Children will recognize the importance of rest, exercise, and regular visits to the doctor and dentist.

Stop and Go

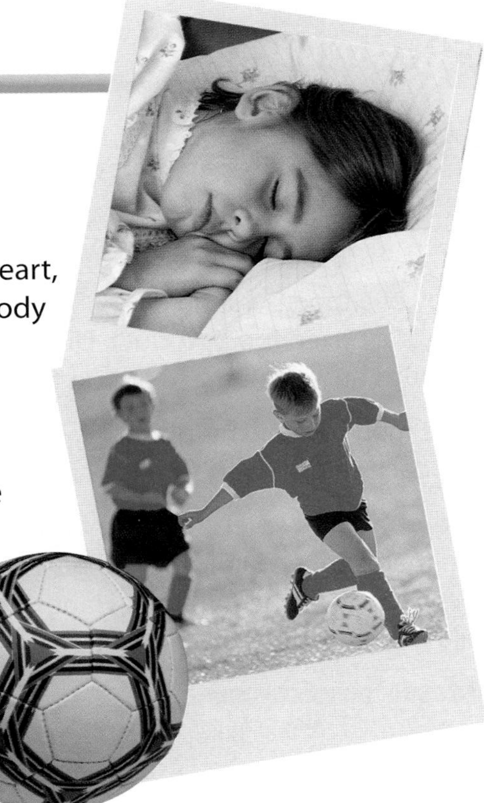

You need **old magazines, glue sticks, manila folders**

Inquiry Skills: communicate, classify

- Discuss with children that being active every day keeps your heart, bones, and muscles healthy. Explain also that rest allows the body to restore its energy.
- Have children look through magazines to find and cut out pictures that show being active and being restful.
- Open a folder and have children sort the pictures and glue the "active" pictures on the left side and the "rest" pictures on the right side. Label each side appropriately.

Doctor Tools

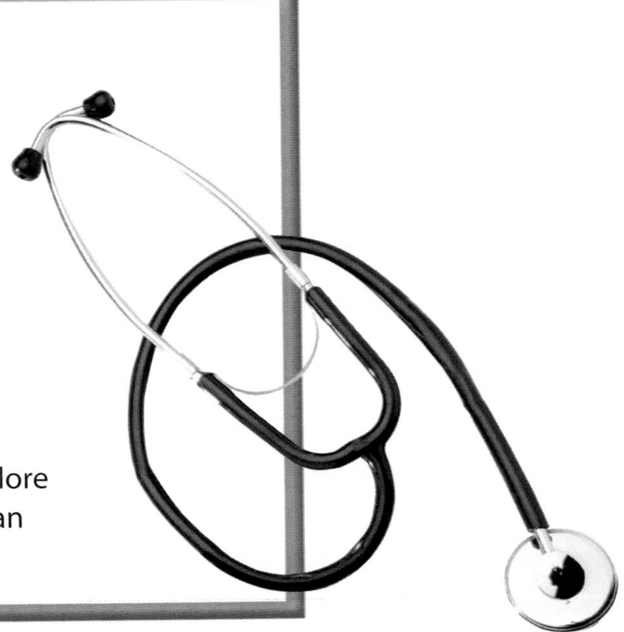

You need **pretend doctor's kit, stethoscope, large craft sticks, toothbrush, paper, markers**

Inquiry Skills: communicate, measure

- Collect a variety of doctor's and dentist's tools (or a child's doctor/dentist kit) to share with the class. As a group, pass each tool around, inviting children to share what they know about each tool.
- Explain that doctors and dentists help us stay healthy and use tools to do this.
- Place the tools at the Science Center and invite children to explore the tools more closely and draw a picture of how these tools can be used.

Eat Healthful Foods

Eating habits and food choices are often made for children. As they mature, children will begin to make their own food decisions. Therefore, it is important for children to learn about healthful foods, the basic food groups, and why it is important to eat healthfully.

Food Pyramid

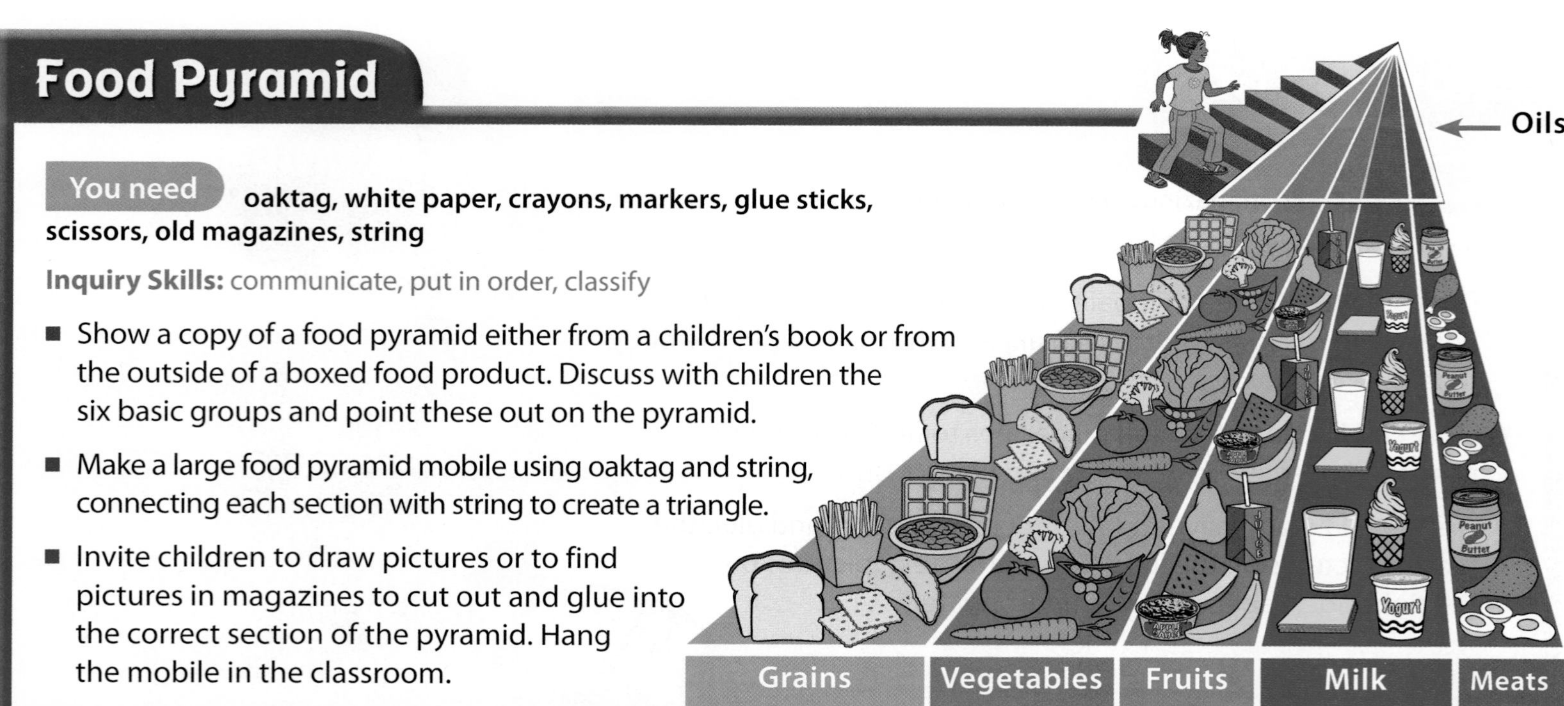

You need **oaktag, white paper, crayons, markers, glue sticks, scissors, old magazines, string**

Inquiry Skills: communicate, put in order, classify

- Show a copy of a food pyramid either from a children's book or from the outside of a boxed food product. Discuss with children the six basic groups and point these out on the pyramid.
- Make a large food pyramid mobile using oaktag and string, connecting each section with string to create a triangle.
- Invite children to draw pictures or to find pictures in magazines to cut out and glue into the correct section of the pyramid. Hang the mobile in the classroom.

Healthy Menu

You need **pretend food, paper plates, plastic forks and spoons, paper, markers, old magazines**

Inquiry Skills: observe, make a model

- Invite children to create a restaurant in the Dramatic Play Center. Provide them with plates, utensils, and pretend food.
- Discuss with children what types of healthful foods they would like to offer at the restaurant. Review the food pyramid and remind children that a balance of fruits, vegetables, meats, dairy, and grains is healthful.
- Help children create menus listing these foods. Invite children to illustrate each food or to cut out pictures from magazines for each food.

Getting Along

Getting along, sharing feelings, and working with others are important social skills for Kindergartners to learn. Children will recognize that people have different feelings and express them in a variety of ways.

Our Feelings

You need **chart paper, markers, paper plates, yarn, craft sticks, tape**

Inquiry Skills: communicate, compare

- As a class, brainstorm different feelings that people can have. Encourage children to share experiences when they felt sad, happy, surprised, upset, scared, angry, worried, or excited. Create a class chart listing the feelings children shared.
- Invite volunteers to act out each emotion using their facial expressions and have others try to guess the emotion they are acting out.
- Have children create face masks of different emotions and invite them to perform a puppet show about different feelings.

Bill of Rights

You need **markers, chart paper**

Inquiry Skill: communicate

- Explain to the class that they are going to help you create a class Bill of Rights. Teach children that a Bill of Rights is something that lists all of the things that people should be able to do, feel, and say.
- Begin the document with three main heads: "We have the right to be heard," "We have the right to feel safe," and "We have the right to have our work and feelings respected." Invite volunteers to list examples of each "right" and add their suggestions to the chart.
- Have every child sign the bottom of the document and post your Bill of Rights in the classroom for the year.

Bill of Rights

We have the right to be heard.

We have the right to have our work and feelings respected.

Be Safe Indoors

Indoor safety is an important topic to address with children. They are curious about their surroundings and need to be aware of general safety. Children will recognize what is safe and what is not safe to do and to touch indoors.

Keep It Safe

You need **oaktag, markers, tape, glue**

Inquiry Skills: observe, communicate

- Discuss with children that you will be adding a new job to your job chart called "the safety monitor." This person will be in charge of checking certain things in the classroom to make sure that they are safe.
- Have the class help to create a list of things for the safety monitor to watch for, such as careful use of scissors, not running in the hallways, staying away from hot surfaces (such as a stovetop, oven, or radiator). Post the list in the classroom.
- Create a safety monitor badge or hat for the monitor to wear and have children do this job during a clean-up time each day.

Safety Book

You need **paper, crayons, markers, yarn**

Inquiry Skill: communicate

- Ask children how they keep themselves safe at home. Record their ideas and discuss how they are similar and different to staying safe in school.
- Have children draw a picture of how they work to stay safe at home. Help them to write words to describe their work.
- Collect the pages and bind them together into a classroom book about safety. Encourage children to be "safety monitors" at home.

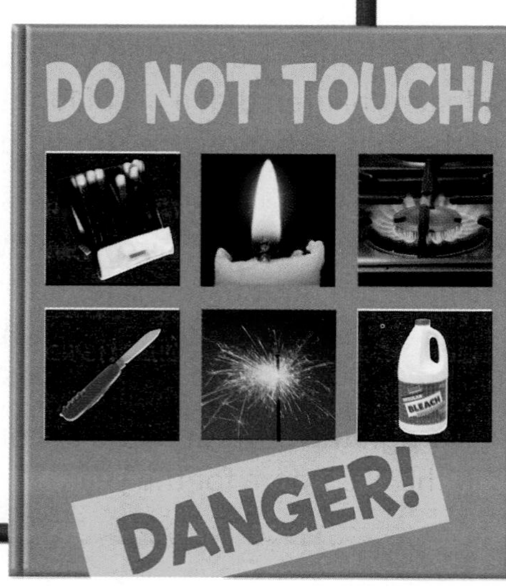

Be Safe Outdoors

It is important to discuss outdoor safety with children before working and playing outside. Children will explore outdoor safety and the rules necessary to follow in order to stay safe.

Play It Safe

You need **outdoor safety equipment such as knee pads, helmets, elbow pads, sunglasses**

Inquiry Skills: observe, communicate

- Show children the different protective gear that you have collected. If you are unable to collect these materials, find pictures of these items in magazines or books.
- Invite children to explore these materials and share any experiences they have had with them.
- Place the protective gear in the Dramatic Play Center for children to use during their pretend play. You can also bring these items outside with you during an outdoor playtime and allow children to use them.

Safety Mural

You need **butcher paper, markers, crayons**

Inquiry Skills: communicate, make a model

- Discuss with children the different ways they stay safe outdoors: wear safety gear on bikes, in-line skates, boats; cross the street at a crosswalk; wear seat belts in the car; do not talk to strangers.
- Invite children to help illustrate a class mural about outdoor safety. Label each section of the mural and have volunteers illustrate each safety rule.
- Hang the mural in the classroom and invite other classes in the school to come to your room for a safety discussion.

Name ______________________ Date ______________________

Venn Diagram

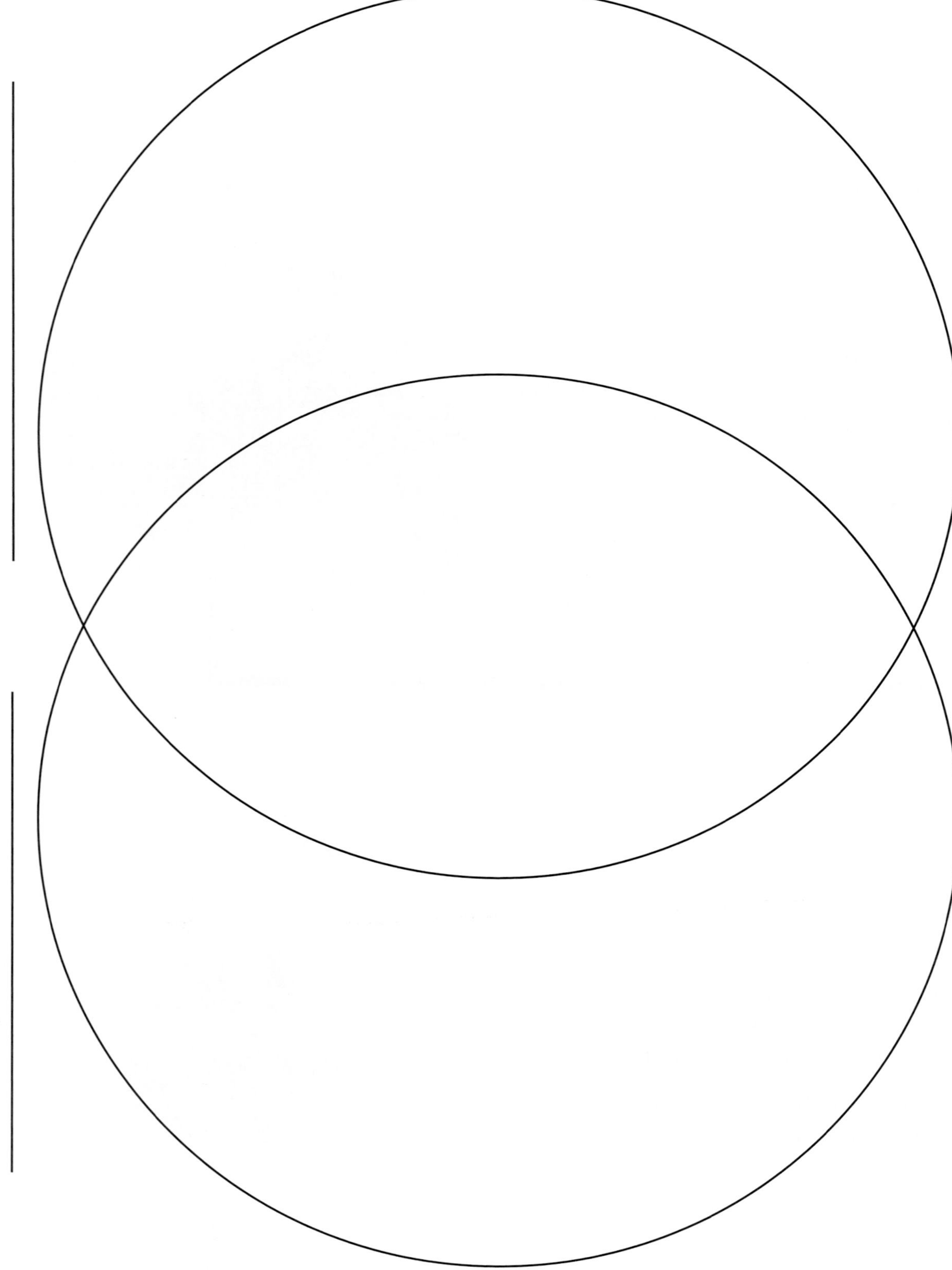

Name ______________________ Date ______________________

4-Row Graph

Name ______________________ Date ______________________

Word Web

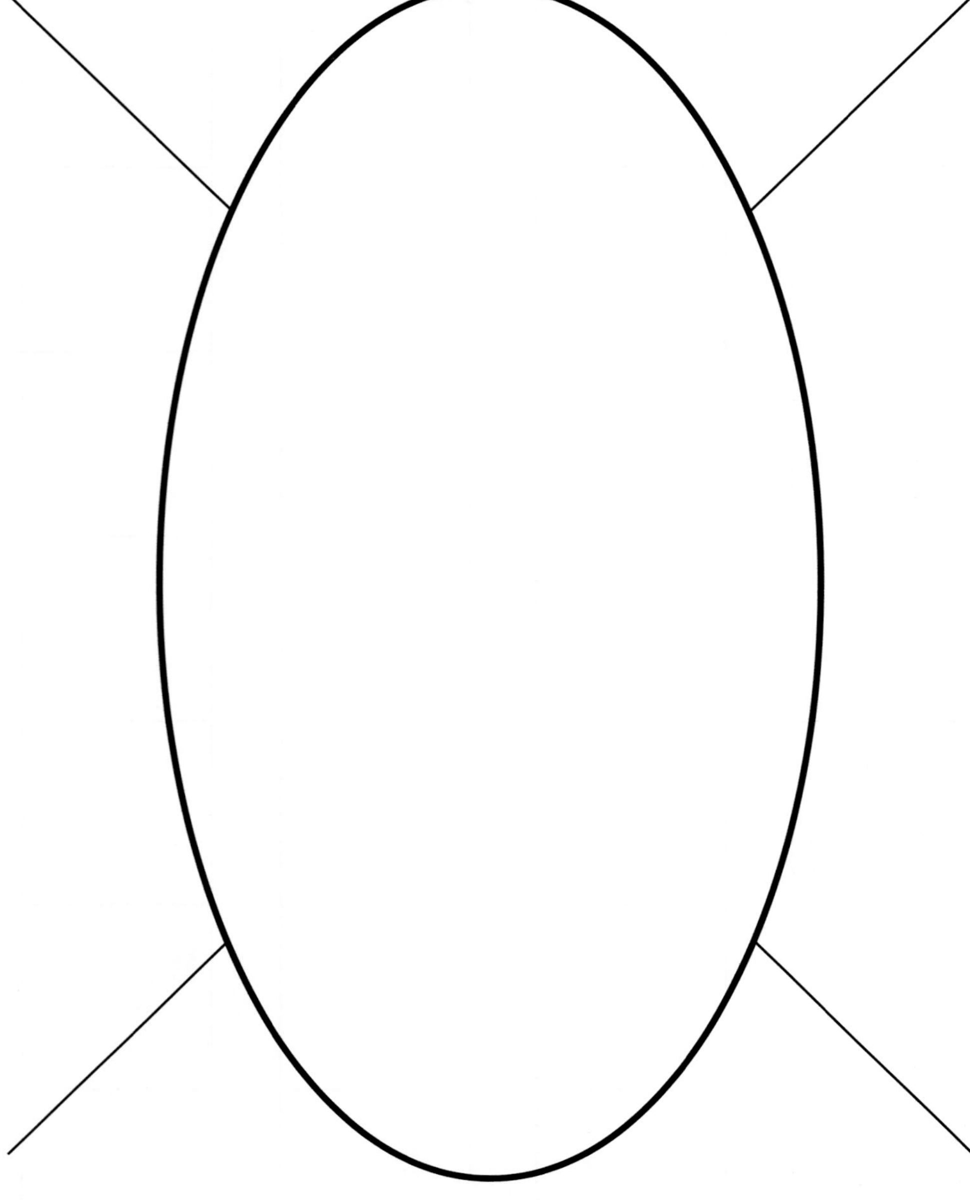

Name ______________________ Date ______________________

KWL Chart

K What You Know	W What You Want To Know	L What You Learned

Name ______________________ Date ______________

Calendar

Sunday	Monday	Tuesday	Wednesday	Thursday	Friday	Saturday

Name ______________________ Date ______________________

Inch Graph Paper

Apples

sung to the tune of **"I've Been Working on the Railroad"**

March Tempo

lyrics by Becky Manfredini and Jenny Reznick

In My Garden

sung to the tune of **"My Bonnie Lies Over the Ocean"**

Jazzy Feel

lyrics by Becky Manfredini and Jenny Reznick

F Gm7 Am7 C7
Oh, I plant some seeds in my gar - den.______ I

F G7 C7 C7
plant the seeds all in a row.______ I

FM7 Gm7 Am7 D7
wa - ter the seeds ev' - ry morn - ing.______ The

Gm7 C7 F
sun shines on them so they grow.______

F F7 B♭ G7 C C7 F
In my gar - den, I grow pink flow-ers and red ones too.

F F7 B♭ G7 C C7 F
In my gar - den, I grow some flow-ers for you!______

Take a Look

sung to the tune of **"Skip to My Lou"**

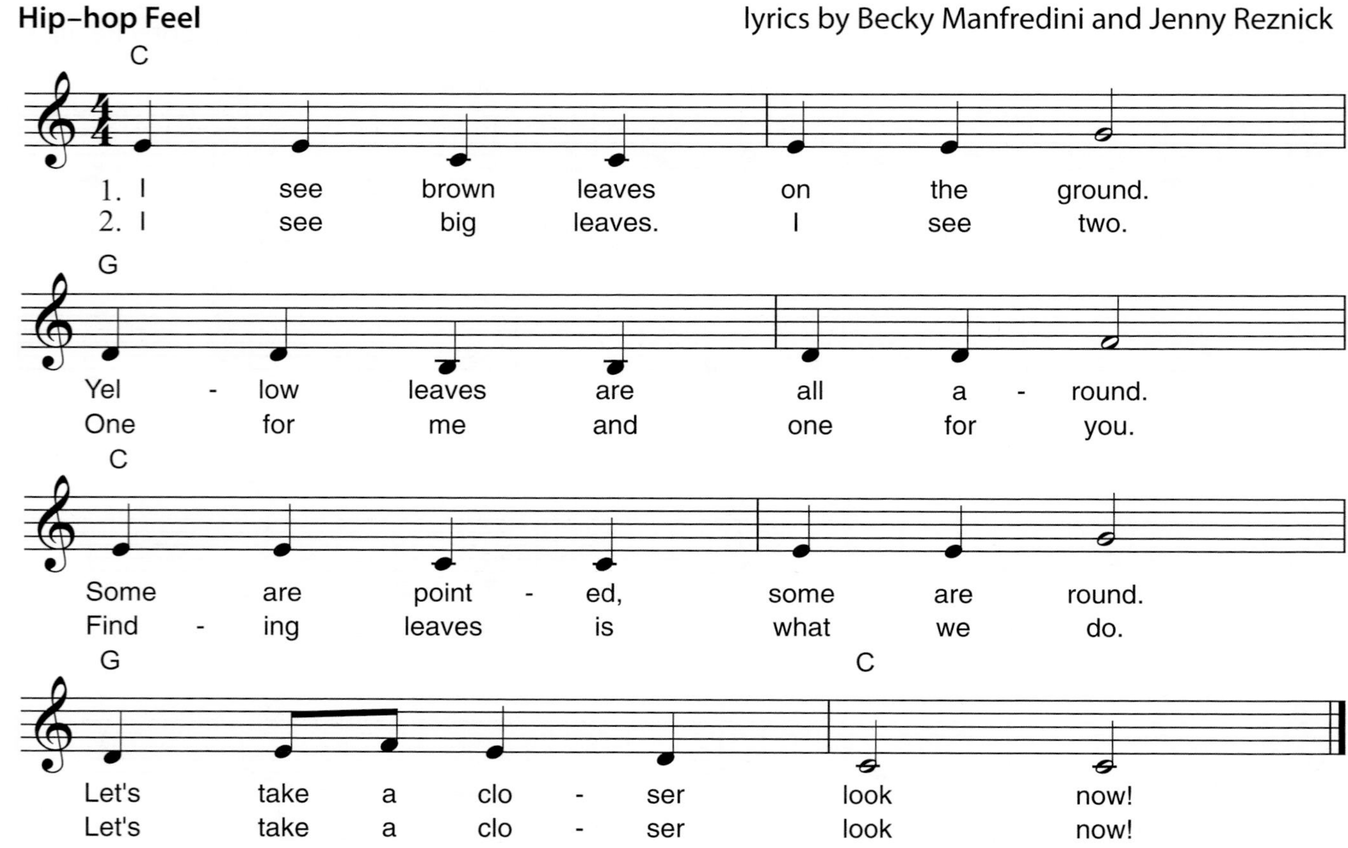

Animals Need

sung to the tune of **"Here We Go 'Round the Mulberry Bush"**

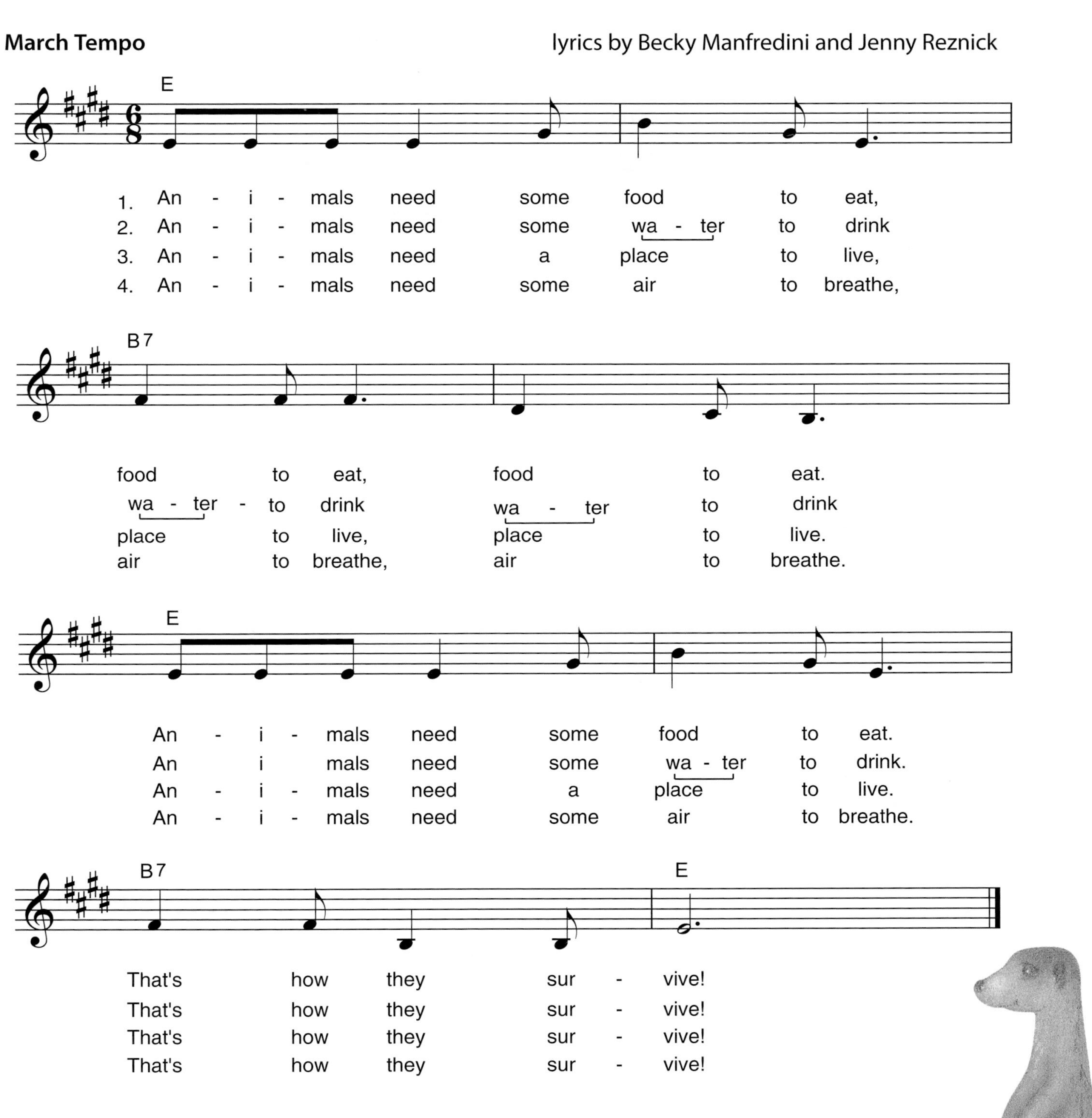

Animals in Motion!

an original song

March Tempo

lyrics by Becky Manfredini and Jenny Reznick
music by Harry Manfredini

Let's Look

sung to the tune of **"On Top of Old Smokey"**

Country Waltz Tempo

lyrics by Becky Manfredini and Jenny Reznick

B E

1. Let's look at the moun - tain. It's
2. Let's look at the large plain. This
3. Let's look at the hill now. This

E B

Earth's high - est land. I
land is so flat. Let's
land is not flat. A

B F♯

wish we could climb it. This
look at the corn grow. Let's
hill's not a moun - tain. It's

F♯7 B

moun - tain is grand!
eat some of that!
smal - ler than that!

Water All Around

an original song

Calypso Tempo

lyrics by Becky Manfredini and Jenny Reznick
music by Harry Manfredini

G F G
Wa - ter's all___ a - round us.___

G F G
Is there wa - ter right___ near you?

G F G
Find an o - cean, lake___ and ri - ver.

C D7 G *Fine*
Now lis - ten to___ each clue.___

C D
1. An o - cean has salt wa - - ter.___
2. A lake has land a - round it.___

C D
It's where you see ships sail.___
There's ve - ry lit - tle mo - tion___

C B7 Em
It's large and has big waves.___
A ri - ver flows a - cross the land. -

A7 D7 *2nd time: D. C. al Fine*
You might just see a whale!___
It some - times meets an o - cean!

What Do You See?

sung to the tune of **"Twinkle, Twinkle, Little Star"**

Moderate Tempo

lyrics by Becky Manfredini and Jenny Reznick

C F C
In my tel - e - scope I see

G C G C
stars that twin - kle down on me.

C F C G
They light up the sky at night.

C F C G
They are hot and ve - ry bright.

C F C
Stars make pat - terns in the sky.

G C G C
What do you see way up high?

an original song

lyrics by Becky Manfredini and Jenny Reznick
music by Harry Manfredini

A F♯m D E A F♯m
1. Tell me what is that I see. It looks just like a -
2. morn - ing it's long - er than me. But by noon, it's

D E A F♯m D E
no - ther me. Some - times it's short and then it grows. It
small you see. Then it grows much long - er through the day When the

Bm D A E A
seems to meet me at my toes.
sun goes down it goes a - way.

Refrain

D G♯7 C♯m F♯m
It's my sha - dow. This I know. It

Bm C♯m D B7 E
fol - lows me where e - ver I go. It

D G♯7 C♯m F♯m
jumps and runs and moves with me. It

1. Bm D Bm E7 A **Verse 2**
likes to keep me com - pa - ny.
2. In the

2. Bm D Bm E7 A
likes to keep me com - pa - ny.

Water, Steam, and Ice

sung to the tune of **"I'm a Little Teapot"**

Moderate Rock Tempo

lyrics by Becky Manfredini and Jenny Reznick

A D A

1. Watch Mom bo - il wa - ter for some tea.
2. Watch Mom make some ice cubes in the tray.

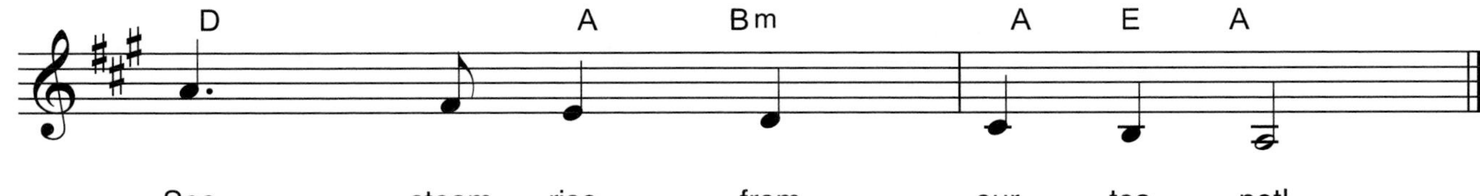

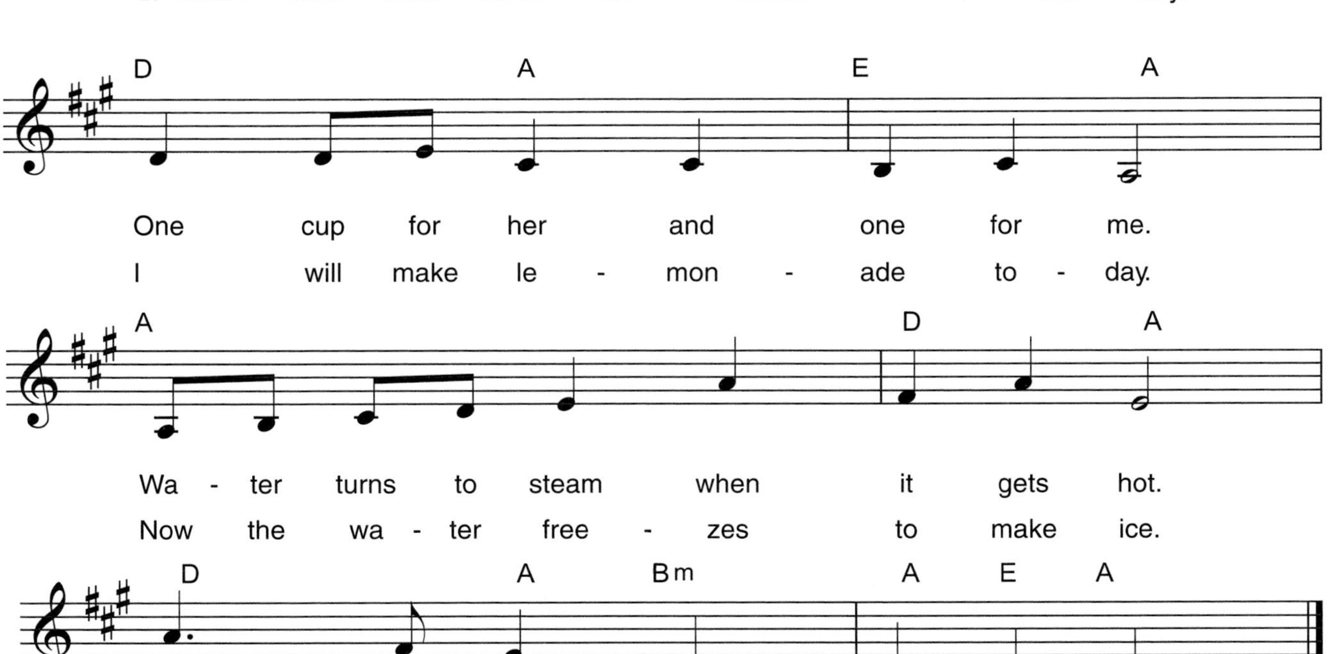
D A E A
One cup for her and one for me.
I will make le - mon - ade to - day.
A D A
Wa - ter turns to steam when it gets hot.
Now the wa - ter free - zes to make ice.
D A Bm A E A

See steam rise from our tea - pot!
Ice makes le - mon - ade taste nice!

Roll Along!

an original song

Here Comes the Band

an original song

Stately March Tempo

lyrics by Becky Manfredini and Jenny Reznick
music by Harry Manfredini

A
D
A
(bass drum solo)
Lis - ten to the bass drum keep the beat.
D
E7
A
bass drum solo
Boom, boom boom, down the street!
A
D
A
(tuba solo)
Lis - ten to the tu - ba play out loud.
D
E7
A
(tuba solo)
oom - pah oom - pah for the crowd!

Materials

Materials required to complete activities in the Teacher's Edition are listed below.

Consumable Materials

Based on 6 groups

MATERIALS	QUANTITY NEEDED PER GROUP	KIT QUANTITY	UNIT / LESSON
Aluminum foil			E-2
Apple peels, oatmeal or corn meal			B-Unit Project
Bag			D-1
Bag, resealable	1 or 2	18	A-3; F-4
Battery, D-cell	2	12	D-5
Bulletin-board paper			C-4
Celery stalk w/ leaves			A-1
Chart paper			A-2, 4; B-3, 4, 5; C-2, 4, 5, 6; D-2, 3, 5
Chenille stems	1	100	E-2
Construction paper, dark			B-Unit Project
Cracker, square			B-3
Craft sticks	1	30	E-4
Crayon			B-4; C-2, 4; D-4; F-2, 5
Cup, paper	1	25	A-3; E-4; F-1
Cup, plastic	1	50	B-4; C-3, 4, 5, 6; E-4
Drawing paper			B-4
Drink mix	1	1 packet	E-4
Dry erase markers			D-3, 4
Eggs, Uncooked			B-4
Food coloring, blue	1	1 bottle	A-1
Fresh flowers			A-4
Fruits and vegetables	5		A-5
Glue			C-4, 6; D-3; E-2
Grass and dead leaves			B-Unit Project
Honey			C-6
Ice cubes			E-4
Index cards	10		A-1
Leaves			A-4
Live Coupon - Earthworms (12)	1	1	B-Unit Project
Marker			A-1, 2, 4; B-3, 4, 5; C-2, 4, 5, 6; D-2, 3, 5; E-1; F-1, 3, 4
Masking tape			A-3; B-Unit Project; E-2; F-2
Milk cartons			A-3
Modeling clay, 4 color	1	2 boxes	C-3; E-4
Modeling clay, brown	1	4 boxes	C-3
Modeling clay, cream	1	4 boxes	C-3
Newspaper			A-3; C-1; D-Unit Project; E-1

Materials

Materials required to complete activities in the Teacher's Edition are listed below.

Non-Consumable Materials

Based on 6 groups

MATERIALS	QUANTITY NEEDED PER GROUP	KIT QUANTITY	UNIT / LESSON
Attribute blocks	1	1 set	B-3
Ball, rubber	1	6	F-2
Bean bag			F-3
Blender			E-1
Block, curvy			D-5
Book			F-2, 5
Bowl			C-5
Box, small			A-5
Cardboard			C-3; E-2; F-1, 5
Cardboard box			E-2
Cardboard tubes			F-4
Classroom objects			D-5; E-4; F-1, 3, 5
Clipboard			C-4; D-3
Cloth swatches			E-1
Connecting cubes	6	1 set	A-2, 3; B-4, 6, 8; C-3; D-5
Dropper	1	6	C-3, 4
Fabric			F-5
Flashlight	1	6	D-5
Freezer			E-4
Goggles	5		F-4
Hand lens	1	6	B-2, 3; C-1, 2
Jar, plastic	1	6	A-1; C-5
Magnet, bar	1	6	F-5
Magnet, disk	1	6	F-5
Magnetic strips	1	1 roll	F-5
Magnets, horseshoe	1	6	F-5
Measuring cup	1	6	C-5
Metal objects			E-2
Nuts, screws, and washers			E-2
Overhead			D-5
Paintbrushes			F-4
Pan, aluminum foil	1	6	E-1
Paper clips			E-2; F-5
Photo, bug			B-3
Photo sorting cards	1	1 packet	A-4; B-4; C-2, 4; D-1, 4; E-1; F-1
Plastic insects	1	6	B-3
Plastic lizards	1	6	B-4
Pulley	1	6	F-1

Materials required to complete activities in the Teacher's Edition are listed below.

Consumable Materials

Based on 6 groups

MATERIALS	QUANTITY NEEDED PER GROUP	KIT QUANTITY	UNIT/ LESSON
Paint			C-4, 6; D-3; F-4
Paper			C-Unit Project, 4, 6; D-3, 4, 5; E-1, 2, 3; F-1, 2, 4, 5
Paper, colored			F-5
Paper towel			A-3
Paper towel rolls			C-4
Pencil			A-1, 4, 5; B-2, 3, 4, 5, 6, 8; C-Unit Project, 2, 4, 5, 6; D-Unit Project, 1, 3, 5; E-3; F-1, 2, 3, 5
Plant			A-2
Plants with roots	8		A-2
Play dough			E-3
Potter's clay			E-3
Rubber bands, assorted	1	1 box	F-4
Salt	1	1	A-1; E-4
Sand	1	1 bag	C-3
Scrap paper			E-1
Seeds, lima beans	2	1 packet	A-3
Seeds, radish	4	1 packet	A-3
Soil, potting	1	1 bag	A-2, 3; B-Unit Project; C-1, 3
Spoons, plastic	1	24	C-6
Straws	1	50	B-4
String	1	1 roll	A-4; C-3; F-1
Tissue boxes, empty			F-4
Vegetables			A-5
Water			C-3, 4
Waxed paper			C-4
Wire, copper	1	1 roll	E-2
Yarn	1	1 roll	A-4, F-1
Yogurt			C-6

Non-Consumable Materials

Based on 6 groups

MATERIALS	QUANTITY NEEDED PER GROUP	KIT QUANTITY	UNIT/ LESSON
Recyclable items			C-6
Refrigerator magnets			F-5
Rocks			C-Unit Project, 2
Ruler			C-3; D-5
Sand table			C-3
Scissors			C-3, 6; D-5
Sorting trays			A-4
Spoon			D-5
Stapler			E-2
Stick			D-5
Stormy music			D-2
Strainer, 6.9 cm	1	3	C-1
Strainer, 6.9 cm (fine mesh)	1	3	C-1
Terrarium	1	1	B-Unit Project
Thermometer	1	6	D-Unit Project
Tools			E-2
Toy car	1	6	F-1, 2
Toy truck	1	6	F-1
Tub, plastic	1	3	E-4
Tunnels			F-1
Venn Diagram			D-3
Watering can	1	1	C-3
Window box or flower pots			A-2
Window screen	1	3	E-1
Wire cutter	1	1	E-2
Wood block	1	6	E-2; F-5
Wood scraps			E-2
Wooden objects			E-2

Bibliography

BE A SCIENTIST

Apples. Gibbons, Gail. Holiday House, 2000.

Apples. Robbins, Ken. Atheneum, 2002.

Apples, Apples, Apples. Wallace, Nancy Elizabeth. Winslow, 2000.

Cloudy Day, Sunny Day. Crews, Donald. Harcourt Brace, 1999.

How Tall, How Short, How Faraway. Adler, David A. Holiday House, 1999.

Nature's Paintbrush: The Patterns and Colors Around You. Stockdale, Susan. Simon & Schuster, 1999.

Out and About. Hughes, Shirley. Lothrop Lee & Shephard, 1988.

Picking Apples. Hutchings, Amy and Richard. Scholastic, 1994.

The Seasons of Arnold's Apple Tree. Gibbons, Gail. Harcourt, Brace, 1984.

Teaching Children About Life and Earth Sciences: Ideas and Activities Every Teacher and Parent Can Use. Levenson, Elaine. Tab Books, 1994.

Teaching Children About Physical Science: Ideas and Activities Every Teacher and Parent Can Use. Levenson, Elaine. Tab Books, 1994.

Worms, Shadows, and Whirlpools: Science in the Early Childhood Classroom. Grollman, Sharon, and Karen Worth. Heinemann, 2003.

UNIT A: PLANTS

Carrot Seed. Krauss, Ruth. HarperCollins, 2005.

Compost! Growing Gardens from Your Garbage. Glaser, Linda. Millbrook, 1996.

Dandelions: Stars in the Grass. Posada, Mia. Carolrhoda Books, 2000.

Fruits and Vegetables/Frutas y Vegetales. Rosa-Mendoza, Gladys. Me+Mi Publishing, 2002.

A Fruit Is a Suitcase for Seeds. Richards, Jean. Millbrook, 2002.

Growing Vegetable Soup. Ehlert, Lois. Harcourt, 1990.

A Handful of Sunshine: Growing a Sunflower. Eclasre, Melanie. Ragged Bears, 2001.

Leaves! Leaves! Leaves!. Wallace, Nancy Elizabeth. Marshall Cavendish, 2003.

Market Day. Ehlert, Lois. Harcourt, 2002.

National Audubon Society First Field Guide: Trees. Cassie, Brian. Scholastic, 1999.

Oh Say Can You Seed. Worth, Bonnie. Random House, 2001.

The Pumpkin Book. Gibbons, Gail. Holiday House, 1999.

Pumpkin Circle: The Story of a Garden. Levenson, George. Tricycle Press, 1999.

Red Leaf, Yellow Leaf. Ehlert, Lois. Harcourt, 1991.

Tell Me, Tree: All About Trees for Kids. Gibbons, Gail. Little Brown & Company, 2002.

Ten Seeds. Brown, Ruth. Knopf, 2001.

A Tree Is a Plant. Bulla, Clyde Robert. HarperCollins, 2001.

Zinnia's Flower Garden. Wellington, Monica. Dutton Books, 2005.

UNIT B: ANIMALS

Birds Build Nests. Winer, Yvonne. Charlesbridge, 2002.

Bugs Are Insects. Rockwell, Anne. HarperCollins, 2001.

Butterflies in the Garden. Lerner, Carol. HarperCollins, 2002.

Butterfly: Watch Me Grow. Magloff, Lisa. DK Publishing, 2004.

Central Park Serenade. Root, Barry. HarperCollins, 2002.

Dig, Wait, Listen: A Desert Toad's Tale. Sayre, April. Greenwillow, 2001.

Do You Know the Difference? Bischhoff-Miersch, Andrea and Michael. North-South, 1995.

Duck on a Bike. Shannon, David. Blue Sky Press, 2002.

The Emperor's Egg. Jenkins, Martin. Candlewick, 1999.

Farm Animals. Schlepp, Tammy. Copper Beech Books, 2000.

Five Creatures. Jenkins, Emily. Farrar, Straus & Giroux, 2001.

Giant Pandas. Gibbons, Gail. Holiday House, 2002.

Hello, Fish! Visiting the Coral Reef. Earle, Sylvia. National Geographic, 1999.

Honeybees. Heligman, Deborah. National Geographic Society, 2002.

How Animal Babies Stay Safe. Fraser, Mary Ann. HarperCollins, 2001.

Insects. Bernard, Robin. National Geographic Society, 2001.

I See Animals Hiding. Arnosky, Jim. Scholastic, 2000.

Jellies: The Life of Jellyfish. George, Twig. Millbrook, 2000.

Ladybugs: Red, Fiery, and Bright. Posada, Mia. Carolrhoda, 2002.

Leaving Home. Collard, Sneed. Houghton Mifflin, 2002.

Make Way for Ducklings. McCloskey, Robert. Viking Press, 2001.

A Mother's Journey. Markle, Sandra. Charlesbridge Publishing, 2005.

Over in the Ocean: In a Coral Reef. Berkes, Marianne. Dawn Publications, 1999.

The Mountain That Loved a Bird. McLerran, Alice. Simon & Schuster, 2000.

Pond. Morrison, Gordon. Houghton Mifflin, 2002.

Red Eyes or Blue Feathers: A Book About Animal Colors. Stockland, Patricia M. Picture Window Books, 2005.

Slinky Scaly Slithery Snakes. Patent, Dorothy. Walker, 2000.

Water Hole Waiting. Kurtz, Christopher and Jane. Greenwillow, 2002.

What Do You Do When Something Wants to Eat You? Jenkins, Steven. Houghton Mifflin, 2001.

What Do You Do with a Tail Like This? Page, Robin. Houghton Mifflin, 2003.

The Year at Maple Hill Farm. Provensen, Alice and Martin. Simon & Schuster, 2001.

The Yucky Reptile Alphabet Book. Pallotta, Jerry. Library Binding, 1999.

Bibliography

Unit C: OUR EARTH, OUR HOME

The Amazing Pop-Up Geography Book. Petty, Kate. Dutton Juvenile, 2000.

Danger! Earthquakes. Simon, Seymour. SeaStar, 2002.

Dirt: Jump into Science. Tomecek, Steve. National Geographic, 2002.

Down to Earth. Stewart, Melissa. Compass Point Books, 2004.

The Floating House. Sanders, Scott Russell. Aladdin, 1999.

Follow the Water from Brook to Ocean. Dorros, Arthur. HarperTrophy, 1993.

A Handful of Dirt. Bial, Raymond. Walker & Co., 2001.

Hello, Ocean. Ryan, Pam Munoz. Talewinds, 2001.

How Mountains Are Made. Zoehfeld, Katherine Weidner. HarperCollins Children's Books, 1995.

If You Find a Rock. Christian, Peggy. Harcourt, 2000.

Let's Go Rock Collecting. Gans, Roma. HarperCollins Juvenile Books, 1997.

The Magic Schoolbus: At the Waterworks. Cole, Joanna. Scholastic, 1988.

Recycle: A Handbook for Kids. Gibbons, Gail. 1992.

Rocks: Hard, Soft, Smooth, and Rough. Rosinsky, Natalie. Picture Window Books, 2003. Little Brown & Company, 1996.

Sand. Prager, Ellen J. National Geographic, 2002.

What's Up, What's Down? Schaefer, Lola M. Greenwillow, 2002.

Where Does the Garbage Go? Showers, Paul. HarperTrophy, 1994.

Why Should I Recycle? Green, Jen. Barron's Educational Series, 2005.

Why Should I Save Water? Green, Jen. Barron's Educational Series, 2005.

Unit D: WEATHER AND SKY

Big Bear Ball. Ryder, Joanne. HarperCollins, 2002.

Clouds. Dane Bauer, Marion. Aladdin, 2004.

Common Ground: The Water, Earth, and Air We Share. Bang, Molly. Blue Sky, 1997.

Day Light, Night Light: Where Light Comes From. Branley, Franklyn M. HarperCollins, 1998.

Here Comes the Year. Narahashi, Keiko, and Spinelli, Eileen. Henry Holt, 2002.

It Looked Like Spilt Milk. Shaw, C. G. HarperTropy, 1988.

Light: Shadows, Mirrors, and Rainbows. Rosinsky, Natalie M. Picture Window Books, 2002.

Like a Windy Day. Asch, Devin, and Frank. Harcourt, 2002.

Moon Glowing. Partridge, Elizabeth. Dutton, 2002.

Pieces: A Year in Poems and Quilts. Hines, Anna Grossnickle. Greenwillow, 2001.

Rain Romp: Stomping Away a Grouchy Day. Kurtz, Jane. Greenwillow, 2002.

Shadows. Otto, Carolyn B. Scholastic, 2001.

Snow Day! Lakin, Patricia. Dial, 2002.

Some from the Moon, Some from the Sun: Poems and Songs for Everyone. Zemach, Margot. Farrar, Straus, Giroux, 2001.

Spring Thaw. Schnur, Steven. Viking, 2000.

The Sun: Our Nearest Star. Branley, Franklyn M. HarperCollins, 2002.

Super Storms. Simon, Seymour. SeaStar, 2002.

This Is the Rain. Schaefer, Lola M. Greenwillow, 2001.

Too Many Rabbits and Other Fingerplays. Cooper, Kay. Scholastic, 1995.

Tornadoes. Simon, Seymour. HarperCollins, 1999.

Twister on Tuesday (Magic Tree House #23). Osbourne, Mary Pope. Random House Children's Books, 2001.

The Water, Earth, and Air We Share. Bang, Molly. Blue Sky, 1997.

UNIT E: EXPLORING MATTER

Clay. Firestone, Mary. Capstone Press, 2005.

Clay (Let's Create Series). Gibbs, Dorothy L. Gareth Stevens Publishing, 2004.

Creating with Fimo Acrylic Clay. Nicholson, Libby. Kids Can Press, 1996.

Down Comes the Rain. Branley, Franklyn. HarperCollins, 1997.

How Is Paper Made? Asimov, Isaac, and Elizabeth Kaplan. Gareth Stevens, 1993.

Joseph Had a Little Overcoat. Taback, Simms. Viking, 1999.

The Jumbo Book of Paper Crafts. Lewis, Amanda. Kids Can Press, 2002.

Kid's Clothes: From Start to Finish. Woods, Samuel G. Blackbirch Press, 2001.

Let's Try It Out in the Water: Hands-On Early Learning Science Activities. Fauteux, Nicole, and Seymour, Simon. Simon & Schuster, 2001.

Making Things Float and Sink. Gibson, Gary. Copper Beech Books, 1995.

Metal. Llewellyn, Claire. Sea to Sea, 2006.

Mud Is Cake. Ryan, Pam Munoz. Hyperion, 2002.

Paper. Kras, Sara Louise. Capstone Press, 2003.

Pop! A Book About Bubbles. Bradley, Kimberly. HarperCollins, 2001.

The Pot That Juan Built. Andrews-Goebel, Nancy. Lee & Low Books, 2002.

Rising Up, Falling Down. Hammersmith, Craig. Compass Point Books, 2002.

Screws. Feldmann, Roseann, and Walker, Sally M. Lerner, 2002.

Snowflake Bentley. Martin, Jacqueline Briggs. Houghton Mifflin, 1998.

The Story of Paper. Compestine, Ying Chang. Holiday House, 2003.

Tools. Moris, Ann. HarperCollins, 1998.

Turtle and Snake Fix It. Spohn, Kate. Scholastic, 2001.

What Is the World Made Of? All About Solid, Liquids, and Gases. Zoehfeld, Kathleen Weidner. HarperTrophy, 1998.

Wood. Llewellyn, Clair. Sea to Sea, 2006.

Workshop. Clements, Andrew. Clarion Books, 1999.

UNIT F: MOVING RIGHT ALONG

Cinnamon's Day Out. Roth, Susan L. Dial Books, 1998.

Construction Trucks. Dussling, Jennifer. Grosset & Dunlap, 1998.

Dig Dig Digging. Mayo, Margaret. Henry Holt, 2002.

I Fall Down. Cobb, Vicki. HarperCollins, 2004.

In the Spin of Things. Dotlich, Rebecca Kai. Boyds Mill Press, 2003.

Inclined Planes and Wedges. Feldmann, Roseann, and Walker, Sally M. Lerner, 2002.

Kite Flying. Lin, Grace. Dragonfly Books, 2004.

Levers. Feldmann, Roseann, and Walker, Sally M. Lerner Publications, 2002.

The Listening Walk. Showers, Paul. HarperCollins, 1993.

The Little Engine That Could. Piper, Watty. Grosset and Dunlap, 1978.

Mama Zooms. Cowen-Fletcher, Jane. Scholastic, 1996.

Mickey's Magnet. Branley, Franklyn M., and Vaughan, Eleanor K. Scholastic, 1999.

Mike Mulligan and His Steam Shovel. Burton, Virginia Lee. Houghton Mifflin, 1977.

Night in the Country. Ryland, Cynthia. Simon and Schuster, 1991.

Polar Bear, Polar Bear, What Do You Hear? Martin Jr., Bill. Henry Holt & Company Inc., 1997.

Science All Around Me: Magnets. Bryant-Mole, Karen. Heinemann, 1998.

The Science of Gravity. Stringer, John. Raintree, 2000.

Screws. Feldmann, Roseann, and Walker, Sally M. Lerner Publications, 2002.

Snow Music. Perkins, Lynne Rae. HarperCollins Children's Books, 2003.

The Way Things Move. Dwordin, Heidi Gold, and Ullman, Robert K. McGraw-Hill, 2001.

What Does a Wheel Do? Pipe, Jim. Copper Beech Books, 2002.

What Makes a Magnet? Branley, Franklyn M. HarperCollins, 1996.

Wheels and Axles. Feldmann, Roseann, and Walker, Sally M. Lerner, 2002.

The Wheels on the Bus. Moore, Mary-Alice. HarperCollins Children's Books, 2006.

Scope and Sequence

Inquiry, Technology, History and Nature of Science

Inquiry

Inquiry Skills	K	1	2	3	4	5	6
Observe	I	D	D	D	D	D	D
Infer	I	D	D	D	D	D	D
Predict	I	D	D	D	D	D	D
Communicate	I	D	D	D	D	D	D
Measure	I	D	D	D	D	D	D
Put things in order	I	D	D				
Compare	I	D	D				
Classify	I	D	D	D	D	D	D
Investigate		I	I	D	D	D	D
Make models		D	D	D	D	D	D
Draw conclusions	I	D	D				
Use numbers				D	D	D	D
Interpret data				D	D	D	D
Form a hypothesis				D	D	D	D
Experiment				D	D	D	D

Special Abilities for Inquiry	K	1	2	3	4	5	6
Critical thinking	I	I	I	D	D	D	D
Use of mathematics	I	I	I	D	D	D	D
Use of technology	I	I	I	D	D	D	D

Understandings About Inquiry	K	1	2	3	4	5	6
Different questions suggest different kinds of investigations.	I	I	I	D	D	D	D
Current scientific knowledge guides investigations.				D	D	D	D
Investigations may lead to new ideas.				D	D	D	D
Explanations are based on evidence.	I	I	I	D	D	D	D

Science and Technology

Abilities of Technological Design	K	1	2	3	4	5	6
Identify appropriate problems.	I	I	D	D	D	D	D
Design a solution.	I	I	D	D	D	D	D
Implement the solution.	I	I	D	D	D	D	D
Evaluate/communicate the solution.	I	I	D	D	D	D	D

KEY		
I = Introduced	**D** = Developed	**R** = Reinforced

Science and Technology *cont.*

Understandings about Science and Technology	K	1	2	3	4	5	6
Contributions made by people in different cultures	D	I	I	D	D	D	D
Trade-offs: benefits and risks, consequences	I	I	I	D	D	D	D
Reciprocity of science and technology			I	D	D	D	D
Distinguishing natural from human-made objects	I	D	D	R			

Personal and Social Perspectives

Personal Health	K	1	2	3	4	5	6
Practice of healthful behaviors	I	I	I	D	D	D	D
Injury preventions	I	I	I	D	D	D	D

Populations, Resources, Environments	K	1	2	3	4	5	6
Effect of population density on resources and environment	I	I	I	D	D	D	D
Limited supply of many resources	I	I	D	D	D	D	D
Natural and human causes of environmental change	I	I	D	D	D	D	R
Rapid and gradual environmental change	I	I	D	D	R	R	R

Natural Hazards	K	1	2	3	4	5	6
Resulting from Earth processes	I	I	D	D	D	D	D
Induced by human activities			I	D	D	D	D
Challenges to society			I	D	D	D	D
Risks of natural hazards	I	I	D	D	D	D	D

Science, Technology, and Society	K	1	2	3	4	5	6
Effects of science and technology on society	I	I	I	D	D	D	D
Effect of society on science investigation					I	D	D
Many settings for carrying out science and technology					I	D	D
Inability of science to answer all questions					I	I	D

History and Nature of Science

Science as Human Endeavor	K	1	2	3	4	5	6
Carried out by men and women of various backgrounds		I	I	D	D	D	D
The need for different abilities		I	I	I	D	D	D
Varied careers in science	I	I	I	D	D	D	D

Nature of Science	K	1	2	3	4	5	6
Explanations based on observation and experiments	I	I	D	D	D	D	D
Conflicting scientific interpretations					I	D	D
Evaluation of scientific ideas				I	D	D	D

History of Science	K	1	2	3	4	5	6
The change of accepted ideas because of scientific explanations					I	D	D
Contribution of individuals throughout history				I	D	D	D

Scope and Sequence

Life Science

Plants

Needs	K	1	2	3	4	5	6
Identified	I	I	I	I	D	D	R
Related to habitat	I	D	D	D	R	R	R
Related to structures	I	I	I	I	D	D	R

Photosynthesis	K	1	2	3	4	5	6
Raw materials/products		I	I	I	D	D	R
Chemical formula					I	D	D
Producers							

Structures and Functions	K	1	2	3	4	5	6
Roots, stems, leaves	I	I	I	I	D	D	R
Flowering plants	I	I	I	D	D	D	R
▸ Parts of a flower	I	I	I	I	D	D	R
▸ Monocots/dicots					I	D	D
Nonflowering plants				I	D	D	R
▸ Mosses (parts)					D	D	R
▸ Ferns (parts)				I	D	D	R
▸ Gymnosperms (parts)				I	D	D	R

Reproduction (Life Cycles)	K	1	2	3	4	5	6
Flowering plants	I	I	I	I	D	D	R
Gymnosperms				I	I	D	R
Mosses/ferns				I	I	D	R
Bulbs/cuttings/tubers (asexual means)		I		I	D	R	

Regulation and Responses	K	1	2	3	4	5	6
Tropisms				I	I	D	D
Other responses to environment	I	I	I	D	D	R	R

Animals

Needs	K	1	2	3	4	5	6
Identified	I	I	I	I	D	D	R
Related to habitat	I	D	D	D	R	R	R
Related to structures	I	I	I	I	D	D	R

Structures and Functions	K	1	2	3	4	5	6
External characteristics	I	I	I	I	D	D	R
Organ systems					D	D	D

KEY		
I = Introduced	D = Developed	R = Reinforced

Animals *cont.*

Life Cycles	K	1	2	3	4	5	6
Basic examples (e.g., birds, mammals)	I	I	I	I	D	D	R
Complete metamorphosis (insect, amphibian)	I	I	I	D	D	D	R
Incomplete metamorphosis				I	D	D	R
Asexual reproduction (regeneration)					I	D	R

Regulation and Behavior	K	1	2	3	4	5	6
Response to environment	I	I	D	D	D	R	R
Internal stimuli				I	D	R	R

Six Kingdom Classification

Binomial Classification	K	1	2	3	4	5	6
6 kingdoms defined					I	D	R
Kingdom to species delineations					I	D	R
Species names					I	D	R

Plant Kingdom	K	1	2	3	4	5	6
Divisions identified					I	D	R
Nonvascular vs. vascular					D	D	R
Seedless vs. seed				I	D	D	R
▶ Nonflowering vs. flowering				I	D	D	R
▶ Monocots/dicots					I	D	R

Animal Kingdom	K	1	2	3	4	5	6
Basic groupings (mammals, insects, reptiles, birds)	I	I	I	I	D	D	R
Invertebrates				I	D	D	R
▶ Basic phyla				I	D	D	R
Vertebrates				I	D	D	R
▶ Classes				I	D	D	R

Microorganisms	K	1	2	3	4	5	6
Bacteria				I	I	D	D
▶ Two kingdoms					I	D	D
Protist Kingdom				I	I	D	D
Fungus Kingdom				I	I	D	D
Reproduction					I	D	D
▶ Asexual vs. sexual						D	D

Scope and Sequence

Cells (Cell Theory of Life)

Cell Structure	K	1	2	3	4	5	6
Cell as unit of life				I	I	D	D
Plant cells vs. animal cells					I	D	R
Multicellular vs. single-celled organism					D	D	R
Levels of organization (cell-tissue-organ-system)					D	D	R

Cell Activities	K	1	2	3	4	5	6
Photosynthesis/respiration		I	I	D	D	D	D
Cell division						D	D
▸ Mitosis/meiosis						I	D
Diffusion/osmosis							D

Heredity and Genetics

Concepts of Heredity	K	1	2	3	4	5	6
Inherited traits	I	I	I	D	D	D	D
▸ In animals	I	I	I	D	D	D	D
▸ In plants	I	I	I	D	D	D	D
Learned behavior/acquired traits	I	I	I	D	D	D	D
Variation	I	I	I	I	D	D	D
Effect of environment	I	I	I	D	D	D	D
Natural selection							D

Mechanisms of Heredity	K	1	2	3	4	5	6
Mendel's patterns						I	D
Chromosomes						I	D
Genes						I	D
DNA							D

Genetics	K	1	2	3	4	5	6
Pedigree						I	D
Sex-linked traits							D
Genetic engineering						I	D

KEY		
I = Introduced	**D** = Developed	**R** = Reinforced

Ecology/Ecosystems

Organization of Ecosystems	K	1	2	3	4	5	6
Habitats	I	I	I	D	D	R	R
Populations and communities			I	D	D	R	R
Niches				I	D	D	R
Land vs. water	I	I	D	D	D	D	R
Biomes					I	D	D
▶ Latitude					I	D	D
▶ Comparisons					I	D	D
▶ Adaptations					D	D	R

Dynamics of Ecosystems	K	1	2	3	4	5	6
Energy transfer		I	I	D	D	D	D
▶ Food chains		I	I	D	D	D	D
▶ Food webs			I	I	D	D	D
▶ Energy pyramids					I	D	D
Cycles of matter				I	I	D	R
▶ Carbon				I	I	D	R
▶ Nitrogen						D	R
Competition				I	D	D	D
Symbiotic relationships					I	D	D

Adaptation and Survival	K	1	2	3	4	5	6
Adaptations	I	I	I	D	D	D	D
▶ Animal behaviors	I	I	I	D	D	D	D
▶ Animal structures	I	I	I	D	D	D	D
▶ Plant responses				I	D	D	R
▶ Plant structures		I	I	D	D	D	R
Changes in ecosystems			I	D	D	D	D
▶ Destructive change			I	D	D	D	D
▶ Effect on populations			I	D	D	D	D
▶ Regrowth/succession						D	D
Extinction			I	D	D	D	D
▶ Endangered species			I	D	D	D	D
▶ Extinct species			I	D	D	D	D
▶ Comparison of present and past species			I	D	R	R	D

Scope & Sequence

Scope and Sequence

Earth and Space Science

Earth Processes and Materials

Earth Structure	K	1	2	3	4	5	6
Landforms	I	I	I	D	D	R	R
▸ Comparisons	I	I	I	D	D	R	R
▸ Classified by Earth processes				I	I	D	D
Features of ocean bottom				I	I	D	D
▸ Spreading zone				I	I	D	D
▸ Subduction						I	D
Crust, mantle, core			I	I	D	D	R
▸ Lithosphere					I	I	D
Hydrosphere	I	I	I	D	D	D	D
▸ Sources of fresh water	I	I	I	D	D	D	D
Glaciers				I	D	D	D

Processes of Change	K	1	2	3	4	5	6
Weathering		I	I	D	D	D	D
▸ Causes		I	I	D	D	D	D
▸ Chemical vs. physical					I	D	D
▸ Soil as product			I	I	I	D	D
Erosion/deposition		I	I	D	D	D	D
▸ Basic agents		I	I	D	D	D	D
▸ River systems					I	D	D
▸ Glacier systems				I	D	R	R
Plate tectonics					I	D	D
▸ Evidence						D	D
▸ Plate model						D	D
▸ Earthquakes				I	I	D	D
▸ Volcanoes				I	I	D	D
▸ Mountain building					I	D	D

Earth History	K	1	2	3	4	5	6
Fossil evidence			I	D	D	D	D
Relative dating					I	D	D
Absolute dating						I	D
Eras of time						I	D
Change of environments over time			I	D	D	D	D

Minerals	K	1	2	3	4	5	6
Examples			I	I	I	D	D
Comparison/observed properties			I	I	I	D	D
Identification via key					I	D	D

KEY		
I = Introduced	D = Developed	R = Reinforced

Earth Processes and Materials *cont.*

Rocks	K	1	2	3	4	5	6
Examples	I	I	I	D	D	R	R
Comparison/observed properties	I	I	I	D	D	R	R
Classification (igneous, sedimentary, metamorphic)				I	D	D	D
Rock cycle					I	D	D

Soil	K	1	2	3	4	5	6
Formation			I	I	D	D	D
Components	I	I	I	D	D	R	R
Horizons				I	D	D	R
Properties	I	I	I	D	D	R	R
▶ Water capacity		I	I	I	D	D	R
▶ Permeability			I		I	D	R
Pollution/conservation				I	D	D	D

Renewable and Nonrenewable	K	1	2	3	4	5	6
Air		I	I	D	D	D	D
▶ Pollution/conservation		I	I	D	D	D	D
▶ Natural recycling				I	D	D	R
Water	I	I	I	D	D	R	R
▶ Pollution/conservation	I	I	I	D	D	D	D
▶ Water cycle		I	I	D	D	R	R
Fossil fuels				I	D	D	R
▶ Formation				I	D	D	D
▶ Alternative energy sources				I	D	D	D

Weather and Climate

The Water Cycle	K	1	2	3	4	5	6
Processes		I	I	D	D	R	R
▶ Evaporation/condensation/precipitation		I	I	D	D	R	R
▶ Runoff/infiltration/transpiration				I	D	D	R
Clouds	I	I	D	D	D	D	R
▶ Formation	I	I	D	D	D	D	R
▶ Classification	I	I	D	D	D	D	R
▶ Shapes	I	I	D	D	R	R	R
▶ Cloud cover (sky conditions)			I	D	D	R	R
Precipitation	I	I	I	I	D	D	D
▶ Forms	I	I	I	I	D	D	D
▶ Formation				I	D	D	D
▶ Relation to clouds			I	I	D	D	R

Weather and Climate *cont.*

Weather Variables & Measurement	K	1	2	3	4	5	6
Air temperature	I	I	I	D	D	R	R
▸ Conceptual model			I	I	D	D	R
▸ Thermometer—Celsius/Fahrenheit	I	I	D	D	D	R	R
Air pressure				I	D	D	D
▸ Conceptual model					I	D	D
▸ Barometer				I	D	D	R
Wind	I	I	I	D	D	R	R
▸ Relation to air pressure				I	I	D	D
▸ Direction—vanes	I	I	D	D	D	R	R
▸ Speed—anemometer		I	I	D	D	R	R
▸ Global patterns					I	D	D
▸ Coriolis effect						D	R

Weather Prediction	K	1	2	3	4	5	6
Air masses/fronts					I	D	D
Station models					I	D	R
Weather maps				I	D	D	D
Barometer readings, predictions					I	D	R
Cloud types, trends			I		D	R	R
Severe weather	I	I	D	D	D	D	R
▸ Thunderstorms	I	I	D	D	D	D	D
▸ Hurricanes	I	I	D	D	D	D	D
▸ Tornadoes			I	I	D	D	D
▸ Hailstorms			I	I	D	D	D

Climate	K	1	2	3	4	5	6
Components of				I	D	D	D
Comparison/classification				I	D	D	D
Seasons	I	D	D	R	R	R	R
Factors of				I	D	D	D
▸ Latitude				I	D	D	D
▸ Altitude				I	D	D	D
▸ Coastal/inland				I	D	D	D
▸ Mountain ranges				I	D	D	D
▸ Ocean currents					I	D	R
▸ Global winds					I	D	R

KEY		
I = Introduced	**D** = Developed	**R** = Reinforced

Astronomy

Earth-Sun Moon System	K	1	2	3	4	5	6
Day and night	I	I	D	D	D	R	R
▶ Daily observations	I	I	D	D	D	R	R
▶ Rotational model		I	I	D	D	D	R
Year/seasons	I	I	D	D	D	R	R
▶ Orbital model		I	I	D	D	R	R
▶ Solar radiation variations				I	D	R	R
Earth's Moon	I	I	I	D	D	D	R
▶ Basic observations/phases	I	I	I	D	D	D	R
▶ Earth-Sun-Moon model			I	I	D	D	R
▶ Eclipses					I	D	D
▶ Tides						D	D

Solar System	K	1	2	3	4	5	6
Planets		I	I	I	D	D	D
▶ Inner vs. outer			I		I	D	D
▶ Asteroids					I	D	D
Effects of gravity/inertia						D	D
▶ Shape of orbital path					I	D	R
▶ Speed of planets						D	R
Comets					I	D	R
▶ Explanation					I	D	R
▶ Historical sightings						D	R
Meteors/meteorites					I	D	D
▶ Explanation					I	D	D
▶ Impact on Earth					I	D	D
Formation of solar system						D	D

Stars/Universe	K	1	2	3	4	5	6
Star properties/comparisons/constellations	I	I	I	D	D	D	D
▶ Life cycle						D	D
▶ H-R diagram						I	D
Galaxies						I	D
▶ Examples/classification						I	D
▶ Motion/red- vs. blueshift							D
Universe						D	D
▶ Structure						I	D
▶ Change						I	D
▶ Formation						I	D

Physical Science

Matter

Properties of Objects/Materials	K	1	2	3	4	5	6
Observations	I	I	D	D	D	R	R
Classification (physical vs. chemical)					I	D	D
Measurements	I	I	I	D	D	R	R
▶ Length (ruler)	I	I	I	D	D	R	R
▶ Mass (balance)	I	I	I	D	D	R	R
▶ Volume/capacity (containers)	I	I	I	D	D	R	R
▶ Temperature (thermometer)	I	I	I	D	D	R	R

Properties of Matter	K	1	2	3	4	5	6
Density					I	D	R
Conductivity				I	D	D	R
Magnetism	I	D	D	R	R	R	R
▶ Poles	I	D	D	R	R	R	R
▶ Fields			I	I	D	D	R
▶ Interactions	I	D	D	R	R	R	R

States of Matter	K	1	2	3	4	5	6
Descriptions/properties		I	I	D	D	R	R
Particulate models					I	D	D
Melting/freezing point				I	I	D	R
Boiling/condensing point					I	D	R

Classification of Matter	K	1	2	3	4	5	6
Elements				I	I	D	D
▶ Metals				I	I	D	D
▶ Nonmetals					I	D	D
▶ Metalloids						I	D
▶ Periodic table					I	D	R
Mixtures		I	I	I	D	D	R
▶ Solid/liquid/gas combinations		I	I	I	D	D	R
▶ Suspensions			I	I	D	D	R
▶ Solutions					I	D	R
Compounds				I	D	D	R
▶ Compared with mixtures				I	D	D	R
▶ Names for						D	R
▶ Carbon compounds compared							D

KEY		
I = Introduced	D = Developed	R = Reinforced

Matter *cont.*

Changes of Matter	K	1	2	3	4	5	6
Physical changes	I	I	D	D	D	R	R
▶ Tearing/cutting/molding	I	I	D	D	D	R	R
▶ Change of state	I	I	D	D	D	R	R
▶ Particle/heat model					I	D	R
▶ Temperature/time graph						D	R
▶ Formation of mixtures		I	I	I	D	D	R
▶ Separation of mixtures			I	I	D	D	R
▶ Solubility (means of increasing)						D	D
Chemical Changes			I	I	D	D	D
▶ Chemical reactions					I	D	D
▶ Signs of chemical change					I	D	D
▶ Formation of compounds				I	D	D	D
▶ Acid/base/salts					I	D	D
Nuclear changes							D
▶ Radioactivity							D
▶ Fission vs. fusion							D

Atomic Model of Matter	K	1	2	3	4	5	6
Atoms					I	D	D
▶ Parts of atoms						D	D
▶ Models of atoms of elements						D	D
▶ Model through history						D	D
Molecules						D	D
▶ Bonding within						I	D
▶ Models of compounds						D	D
▶ Models of diatomic elements						D	D

Scope and Sequence

Position, Motion, Forces

Position	K	1	2	3	4	5	6
Position words (*above, below*, etc.)	I	D	D	R			
Relative position, fixed/moving				I	D	D	R

Motion	K	1	2	3	4	5	6
Distance			I	D	R	R	R
Speed	I	I	I	I	D	D	R
▸ Comparing fast and slow	I	I	D	R			
▸ Using mathematical formula						I	D
▸ Compared with velocity					I	D	R
Velocity/acceleration					I	D	R

Forces	K	1	2	3	4	5	6
Pushes and pulls	I	I	D	D	R	R	R
Gravity			I	I	D	D	D
▸ Weight change with distance				I	D	D	D
▸ Universal gravitation						D	D
▸ Weight on other planets					I	D	D
Friction		I	I	D	D	D	R
▸ Reduction of, lubrication			D	D	D	R	R
▸ Source of heat		I	I	D	D	R	R
▸ Kinds (rolling, standing, sliding)						I	D
Magnetism	I	I	D	R	R	R	R
▸ Poles	I	I	D	R	R	R	R
▸ ▸ Attraction vs. repulsion	I	I	D	R	R	R	R
▸ ▸ Magnetic vs. geographic					I	D	R
▸ Magnetic fields			I	I	D	D	R
▸ Magnetic domains						D	R
Effect on position and motion		I	I	D	D	D	D
▸ Changes in speed and direction (acceleration)		I	I	D	D	D	D
▸ Balanced vs. unbalanced forces				I	D	D	D
▸ Inertia					I	D	D
▸ Circular motion						D	D
▸ Newton's First Law			I	I	I	D	D
Newton's Second Law			I	I	I	D	D
▸ Effect of changing mass on constant force			I	I	I	D	D
▸ Effect of changing force on constant mass			I	I	I	D	D
Newton's Third Law (action-reaction)						D	R
Momentum						I	D

KEY		
I = Introduced	D = Developed	R = Reinforced

Work and Energy

Work	K	1	2	3	4	5	6
Examples of				I	D	D	R
Mathematical formula	I	I			I	D	D
Calculations in joules						I	D

Energy	K	1	2	3	4	5	6
Potential				I	D	D	R
▸ Examples of				I	D	D	R
▸ Gravitational potential				I	D	D	R
Kinetic				I	D	D	R
▸ Examples of				I	D	D	R
▸ Relation to temperature					I	D	D
Transformation of potential and kinetic				I	I	D	R
Law of Conservation of Energy					I	D	R

Simple Machines	K	1	2	3	4	5	6
Relation to needs, found in nature		I	I	D	D	R	R
Examples of		I	I	D	D	R	R
▸ Levers		I	I	D	D	D	R
▸ Fulcrum			I	I	D	D	R
▸ Classification			I	I	D	D	R
▸ Pulleys	I	I	I	I	D	D	R
▸ Fixed vs. movable				I	D	D	R
▸ Wheel-and-axles	I	I	I	I	D	D	R
▸ Inclined planes		I	I	D	D	D	R
▸ Compared with lifting		I	I	D	D	D	R
▸ Effects of texture, angle				D	D	D	R
▸ Wedges		I	I	I	D	D	R
▸ Screws		I	I	I	D	D	R
Mechanical advantage						I	D
Friction/efficiency						I	D
Relation to compound machines				D	D	D	R

Scope & Sequence

Scope and Sequence

Transfer of Energy

Heat	K	1	2	3	4	5	6
Sources		I	I	D	D	R	R
Transfer				I	D	D	D
▸ Conduction				I	D	D	D
▸ Convection				I	D	D	D
▸ Radiation				I	D	D	D
Related to temperature		I	I	I	D	D	D
Kinetic energy model					I	D	D
Heating of different surfaces, trends					I	D	D

Sound	K	1	2	3	4	5	6
Sources (vibrations)	I	I	I	D	D	D	R
Transfer of sound			I	D	D	D	R
▸ Through solids, liquids, gases			I	D	D	D	R
▸ Compression wave model					I	D	D
Pitch variations	I	I	I	D	D	D	R
▸ Frequency					I	D	R
▸ Hertz measurements						I	D
▸ Doppler effect							D
Volume (loudness) variations	I	I	I	D	D	D	R
▸ Amplitude					I	D	D
▸ Decibel measurements						I	D
Absorption (insulation)						D	D
Reflection (echoes)					I	D	R

Light	K	1	2	3	4	5	6
Sources		I	I	D	D	D	R
▸ Natural vs. artificial		I	I	I	D	D	R
Straight-line path		I	I	D	D	D	R
▸ Opaque, transparent, translucent materials				I	D	D	R
▸ Shadows	I	I	I	D	D	D	R
Reflection			I	I	D	D	D
▸ Law of reflection					I	D	D
▸ Mirrors			I	I	D	D	D
▸ ▸ Convex vs. concave					I	D	D
Refraction			I	I	D	D	D
Lenses			I	I	D	D	D
▸ Convex vs. concave					I	D	D

KEY		
I = Introduced	D = Developed	R = Reinforced

Transfer of Energy *cont.*

Light *cont.*	K	1	2	3	4	5	6
Prism			I	I	D	D	D
▶ Visible spectrum (colors)			I	I	I	D	D
▶ Mixing colors						D	R
▶ Mixing pigments						D	R
Wave model						D	D
▶ Parts (crest, trough)						D	D
▶ Photons						D	D
Electromagnetic spectrum						I	D
▶ Descriptions, uses						I	D
▶ Wavelengths/frequencies compared						I	D

Electricity	K	1	2	3	4	5	6
Static			I		D	D	R
▶ Attraction vs. repulsion			I		D	D	R
▶ Induced charge					I	D	D
▶ Discharge					I	D	D
▶ Insulators/conductors				I	D	D	R
Current/circuits			I	I	D	D	R
▶ Closed vs. open				I	D	D	R
▶ Parts of		I	I	D	D	D	R
▶ Batteries/wet-dry cells		I	I	I	D	D	R
▶ Transformation of energy in		I	I	I	D	R	R
▶ Series vs. parallel					D	D	D
Electromagnetism					D	D	D
▶ Making electromagnets					D	D	D
▶ Increasing strength of					D	D	D
▶ Common uses				I	D	D	D
▶ Motors				I	D	D	R
▶ Generating current					D	D	R
▶ AC vs. DC					D	D	R
▶ Power transmission					D	D	R
Household uses	I	I	I	D	D	R	R
▶ Safety					D	D	D

Scope & Sequence

Teacher's Glossary

The National Science Education Standards state that all teachers of science must have a strong, broad base of scientific knowledge extensive enough for them to understand the fundamental facts and concepts in major science disciplines. This glossary is designed to help the Kindergarten teacher understand the basic facts and concepts of science.

acceleration The rate of the speed or direction of a moving object. The acceleration of something falling due to Earth's gravity is about 32 feet per second.

adaptation Changes that take place over many generations in an animal or a plant that increases its chances of survival. A polar bear's thick, white coat is an example of an adaptation to arctic conditions.

adhesion A force that attracts molecules of unlike substances to each other. For example, the graphite from a pencil sticks to paper when you write.

aerobic A term used to describe animals and plants that live in an oxygen environment. Most animals and plants are aerobic.

AIDS A fatal viral disease that is transmitted by the exchange of bodily fluids. AIDS cannot be contracted through casual contact.

air A mixture of nitrogen, oxygen, water vapor, and small amounts of other gases that is colorless and usually odorless.

air mass A body of air that has about the same humidity and temperature at any altitude. Air masses cause changes in the weather as they move.

air pressure The force or weight of the atmosphere on a unit of area.

algae Small green water plants with no roots, stems, or leaves. Algae plants turn pond water green.

alloy A mixture of one metal with another metal or a nonmetal, such as carbon. For example, brass is an alloy of copper and zinc.

altitude The measurement of height from sea level. Mount Everest is the highest mountain, with an altitude of about 29,028 feet.

aluminum A lightweight metal that is the most common metallic element found on Earth. It conducts electricity and is used to make overhead electric cables where having lightweight metal is important.

amphibian An animal that is cold-blooded and lives part of its life in water and part on land. Amphibians do not have hair, scales, or feathers, but they do have moist, smooth skin. They are often born in water and have gills. As they grow, they develop lungs and can live on land. They lay their eggs in water. Frogs, toads, and salamanders are amphibians.

animal A living thing that moves around, takes in other animals or plants as food, and is sensitive to its surroundings.

antennae Moveable, hair-like feelers found on an insect's head. They are also found on some shelled animals.

arthropods Animals that have a hard outer shell and jointed legs. Some, like shrimp and crabs, live in water. Others, like insects and spiders, live on land.

astronomer A scientist who studies celestial objects and their movements.

autumn The autumnal equinox, where day and night are equal in length, marks the beginning of the autumn season.

axis A real or imaginary straight line that passes through an object and around which the object spins.

bark The tough, waterproof, outer layer of wood on a tree or woody plant.

barograph An instrument that records changes in air pressure.

barometer An instrument that measures air pressure and is used to help predict weather changes. Rising air pressure usually means fair weather, falling air pressure usually means rain or snow.

basalt A hard, dark-gray to black igneous rock formed by the cooling and hardening of lava from a volcano.

biologist A scientist who studies the habits and characteristics of living things.

birds Animals that have feathers, are warm-blooded, lay eggs, and have wings. Most, but not all birds, can fly. Birds have hollow, lightweight bones to help make it possible for them to fly.

bubble A thin skin of a liquid that is filled with a gas. For example, a soap bubble is filled with air. When it breaks, there is only a tiny bit of soap left.

bugs Insects or other creeping or crawling invertebrates that feed with sharp, tube-like mouths, which they use to suck their food.

bulb The stem of a small underground plant covered by leaves. Plant food is stored in the bulb. Tulips and daffodils are examples of plants that grow from bulbs.

C

camouflage A way some animals protect themselves by seeming to blend into their surroundings.

canyon A deep, narrow passage between hills. There is often a river at the bottom of a canyon.

carbon A common, nonmetallic, chemical element found in all animals and plants. Graphite and diamonds are different forms of carbon.

carbon dioxide Formed when carbon combines with oxygen during burning or respiration; carbon dioxide is one of the gases in air.

cave An open space in rock that is naturally formed. A cave usually opens to the surface through a passageway.

cement A gray powder made from crushed limestone and clay. It hardens into concrete when mixed with water and sand.

center of gravity The point in an object where the weight seems to be concentrated. When an object is supported beneath its center of gravity, it is balanced.

characteristic A feature or trait of an animal or plant that is innate. For example, a characteristic of owls is that they have keen night vision.

cirrus The name given to clouds that are usually high, thin, and made of ice crystals.

classify To classify is to put things into groups to show how they are alike.

claw The curved, pointed nail used to grab and hold an object. Claws are found on the toes of animals such as birds and cats.

clay A very small particle of soil that is smaller than sand or silt particles. Clay is used in ceramics and pottery and to make bricks and tiles.

clouds Clouds are formed by millions of tiny water droplets or ice crystals floating in air. Clouds are classified into three main types: cumulus (puffy) clouds are mountain or mushroom shaped; cirrus (wispy) clouds are high and feather shaped; stratus (layered) clouds are low-lying clouds that make the sky look gray. Clouds along the ground are called fog.

coal Hard, black carbon that is burned as fuel. Coal was formed many millions of years ago from the remains of tropical and subtropical plants.

cold-blooded This term refers to the body temperature of reptiles, amphibians, and fish. Their body temperatures change when the surrounding water or air temperature changes.

color When light enters the eye, the brain senses the wavelengths, adds them together, and produces the sensation of different colors. Only a few mammals are able to see colors.

communicate To share information with others.

compare To tell how things are alike and different.

compound eye An eye composed of several thousand individual eyes that enable an insect to see a wide area around it.

conclusion To use information collected to explain a given phenomenon.

condensation The process by which a gas cools and changes into a liquid. The dew that appears on windows and mirrors is the result of the condensation of water vapor on cold objects.

conduction The process by which sound, electricity, heat, or some other form of energy travels through a substance.

conifer Conifers are trees that produce cones and usually have needle-shaped leaves. Pines, spruces, and firs are all conifers.

constellation A group of stars that forms a pattern in the night sky.

continent The seven large land areas on Earth's surface (North America, South America, Europe, Asia, Africa, Australia, Antarctica) are called continents.

coral Coral are small sea animals found in many different shapes and colors. When they die, their limestone skeletons form coral rock. Coral reefs and tropical islands are made from coral rock.

cotton A plant whose soft, white fibers attached to its seeds are used to make thread. Cotton thread is used to make cotton cloth.

crescent moon The thin, curved shape of the Moon that appears in the west at sunset.

crystal Different chemical substances whose atoms are arranged in a definite geometrical shape. They come in a variety of shapes and colors.

cumulonimbus Clouds that are often called "thunderheads" that frequently bring rain, snow, hail, or thunderstorms.

cycle A chain of events in nature that repeats itself in the same order.

decay The rotting and breakdown of dead materials caused by bacteria and other microorganisms.

decibel A unit used to measure sound or noise levels.

deciduous Trees and other plants that lose their leaves before the onset of cold weather.

dehydrate To remove water from a substance.

density The mass of a substance compared to its volume. A block of lead weighs more than an equal volume of wood because the lead is denser than the wood.

desert A large area of land that has very little rain. Deserts can be hot or cold. Parts of the Arctic, for example, are cold deserts where water is trapped in the form of ice.

dew point The temperature of the air when it cannot hold any more water vapor. When air becomes colder than the dew point, water vapor forms clouds, fog, frost, or dew.

dormant An inactive plant or animal, for example, a seed in winter.

dune A mound or ridge of sand formed by the wind and found in deserts or along the sea.

earthquake Caused by the sudden shifting of rocks beneath the surface of Earth along a fault line.

earthworm A large worm found in forest or grassland soils. Its long, narrow body is made of many rings joined together by a softer material. It eats plant materials in the soil and has tiny bristles on the bottom of each segment.

effort A force needed to do work. Effort is the force that produces motion when levers, inclined planes, or pulleys are used.

electricity Energy formed by the interaction of charged particles, particularly protons and electrons. When the particles are in motion, it is called an electrical current. At rest it is called static electricity.

electromagnet A magnet produced by an electric current that flows through a conductor and forms a magnetic field. When the current stops, the magnetic field collapses.

energy There are many forms of energy: electrical, chemical, mechanical, nuclear, light, and heat. Energy cannot be created or destroyed, only changed from one form into another.

engine A machine that transforms energy from one form to another. For example, a car's combustion engine turns heat from gasoline into movement.

environment The conditions around a plant or animal that influence its life: temperature, light, living space, water, soil, and food source.

erosion The gradual wearing away of Earth's surface. It is caused by wind, waves, running water, rain, ice, or other natural forces.

evaporate When a liquid changes into a vapor or gas it evaporates. Heat increases the rate of evaporation.

evergreen A tree or plant that keeps its leaves during the winter.

family Biologists use this term for a group of living things with similar characteristics.

feather Only birds have feathers. They usually have a partly hollow central shaft and many barbs. Feathers help insulate and waterproof the bird's body. When in flight, the bird's muscles move the feathers.

feelers The long, jointed growths on the head of insects and some other animals. They are used for touch and smell and are sometimes called antennae.

fern A green plant that does not produce flowers or seeds. The tiny spores found on the underside of the feathery leaves are used for reproduction. There are over 10,000 varieties of ferns.

file A steel tool with a rough surface used for smoothing or grinding a hard material, for example, a nail file.

fin A moveable part of the body of a fish or other water animal. Fins are used for steering, swimming, and balancing in water.

fish A cold-blooded animal that has a backbone and lives in water. Most fish lay eggs in water, but some give birth to live young. There are about 30,000 different kinds of saltwater and freshwater fish.

float Objects that stay up or at the water's surface float. They can only float if they weigh less than an equal volume of the water. Because salt water weighs more, objects that sink in fresh water may be able to float in salt water.

flower The part of some plants that contains its reproductive parts—stamens and pistils. Tiny grains of pollen from a stamen fertilize a pistil. Fruits and seeds come from fertilized pistils.

fog A cloud near the ground.

force A push or a pull that makes something move, or if it is already moving, changes an object's speed or direction.

fossil The naturally preserved remains of a plant or animal. Fossils can be the remains of a leaf, bone, skeleton, or even the entire plant or animal. Fossils help us learn about the living things of the past.

friction The resistance caused by moving the surface of one object over the surface of another. Friction slows down movement and produces heat. For example, rubbing your hands together will warm them up.

frost Ice crystals that form when water vapor condenses on a surface colder than 32 degrees Fahrenheit.

fruit The part of a flowering plant that stores seeds.

fur The soft, thick hair that covers the skin of many mammals and provides insulation.

G

galaxy Star systems, each containing billions of stars, planets, and other objects. Our Sun is in the Milky Way galaxy.

gas One of the three states of matter. Gas molecules move rapidly. Gases expand to fill up and take the shape of their containers. When the temperature of a gas is lowered enough, it becomes a liquid. When lowered even more, it often becomes a solid.

gear A wheel with teeth along the edge. When two gears are placed together, the teeth fit into one another. As one gear turns, the other turns in the opposite direction. The metal rod connected to each gear is called the axle. As a gear turns, the axle also turns.

gem A hard, usually transparent mineral that can be cut or polished. Gems come in many different colors and are often used for jewelry or decoration.

geologist A scientist who studies rocks and minerals, and the ways in which Earth has changed over the course of millions of years.

germinate Seeds need water, sunlight, and air to germinate or start to grow.

gills The part of the body of many water animals that enables them to get the oxygen they need in order to live.

glacier A huge sheet of ice that moves slowly over land. Glaciers form when, over many years, water freezes more than it melts.

glass Glass is made by heating a mixture of sand, sodium, and lime until it becomes a liquid. That liquid can then be shaped into bottles, sheets, or other forms. As it cools, the glass hardens and becomes a transparent solid.

gold A metal that resists corrosion and is used to make coins, jewelry, and other items. It can be hammered into a thin sheet or drawn into a fine wire.

grain A single seed or fruit of crops such as corn, wheat, rice, or oats.

granite An igneous rock that is hard and ranges in color from gray to pink or even red. It contains quartz, feldspar, and a small amount of other minerals.

graph A diagram that shows the relationship between two or more different things.

graphite A soft, shiny, black mineral that is a form of carbon. Graphite, not lead, is used in pencils.

grass A green plant with long, narrow leaves and jointed stems. Wheat, corn, oats, bamboo, and rye are all examples of grasses.

gravel A mixture of pebbles and small bits of rocks. It is used in various ways in the construction industry. For example, gravel is used to make concrete.

gravity The force that attracts things to move toward the center of Earth. The larger the object, the greater its gravitational force. Gravity is what causes people and objects to have weight on Earth and what makes objects fall towards Earth.

grub The worm-like form of an insect. A grub hatches and then develops into a pupa or chrysalis and then into an adult.

H

habit An instinctive way that animals or plants behave or grow.

habitat The natural surroundings of a particular area where a plant or animal naturally lives and grows.

hail Frozen raindrops that grow larger as they are tossed around by strong winds in thunderclouds.

hair A filament that grows from the skin of mammals that provides warmth and protection.

hardwood The wood that comes from trees with broad, flat leaves. Oak, maple, and cherry trees produce hardwood. Trees with needles, such as pine and fir, produce soft wood—wood that is less dense and lighter than hardwood.

hearing The process of sound waves passing into the ear, which causes the bones in the ear to vibrate. The vibrations are then picked up by nerve cells and sent to the brain, which interprets the messages.

heat A form of energy caused by the movements of molecules in a substance. Heat can change a solid into a liquid and a liquid into a gas.

herb A flowering plant without a woody stem, often used to flavor foods.

heredity The process in which plant or animal characteristics are passed from one generation to the next by means of genes. Such traits as eye color and size are hereditary.

horizon The apparent line where the land or sea seems to meet the sky. Because Earth is curved, it is impossible to see beyond the horizon.

horn Often found on the heads of grazing animals, horns are hard, curved growths.

hull The outer covering of a seed or fruit.

hurricane Large, intense tropical storms that usually form in the late summer or early fall.

hydrogen The simplest, lightest, and most prevalent gas in our universe. When combined with oxygen it produces water (H_20).

hypothesis A possible explanation for why something occurs. Until tested, a hypothesis cannot be proved nor disproved.

ice A colorless solid that is formed when water reaches 32 degrees Fahrenheit. Because ice is less dense than water, it floats on top of bodies of water.

igneous rock One of the three main kinds of rock. Igneous rocks are formed from the cooling and subsequent hardening of lava after a volcanic eruption. Granite and basalt are both igneous rocks.

inclined plane A simple machine that makes it easier to move a heavy load. It is set at an angle to the ground.

infer To use what is known to figure something out.

insects Small animals with three pairs of legs (6 total) and three segmented body parts: head, thorax, and abdomen. Most insects have one or two pairs of wings. Flies, bees, beetles, ants, grasshoppers, to name just a few, are all insects. Spiders, however, are not. They have eight, not six, legs.

instinct An animal behavior that is inherited, not learned.

investigate To make a plan to explore a question.

iron The most common metal on Earth. It is hard, gray, and very magnetic. When mixed with carbon and other elements, it becomes steel.

island A body of land smaller than a continent that is completely surrounded by water.

jade A gem, usually green, but also white, brown, or yellow. It is often carved and used in jewelry.

Jupiter The largest planet in our solar system. Jupiter has at least 16 moons. It takes 12 years for Jupiter to travel once around the Sun. It is more than 1,300 times the size of Earth.

kernel The soft part of a plant seed that can grow into a plant. A kernel is protected by the husk.

lake A large body of water that is entirely or almost completely surrounded by land.

larva The early stage of insects that will change greatly in shape when they become adults. For example, a tadpole is the larva of a frog or toad.

lava Rock that is so hot that it has become liquid and flows through volcanoes and cracks in Earth.

leaf A thin, green outgrowth of a plant that grows on a stem or comes up from the roots. Leaves absorb carbon dioxide from air or water and use sunlight to carry on photosynthesis, which creates the food plants need.

legume A plant that creates pods that contain seeds. Peas, beans, lentils, and soybeans are examples of legumes.

lever A bar that turns on a fixed support called a fulcrum. For example, a seesaw is a lever with the fulcrum in the middle. When you push down on one end, it raises the other.

life A term that has no exact definition, but some of the characteristics or properties include: something that reproduces, grows, and responds to stimuli.

life cycle A series of changes that living things pass through: born, grow older, reproduce, and die.

light A form of electromagnetic radiation that humans and other animals can see because of the light-sensitive cells in their eyes.

lightning When electricity in the atmosphere is discharged, a powerful flash can be seen. Lightning can appear between two clouds or between a cloud and the ground.

limestone A sedimentary rock made mostly of calcite.

liquid One of the three states of matter. A liquid has a definite volume, but not a definite shape. It can be poured, and it takes the shape of the container into which it is poured.

lizard Lizards live in tropical or semitropical climates and have a long body with a slender tail and four legs.

lodestone A magnetic rock, sometimes called magnetite, that was used as a primitive compass.

machine Something that transmits or changes energy or motion. Simple machines have one part that provides a particular type of work. More complex machines consist of many moving parts that work together.

magma The very hot, molten rock that lies beneath Earth's surface. During volcanic eruptions, magma is forced out of Earth and then it is called lava.

magnet An object with a powerful magnetic field that will attract iron, steel, nickel, and cobalt.

magnetic field The space around a magnet that can attract objects. The Sun, Earth, and other planets all have magnetic fields that extend for millions of miles.

magnetic pole Magnets have two poles—a north pole and a south pole. There is an invisible line that extends between the two poles. If you place two magnets near one another, the like poles will repel one another, opposite poles will attract.

mammals Warm-blooded animals with backbones that usually have hair or fur. They give birth to living young, and a female mammal produces milk to feed her young.

marble A hard metamorphic rock that can be polished. Marble can be white, streaked, or a variety of other colors.

Mars The fourth planet from the Sun. Mars is smaller than Earth and has two moons.

marsupial A specific group of mammals; female marsupials have a pouch that covers their milk-producing mammary glands. Kangaroos, koalas, and wombats are examples of marsupials.

mass The amount of matter in an object. An object's mass is not the same as its weight. Rather, it is a measure of its ability to resist changes in its motion. The greater an object's mass, the greater its ability to resist changes in its motion.

matter Matter can exist in three states; solid, liquid, gas. It is made of atoms, has mass, and occupies space.

measure To find out the length, width, height, temperature, capacity, or volume of something.

melt To change from a solid to a liquid. Ice melts at 32 degrees Fahrenheit. Metals also can melt when heated at very high temperatures.

Mercury The planet closest to the Sun. It takes only about 88 days for Mercury to travel around the Sun. It is the second smallest planet.

metamorphic A type of rock that has been changed by heat, pressure, or water.

meteorologist A scientist who studies weather patterns and forecasts the weather.

migration The movement of animals from one place to another. Migrations usually take place with the change of the seasons. For example, many North American birds migrate south before winter and return in the spring.

mineral An object found in the ground that does not come from living things and has a crystal form. For example, quartz and mica are both minerals.

mist Mist is a cloud composed of very tiny water droplets.

model Something made to show how something does or might look.

Moon The Moon travels in an orbit around Earth. It takes 29.5 days to completely circle Earth. Like Earth, it has a gravitational field. It is the pull of the Moon's gravitational force that helps create ocean tides.

moss A group of very small plants that grow close to the ground. They can be green, brown, or gray, and have no flowers.

moth A small flying insect with four wings and antennae. Moths usually fly at night and are attracted to light sources.

motor A machine that changes one form of energy into motion (mechanical energy). For example, the motor in a food mixer transforms electricity into motion that turns the beaters.

mountain A very high outcropping of land. There are large, long mountains both above ground and along the ocean floor.

nail A thin, pointed piece of metal that has a flat head at one end and can be hammered into wood. The term nail is also used to refer to the hard plate that protects the fingers and toes of animals with backbones.

natural A substance found in nature that is not human-made.

Neptune The fourth largest planet in our solar system. It is the eighth planet from the Sun. It takes 165 years for Neptune to orbit the Sun. Neptune has two large moons and at least six small ones.

nickel A hard, silver-colored metal that is often combined with other metals because it resists corrosion.

nimbostratus A dark gray layer of clouds that produce lengthy rain or snow showers.

nitrogen A gas that is found in Earth's atmosphere. It is colorless, tasteless, and odorless. It is found in all living things. Unlike some other gases, nitrogen does not burn.

node The place on the stem of a plant where leaves grow.

nut A fruit with a shell that contains a seed or kernel. Nuts can be as large as a coconut or as small as a hazelnut.

nutrient Something that provides nourishment for a plant or an animal.

nutritionist A scientist who studies the process by which living things eat food, digest it, and use the nutrients contained in the food.

nylon A synthetic (not natural) material that is strong and lightweight. It is used to make cloth and ropes, or molded to make a variety of other objects.

nymph A stage in the development of certain insects. A nymph looks much like a grown insect, but its wings are not yet fully developed.

observation The process of looking at something very carefully. Scientists take notes or make records of the things they observe.

ocean The large connected body of salt water that covers almost three fourths of Earth's surface.

oil A liquid substance that does not mix easily with water. Oil comes from three sources: from the ground (mineral oils) and from vegetables and animals (olive oil, animal fat).

opaque A characteristic of substances that does not allow many or any light rays to pass through.

orbit The path of one celestial object around another.

order To place objects or events in a sequential arrangement.

organism Every living thing can be called an organism. Organisms reproduce, take in and use nutrients, and die.

oxygen One of the most abundant gases on Earth, it is required for most organisms to live. It is produced through the process of photosynthesis. It can be combined with many other elements.

paleontologist A scientist who studies fossils (the remains of living things) to learn more about life in past times.

peninsula A large piece of land that is surrounded by water on three sides. Florida is an example of a peninsula.

perennial Any plant whose roots grow for two or more years. Trees, tulips, and roses are examples of perennials.

phase Astronomers refer to the apparent shape of the Moon by describing four phases: new moon, first quarter, full moon, and last quarter.

photosynthesis The process by which green plants produce carbohydrates, which in turn provide nutrients for the plant. This is done using energy from sunlight acting on the combination of water and carbon dioxide with chlorophyll and enzymes in the plant. During this process, oxygen is released. Most of the oxygen in our atmosphere comes from this process.

physical change A change in a substance that does not result in the creation of a new substance. For example, when water turns to ice, it changes form, but it is still the same chemical substance.

physiologist A scientist who studies the way living things function. Physiologists who study the human body help doctors learn how to treat patients more effectively.

pistil A flower's female reproductive part that produces seeds.

plain A large span of nearly level land.

plant A many-celled living thing that usually makes its own food from carbon and water, using the energy from sunlight. Although scientists now know that some one-celled living things could be classified as plants or animals, they usually agree that the term plant should be reserved for multicelled organisms that can create their own food.

plastic A substance made from petroleum. Acrylics, nylon, and polyurethane are all examples of a plastic substance.

plate One of the huge blocks that make up the crust of Earth. Earth's crust is formed by many plates that sometimes push against one another to produce changes in the shape of Earth's surface.

Pluto This dwarf planet takes approximately 249 years to orbit the Sun. Pluto is actually smaller than our own Moon.

pod The casing or sac that contains the seeds in some kinds of plants. Peas and beans have pods.

pollen The tiny, yellow grains produced by the male parts of flowering or cone-bearing plants. Each tiny grain contains one male reproductive cell. When there is a great deal of pollen floating in the air, it can cause allergic reactions in some people and animals.

pollute To dirty or poison the natural environment by introducing toxic substances or things that do not break down in the natural environment.

pond A shallow body of fresh water that is smaller than a lake. Generally, sunlight can reach the bottom of a pond.

potential energy The energy present in an object because of its position. For example, a toy car at the top of a block ramp has potential energy. When it is pushed and travels down the ramp, the potential energy becomes kinetic energy (the energy of movement).

prairie A flat, grassy plain with few trees.

precipitation The process by which water vapor condenses and becomes water or ice and falls to Earth as rain, hail, snow, sleet, dew, fog, or mist.

predict To predict is to use what you know to tell what might happen.

pressure A constant weight or force that acts on a surface. The air in our atmosphere creates pressure on Earth and is reported in units of force per unit of area. For example, air pressure at sea level is about 14.7 pounds per square inch. At the bottom of the ocean, the pressure of the water on the ocean floor is more than 8 tons per square inch.

primary color One of three colors (red, blue, yellow) that, when mixed together in different amounts, can make all other colors. For example, blue and yellow produce green. If you use more yellow, the tint of green changes.

prism Light is made up of different wavelengths. A glass prism is made in such a way that when light passes through it, the light rays are bent and the white light is split into the colors of the spectrum.

propagate To reproduce plants either by seeds or from the cuttings (pieces) of other plants.

pulley A machine that eases work. It is made by wrapping a rope or chain around a grooved wheel that has been mounted on a block. Pulleys change the direction of a force to lift a load. By creating a system of several pulleys working together, a small amount of effort can lift a heavy load.

pupa A stage in the development of many insects. A butterfly is in the pupa stage while encased in its chrysalis.

Q

quartz A very hard mineral that, in its crystal form, has six sides and comes to a point. Quartz is often embedded in granite and sandstone. Agate, jasper, and amethyst are different forms of quartz.

queen A female in a colony of insects. There is usually only one queen in a colony. The queen lays eggs and produces young.

R

radiation Energy in the form of waves or rays. Light and heat rays radiate heat and warm Earth. Radiation is also the particles or rays that come from atoms in a radioactive substance.

radicle The part of a seed that becomes the root of a plant. When seeds germinate, the radicle is the first part to grow.

rain Precipitation that falls from clouds as drops of water. Rain is formed when water vapor in the air condenses.

rain forest A forest of tall trees where the rainfall is particularly heavy. Because of the lush foliage in these forests, very little sunlight penetrates to the forest floor.

ray A beam of light, heat, or other forms of radiant energy. A ray is also a group of large, flat, saltwater fish (stingray).

recycling The process of breaking down, changing, or reducing human-made materials into something new.

reef A large piece of rock or coral that lies just below the surface of the water. Coral reefs provide a rich source of food for a large variety of tropical fish and other water animals.

reflection Radio, light, or heat waves that hit a surface and bounce back. When a sound wave reflects off a hard surface, you hear an echo.

refract To bend light, heat, or sound waves from a straight path. A curved lens refracts light.

reproduction The process of producing offspring similar to the parents. Both plants and animals reproduce.

reptile A cold-blooded animal with a backbone that also has lungs. Reptiles have dry, scaly skin and lay eggs. Snakes, lizards, crocodiles, and turtles are all reptiles. Dinosaurs were prehistoric reptiles.

rhizome Some plants have scaly stems that grow underground called rhizomes. Ferns, grasses, and some flowers have rhizomes.

ridge A long, narrow chain of hills or mountains.

river A long body of water that flows in a channel and empties into a lake, pond, or ocean. Water for rivers comes from springs, melting glaciers, and lakes. On a steep hillside, the water can flow rapidly. Near the mouth of a river, it often flows much more slowly and is filled with mud and silt that has been washed down the river.

rock Most rocks are made up of two or more minerals. They can be very hard or very soft. Rocks are classified according to the way they were formed. There are three types: igneous, sedimentary, and metamorphic.

rodent A group of mammals that are small and have two long teeth that continue to grow throughout the animal's life. Mice, guinea pigs, squirrels, and beavers are all rodents.

root The part of a plant that grows downward, absorbs water, holds the plant in place, and may also store food.

rubber A substance made from the sap of rubber trees. It can also be made synthetically. Rubber is impervious to water and air.

salt Salt is found in underground deposits and seawater that consists of sodium chloride. It flavors food, and because it kills bacteria, salt is also used to preserve foods.

sand Grains of rock, often quartz or feldspar, that are smaller than gravel and can be many different colors.

sandstone A rock that is formed when grains of sand bond together with minerals. Sandstone is a sedimentary rock and varies in color from yellow or red to gray or brown.

Saturn The sixth planet from the Sun, Saturn is the second largest planet. Made of gases, Saturn has more than 20 moons and is encircled by hundreds of rings made up of tiny ice particles.

savanna A grassland of tropical Africa. There are some trees and shrubs that provide protection and shade for animals that graze on the savanna.

scale The thin, flat, hard plates that cover many fish, snakes, and lizards. Also, a tool used to measure the weight of objects.

screw A thick inclined plane wrapped around a cylinder, most often used as a fastener.

sedimentary Rocks formed by sediment, such as sand or mud, that is deposited and becomes compacted. Sandstone and limestone are examples of sedimentary rocks.

seed A part of a flowering plant that contains an embryo and the necessary food to enable the embryo to grow into a new plant.

shale A sedimentary rock made of compacted and hardened clay, silt, or mud. It forms in distinct layers and can be split easily. It varies in color from black or gray to brown or red.

silt Small particles of rock that are smaller than sand and larger than clay. Silt is often found at the bottom of lakes and ponds, where it accumulates slowly.

silver A soft, shiny metal that conducts heat and electricity better than any other metal.

slate A metamorphic rock that forms from shale. Slate also splits into thin layers. It ranges in color from gray to black or from red to green.

sleet Partially frozen or frozen raindrops that fall from clouds.

snakes Meat-eating reptiles that have long, narrow bodies, often just one lung, and a forked tongue. Some, but not all, snakes have venom glands and sharp fangs that can give a poisonous bite.

snow Ice crystals that form from water vapor in clouds and then fall to Earth.

soapstone A soft metamorphic rock composed mostly of the mineral talc.

soil The top layer of Earth's surface. It is composed of rock and mineral particles and decaying organic matter (humus). Soil provides the nutrients that many plants need to grow.

solid One of the three basic forms of matter. The molecules in a solid have little or no ability to change places. Solids also have a fixed shape and volume.

sound Sound is produced when an object, such as a vocal cord or guitar string, vibrates. These vibrations travel in waves as they move through a substance.

space The part of the universe that lies outside Earth's atmosphere. It can also mean the three-dimensional volume in which matter exists.

spectrum The colors of white light when it has been split by passing through a prism or some other object that refracts light. A rainbow, which is made when light passes through water droplets, has all the colors of the spectrum.

spiders A group of small animals with a body divided into two parts (unlike insects which have three body parts). Spiders have a head and thorax, four pairs of jointed legs, usually eight eyes, and no antennae.

spinneret The part of the body that enables spiders and some other animals (silkworms) to spin threads for webs or cocoons.

spore The tiny reproductive body found on flowerless plants. Given the necessary conditions, a spore can grow into a complete new plant. Spores can often survive for many years, under unfavorable conditions, until they get what they need to grow.

stalk The main part of a plant that grows upward. Leaves and flowers are supported by a plant's stalk.

stamen The male part of a flower that produces the pollen the plant needs to reproduce. It is a slender stalk with an anther (pollen sac) at the end. The stamen is usually in between the petals of the flower on the plant.

star A large, glowing body in space composed of very hot gases, mainly hydrogen and helium. The Sun is our closest star. The stars in the night sky look like points of light because they are so far away from Earth.

state The condition of a substance. Water, for example, can be in a liquid, solid, or gaseous state.

steam When water boils, it produces steam. The steam is actually water vapor. When the water vapor molecules cool and join together to produce water droplets, the steam becomes visible.

steel An alloy made of iron and small amounts of carbon and other elements. Steel is much stronger than iron, resists corrosion, and is used to build many things.

stem The main stalk of a plant that bears leaves. The stem carries water and nutrients from the roots to the other parts of the plant and carries the food the leaves make back down.

stone A small piece of a rock.

stream A long, narrow body of water that runs along a channel or bed. Stream is used to describe bodies of water that are larger than a brook and smaller than a river.

Sun A ball of boiling hot gases with a temperature of millions of degrees at its core. Earth orbits around the Sun once a year. The Sun supplies Earth with heat and light.

surface tension Surface tension is caused by the attraction of molecules in a liquid. You can see the effects of this tension by placing drops of water on waxed paper and pulling drops across the paper to join with other drops.

T

tadpole The larval stage of a young frog or toad. Tadpoles live in water and have gills that enable them to get oxygen from the water. Gradually, tadpoles lose their long tails and gills, develop lungs and legs, and move from the water to the land.

talc A soft, smooth mineral used to make chalk, cosmetics, and other substances.

taste The sense by which the flavor of foods and other substances is perceived. There are four basic types: sweet, salty, sour, and bitter.

theory An explanation, based on observations and investigations, of how or why something happens. Theories can help us predict events or behavior.

thermometer A tool used for measuring temperature. Most thermometers are made of a thin glass tube containing a colored liquid. As the temperature rises, the liquid expands and the column of liquid rises in the thermometer. When the liquid cools, the column falls.

thorax In insects, the middle section of the body between the head and the abdomen. The legs and wings are attached to the thorax.

thunder The loud sound that follows a flash of lightning. Thunder is caused by the shock wave from the sudden heating of the atmosphere that happens when lightning strikes.

tide The regular rise and fall of the ocean. There is a high tide about every 12 hours, so there are usually two tides in a day. Tides are caused by the gravitational pulls of the Moon and the Sun.

tin A soft metal that can be molded and bent and has a low melting point.

tornado A small, violent, funnel-shaped storm. The swirling winds of a tornado can reach 320 miles per hour. Tornadoes can cause great damage as they roll across the countryside.

touch The sense that enables us to perceive how an object feels. Sense receptors are concentrated in our fingertips, the rest of the hand, and the face.

translucent This term is used to describe an object that lets in some light but is not transparent. For example, tissue paper and waxed paper are translucent.

transparent This term is used to describe an object that lets light pass freely through it. Air that is not polluted, clean glass, and clean water are all transparent.

tree A large, woody plant with one main stem called a trunk. Trees also have leaves, branches, and stems, and are at least several feet above the ground. They are divided into two groups, those that keep their leaves all year (evergreen) and those that lose their leaves in the winter (deciduous).

trunk The main stem of a tree. The roots and branches are attached to the trunk.

tuber The underground part of a root or stem that contains stored food. Potatoes are tubers.

U

universe All the matter and energy in space, including the planets, moons, and stars, now and in the past.

Uranus The seventh planet from the Sun and the third largest. It takes Uranus 84 Earth years to travel once around the Sun. It is made up of gases. Astronomers have discovered 15 moons as well as rings circling Uranus.

Teacher's Glossary

valley A long, narrow strip of land lying between hills or mountains. Because valleys are usually formed when land is eroded by rivers or glaciers, they often have fertile soil.

vapor A gas formed from something that is usually in a solid or a liquid state. Water vapor is water in its gaseous state.

vegetable Plants we eat that are not fruits. For example, lettuce, carrots, asparagus. Fruits are the ripened ovaries of plants, so some foods commonly termed as vegetables (tomatoes, corn, string beans) are technically fruits.

Venus The second planet from the Sun. It orbits the Sun every 225 days and is the planet that comes closest to Earth. However, its atmosphere is very different. Temperatures reach 900 degrees Fahrenheit, and the clouds are formed not from water vapor, but from sulfuric acid.

vertebrate An animal with a backbone. Mammals, birds, reptiles, amphibians, and fish are all vertebrates.

vibrate To move quickly back and forth at a steady rate. For example, when a harp or guitar string is plucked, it vibrates back and forth and sets the air in motion, which we then hear as sound at a particular pitch (depending on the length of the string).

vision The sense of sight. Vision is the ability to receive and interpret light waves.

volcano An opening or vent in Earth's surface through which lava, hot gases, and ash flow. Volcanoes can be cone-shaped or an opening in Earth in which lava flows along the ground.

warm-blooded This term refers to an animal whose body temperature stays about the same regardless of the temperature of its surroundings. Birds and mammals are warm-blooded.

water A liquid that covers almost three fourths of Earth's surface. At normal atmospheric pressure, it freezes at 32 degrees Fahrenheit and boils at 212 degrees Fahrenheit. Without water, living things eventually die.

water cycle The movement of water from the atmosphere (water vapor) to Earth as it falls as rain or snow. This liquid ends up in bodies of water or the soil, where it slowly evaporates and returns to the atmosphere as water vapor.

wave A rapidly vibrating motion that passes through a medium, such as water or air, that does not change what it passes through. Light, heat, electricity, and sound all travel in waves.

wavelength The distance between two waves of energy, usually measured from the crest of one to the next. The wavelength of radio waves is much longer than the wavelength of light.

weight The force with which an object is attracted to Earth by gravity, or pulled vertically downward.

wetland An area of land that often floods and where you can often see standing water for at least part of the year.

whale A large mammal that lives in the ocean. It is not a fish. It has lungs and must come up for air in order to breathe. Whales give birth to live young. They vary greatly in size. The blue whale is the largest animal that has ever lived.

wheel A round frame that turns on a central axle. Wheels are used to transmit power, move heavy loads, and steer vehicles on land and water. They are also used in a variety of mechanical toys, clocks, and other machines.

wind A stream of air that moves over Earth's surface. It is caused by the unequal heating of Earth by the Sun and by Earth's orbit. When air comes into contact with the surface, it expands, becomes lighter, and rises. Cooler air rushes in to fill the lower space and this creates wind.

wire Metal that has been drawn into a thread. Copper, aluminum, iron, and steel can all be stretched into a wire.

wood The hard, fibrous part of the trunk of a tree or shrub.

wool The soft, curly hair of sheep, goats, camels, and alpacas. This hair can be spun into thread and used to make cloth or yarn.

worms Long, thin, soft-bodied, legless animals without a backbone (invertebrate) that can live in soil, freshwater lakes or ponds, the ocean, or as parasites on other animals.

year The time it takes for Earth to make one revolution around the Sun. It takes 365 days, 5 hours, 48 minutes, and 45.5 seconds, and that is why every four years we add one extra day to our calendar (leap year).

yeast Tiny one-celled fungi that produce alcohol as they eat and digest sugar. Some forms of yeast are used to bake bread, make wine, and brew beer.

Z

zoologist A scientist who studies animal behavior, genetics, evolution, and ecology.

For additional glossary support, see:

Blackbirch Encyclopedia of Science and Invention, by Jenny Tesar and Bryan Bunch (Blackbirch Press, 2001)

International Encyclopedia of Science and Technology (Oxford University Press, 1999)

Science Dictionary, by Seymour Simon (HarperCollins, 1994)

Index

Page references followed by an asterisk * indicate activities.

D

E

Page references followed by an asterisk * indicate activities.

F

G

H

I

Page references followed by an asterisk * indicate activities.

Page references followed by an asterisk * indicate activities.

N

O

P

Page references followed by an asterisk * indicate activities.

Q

R

S

Page references followed by an asterisk * indicate activities.

T

Page references followed by an asterisk * indicate activities.

U

V

W

Page references followed by an asterisk * indicate activities.

Correlation to National Science Education Standards

Grades K-4

The Science content standards are presented in the *National Science Education Standards*. These standards offer a structured guide to what scientifically literate students should know, understand, and be able to do at different grade levels. The standards are clustered for Grades K–4, 5–8, and 9–12.

The following table shows where the content standards are met in *Science A Closer Look,* Grades K–4. Grade K page numbers refer to the Teacher's Edition. Grades 1–2 page numbers refer to the Student Edition or, when preceded by AF, to the Activity Flipchart. Grades 3–4 page numbers refer to the Student Edition.

Science As Inquiry	
Abilities Necessary to Do Scientific Inquiry	
Ask a question about objects, organisms, and events in the environment. This aspect of the standard emphasizes students asking questions that they can answer with scientific knowledge, combined with their own observations. Students should answer their questions by seeking information from reliable sources of scientific information and from their own observations and investigations.	**Grade K:** 12, 52, 65, 82, 89, 96, 97, 102, 116, 127, 130, 131, 136, 144, 150, 158, 161, 164, 202, 218, 237, 254, 255, 257, 260, 266, 269, 272, 273 **Grade 1:** 3, 11, 23, 29, 37, 53, 59, 67, 87, 95, 103, 109, 127, 133, 141, 163, 171, 179, 195, 201, 209, 229, 235, 241, 249, 265, 271, 279, 299, 307, 315, 329, 333, 341, 361, 367, 373, 381, 397, 403, 411, 419, AF4, AF11, AF16, AF27, AF30, AF34, AF45, AF50, AF54 **Grade 2:** 3, 11, 23, 29, 39, 55, 61, 69, 89, 95, 103, 121, 129, 135, 155, 163, 171, 187, 195, 201, 223, 231, 237, 253, 259, 267, 275, 295, 301, 309, 325, 331, 339, 361, 367, 377, 385, 399, 405, 415, 421, AF4, AF10, AF16, AF19, AF22, AF26, AF28, AF34, AF38, AF41, AF46, AF50, AF54, AF56, AF59 **Grade 3:** 3, 21, 31, 43, 53, 69, 81, 91, 107, 119, 133, 151, 161, 173, 191, 203, 213, 227, 239, 249, 259, 279, 289, 303, 317, 327, 337, 347, 363, 373, 383, 397, 407, 417, 433, 443, 453, 463, 479, 489, 499, 511 **Grade 4:** 3, 21, 33, 45, 59, 77, 89, 99, 109, 129, 137, 149, 165, 175, 183, 203, 213, 225, 237, 251, 263, 273, 285, 295, 313, 323, 335, 345, 359, 369, 379, 393, 411, 421, 431, 445, 457, 467, 483, 493, 503, 513, 529, 539, 551, 563, 575
Plan and conduct a simple investigation. In the earliest years, investigations are largely based on systematic observations. As students develop, they may design and conduct simple experiments to answer questions. The idea of a fair test is possible for many students to consider by fourth grade.	**Grade K:** 30, 38, 44, 52, 58, 68, 76, 82, 88, 96, 102, 110, 116, 130, 136, 144, 150, 158, 164, 176, 182, 188, 196, 202, 216, 222, 228, 236, 246, 254, 260, 266, 272 **Grade 1:** 59, 99, 195, 201, 209, 271, 279, 329, 331, 333, 335, 341, 361, 367, 370, 373, 381, 397, 403, 419, AF4, AF11, AF27, AF30, AF34, AF50, AF54 **Grade 2:** 39, 73, 132, 137, 163, 325, 339, 344, 380, 388, AF22, AF28, AF34, AF46, AF50, AF54, AF56 **Grade 3:** 40–41, 144–145, 268–269, 334–335, 422–423, 450–451, 496–497 **Grade 4:** 56–57, 106–107, 244–245, 270–271, 282–283, 352–353, 400–401, 472–473, 522–523, 560–561, 572–573

Abilities Necessary to Do Scientific Inquiry continued	
Employ simple equipment and tools to gather data and extend the senses. In early years, students develop simple skills, such as how to observe, measure, cut, connect, switch, turn on and off, pour, hold, tie, and hook. Beginning with simple instruments, students can use rulers to measure the length, height, and depth of objects and materials; thermometers to measure temperature; watches to measure time; beam balances and spring scales to measure weight and force; magnifiers to observe objects and organisms; and microscopes to observe the finer details of plants, animals, rocks, and other materials. Children also develop skills in the use of computers and calculators for conducting investigations.	**Grade K:** 52, 58, 68, 76, 82, 102, 110, 117, 127, 130, 131, 133, 136, 139, 144, 150, 151, 153, 164, 165, 168F, 176, 177, 197, 199, 202, 203, 222, 223, 229, 231, 243, 246, 255, 261, 267, 268, 269, 272, 273, TR7, TR8, TR9, TR10, TR11 **Grade 1:** 29, 53, 87, 95, 127, 171, 229, 235, 249, 271, 307, 311, 315, 329, 341, 343, 376, 381, 384, 397, 403, 411, 419, R3, R4, R5, R6, AF4, AF27, AF34, AF45, AF50, AF54 **Grade 2:** 3, 29, 39, 69, 89, 95, 103, 129, 135, 171, 187, 195, 223, 231, 253, 259, 267, 275, 301, 309, 325, 339, 367, 377, 385, 399, 405, 415, 421, R3, R4, R5, R6, AF4, AF10, AF16, AF22, AF28, AF34, AF46, AF50, AF54, AF56, AF59 **Grade 3:** 26, 31, 40, 41, 53, 69, 81, 107, 119, 133, 144, 145, 161, 173, 200–201, 213, 231, 239, 243, 259, 268, 269, 307, 317, 334–335, 339, 347, 349, 363, 373, 380, 381, 383, 397, 407, 417, 422, 423, 438, 450, 460–461, 481, 486–487, 489, 496–497, 499, 511, 516, R2, R4, R5, R6, R7, R8, R9 **Grade 4:** 21, 30, 33, 39, 45, 56, 93, 99, 106–107, 109, 129, 134–135, 137, 143, 149, 172–173, 175, 237, 241, 251, 263, 267, 270–271, 285, 289, 295, 332–333, 352–353, 359, 363, 369, 379, 393, 400, 411, 421, 425, 431, 445, 461, 464–465, 467, 483, 490–491, 493, 503, 513, 522–523, 536–537, 539, 551, 560–561, 575, 580, R4, R5, R6, R7, R8, R9
Use data to construct a reasonable explanation. This aspect of the standard emphasizes the students' thinking as they use data to formulate explanations. Even at the earliest grade levels, students should learn what constitutes evidence and judge the merits or strength of the data and information that will be used to make explanations. After students propose an explanation, they will appeal to the knowledge and evidence they obtained to support their explanations. Students should check their explanations against scientific knowledge, experiences, and observations of others.	**Grade K:** 7, 30, 58, 76, 79, 99, 113, 158, 176, 203, 228, 243, 246, 260 **Grade 1:** 11, 141, 397, 403, 411, 419, AF4, AF8, AF27, AF30, AF34, AF38, AF40, AF45, AF50, AF54 **Grade 2:** 11, 129, 171, 201, 237, 259, 267, 275, 309, 325, 361, AF4, AF10, AF22, AF26, AF28, AF34, AF38, AF41, AF46, AF50, AF54, AF56, AF59 **Grade 3:** 21, 31, 41, 43, 53, 69, 81, 91, 107, 119, 133, 145, 151, 161, 173, 191, 203, 213, 227, 239, 249, 259, 269, 279, 289, 303, 317, 327, 335, 337, 347, 363, 373, 383, 397, 407, 417, 422, 423, 433, 443, 450, 453, 463, 479, 489, 497, 499, 511 **Grade 4:** 21, 33, 45, 57, 59, 77, 89, 99, 107, 109, 129, 137, 149, 165, 175, 183, 203, 213, 225, 237, 245, 251, 263, 271, 273, 283, 285, 295, 313, 323, 335, 345, 352, 359, 369, 379, 393, 401, 411, 421, 431, 445, 457, 467, 473, 483, 493, 503, 513, 523, 529, 539, 551, 561, 563, 573, 575

Abilities Necessary to Do Scientific Inquiry continued	
Communicate investigations and explanations. Students should begin developing the abilities to communicate, critique, and analyze their work and the work of other students. This communication might be spoken or drawn as well as written.	**Grade K:** 26, 30, 32, 38, 40, 44, 46, 52, 54, 58, 64, 68, 70, 76, 78, 82, 84, 88, 90, 96, 98, 102, 104, 110, 112, 116, 126, 130, 132, 136, 138, 144, 146, 150, 152, 158, 160, 164, 170, 176, 178, 182, 184, 188, 190, 196, 198, 202, 212, 216, 218, 222, 224, 228, 230, 236, 242, 246, 248, 254, 256, 260, 262, 266, 268, 272 **Grade 1:** 3, 29, 32, 37, 56, 59, 67, 95, 103, 127, 133, 141, 163, 195, 235, 271, 279, 299, 315, 329, AF4, AF11, AF16, AF27, AF30, AF34, AF50, AF54, AF58 **Grade 2:** 11, 29, 55, 59, 63, 89, 93, 95, 97, 106, 163, 187, 195, 201, 261, 331, 361, AF4, AF10, AF16, AF22, AF28, AF34, AF38, AF45, AF46, AF50, AF54, AF56, AF59 **Grade 3:** 26, 35, 43, 47, 53, 85, 91, 93, 114, 116–117, 144, 151, 173, 203, 207, 239, 251, 265, 268, 327, 334, 349, 363, 367, 373, 387, 397, 423, 433, 438, 453, 463, 479, 499, 511 **Grade 4:** 12, 21, 45, 57, 77, 129, 151, 203, 225, 251, 260, 261, 273, 277, 295, 299, 323, 339, 349, 352, 359, 467, 472, 523, 539, 561, 563, 573, 575
Understanding About Scientific Inquiry	
Scientific investigations involve asking and answering a question and comparing the answer with what scientists already know about the world.	**Grade K:** 30, 38, 44, 52, 58, 68, 76, 82, 88, 96, 102, 110, 116, 130, 136, 144, 150, 158, 164, 176, 182, 188, 196, 202, 216, 222, 228, 236, 246, 254, 260, 266, 272 **Grade 1:** 3, 11, 23, 29, 37, 53, 59, 67, 87, 95, 103, 109, 127, 133, 141, 163, 171, 179, 195, 201, 209, 229, 235, 241, 249, 265, 271, 279, 299, 307, 315, 329, 333, 341, 361, 367, 373, 381, 397, 403, 411, 419, AF4, AF30, AF34, AF50, AF54, AF55 **Grade 2:** 3, 11, 23, 29, 39, 55, 61, 69, 89, 95, 103, 121, 129, 135, 155, 163, 171, 187, 195, 201, 223, 231, 237, 253, 259, 267, 275, 295, 301, 309, 325, 331, 339, 361, 367, 377, 385, 399, 405, 415, 421, AF4, AF10, AF26, AF28, AF34, AF46, AF50, AF54, AF56, AF59 **Grade 3:** 3, 21, 31, 40–41, 43, 53, 69, 81, 91, 107, 119, 133, 144–145, 151, 161, 173, 191, 203, 213, 227, 239, 249, 259, 268–269, 279, 289, 303, 317, 327, 334–335, 337, 347, 363, 373, 383, 397, 407, 417, 422–423, 433, 443, 450–451, 453, 463, 479, 489, 493, 496–497, 499, 511 **Grade 4:** 3, 21, 33, 45, 56–57, 59, 77, 89, 99, 106–107, 109, 129, 137, 149, 165, 175, 183, 203, 213, 225, 237, 244–245, 251, 263, 270–271, 273, 282–283, 285, 295, 313, 323, 335, 345, 352–353, 359, 369, 379, 393, 400–401, 411, 421, 431, 445, 457, 467, 472–473, 483, 493, 503, 513, 522–523, 529, 539, 551, 560–561, 563, 572–573, 575

Understanding About Scientific Inquiry continued	
Scientists use different kinds of investigations depending on the questions they are trying to answer. Types of investigations include describing objects, events, and organisms; classifying them; and doing a fair test (experimenting).	**Grade K:** 30, 38, 44, 52, 58, 68, 76, 82, 88, 96, 102, 110, 116, 130, 136, 144, 150, 158, 164, 176, 182, 188, 196, 202, 216, 222, 228, 236, 246, 254, 260, 266, 272 **Grade 1:** 3, 11, 23, 25, 29, 32, 37, 40, 53, 56, 59, 63, 67, 70, 87, 91, 95, 99, 103, 107, 109, 113, 127, 130, 133, 135, 141, 144, 163, 166, 171, 173, 179, 183, 195, 198, 201, 204, 209, 211, 229, 232, 235, 238, 241, 243, 249, 251, 265, 267, 271, 273, 279, 281, 299, 301, 307, 311, 315, 319, 329, 331, 333, 335, 341, 343, 361, 364, 367, 370, 373, 376, 381, 384, 397, 399, 403, 405, 411, 414, 419, 421, AF8, AF27, AF30, AF34, AF38, AF50, AF54 **Grade 2:** 3, 11, 23, 26, 29, 32, 39, 42, 55, 59, 61, 63, 69, 73, 89, 93, 95, 97, 103, 106, 121, 124, 129, 132, 135, 137, 155, 161, 163, 166, 171, 173, 187, 191, 195, 199, 201, 207, 223, 226, 231, 234, 237, 240, 253, 256, 259, 261, 267, 272, 275, 278, 295, 298, 301, 304, 309, 313, 325, 328, 331, 335, 339, 344, 361, 364, 367, 371, 377, 380, 385, 388, 399, 403, 405, 407, 415, 418, 421, 424, AF4, AF8, AF22, AF26, AF28, AF34, AF38, AF50, AF54, AF56, AF59 **Grade 3:** 3, 21, 26, 31, 35, 43, 47, 53, 55, 69, 73, 81, 85, 91, 93, 107, 114, 119, 127, 133, 137, 151, 155, 161, 167, 173, 177, 191, 195, 203, 207, 213, 217, 227, 231, 239, 243, 249, 251, 259, 265, 279, 283, 289, 293, 303, 307, 317, 319, 327, 331, 337, 339, 347, 349, 363, 367, 373, 377, 383, 387, 397, 401, 407, 412, 417, 419, 433, 438, 443, 447, 453, 457, 463, 469, 479, 481, 489, 493, 499, 505, 511, 516 **Grade 4:** 3, 21, 27, 33, 39, 45, 53, 59, 64, 77, 80, 89, 93, 99, 102, 109, 116, 129, 131, 137, 143, 149, 151, 165, 169, 175, 177, 183, 186, 203, 207, 213, 219, 225, 231, 237, 241, 251, 255, 263, 267, 273, 277, 285, 289, 295, 299, 313, 317, 323, 328, 335, 339, 345, 349, 359, 363, 369, 373, 379, 384, 393, 397, 411, 415, 421, 425, 431, 435, 445, 449, 457, 461, 467, 470, 483, 487, 493, 498, 503, 507, 513, 515, 529, 533, 539, 544, 551, 557, 563, 569, 575, 580

Understanding About Scientific Inquiry continued	
Simple instruments, such as magnifiers, thermometers, and rulers, provide more information than scientists obtain using only their senses.	**Grade K:** 52, 58, 68, 76, 82, 102, 110, 117, 127, 130, 131, 133, 136, 139, 144, 150, 151, 153, 164, 165, 168F, 176, 177, 197, 199, 202, 203, 222, 223, 229, 231, 243, 246, 255, 261, 267, 268, 269, 272, 273, TR7, TR8, TR9, TR10, TR11 **Grade 1:** 29, 53, 87, 95, 127, 171, 229, 235, 249, 271, 307, 311, 315, 329, 341, 343, 376, 381, 384, 397, 403, 411, 419, R3, R4, R5, R6, AF4, AF34 **Grade 2:** 3, 29, 39, 69, 89, 95, 103, 129, 135, 171, 187, 195, 223, 231, 253, 259, 267, 275, 301, 309, 325, 339, 367, 377, 385, 399, 405, 415, 421, R3, R4, R5, R6, AF28, AF34, AF46, AF50, AF54, AF56, AF59 **Grade 3:** 26, 31, 40, 41, 53, 69, 81, 107, 119, 133, 144, 145, 161, 173, 200–201, 213, 231, 239, 243, 259, 268, 269, 307, 317, 334–335, 339, 347, 349, 363, 373, 380, 381, 383, 397, 407, 417, 422, 423, 438, 450, 460–461, 481, 486–487, 489, 496–497, 499, 511, 516, R2, R4, R5, R6, R7, R8, R9 **Grade 4:** 21, 30, 33, 39, 45, 56, 93, 99, 106–107, 109, 129, 134–135, 137, 143, 149, 172–173, 175, 237, 241, 251, 263, 267, 270–271, 285, 289, 295, 332–333, 352–353, 359, 363, 369, 379, 393, 400, 411, 421, 425, 431, 445, 461, 464–465, 467, 483, 490–491, 493, 503, 513, 522–523, 536–537, 539, 551, 560–561, 575, 580, R4, R5, R6, R7, R8, R9
Scientists develop explanations using observations (evidence) and what they already know about the world (scientific knowledge). Good explanations are based on evidence from investigations.	**Grade K:** 7, 30, 58, 76, 79, 99, 113, 158, 176, 203, 228, 243, 246, 260 **Grade 1:** 11, 141, 397, 403, 411, 419, AF4, AF11, AF16, AF30, AF50, AF54 **Grade 2:** 11, 129, 171, 201, 237, 259, 267, 275, 309, 325, 361, AF4, AF10, AF22, AF26, AF28, AF34, AF38, AF46, AF50, AF54, AF56, AF59 **Grade 3:** 21, 31, 41, 43, 53, 69, 81, 91, 107, 119, 133, 145, 151, 161, 173, 191, 203, 213, 227, 239, 249, 259, 269, 279, 289, 303, 317, 327, 335, 337, 347, 363, 373, 383, 397, 407, 417, 422, 423, 433, 443, 450, 453, 463, 479, 489, 497, 499, 511 **Grade 4:** 21, 33, 45, 57, 59, 77, 89, 99, 107, 109, 129, 137, 149, 165, 175, 183, 203, 213, 225, 237, 245, 251, 263, 271, 273, 283, 285, 295, 313, 323, 335, 345, 352, 359, 369, 379, 393, 401, 411, 421, 431, 445, 457, 467, 473, 483, 493, 503, 513, 523, 529, 539, 551, 561, 563, 573, 575

Understanding About Scientific Inquiry continued	
Scientists make the results of their investigations public; they describe the investigations in ways that enable others to repeat the investigations.	**Grade K:** 6, 7, 12, 13, 19, 26, 27, 30, 31, 32, 33, 38, 39, 41, 44, 45, 46, 47, 52, 53, 54, 55, 58, 59, 64, 65, 68, 69, 70, 71, 76, 77, 79, 83, 88, 90, 91, 97, 98, 99, 102, 103, 104, 105, 110, 111, 113, 116, 117, 126, 127, 132, 133, 136, 137, 138, 139, 146, 147, 150, 151, 152, 153, 158, 159, 160, 161, 164, 165, 170, 171, 177, 178, 179, 182, 183, 184, 185, 188, 189, 190, 191, 196, 197, 198, 199, 203, 212, 213, 217, 219, 222, 223, 224, 225, 229, 231, 236, 237, 242, 243, 246, 247, 248, 255, 256, 257, 261, 262, 263, 266, 267, 269, 273 **Grade 1:** 3, 29, 32, 37, 56, 59, 67, 95, 103, 127, 133, 141, 163, 195, 235, 271, 279, 299, 315, 329, AF11, AF16, AF30, AF34, AF50 **Grade 2:** 11, 29, 55, 59, 63, 89, 93, 95, 97, 106, 163, 187, 195, 201, 261, 331, 361, AF10, AF54 **Grade 3:** 7, 26, 35, 43, 47, 53, 85, 91, 93, 114, 116–117, 144, 151, 173, 203, 207, 239, 251, 265, 268, 327, 334, 349, 363, 367, 373, 387, 397, 423, 433, 438, 453, 463, 479, 499, 511 **Grade 4:** 12, 21, 45, 57, 77, 129, 151, 203, 225, 251, 260, 261, 273, 277, 295, 299, 323, 339, 349, 352, 359, 467, 472, 523, 539, 561, 563, 573, 575
Scientists review and ask questions about the results of other scientists' work.	**Grade 2:** 280–281, AF16 **Grade 3:** 10–11, 180–181 **Grade 4:** 106–107, 387

Physical Science

Properties of Objects and Materials	
Objects have many observable properties, including size, weight, shape, color, temperature, and the ability to react with other substances. Those properties can be measured using tools, such as rulers, balances, and thermometers.	**Grade K:** 210E, 211, 212, 213, 214, 215, 218, 223, 224, 226, 227, 228, 230, 231, 232, 233, 234, 235, 236, 237, 238, 239, 240J, 249, 254, 255, 260, 268, 269, 270, 271, 273, 274, 275, 276 **Grade 1:** 298, 299, 300, 301, 302, 303, 304, 305, 306, 307, 308–311, 314, 315, 316–319, 320–323, 324, 325, 328, 329, 330–331, 332–337, 338, 339, 341, 342–345, 346, 347, 348–351, 352, 353, AF45, AF54 **Grade 2:** 295, 296, 297, 298, 299, 300, 301, 302, 303, 304, 305, 306–307, 308, 309, 310, 311, 312, 313, 314, 315, 316–319, 320, 321, 324, 325, 326–329, 331, 332–335, 336–337, 348–351, 352, 353, AF45 **Grade 3:** 358–359, 362, 363, 364, 365, 366, 367, 368, 369, 372, 373, 374, 375, 376, 377, 378, 379, 380–381, 383, 384–389, 390, 391, 392, 393, 406, 407, 408–413, 414–415, 416, 417, 418–421, 422–423, 424, 425, 479, 482, 483, 485 **Grade 4:** 411, 412, 413, 414, 415, 416, 417, 418, 419, 421, 422, 423, 424, 425, 426, 427, 428–429, 431, 432, 433, 434, 435, 436, 437, 440, 441, 445, 446, 447, 448, 449, 450, 451, 452, 453, 454–455, 456, 463, 467, 468, 469, 470, 471, 472–473, 474, 475

Correlation to National Standards

Properties of Objects and Materials continued	
Objects are made of one or more materials, such as paper, wood, and metal. Objects can be described by the properties of the materials from which they are made, and those properties can be used to separate or sort a group of objects or materials.	**Grade K:** 210E, 210J, 210, 212, 214, 215, 216, 217, 218, 219, 220, 221, 222, 225, 226, 229, 239, 271 **Grade 1:** 312–313, 325, 346–347 **Grade 2:** 306–307, 336–337, 339, 340, 341, 342, 343, 344, 345, 346, 347, 348–351, 352, 353 **Grade 3:** 366, 367, 368, 369, 370–371, 409, 410, 411, 412, 413, 414–415, 424, 425 **Grade 4:** 54, 406–407, 444, 456, 457, 458, 459, 460, 461, 462, 463, 474, 475
Materials can exist in different states—solid, liquid, and gas. Some common materials, such as water, can be changed from one state to another by heating or cooling.	**Grade K:** 230, 231, 232, 233, 237, 238, 239 **Grade 1:** 294, 295, 306, 307, 308, 309, 310, 311, 314, 315, 316, 317, 318, 319, 320–323, 324, 325, 341, 342, 343, 344, 345, 352, 353, AF50, AF54 **Grade 2:** 299, 300–305, 308, 309, 310–313, 314, 315, 316–319, 320, 321, 327, 330, 331, 332–335, 336–337, 348–351, 352, 353, 400, AF41, AF46 **Grade 3:** 380–381, 382, 383, 384, 385, 386, 387, 388, 389, 392, 393, 396, 397, 398, 399, 400, 401, 402, 403, 404–405, 408, 424, 425, 483, 485 **Grade 4:** 323, 324, 325, 331, 410, 414, 415, 417, 446, 447, 448, 449, 453, 534, 535
Position and Motion of Objects	
The position of an object can be described by locating it relative to another object or the background.	**Grade 1:** 361, 362, 363, 365, 392, 393 **Grade 2:** 360, 361, 362, 363, 365, 394, 395 **Grade 3:** 432, 433, 434, 435, 439 **Grade 4:** 484, 485, 503
An object's motion can be described by tracing and measuring its position over time.	**Grade K:** 240F, 242, 243, 250, 251, 252, 253, 255 **Grade 1:** 356–357, 364, 365, 392 **Grade 2:** 363, 364, 365, 367, 372, 373, 383, 390–393, 394, 395 **Grade 3:** 429, 436, 437, 438, 439, 443, 449 **Grade 4:** 482, 483, 484, 485, 486, 487, 488, 489, 490–491, 496, 497, 498, 499, 503, 524, 525
The position and motion of objects can be changed by pushing or pulling. The size of the change is related to the strength of the push or pull.	**Grade K:** 240, 241, 244, 245, 246, 247, 248, 249, 252, 253, 255, 274, 275, 277 **Grade 1:** 366, 367, 368, 369, 370, 371, 373, 388–391, 392, 393 **Grade 2:** 367, 368, 369, 370, 371, 372, 373, 390–393, 394, 395 **Grade 3:** 442, 443, 444, 445, 446, 447, 448, 449, 450–451 **Grade 4:** 486, 487, 488, 489, 490–491, 492, 493, 494, 495, 496, 497, 498, 499, 500, 501, 503, 504, 505, 521, 524, 525

Position and Motion of Objects continued	
Sound is produced by vibrating objects. The pitch of the sound can be varied by changing the rate of vibration.	**Grade K:** 264, 265, 266, 267, 274, 275 **Grade 1:** 402, 403, 404, 405, 406, 407, 428, 429 **Grade 2:** 356–357, 405, 406, 407, 409, 410, 411, 412, 429, 432 **Grade 3:** 489, 490, 491, 492, 493, 494, 495, 496–497, 520, 521 **Grade 4:** 538, 539, 540, 541, 542, 543, 544, 545, 546, 547, 548, 549, 588, 589
Light, Heat, Electricity, and Magnetism	
Light travels in a straight line until it strikes an object. Light can be reflected by a mirror, refracted by a lens, or absorbed by the object.	**Grade K:** 198, 199, 200, 201, 202, 203 **Grade 1:** 411, 412, 413, 414, 415, 416, 417, 426, 429 **Grade 2:** 414, 416, 417, 419, 430, 432 **Grade 3:** 498, 499, 500, 501, 502, 503, 504, 505, 506, 507, 508–509, 520, 521 **Grade 4:** 506, 509, 550, 551, 552, 553, 554, 556, 557, 558, 559, 560–561, 588, 589
Heat can be produced in many ways, such as burning, rubbing, or mixing one substance with another. Heat can move from one object to another by conduction.	**Grade 1:** 396, 397, 400, 401, 428, 429, AF47, AF58 **Grade 2:** 399, 400, 401, 403, 428, 433 **Grade 3:** 478, 479, 480, 481, 482, 483, 484, 485, 486–487, 520, 521 **Grade 4:** 507, 509, 530, 531, 532, 533, 534, 535, 536–537, 588, 589
Electricity in circuits can produce light, heat, sound, and magnetic effects. Electrical circuits require a complete loop through which an electrical current can pass.	**Grade 1:** 418, 419, 420, 421, 422, 423, 427, 428, 429 **Grade 2:** 420, 421, 422, 423, 425, 426–427, 431, 432, 433 **Grade 3:** 511, 514, 515, 516, 517, 518, 520, 521 **Grade 4:** 508, 509, 562, 566, 567, 568, 569, 570, 571, 580, 581, 582, 583, 584, 585, 586–587, 588
Magnets attract and repel each other and certain kinds of other materials.	**Grade K:** 240J, 268, 269, 270, 271, 272, 273, 274, 275 **Grade 1:** 380, 381, 382, 383, 384, 385, 386, 387, 392 **Grade 2:** 384, 385, 386, 387, 388, 389, 393, AF54 **Grade 3:** 367, 446, 450–451 **Grade 4:** 574, 575, 576, 577, 578, 579, 580, 581, 582, 583, 585, 588, 589

Life Science

The Characteristics of Organisms	
Organisms have basic needs. For example, animals need air, water, and food; plants require air, water, nutrients, and light. Organisms can survive only in environments in which their needs can be met. The world has many different environments, and distinct environments support the life of different types of organisms.	**Grade K:** 26, 32, 33, 34, 35, 37, 38, 39, 43, 44, 45, 60, 61, 67, 70, 71, 72, 73, 74, 75, 77, 82, 83, 92, 93, 98, 120, TR3, TR4, TR5, TR6 **Grade 1:** 24, 26, 27, 30, 42–43, 49, 50, 56, 58, 59, 67, 95, 96, 97, 99, 100, 104, 118, 135, 202, AF11 **Grade 2:** 23, 24, 25, 26, 27, 47, 51, 56, 59, 77, 89, 90, 93, 94 **Grade 3:** 20, 24, 25, 27, 32, 33, 36, 37, 39, 40–41, 43, 46, 47, 49, 64, 65, 69, 70, 71, 78–79, 82, 119, 120, 121, 122–128 **Grade 4:** 22, 46, 48, 49, 50, 51, 55, 59, 64, 72

The Characteristics of Organisms continued	
Each plant or animal has different structures that serve different functions in growth, survival, and reproduction. For example, humans have distinct body structures for walking, holding, seeing, and talking.	**Grade K:** 5, 10–11, 26, 27, 28, 29, 30, 31, 42, 43, 60, 61, 78, 80, 81, 86, 87, 88, 90, 91, 92, 93, 94, 95, 97, 99, 100, 101, 103, 105, TR9 **Grade 1:** 19, 28, 29, 31, 32, 33, 35, 49, 52, 54, 55, 56, 57, 59, 60, 62, 63, 64–65, 68, 69, 70, 71, 78, 79, 82–83, 89, 90, 92, 93, 95, 98, 99, 102, 103, 105, 106, 107, 114, 115, 123, 129, 130, 131, 133, 136, 137, 138, 139, 154, AF13, AF16, AF52 **Grade 2:** 11, 12, 13, 26, 27, 30, 31, 32, 33, 34, 35, 36, 39, 42, 46–49, 50, 57, 58, 59, 69, 70, 71, 72, 73, 74, 77, 78, 80, 89, 93, 123, 125, 129, 130, 131, 132, 133, 137, 139, R8, R9, R10, R11, AF4 **Grade 3:** 28–29, 31, 32, 33, 34, 35, 36, 37, 38, 39, 42, 43, 44, 45, 46, 47, 48, 49, 50–51, 53, 54, 55, 56, 57, 58, 59, 60, 61, 62, 64, 65, 70, 72, 73, 75, 76, 77, 102–103, 132, 133, 134, 135, 136, 137, 138, 139, 140, 141, 142, 143, 144–145, 146, 147, 449, 494, 506, R14, R15, R16, R17, R18, R19, R20, R21, R22 **Grade 4:** 17, 27, 45, 46, 47, 49, 50, 51, 52, 53, 55, 56–57, 60, 61, 62, 63, 70, 71, 72, 82, 83, 84, 85, 88, 89, 90, 91, 92, 93, 95, 98, 100, 101, 102, 103, 104, 105, 106–107, 110, 120, 121, 124–125, 164, 165, 166, 167, 168, 169, 171, 172–173, 174, 178, 179, 194, 195, 528, 529, R14, R15, R16, R17, R18, R19, R20, R21, R22
The behavior of individual organisms is influenced by internal cues (such as hunger) and by external cues (such as a change in the environment). Humans and other organisms have senses that help them detect internal and external cues.	**Grade 1:** 17, 19 **Grade 2:** 42, 58, 72, 73, R9, AF4 **Grade 3:** 24, 25, 44, 49, R21 **Grade 4:** 99, 100, 101, 105, 120, 478–479
Life Cycles of Organisms	
Plants and animals have life cycles that include being born, developing into adults, reproducing, and eventually dying. The details of this life cycle are different for different organisms.	**Grade K:** 40, 41, 42, 43, 44, 47, 104, 105, 106, 107, 108, 109, 110, 111, 119 **Grade 1:** 54, 57, 60, 61, 62, 63, 78, 79, 91, 110, 111, 112, 113, 114, 115, 116, 117, 123 **Grade 2:** 8, 9, 25, 30, 31, 34, 35, 48, 49, 50, 62, 63, 64, 65, 67, 78, 80, 81, AF10, AF13 **Grade 3:** 17, 23, 27, 64, 68, 70, 71, 72, 73, 74, 75, 76, 77, 80, 81, 82, 83, 84, 85, 86, 87, 88, 98, 99 **Grade 4:** 47, 52, 53, 58, 59, 62, 63, 64, 65, 68, 69, 72, 94, 108, 109, 110, 111, 112, 113, 114, 115, 117, 118–119, 120, 158
Plants and animals closely resemble their parents.	**Grade K:** 104, 106, 107, 108, 109, 115 **Grade 1:** 108, 109, 110, 111, 112, 113, 123 **Grade 2:** 24, 35, 40, 41, 43, 47, 61 **Grade 3:** 22, 84, 85, 86, 88, 90, 91, 92, 93 **Grade 4:** 66, 69, 112

Life Cycles of Organisms continued	
Many characteristics of an organism are inherited from the parents of the organism, but other characteristics result from an individual's interactions with the environment. Inherited characteristics include the color of flowers and the number of limbs of an animal. Other features, such as the ability to ride a bicycle, are learned through interactions with the environment and cannot be passed on to the next generation.	**Grade K:** 109 **Grade 1:** 113 **Grade 2:** 42, 43, 47, 75, 78 **Grade 3:** 22, 23, 86, 91, 92, 93, 94, 95, 96–97, 98, 99 **Grade 4:** 66, 67, 69, 114, 116, 117, 478–479
Organisms and Their Environments	
All animals depend on plants. Some animals eat plants for food. Other animals eat animals that eat the plants.	**Grade K:** 56, 57, 58, 59 **Grade 1:** 12, 13, 14, 104, 105, 142, 144, 145, 154, 155, 198 **Grade 2:** 27, 92, 94, 95, 96, 97, 98, 99, 100, 101, 116, 117, 136, 138 **Grade 3:** 106, 107, 108, 109, 110, 111, 112, 113, 114, 115, 116–117, 146, 147 **Grade 4:** 44, 54, 124–125, 131, 148, 149, 150, 151, 152, 153, 154, 155, 156, 157, 158, 159, 160, 161, 164, 198–199
An organism's patterns of behavior are related to the nature of that organism's environment, including the kinds and numbers of other organisms present, the availability of food and resources, and the physical characteristics of the environment. When the environment changes, some plants and animals survive and reproduce, and others die or move to new locations.	**Grade K:** 89, 102, 103 **Grade 1:** 11, 12, 13, 14, 15, 66, 67, 68, 69, 70, 71, 97, 126, 127, 128, 130, 131, 132, 133, 134, 135, 136, 137, 138, 141, 144, 145, 150–153, 154, 155, 241, AF19 **Grade 2:** 42, 43, 90, 91, 92, 104, 105, 106, 108, 109, 110–111, 112–115, 116, 117, 123, 130, 131, 132, 133, 135, 136, 137, 138, 139, 142–145, 147, 150–151 **Grade 3:** 107, 108, 109, 115, 123, 125, 126, 128, 136, 137, 138, 139, 141, 142, 143, 146, 147, 152, 153, 164, 165, 166, 167, 168, 169, 170, 171, 174, 175, 179, 182, 183, 274–275 **Grade 4:** 23, 96, 98, 99, 118–119, 121, 124–125, 141, 142, 143, 144, 155, 158, 168, 170, 171, 175, 176, 177, 179, 180, 181, 185, 188, 189, 194, 195, 238
All organisms cause changes in the environment where they live. Some of these changes are detrimental to the organism or other organisms, whereas others are beneficial.	**Grade K:** 124–125, 154, 155, 160, 161, 162, 163, 164, 165 **Grade 1:** 143, 146, 147, 195, AF30 **Grade 2:** 105, 106, 107, 109, 126–127 **Grade 3:** 28–29, 114, 115, 128, 150, 151, 152, 153, 154, 155, 156, 157, 182, 183 **Grade 4:** 42–43, 124–125, 185, 191
Humans depend on their natural and constructed environments. Humans change environments in ways that can be either beneficial or detrimental for themselves and other organisms.	**Grade K:** 124–125, 150, 152, 153, 154, 155, 156, 157, 158, 159, 160, 161, 162, 163, 164, 165 **Grade 1:** 64–65, 146, 147, 195, 204, 205, 246–247 **Grade 2:** 44, 105, 106, 107, 109, 126–127 **Grade 3:** 28–29, 154, 155, 156, 157, 158–159, 170–171, 182, 183, 184, 210–211, 218, 219 **Grade 4:** 54, 96, 118–119, 146–147, 159, 186, 187, 190, 191, 232, 233

Correlation to National Standards

Earth and Space Science

Properties of Earth Materials

Standard	Pages
Earth materials are solid rocks and soils, water, and the gases of the atmosphere. The varied materials have different physical and chemical properties, which make them useful in different ways, for example, as building materials, as sources of fuel, or for growing the plants we use as food. Earth materials provide many of the resources that humans use.	**Grade K:** 124E, 126, 127, 128, 129, 130, 131, 132, 133, 134, 135, 136, 137, 146, 147, 148, 149, 151, 166, 167 **Grade 1:** 64–65, 158–159, 162, 163, 164, 165, 166, 167, 168, 169, 170, 171, 172, 173, 174, 175, 176–177, 179, 185, 186–189, 190, 191, 196, 197, 198, 199, 200, 201, 202, 203, 204, 205, 206–207, 312–313, AF23 **Grade 2:** 155, 162, 163, 164, 165, 166, 167, 168, 169, 171, 172, 173, 178–181, 182, 183, 186, 187, 188, 189, 190, 191, 192, 193, 194, 195, 196, 197, 198, 199, 200, 201, 202, 203, 210–213, 214, 215, 244–247, 401 **Grade 3:** 190, 191, 192, 193, 194, 195, 196, 197, 199, 200–201, 213, 214, 215, 216, 219, 220, 221, 222, 223, 226, 227, 228, 229, 230, 231, 232, 233, 234, 235, 236, 238, 239, 240, 241, 242, 243, 244, 245, 251, 258, 259, 260, 261, 262, 263, 267, 268–269, 270, 271, 278, 279, 280, 281 **Grade 4:** 250, 251, 252, 253, 254, 255, 256, 257, 258, 259, 260–261, 262, 263, 264, 265, 266, 267, 268, 269, 270–271, 284, 285, 286, 287, 288, 289, 290, 291, 304, 305, 314, 315, 319, 428–429, 438–439
Soils have properties of color and texture, capacity to retain water, and ability to support the growth of many kinds of plants, including those in our food supply.	**Grade K:** 126, 127, 128, 129, 130, 131 **Grade 1:** 174, 175, 190, 198, 199, AF27 **Grade 2:** 194, 195, 196, 197, 198, 199, 203, 210, 211, 214, 215, AF28 **Grade 3:** 238, 239, 240, 241, 242, 243, 244, 245, 246, 247 **Grade 4:** 262, 263, 264, 265, 266, 267, 268, 269, 270–271, 304, 305
Fossils provide evidence about the plants and animals that lived long ago and the nature of the environment at that time.	**Grade 1:** 148–149 **Grade 2:** 108, 109, 110, 111, AF16 **Grade 3:** 172, 173, 174, 175, 176, 177, 178, 179, 180–181, 182, 183, 248, 249, 250, 251, 255, 270 **Grade 4:** 189, 255, 272, 273, 274, 275, 276, 277, 281, 282–283, 304

Objects in the Sky	
The sun, moon, stars, clouds, birds, and airplanes all have properties, locations, and movements that can be observed and described.	**Grade K:** 190, 191, 192, 193, 194, 195, 196, 197, 203, 205, 256, 257, 258, 259, 261 **Grade 1:** 235, 237, 238, 239, 262, 264, 265, 266, 267, 268, 269, 270, 272, 273, 276, 278, 279, 280, 281, 286–289, 290, 291, AF38, AF40 **Grade 2:** 232, 233, 234, 236, 237, 238, 239, 248, 249, 253, 254–257, 262, 263, 265, 266, 267, 268, 269, 270, 271, 272, 273, 280–281, 286, 287, 374–375, AF38 **Grade 3:** 273, 290, 291, 292, 293, 316, 317, 318, 319–321, 322, 323, 326, 327, 328, 329, 330, 331, 332, 333, 334–335, 346, 347, 348, 349, 350, 351, 352–353, 354, 355, 370–371 **Grade 4:** 326, 328, 329, 331, 343, 358, 361, 363, 364, 365, 368, 369, 370, 371, 372, 373, 374, 375, 390–391, 392, 393, 394, 395, 396, 397, 398, 399, 400–401, 402
The sun provides the light and heat necessary to maintain the temperature of the earth.	**Grade K:** 198, 199, 200, 201, 202, 203 **Grade 1:** 243, 245, 267, 268, 269, 271, 273, 286–289, 290, 291, AF40, AF58 **Grade 2:** 254–257, 262, 263, 265, 273, 400, 416, AF59 **Grade 3:** 294, 305, 316, 319, 320, 321, 322, 323, 480 **Grade 4:** 48, 50, 360, 366, 367, 398, 506
Changes in the Earth and Sky	
The surface of the earth changes. Some changes are due to slow processes, such as erosion and weathering, and some changes are due to rapid processes, such as landslides, volcanic eruptions, and earthquakes.	**Grade K:** 138, 140, 141, 142, 143, 144, 148, 149 **Grade 1:** 179, 180, 181, 182, 183, 184, 191 **Grade 2:** 170, 171, 172, 173, 174, 175, 176–177, 178–181, 182, 183 **Grade 3:** 202, 203, 204, 205, 206, 207, 208, 209, 210–211, 212, 213, 214, 215, 216, 217, 219, 220, 221, 222, 223 **Grade 4:** 3–11, 184, 197, 198–199, 202, 204, 206, 210–211, 213, 214, 215, 216, 217, 218, 220, 221, 224, 225, 226, 227, 228, 229, 230, 231, 233, 234, 235, 236, 237, 238, 242, 243, 244–245, 246, 247

Correlation to National Standards

Changes in the Earth and Sky continued	
Weather changes from day to day and over the seasons. Weather can be described by measurable quantities, such as temperature, wind direction and speed, and precipitation.	**Grade K:** 4–5, 6, 7, 168E, 168F, 168J, 168, 169, 170, 171, 172, 173, 174, 175, 176, 177, 178, 179, 180, 181, 182, 183, 184, 185, 186, 187, 188, 189, 204, 205, 206, 207 **Grade 1:** 228, 229, 230, 231, 232, 233, 235, 236, 237, 238, 239, 240, 241, 242, 243, 244, 246–247, 248, 249, 250, 251, 252, 253, 254, 255, 256–259, 260, 261, 275, AF32, AF34 **Grade 2:** 176–177, 220, 222, 223, 224, 225, 226, 227, 228, 229, 236, 237, 240, 241, 242–243, 248, 249, 258, 259, 260, 261, 262, 263, 264, 265, AF32 **Grade 3:** 274–275, 278, 280, 281, 282, 283, 284, 285, 286–287, 288, 289, 290, 291, 292, 293, 294, 295, 296, 297, 298, 299, 300–301, 303, 304, 305, 306, 307, 308, 309, 312, 313, 321, 324 **Grade 4:** 240, 241, 243, 312, 313, 316, 317, 318, 319, 320, 321, 325, 327, 328, 329, 330, 331, 334, 335, 336, 337, 338, 339, 340, 341, 342–343, 345, 346, 350, 351, 352–353, 354, 355
Objects in the sky have patterns of movement. The sun, for example, appears to move across the sky in the same way every day, but its path changes slowly over the seasons. The moon moves across the sky on a daily basis much like the sun. The observable shape of the moon changes from day to day in a cycle that lasts about a month.	**Grade K:** 190, 192, 193, 194, 195, 196, 197, 199, 200, 201, 203, 205 **Grade 1:** 272, 273, 274, 275, 276, 279, 280, 281, 282, 283, 284–285, 286–289, 290, 291, AF38 **Grade 2:** 253, 254, 255, 256, 257, 262, 263, 268, 269, 270, 271, 273, 275, 276–279, 280–281, 282–285, 286, 287, AF34, AF38 **Grade 3:** 316, 317, 318, 319, 320, 321, 323, 324, 327, 328, 329, 330, 331, 332, 333, 334–335, 350, 354, 355 **Grade 4:** 358, 359, 360, 361, 362, 363, 364, 365, 372, 373, 374, 375, 376–377, 402, 403

Science and Technology

Abilities of Technological Design	
Identify a simple problem. In problem identification, children should develop the ability to explain a problem in their own words and identify a specific task and solution related to the problem.	**Grade 1:** 229, 249, 271, 403, 421 **Grade 2:** 377, 380, 405, 421 **Grade 3:** 144–145, 289, 443, 460–461, 463, 511 **Grade 4:** 106–107, 210–211, 335, 483, 503 See also *Technology A Closer Look*, **Grades K–2 and 3–4.**
Propose a solution. Students should make proposals to build something or get something to work better; they should be able to describe and communicate their ideas. Students should recognize that designing a solution might have constraints, such as cost, materials, time, space, or safety.	**Grade K:** 247 **Grade 1:** 271, 373, 376, 405 **Grade 2:** 234, 367, 380, 405, 411 **Grade 3:** 203, 289, 443, 450–451, 463, 475, 486–487, 489, 511 **Grade 4:** 89, 102, 335, 483, 493, 503, 525 See also *Technology A Closer Look*, **Grades K–2 and 3–4.**

Abilities of Technological Design continued	
Implementing proposed solutions. Children should develop abilities to work individually and collaboratively and to use suitable tools, techniques, and quantitative measurements when appropriate. Students should demonstrate the ability to balance simple constraints in problem solving.	**Grade K:** 246 **Grade 1:** 204, 249, 373, 376, 403, 405 **Grade 2:** 171, 267, 380, 421 **Grade 3:** 55, 293, 463, 469, 486–487 **Grade 4:** 102, 210–211, 483, 503, 522–523 See also *Technology A Closer Look*, **Grades K–2 and 3–4.**
Evaluate a product or design. Students should evaluate their own results or solutions to problems, as well as those of other children, by considering how well a product or design met the challenge to solve a problem. When possible, students should use measurements and include constraints and other criteria in their evaluations. They should modify designs based on the results of evaluations.	**Grade 1:** 229, 249, 373, 379, 403 **Grade 2:** 129, 367, 411 **Grade 3:** 203, 283, 443, 463, 469, 486–487, 495, 511 **Grade 4:** 89, 106–107, 210–211, 483, 493, 503, 522–523 See also *Technology A Closer Look*, **Grades K–2 and 3–4.**
Communicate a problem, design, and solution. Student abilities should include oral, written, and pictorial communication of the design process and product. The communication might be show and tell, group discussions, short written reports, or pictures, depending on the students' abilities and the design project.	**Grade K:** 150 **Grade 1:** 95, 229, 249, 271, 373, 379, 403 **Grade 2:** 95, 421 **Grade 3:** 289, 443, 460–461, 463 **Grade 4:** 106–107, 490–491, 522–523 See also *Technology A Closer Look*, **Grades K–2 and 3–4.**
Understanding About Science and Technology	
People have always had questions about their world. Science is one way of answering questions and explaining the natural world.	**Grade K:** 4–5, 6, 7, 10–11, 12–13, 16–17, 18–19 **Grade 1:** 206–207, 284–285, 292 **Grade 2:** 110–111, 242–243, 276, 277, 280–281 **Grade 3:** 4, 130–131, 352–353, 370–371 **Grade 4:** 3–11, 38, 276–277, 282–283, 381, 382, 406–407, 438–439, 552
People have always had problems and invented tools and techniques (ways of doing something) to solve problems. Trying to determine the effects of solutions helps people avoid some new problems.	**Grade K:** 150 **Grade 1:** 374–377, 378–379, 398, 399, 408–409, 414, 415, 421, 422 **Grade 2:** 176–177, 189, 242–243, 280–281, 354, 426–427 **Grade 3:** 28–29, 210–211, 254, 262, 263, 300–301, 342, 440–441, 471 **Grade 4:** 43, 218, 219, 279, 280, 288, 289, 302–303, 382, 383, 389, 390–391, 397, 477, 510–511, 586–587 See also *Technology A Closer Look*, **Grades K–2 and 3–4.**
Scientists and engineers often work in teams with different individuals doing different things that contribute to the results. This understanding focuses primarily on teams working together and secondarily, on the combination of scientist and engineer teams.	**Grade 2:** 82, 110–111, 288 **Grade 3:** 3–11, 180–181, 187 **Grade 4:** 3–11, 119 See also *Technology A Closer Look*, **Grades K–2 and 3–4.**

Correlation to National Standards

Understanding About Science and Technology continued	
Women and men of all ages, backgrounds, and groups engage in a variety of scientific and technological work.	**Grade K:** 116, 252 **Grade 1:** 80, 116–117, 148–149, 156, 176–177, 206–207, 222, 284–285, 292, 430 **Grade 2:** 66–67, 82, 110–111, 126–127, 148, 216, 288, 354, 374–375, 434 **Grade 3:** 96–97, 100, 130–131, 184, 272, 352–353, 356, 370–371, 426, 522 **Grade 4:** 118–119, 122, 196, 222–223, 306, 404, 438–439, 476, 590 See also *Technology A Closer Look*, **Grades K–2 and 3–4.**
Tools help scientists make better observations, measurements, and equipment for investigations. They help scientists see, measure, and do things that they could not otherwise see, measure, and do.	**Grade K:** TR7, TR8 **Grade 1:** 266, 284–285, 292, R2, R3, R4, R5, R6, R7 **Grade 2:** 224, 225, 226, 227, 242–243, 374–375, R3, R4, R5, R6 **Grade 3:** 6, 7, 26, 209, 281, 282, 283, 284, 300–301, 342, 344–345, 356, 379, R2, R4, R5, R6, R7, R8, R9 **Grade 4:** 28, 218, 219, 276, 318, 319, 342–343, 381, 382, 383, 390–391, 392, 397, 531, R2, R3, R4, R5, R6, R7, R8, R9 See also *Technology A Closer Look*, **Grades K–2 and 3–4.**
Abilities to Distinguish Between Natural Objects and Objects Made by Humans	
Some objects occur in nature; others have been designed and made by people to solve human problems and enhance the quality of life.	**Grade K:** 276–277 **Grade 1:** 195, 312–313, 325, 414 **Grade 2:** 188, 189, 190, 202, 203, 208–209, 216, 306–307 **Grade 3:** 210–211, 252, 253, 262, 263, 414–415 **Grade 4:** 54, 258, 406–407, 459, 463, 474, 476, 506 See also *Technology A Closer Look*, **Grades K–2 and 3–4.**
Objects can be categorized into two groups, natural and designed.	**Grade K:** 276–277 **Grade 1:** 312–313, 414 **Grade 2:** 190 **Grade 3:** 409, 414–415 **Grade 4:** 54, 463, 476 See also *Technology A Closer Look*, **Grades K–2 and 3–4.**

Science in Personal and Social Perspectives

Personal Health	
Safety and security are basic needs of humans. Safety involves freedom from danger, risk, or injury. Security involves feelings of confidence and lack of anxiety and fear. Student understandings include following safety rules for home and school, preventing abuse and neglect, avoiding injury, knowing whom to ask for help, and when and how to say no.	**Grade K:** TR13, TR14 **Grade 1:** 16, 103, 127, 163, 403, R16, R17 **Grade 2:** R16, R17, R18 **Grade 3:** 14, 49, 299, 323, 517 **Grade 4:** 239, 340, 341, 398, 570, 571

Personal Health continued	
Individuals have some responsibility for their own health. Students should engage in personal care—dental hygiene, cleanliness, and exercise—that will maintain and improve health. Understandings include how communicable diseases, such as colds, are transmitted and some of the body's defense mechanisms that prevent or overcome illness.	**Grade K:** TR9, TR10 **Grade 1:** R14 **Grade 2:** R9, R14, R15 **Grade 3:** 323, R23, R26 **Grade 4:** 319, 398, R23, R26
Nutrition is essential to health. Students should understand how the body uses food and how various foods contribute to health. Recommendations for good nutrition include eating a variety of foods, eating less sugar, and eating less fat.	**Grade K:** TR11 **Grade 1:** R13 **Grade 2:** R12, R13 **Grade 3:** 27, R23, R24, R25, R26 **Grade 4:** R23, R24, R25, R26
Different substances can damage the body and how it functions. Such substances include tobacco, alcohol, over-the-counter medicines, and illicit drugs. Students should understand that some substances, such as prescription drugs, can be beneficial, but that any substance can be harmful if used inappropriately.	**Grade 1:** R14, R15 **Grade 2:** R15 **Grade 3:** R23, R26 **Grade 4:** R26
Characteristics and Changes in Populations	
Human populations include groups of individuals living in a particular location. One important characteristic of a human population is the population density—the number of individuals of a particular population that live in a given amount of space.	**Grade 4:** 145, 186
The size of a human population can increase or decrease. Populations will increase unless other factors such as disease or famine decrease the population.	This standard is covered in the Macmillan/McGraw-Hill *Health & Wellness* program.
Types of Resources	
Resources are things that we get from the living and nonliving environment to meet the needs and wants of a population.	**Grade K:** 146, 152, 153, 154, 155, 156, 157, 158, 159, 160, 161, 162, 163, 164, 165 **Grade 1:** 176–177, 195, 196, 197, 198, 199, 200, 201, 202, 203, 216–219, 220, 221, 222 **Grade 2:** 162, 163, 164, 165, 167, 168, 184, 188, 189, 190, 191, 200, 201, 202, 203, 204, 206, 207, 210–213, 214, 215 **Grade 3:** 152, 153, 244, 245, 252, 253, 254, 255, 256–257, 258, 259, 260, 261, 262, 263, 264, 265, 267, 268–269, 270, 271, 518 **Grade 4:** 248, 258, 259, 278, 279, 280, 281, 290, 291, 292, 293, 304, 305 See also *Technology A Closer Look*, **Grades K–2 and 3–4.**
Some resources are basic materials, such as air, water, and soil; some are produced from basic resources, such as food, fuel, and building materials; and some resources are nonmaterial, such as quiet places, beauty, security, and safety.	**Grade K:** 146, 152, 153, 154, 155, 156, 157, 158, 159, 160, 161, 162, 163, 164, 165, 276, 277 **Grade 1:** 64–65, 194, 195, 196, 197, 198, 199, 200, 201, 202, 203, 216–219, 220, 221, 222, 312–313 **Grade 2:** 162, 163, 164, 165, 168, 188, 189, 190, 191, 194–199, 200, 201, 202, 203, 204, 208–209, 210–213, 214, 401, 426 **Grade 3:** 26, 27, 244, 252, 253, 254, 255, 256–257, 259, 260, 261, 262, 263, 267, 270, 271, 518 **Grade 4:** 54, 278, 279, 280, 281, 290, 291, 294, 296, 304, 305 See also *Technology A Closer Look*, **Grades K–2 and 3–4.**

Types of Resources continued	
The supply of many resources is limited. If used, resources can be extended through recycling and decreased use.	**Grade K:** 154, 155, 158, 159, 160, 161, 162, 163, 164, 165 **Grade 1:** 192, 197, 201, 203, 205, 209, 210, 211, 212, 213, 214, 215, 216–219, 220, 221 **Grade 2:** 200, 201, 203, 204, 205, 206, 207, 210–213, 215 **Grade 3:** 156, 157, 158–159, 183, 253, 256–257, 264, 265, 266, 267, 270, 271, 415, 426, 518, 519 **Grade 4:** 187, 292, 293, 298, 299, 300, 301, 302–303, 304, 305, 416 See also *Technology A Closer Look*, **Grades K–2 and 3–4.**
Changes in Environments	
Environments are the space, conditions, and factors that affect an individual's and a population's ability to survive and their quality of life.	**Grade K:** 87, 89, 124–125 **Grade 1:** 11, 66, 68, 69, 70, 71, 126, 128, 129, 130, 131, 132, 133, 134, 135, 136, 137, 141, 150–153 **Grade 2:** 90–93, 120, 121, 122–125, 128, 130–133, 134, 135, 136–139, 140, 141, 142–145, 146 **Grade 3:** 24, 25, 28–29, 108, 109, 118, 120, 121, 122, 123, 124, 125, 126, 128, 129, 146, 147, 184 **Grade 4:** 128, 129, 130, 131, 132, 133, 144, 145, 146–147, 159, 176, 177, 179, 180, 181, 192–193
Changes in environments can be natural or influenced by humans. Some changes are good, some are bad, and some are neither good nor bad. Pollution is a change in the environment that can influence the health, survival, or activities of organisms, including humans.	**Grade K:** 138, 140, 141, 142, 143, 144, 148, 149, 154, 155, 162, 163, 164, 165 **Grade 1:** 131, 146, 147, 155, 184, 204, 205, 206–207, 209, 220 **Grade 2:** 44, 102, 104, 105, 106, 107, 108, 109, 112–115, 116, 117, 126–127, 176–177, 204, 205, 207, 215 **Grade 3:** 28–29, 142, 150, 151, 152, 153, 154, 155, 156, 157, 162, 163, 168, 169, 170, 174, 175, 182, 183, 218, 264, 265, 266, 267, 426 **Grade 4:** 42–43, 96, 118–119, 124–125, 146–147, 159, 183, 184, 185, 186, 187, 188, 189, 190, 191, 192–193, 194, 195, 232, 233, 244–245, 294, 295, 296, 297, 298, 301, 302–303
Some environmental changes occur slowly, and others occur rapidly. Students should understand the different consequences of changing environments in small increments over long periods as compared with changing environments in large increments over short periods.	**Grade K:** 138, 140, 141, 142, 143, 144, 148, 149, 154, 155, 162, 163, 164, 165 **Grade 1:** 131, 146, 147, 155, 182 **Grade 2:** 104, 105, 108, 109, 112–115, 176–177, 178–181 **Grade 3:** 28–29, 114, 150, 151, 152, 153, 154, 155, 162, 163, 168, 169, 170, 174, 175, 179, 182, 183, 202–223 **Grade 4:** 42–43, 118–119, 124–125, 146–147, 190, 191, 192–193, 232, 233, 244–245

Science and Technology in Local Challenges	
People continue inventing new ways of doing things, solving problems, and getting work done. New ideas and inventions often affect other people; sometimes the effects are good and sometimes they are bad. It is helpful to try to determine in advance how ideas and inventions will affect other people.	**Grade K:** 244, 245 **Grade 1:** 284–285, 346–347, 378–379, 398, 399, 408–409, 414–415, 422 **Grade 2:** 126–127, 148, 176–177, 189, 242–243, 374–375, 426–427 **Grade 3:** 210–211, 254, 256–257, 300–301, 342, 440–441, 508–509, 518 **Grade 4:** 218, 219, 279, 280, 288, 289, 290, 342–343, 375, 381, 382, 383, 389, 390–391, 397, 501, 510–511, 586–587 See also *Technology A Closer Look*, **Grades K–2 and 3–4.**
Science and technology have greatly improved food quality and quantity, transportation, health, sanitation, and communication. These benefits of science and technology are not available to all of the people in the world.	**Grade K:** 150, 244, 245 **Grade 1:** 42–43, 64–65, 222, 312–313, 346–347, 378–379, 408–409, 422 **Grade 2:** 242–243, 280–281, 306–307, 336–337, 354, 426–427 **Grade 3:** 28–29, 300–301, 440–441, 508–509 **Grade 4:** 43, 54, 218, 219, 280, 342–343, 489, 500, 501, 506, 507, 508, 510–511, 513, 584, 585, 586–587 See also *Technology A Closer Look*, **Grades K–2 and 3–4.**

History and Nature of Science

Science as a Human Endeavor	
Science and technology have been practiced by people for a long time.	**Grade 1:** 284–285, 346–347, 378–379, 398, 399, 408–409, 414–415, 422 **Grade 2:** 126–127, 148, 176–177, 189, 242–243, 374–375, 426–427 **Grade 3:** 210–211, 254, 256–257, 300–301, 342, 440–441, 508–509, 518 **Grade 4:** 218, 219, 279, 280, 288, 289, 290, 342–343, 375, 381, 382, 383, 389, 390–391, 397, 501, 510–511, 586–587 See also *Technology A Closer Look*, **Grades K–2 and 3–4.**
Men and women have made a variety of contributions throughout the history of science and technology.	**Grade K:** 252 **Grade 1:** 116–117, 148–149, 176–177, 206–207, 284–285, 378–379 **Grade 2:** 66–67, 110–111, 126–127, 280–281, 374–375 **Grade 3:** 96–97, 130–131, 180–181, 256–257, 333, 352–353, 355, 370–371, 440–441 **Grade 4:** 48, 118–119, 222–223, 375, 381, 382, 390–391, 434, 436, 438–439, 552, 586–587

Correlation to National Standards

Science as a Human Endeavor continued	
Although men and women using scientific inquiry have learned much about the objects, events, and phenomena in nature, much more remains to be understood. Science will never be finished.	**Grade K:** 4–5, 6, 7, 10–11, 12–13, 16–17, 18–19 **Grade 1:** 148–149, 284–285 **Grade 2:** 4–9, 44–45, 279, 374–375 **Grade 3:** 4, 11, 96–97, 172, 177, 180–181, 342, 352–353, 355, 370–371 **Grade 4:** 10–11, 34–35, 40, 222–223, 279, 383, 389
Many people choose science as a career and devote their entire lives to studying it. Many people derive great pleasure from doing science.	**Grade K:** 116 **Grade 1:** 80, 116–117, 148–149, 156, 176–177, 206–207, 222, 284–285, 292, 430 **Grade 2:** 66–67, 82, 110–111, 126–127, 148, 216, 288, 354, 374–375, 434 **Grade 3:** 96–97, 100, 130–131, 184, 272, 352–353, 356, 370–371, 426, 522 **Grade 4:** 118–119, 122, 196, 222–223, 306, 404, 438–439, 476, 590

Flipbook Credits

Abbreviation key: MMH=Macmillan/McGraw-Hill

Cover Keith Neale/Masterfile; **1** Jose Luis Pelaez/Photographer's Choice/Getty Images; **2** (tl)Digital Vision/Getty Images, (tr)Photodisc/Getty Images, (c)Dex Image/Getty Images, (bl)Photographer's Choice/Getty Images, (br)CORBIS; **3** (l)Yellow Dog Productions/Riser/Getty Images, (tc)David Young-Wolff/Photo Edit, (tr)CORBIS, (bc)Lifestyle Bokeh/Alamy, (br)D. Hurst/Alamy, (inset)Photodisc/Getty Images; **4** Ken Karp/MMH; **5** Gunter Marx Photography/CORBIS; **6** (tl)Nino Mascardi/Riser/Getty Images, (tcl)Photographer's Choice/Getty Images, (tcr)Randy Wells/Stone/Getty Images, (tr)Stockbyte/PunchStock, (cr)Photodisc/Getty Images, (cl)C Squared Studios/Getty Images, (b)Kevin Summers/Photographer's Choice/Getty Images; **7** (l br)Digital Vision/Getty Images, (c)Jupiter Images/Brand X/CORBIS, (tr)Ngoc Minh & Julian Wass/StockFood Creative/Getty Images; **8** (l)DEA/C.DANI/Getty Images, (c)Gay Bumgarner/Stone/Getty Images, (r)Brand X Pictures/PunchStock; **9** (l)Photodisc/Getty Images, (cl)Jorg Greuel/Photonica/Getty Images, (cr)Ross M. Horowitz/Iconica/Getty Images, (r)Alan Thornton/Photographer's Choice/Getty Images; **10** (tl)Westend61/Getty Images, (tr)David Cavagnaro/Visuals Unlimited/Getty Images, (bl)Paul Harris/Stone/Getty Images, (bcl)Robert J. Erwin/Photo Researchers, Inc., (bcr br)William Leonard/DRK Photo; **11** (cw from top)Rosemary Calvert/Photographer's Choice//Getty Images, (2)Konrad Wothe/Minden Pictures, (3)Frank Krahmer/zefa/CORBIS, (4)Matthew Ward/DK Images, (5)Melba Photo Agency/PunchStock, (6)Photographer's Choice/Getty Images, (7 9)Brand X Pictures/PunchStock, (8)Joseph Van Os/The Image Bank/Getty Images, (10)Darrell Gulin/Photographer's Choice/Getty Images, (11)Siede Preis/Getty Images; **12** (tl)Dorling Kindersley/DK Images, (tr)Matthew Ward/Dorling Kindersley/Getty Images, (c)Darrell Gulin/CORBIS, (bl)Jonathan Buckley/GAP Photos/Getty Images, (br)Image Plan/CORBIS; **13** (cw from top)Westend61/Getty Images, (2)Mark E. Gibson/CORBIS, (3 9)Photodisc/Getty Images, (4)Ingram Publishing/Alamy, (5)Eisenhut & Mayer/Foodpix/JupiterImages, (6)CORBIS, (7)Studio Schmitz/StockFood Munich/StockFood America, (8)Foodcollection/Getty Images; **14** (cw from top)Brand X Pictures/PunchStock, (2)JH Pete Carmichael/The Image Bank/Getty Images, (3 6 8)Photodisc/Getty Images, (4)Digital Zoo/Getty Images, (5)Photographer's Choice/Getty Images, (7)Purestock/PunchStock, (9)Tim Flach/Stone/Getty Images; **15** (cw from top)Jane Burton/Dorling Kindersley/Getty Images, (2)W. Perry Conway/CORBIS, (3 7 11)Brand X Pictures/Getty Images, (4)Photodisc/Getty Images, (5)Mike Dunning/Dorling Kindersley/Getty Images, (6)Rubberball Productions/Getty Images, (8)Kim Taylor/Dorling Kindersley/Getty Images, (9)Ingram Publishing/Alamy, (10)Digital Vision/Getty Images; **16** (l)Benelux/CORBIS, (tr)Michael Gilbert/America 24-7/Getty Images, (c)DLILLC/CORBIS, (br)Visuals Unlimited/CORBIS; **17** (l r)Digital Vision/Getty Images, (cl cr)Photodisc/Getty Images; **18** (cw from top)Westend61 GmbH/Alamy, (2 4 6)Stockbyte, (3 5)Stockbyte/Getty Images, (7)Jerry Young/Dorling Kindersely/Getty Images; **19** (cw from top)IT Stock/PunchStock, (2)Robert Glusic/CORBIS, (3)Digital Vision/PunchStock, (4)Theo Allofs/CORBIS, (5)JH Pete Carmichael/Riser/Getty Images, (6)Jeff Vanuga/CORBIS; **20** blickwinkel/Alamy; **21** Georgette Douwma/Photographer's Choice/Getty Images; **22** (cw from top)Leonard Lee Rue III/DRK Photo, (2)Tim Flach/Stone/Getty Images, (3)Digital Vision/Getty Images, (4)DLILLC/CORBIS, (5)Central Stock/Fotosearch, (6)Visuals Unlimited/CORBIS; **23** (tl)Jonathon Gale/Taxi/Getty Images, (tr br)Dorling Kindersley/Dorling Kindersley/Getty Images, (bl bc)Jane Burton/Dorling Kindersley/Getty Images; **24** (tl)Jeff Foott/Discovery Channel Images/Getty Images, (tc)U.S. Fish & Wildlife Service, (tr)Art Wolfe/Stone/Getty Images, (bl)Gary Vestal/Stone/Getty Images, (br)Image Source/PunchStock; **25** (tl)Digital Vision/Getty Images, (tc)Gary Ombler/Dorling Kindersley/Getty Images, (tr bc)D. Hurst/Alamy, (bl)EIGHTFISH/The Image Bank/Getty Images, (br)Brooke Slezak/Taxi/Getty Images; **26** (cw from top)Steve Maslowski/Visuals Unlimited/Getty Images, (2 5)Photodisc/Getty Images, (3)Gary W. Carter/CORBIS, (4)Brand X Pictures/PunchStock, (6)Darryl Torc[illegible] Photographer's Choice/Getty Images; **27** John Kieffer/Peter Arnold, Inc.; **28** (tl)William Manning/CORBIS, (tc)PCL/Alamy, (tr)Werran/Ochsner/Photographe[illegible] Getty Images, (c)David Frazier/CORBIS, (b)Farrell Grehan/Arcaid/CORBIS; **29** (bl)Neil Fletcher/DK Images, (bc)J.P. Ferrero/Photo Researchers, Inc., (br b[illegible] Streeter/DK Images; **30** (l to r, t to b 4 6)Ken Cavanagh/MMH, (2)Visuals Unlimited/Getty Images, (3)VisionsofAmerica/Getty Images, (5)Photodisc/Getty [illegible], (7)DEA/R.APPIANI/Getty Images, (8)Harry Taylor/Dorling Kindersley/Getty Images; **31** (l)James P. Blair/Stockbyte/Getty Images, (tr)Digital Vision/Getty Images, (br)Photodisc/Getty Images; **32** (l)David Muench/CORBIS, (c)Photodisc/Getty Images, (r)Digital Vision/Getty Images; **33** (l)Digital Vision/Getty Images, (tr)Photographer's Choice/Getty Images, (br)Image Source/Getty Images; **34** (l)Digital Vision/Getty Images, (c)Charles O. Cecil/Alamy, (r)Photodisc/Getty Images; **35** (tl)Jaimie D Travis/DK Stock/Getty Images, (tr)Digital Vision/Getty Images, (bl br)Blend Images/Getty Images; **36** (tl)H. Mark Weidman Photography/Alamy, (tr)Digital Vision/Getty Images, (bl)DEA/M. Borchi/Getty Images, (br)Lars Borges/Stone+/Getty Images; **37** (tl)Tom Bean and Susan/DRK Photo, (tc)Kevin Taylor/Alamy, (tr)Digital Vision/Getty Images, (c)ImageShop/CORBIS, (b)Photodisc/Getty Images; **38** (t)Tim Pannell/CORBIS, (bl)Roine Magnusson/The Image Bank/Getty Images, (br)Photodisc/Getty Images; **39** (tl)Burton McNeely/Stone/Getty Images, (tr)DLILLC/CORBIS, (b)Blend Images/Getty Images; **40** (t)VisionsofAmerica/The Image Bank/Getty Images, (bl)Digital Vision/Getty Images, (bc)National Geographic/Getty Images, (br)Steve Allen/The Image Bank/Getty Images; **41** (t)Blend Images/Alamy, (cl)Blend Images/Getty Images, (cr)Digital Vision/Getty Images, (b)Lifesize/Getty Images; **42** (l)Photodisc/Getty Images, (r)Taxi/Getty Images; **43** (inset)Larry Landolfi/Photo Researchers, Inc., (bkgd)Uvimages/anamaimages/CORBIS; **44** (tl)OJO Images/Getty Images, (tr)Laura Ciapponi/Photonica/Getty Images, (c bl)Photographer's Choice/Getty Images, (br)DAJ/Getty Images; **45** (tl)Jupiter Images/Comstock Images/Alamy, (tr)Crawford A. Wilson III/Flickr/Getty Images, (bl)Digital Vision/Getty Images, (br)Panoramic Images/Getty Images; **46** MMH; **47** (tl)Siede Preis/Getty Images, (tcl)Comstock Images/Alamy, (tcr bl bc)Ken Cavanagh/MMH, (tr)Kari Marttila/Alamy, (c)Nikreates/Alamy, (br)Stock 4B/Getty Images; **48** (l to r, t to b)Medioimages/Photodisc/Getty Images, (2 3 8 9)Photodisc/Getty Images, (4)Santokh Kochar/Photodisc/Getty Images, (5)Ken Cavanagh/MMH, (6 10)D. Hurst/Alamy, (7)Danita Delimont/Alamy; **49** (cw from top)Judith Collins/Alamy, (cl cr 3 5 6 7 8 9)Photodisc/Getty Images, (2)Image Source/PunchStock, (4)Steve Gorton/Dorling Kindersley/Getty Images, (10)Ingram Publishing/Fotosearch, (11)Ingram Publishing/Fotosearch, (12)MaRoDee Photography/Alamy; **50** (tl)Ken Karp/MMH, (tcl)Fancy/Veer/CORBIS, (tcr)Michael S. Yamashita/CORBIS, (tr)Dave Rudkin/Dorling Kindersley/Getty Images, (bl bcr br)Ken Cavanagh/MMH, (bcl)Jupiter Images/BananaStock/Alamy; **51** Roy Mehta/The Image Bank/Getty Images; **52** (l to r, t to b 3)Stockbyte/Getty Images, (2)Dave King/Dorling Kindersley/Getty Images, (4)Brand X Pictures/PunchStock, (5 6 8)Photodisc/Getty Images, (7)United States coin image [or images] from the United States Mint; **53** (tl)Lesley Robson-Foster/Stone/Getty Images, (tc)MMH, (tr)Ben Blankenburg/CORBIS, (cl)Thierry Dosogne/Digital Vision/Getty Images, (cr)Comstock Images/Alamy, (b)Dave King/Dorling Kindersley/Getty Images; **54** (tl)Comstock Images/Alamy, (tr)Purestock/Getty Images, (c)Don Smetzer/Stone/Getty Images, (bl)Caroline De Vries/fStop/Getty Images, (br)Brand X Pictures/PunchStock; **55** (bl tr)Jane Burton/DK Images, (cl)David Hosking/Frank Lane Picture Agency/CORBIS, (c)MMH, (cr)Eric and David Hosking/CORBIS; **56** (tl)Blend Images/Getty Images, (tc br)Ken Cavanagh/MMH, (tr)Andrew Paterson/Photographer's Choice/Getty Images, (bl)Nicholas Eveleigh/Iconica/Getty Images, (bc)Rubberball Productions/Getty Images; **57** (l)Kenneth Gerhardt/Gall Images/Getty Images, (tc)Seanduan.com/Flickr/Getty Images, (r)Larry Gerbrandt/Flckr/Getty Images, (bc)Robert Ross/Gallo Images/Getty Images; **58** (tl)Creatas/PunchStock, (tcl)Comstock Images/Getty Images, (tcr)Ted Streshinsky/CORBIS, (tr)Westend61/Getty Images, (bl)Kris Legg/Alamy, (bcl)Michael Newman/PhotoEdit, (bcr)Jupiter Images/Thinkstock/Getty Images, (br)Photodisc/Getty Images; **59** Ken Cavanagh/MMH; **60** (l)Brand X/Fotosearch, (bkgd)Lester Lefkowitz/The Image Bank/Getty Images.

Teacher's Edition Credits

Cover Photos
Cover Keith Neale/Masterfile.

Illustrations
All illustrations and childrens realia are by Rachel Geswaldo except as noted below.

Marla Baggetta: **TR32**; David Brooks: **TR29**; Carly Castillon: **TR26**; Alan Flinn: **TR31**; April Hartman: **TR28**; Elisa Kleven: **TR23**; Anthony Lewis: **Txiv**, **Txvi–Txxi**, **39** bl., **54**, **69** bl., **89** br., **77** br., **103** bl., **105** b., **111** br., **131** t., **138**, **139** b., **151** br., **177** bl., **177** br., **179** b., **183** bl., **199** b., **197** br., **217** br., **223** bl., **237** bl., **255** bl., **263** b., **269** b., **TR8** t.; Lori Lohstoeter: **TR27**; Jesse Reisch: **TR24**; Marisol Sarrazin: **TR25**; Pam Thompson: **TR22**; Fabricio Vanden Broeck: **TR30**; James Williamson: **TR33**.

Abbreviation key: MMH=Macmillan/McGraw-Hill

Photography
Tiii (cw from top)Erik Stenbakken/stenbakken.com, (2)Scott Stewart, (3)Scott Schauer Photographs, Scottsdale, AZ, (4)Nancy Brown, (5)Joshua P. Roberts, (6)Deborah Attoinese/Overall Productions, (7)Dan Donovan, (8)Ken Cavanagh/MMH, **Tviii** MMH, **Txv** Jane Burton/Dorling Kindersley, **Txxii** MMH; **Txxiii** (t)G.K. Hart & Vikki Hart/Getty Images, (b)American Images, Inc/Getty Images; **Txxix** Harvey Lloyd/Getty Images; **Txxv** (l)Jim Scherer/Stockfood America, (r)Glen Peterson/Stockfood America; **Txxvi** William P. Leonard/DRK PHOTO; **Txxvii** Tim Flach/Getty Images; **Txxviii** MMH; **Txxx** Cathy Melloan/PhotoEdit, Inc.; **Txxxi** MMH; **Txxxii** Siede Preis/Getty Images; **1** Michael Newman/PhotoEdit, Inc.; **2 6 8 9** MMH; **14-15** David Young-Wolff/PhotoEdit, Inc.; **18** Mary Ellen Bartley/Foodpix/JupiterImages; **19 20** MMH; **21** Photodisc/Getty Images; **23** (tl)Stuart Westmorland/Getty Images, (tr)Georgette Douwma/Getty Images, (b)Gunter Marx Photography/CORBIS; **24A** (t)Gunter Marx Photography/CORBIS, (b)Jane Burton/Dorling Kindersley; **24B** Stuart Westmorland/Getty Images; **24D** William P. Leonard/DRK PHOTO; **24F** Siede Preis/Photodisc/Getty Images; **24H** (t to b 2 3 4 5 6 7 11)MMH, (8 10)Siede Preis/Photodisc/Getty Images, (9)Rosemary Calvert/Getty Images; **24J** (l r)Siede Preis/Photodisc/Getty Images, (c)Rosemary Calvert/Getty Images; **30** MMH; **31** (cw from top)MMH, (2 3)Siede Preis/Getty Images, (4)Rosemary Calvert/Getty Images, (5)Dave King/Dorling Kindersley; **33** (t)C Squared Studios/Photodisc/Getty Images, (b)MMH; **38** through **53** MMH; **58** (t)John A. Rizzo/Getty Images, (b)MMH; **59** (t)Renee Comet Photography/Stockfood America, (b)MMH; **62B** Alaska Stock; **62D** (tl)Colin Keates/Dorling Kindersley, (tr)Getty Images, (b)Dan Suzio/Photo Researchers, Inc.; **62H** (fishbowl)Photodisc/Getty Images, (others)MMH; **68** Getty Images; **69** (t)Edwin L. Wisherd/Getty Images, (b)MMH; **76 77** MMH; **78** Dorling Kindersley; **79** Ralph A. Clevenger/CORBIS; **82 83 85** MMH; **88** Richard Hutchings/PhotoEdit, Inc.; **89** through **103** MMH; **111** (t)Getty Images, (b)MMH; **113 116** MMH; **117** (t bl)MMH, (br)Spencer Jones/FOODPIX/JupiterImages; **123** (tl)S.R. Maglione/Photo Researchers, Inc., (tr)Warren E. Faidley/DRK PHOTO, (b)John Kieffer/Peter Arnold, Inc.; **124A** John Kieffer/Peter Arnold, Inc.; **124B** James Randklev/Getty Images; **124D** Richard Hutchings/PhotoEdit, Inc.; **124E** (tl)Michael Durham/DRK PHOTO, (tr)Doug Martin/Photo Researchers, Inc., (c)Wally Eberhart/Visuals Unlimited, (b)Andrew J. Martinez/Photo Researchers, Inc.; **124F** (t to b)Ryan McVay/Photodisc/Getty Images, (2 6)MMH, (3)Joyce Photographics/Photo Researchers, Inc., (4)Stephen J. Krasemann/Photo Researchers, Inc., (5)E. R. Degginger/Photo Researchers, Inc.; **124H**(t to b 2 3 4 5 6 7)MMH, (8)Peter Gardner/Dorling Kindersley, (9)Ryan McVay/Photodisc/Getty Images, (10)David Toase/Photodisc/Getty Images; **127 130** MMH; **131** Horst Schafer/Peter Arnold, Inc.; **132** through **164** MMH; **165** (l)MMH, (r)Spencer Jones/FOODPIX/JupiterImages; **168B** Warren E. Faidley/DRK PHOTO; **168D** Jim Cummins/CORBIS; **168E** Roy Morsch/CORBIS; **168H** (t to b 5 6 7 8)MMH, (4)Dr. Scott Nielsen/DRK PHOTO, (9)Siede Preis/Getty Images; **176 183 185** MMH; **189** (cw from top 6)MMH, (2)Michael Newman/PhotoEdit, Inc., (3)Amy Etra/PhotoEdit, Inc., (4 5)C Squared Studios/Photodisc/Getty Images; **191** MMH; **196** S. Nielsen/DRK PHOTO; **198** Stockbyte/PictureQuest; **202** MMH; **203** (t)Tom Bean/DRK PHOTO, (b)MMH; **203 209 210A** MMH; **210B** Photodisc/Getty Images; **210D** Ken Cavanagh/MMH; **210E** Davies & Starr/Getty Images; **210F** MMH; **210H** (newspapers)Ryan McVay/Photodisc/Getty Images; (others)MMH; **210J** through **237** MMH; **238** (l r)Davies & Starr/Getty Images, (c)David Toase/Getty Images; **240B** Image State/Alamy; **240D 240E** MMH; **240F** (l)Ken Cavanagh/MMH, (r)C Squared Studios/Getty Images; **240H** (letter magnets)C Squared Studios/Photodisc/Getty Images, (others)MMH; **246 247** MMH; **248** ThinkStock/JupiterImages; **254** MMH; **255** (cw from top)Ken Cavanagh/MMH, (2 3)C Squared Studios/Getty Images, (4)PhotoDisc/Punchstock, (5 7)Siede Preis/Getty Images, (8)D. Hurst/Alamy Images; **260** through **273** MMH; **274** (l)Myrleen Ferguson Cate/PhotoEdit, Inc., (r)CORBIS; **TR1** Digital Vision/PunchStock; **TR2** CORBIS; **TR3** MMH; **TR4** C Squared Studios/Getty Images; **TR5** (t)MMH, (b)Ken Karp/MMH; **TR6** (t)Ken Karp/MMH, (b)Photodisc/Getty Images; **TR7** (t b)MMH, (tc bc)C Squared Studios/Getty Images; **TR8** Tom Salyer/Silver Image/MMH; **TR9** MMH; **TR10** (t)Roman Sapecki/MMH, (b)Lew Lause/MMH; **TR11** (t)Laura Dwight/Peter Arnold, Inc., (tc)Lori Adamski Peek/Getty Images, (bc)Brand X Pictures / JupiterImages, (b)Photodisc/Getty Images; **TR12** MMH, **TR13** Myrleen Ferguson Cate/PhotoEdit, Inc.; **TR14** (cw from top 2 3 5)Dave Mager/MMH, (4)Jerry Schad/Photo Researchers, Inc., (6)Photodisc/Getty Images; **TR15** C Squared Studios/Photodisc/Getty Images.